Probability and Statistics for Decision Making

SERIES IN QUANTITATIVE METHODS FOR DECISION MAKING

Robert L. Winkler, Consulting Editor

Chou, Ya-lun PROBABILITY AND STATISTICS FOR DECISION MAKING

Hays, W. L., and R. L. Winkler STATISTICS: PROBABILITY, INFERENCE, AND DECISION (Volumes I and II and combined edition)

Horowitz, I. DECISION MAKING AND THE THEORY OF THE FIRM

LaValle, I. H. **AN INTRODUCTION TO PROBABILITY, DECISION, AND INFERENCE**

Lippman, S. A. ELEMENTS OF PROBABILITY AND STATISTICS

Miller, R. E. MODERN MATHEMATICAL METHODS FOR ECONOMICS AND BUSINESS

Press, S. J. APPLIED MULTIVARIATE ANALYSIS

Spivey, W. A., and R. M. Thrall LINEAR OPTIMIZATION

Winkler, R. L. **AN INTRODUCTION TO BAYESIAN INFERENCE AND DECISION**

INTERNATIONAL SERIES IN DECISION PROCESSES

Ingram Olkin, Consulting Editor

Anderson, T. R., and M. Zelditch, Jr. A BASIC COURSE IN STATISTICS WITH SOCIOLOGICAL APPLICATION, 2d ed.

Bauer, H. PROBABILITY THEORY AND ELEMENTS OF MEASURE THEORY

Chou, Ya-lun PROBABILITY AND STATISTICS FOR DECISION MAKING

Hays, W. L., and R. L. Winkler STATISTICS: PROBABILITY, INFERENCE, AND DECISION (Volumes I and II and combined edition)

Horowitz, I. DECISION MAKING AND THE THEORY OF THE FIRM

Hultquist, R. A. INTRODUCTION TO STATISTICS

Kozlov, B. A., and I. A. Ushakov RELIABILITY HANDBOOK (edited by J. T. Rosenblatt and L. H. Koopmans)

LaValle, I. H. AN INTRODUCTION TO PROBABILITY, DECISION, AND INFERENCE

Lippman, S. A. ELEMENTS OF PROBABILITY AND STATISTICS

Miller, R. E. MODERN MATHEMATICAL METHODS FOR ECONOMICS AND BUSINESS

Press, S. J. APPLIED MULTIVARIATE ANALYSIS

Roscoe, J. T. FUNDAMENTAL RESEARCH STATISTICS FOR THE BEHAVIORAL SCIENCES

Spivey, W. A., and R. M. Thrall LINEAR OPTIMIZATION

Thompson, W. A., Jr. APPLIED PROBABILITY

Walker, H. M., and J. Lev ELEMENTARY STATISTICAL METHODS, 3d ed.

Winkler, R. L. AN INTRODUCTION TO BAYESIAN INFERENCE AND DECISION

Probability and Statistics for Decision Making

Ya-lun Chou

St. John's University
Jamaica, New York

Holt, Rinehart and Winston

New York Chicago San Francisco Atlanta Dallas
Montreal Toronto London Sydney

To Dr. James G. Endicott
the most influential man in my life

Library of Congress Catalog Card Number: 75-167808
ISBN: 0-03-077150-1
Printed in the United States of America
5 4 3 2 038 1 2 3 4 5 6 7 8 9

Preface

This book is designed as a first course in modern business and economic statistics. It is a self-contained text, requiring of the reader only freshman mathematics, serious effort, and, perhaps, a certain degree of maturity.

Textbooks on business and economic statistics published during the past decade fall into three categories: (1) those which stress descriptive measures and classical inferential procedures, paying little or no attention to Bayesian statistics; (2) those which are oriented almost exclusively toward Bayesian methods at the expense of classical analysis and time-series studies; and (3) those which attempt a "balanced" presentation of all elementary statistical methods useful in the study of quantitative economics and management sciences. Texts of the first two kinds seem to view classical and Bayesian theories as substitutes for each other; those of the third kind seem to suggest that they complement each other. In terms of the third approach, methods of one school or another may be more appropriate or useful in certain circumstances and, although both may lead to identical conclusions with the same set of data, the interpretations of results may differ. I have subscribed to this approach in preparing this book.

One difficulty in writing a balanced text, one that includes descriptive measures, analytical statistics—classical as well as Bayesian—and time-series analysis, is in presenting the material within a manageable amount of space without sacrificing rigor, accuracy, and usefulness. I have dealt with this difficulty in two ways.

First, recognizing that one can never learn all the elementary statistical methods in a single course, I have selected only the most useful methods—those which, I believe, constitute the bare minimum of statistical knowledge a student in business or economics must master. To compensate for this limited presentation, I have briefly introduced many other topics, both elementary and advanced, and have included additional reading references in the last chapter.

Second, successful application of any method to concrete problems requires a full understanding of the principle underlying the method. I have therefore placed more stress on statistics as a way of thinking than on the mere mechanical rules employed to solve problems or to make decisions. The treatment of basic concepts and principles, such as probability theory, random variables, theory of sampling, and rationale of decision making under uncertainty, is much more thorough and rigorous in this text than in many elementary statistics texts on the market today. Hopefully, this orientation will both enable the reader to extend the application of principles he has learned here to decision situations beyond those in the text and provide him a sound foundation for later pursuit of more advanced studies in statistical methods, if he is so inspired.

Undoubtedly, the exact coverage of a given topic, or the balance among the different topics introduced in this book, may not be what every instructor might wish. If so, one can extend his lectures either on classical inductive statistics or on Bayesian decision theory or both, moving easily beyond what is offered here, by taking advantage of my rather comprehensive discourse on the foundation of statistics—that is, probability theory and its related concepts. Furthermore, the instructor can make available by omitting some sections in Chapters 9 and 10 and by dropping Chapters 18 and 19 completely from classroom discussion; these materials are mainly descriptive in nature, and the student can study them on his own without much difficulty. Thus, this book can be used with a great deal of flexibility, and the instructor can correct any "imbalance" or "incompleteness" according to his own judgment and the needs of his students.

I am grateful to Professor Robert L. Winkler, of Indiana University, who read all drafts of the manuscript. His comments and suggestions have greatly improved the quality of this book. I also wish to thank a number of Stanford University graduate students, who did an excellent job, under the supervision of Professor Ingram Olkin, in checking computations of illustrative examples in the text. I appreciate the kindnesses of Dr. John C. Alexion, Dean of the College of Business Administration, and Dr. Noel Doherty, Chairman of the Department of Quantitative Analysis, both of St. John's University, who lightened for me the burden of writing this book by giving me a reduced teaching load and convenient class schedules. My greatest appreciation, however, goes to the numerous users of my previous books. Their continuous loyalty and compliments have encouraged and sustained my effort in bringing this book to completion.

Finally, I am indebted to the Literary Executor of the late Sir Ronald A. Fisher, F.R.S., to Dr. Frank Yates, F.R.S., and to Oliver & Boyd, Edinburgh, for permission to reprint Tables II and XXXIII from their book *Statistical Tables for Biological, Agricultural and Medical Research.*

Fresh Meadows, N.Y. — Ya-lun Chou
January 1972

Contents

Preface v
Greek Alphabet xv

Chapter One THE NATURE OF STATISTICS 1

1–1 The Scope of Statistical Investigations 1
1–2 Statistical Definition of the Decision Problem 4
1–3 Populations, Elementary Units, and Observations 5
1–4 Measurements of Elementary Units 8
1–5 Specifying Parameters 10
1–6 Mathematical Prerequisites 11
Exercises 12

Chapter Two PROBABILITY AND PROBABILITY CALCULUS 13

2–1 Random Experiments and Related Concepts 13
2–2 Axioms of Probability for Finite Sample Spaces 15
2–3 The Classical Theory 16
2–4 The Relative-Frequency Theory 18
2–5 The Personalistic Theory 19
2–6 Probabilities Versus Odds 20
2–7 Simple Random Sampling 20
2–8 Mutually Exclusive Events 22
2–9 Joint Events 23
2–10 Complementary Events 25
2–11 Conditional Probability 25
2–12 Dependent Events 29
2–13 Independent Events 31
2–14 The Theorem of Total Probability 35
2–15 Sample Space with Many Outcomes 37
2–16 Infinite Sample Spaces 39
Exercises 42

Chapter Three RANDOM VARIABLES 47

3–1 Random Variables and Their Probability Functions 47
3–2 Cumulative Distribution Functions 51

3–3 Expectation 53
3–4 Properties of Expectation 54
3–5 Variance 56
3–6 Properties of Variance 57
3–7 Standard Random Variables 59
3–8 Chebyshev's Theorem 60
3–9 Continuous Random Variables 61
3–10 A Note on Expectation and Variance of Continuous Random Variables 68
3–11 Additional Properties of Random Variables 69
Exercises 71

Chapter Four SPECIAL DISCRETE PROBABILITY MODELS 75

4–1 The Bernoulli Model 76
4–2 The Binomial Model 76
4–3 The Hypergeometric Model 81
4–4 The Discrete Uniform Model 84
4–5 The Poisson Model 87
4–6 Relationships among Different Probability Models 92
Exercises 95

Chapter Five SPECIAL CONTINUOUS PROBABILITY MODELS 99

5–1 The Continuous Uniform Distribution 100
5–2 The Exponential Model 101
5–3 The General Normal Model 105
5–4 The Standard Normal Model 109
5–5 Summary and Conclusions 115
Exercises 117

Chapter Six INTRODUCTION TO STATISTICAL DECISION THEORY 120

6–1 The Structure of Decisions 120
6–2 Some Decision Strategies 123
6–3 A Decision Model for Perishable Products 129
6–4 Expected Value of Perfect Information 132
6–5 An Inventory Model 133
6–6 Special Forms of Bayesian Strategy and Loss Functions 134
6–7 A Sequential Decision Model 140
6–8 A Note on Utility Theory 143
Exercises 157

Chapter Seven DECISION MAKING WITH SPECIAL PROBABILITY MODELS 159

7–1 Acceptance Sampling: A Sequential Decision Rule 159
7–2 Bayesian Strategy for Nonequivalent State-Action Problems 164
7–3 A Queuing Model 171
7–4 Monte Carlo Simulation 179
Exercises 188

Chapter Eight BAYES' THEOREM AND DECISION MAKING 193

8–1 Bayes' Theorem 194
8–2 Further Comments on Bayes' Theorem 197
8–3 Bayesian Posterior Analysis 201
8–4 Bayesian Preposterior Analysis 204
Exercises 210

Chapter Nine ORGANIZING AND PRESENTING DATA 215

9–1 Introduction 215
9–2 Organizing Qualitative Data 216
9–3 Organizing Univariate Data with Only a Few Distinct Values 218
9–4 The Organization of Univariate Data with Many Distinct Values 220
9–5 Graphic Presentations of Frequency Distributions 226
9–6 Concluding Remarks 236
Exercises 236

Chapter Ten DESCRIPTIVE MEASURES FOR UNIVARIATE DATA 238

10–1 Averages: Measures of Central Tendency 239
10–2 Measures of Dispersion 247
10–3 Descriptive Measures for Grouped Data 253
10–4 Moments, Skewness, and Peakedness 259
Exercises 262

Chapter Eleven THEORY OF SAMPLING 265

11–1 Introduction 265
11–2 Sums of Random Variables 266
11–3 The Law of Large Numbers 267
11–4 The Central Limit Theorem 268
11–5 Sample Observations as Random Variables 270

11–6 Sampling Distribution of the Number of Successes 273
11–7 Sampling Distribution of a Proportion 275
11–8 Sampling Distribution of the Mean 281
11–9 Distribution of the Difference of Two Proportions 285
11–10 Distribution of the Difference of Two Means 286
11–11 Distributions of Median and Related Measures 288
11–12 Significance of Sampling Distributions 289
Exercises 291

Chapter Twelve CLASSICAL DECISION PROCEDURES: ESTIMATION 295

12–1 Statistical Inference 295
12–2 Point Estimation 295
12–3 Maximum Likelihood Function 300
12–4 Interval Estimation 303
12–5 Confidence Limits for μ 305
12–6 Confidence Limits for π 308
12–7 Confidence Limits for $\Delta\mu$ 311
12–8 Confidence Limits for $\Delta\pi$ 312
12–9 Summary and Conclusions 314
Exercises 315

Chapter Thirteen FURTHER CLASSICAL DECISION PROCEDURES: STATISTICAL TESTING 319

13–1 A General Procedure of Testing 319
13–2 Testing the Hypothesis $\mu = \mu_0$ 328
13–3 Testing the Hypothesis $\pi = \pi_0$ 332
13–4 Testing the Hypothesis $\theta_1 = \theta_2$ 334
13–5 Power and OC Functions 339
Exercises 349

Chapter Fourteen CHI-SQUARE DISTRIBUTIONS 354

14–1 Exact-Sampling Theory 354
14–2 Degrees of Freedom 355
14–3 Properties of Chi-Square Distributions 356
14–4 Statistical Inference about Population Variance 358
14–5 The Chi-Square Test on Frequencies 362
14–6 Testing Goodness-of-Fit 365
14–7 Tests of Independence: Contingency-Table Tests 370
14–8 Efficient Formulas for Computing χ^2 375
14–9 Tests of Homogeneity 376
14–10 Summary and Conclusions 379
Exercises 380

Chapter Fifteen STUDENT'S t AND F DISTRIBUTIONS 387

15–1 Properties of Student's Distributions 388
15–2 Properties of F Distributions 390
15–3 Inferences with μ when σ Is Unknown and n Is Small 392
15–4 Inferences concerning Two Population Means with Independent Small Samples 394
15–5 Inferences Concerning Two Population Means with Dependent Samples 397
15–6 F Test of the Equality between Variances 399
15–7 Analysis of Variance 401
15–8 The Least Significant Difference 410
15–9 Relationships among Sampling Distributions 413
Exercises 414

Chapter Sixteen BAYESIAN INFERENCE 420

16–1 Introduction 420
16–2 Estimation as a Bayesian Decision 421
16–3 The Bayesian Estimator of μ 426
16–4 Testing as a Bayesian Decision 430
16–5 Bayesian Inference with Nonequivalent State-Action Problems 434
16–6 Bayesian Versus Classical Inferential Procedures 440
Exercises 442

Chapter Seventeen SIMPLE LINEAR ASSOCIATION ANALYSIS 447

17–1 Association between Variables 447
17–2 The Linear Bivariate Regression Model 449
17–3 Estimating the Population Regression Equation 451
17–4 Computations with Coded Data and Simplified Notations 456
17–5 Estimating the Variance and the Standard Deviation of the Regression 457
17–6 Inference about Population Regression Coefficients 461
17–7 Confidence Limits of Prediction 463
17–8 Population Correlation Coefficient 467
17–9 Sample Correlation Coefficient 470
17–10 Inference about ρ 470
17–11 Population Coefficient of Determination 473
17–12 Sample Coefficient of Determination 477
17–13 Rank-Correlation Coefficient 478
17–14 Concluding Remarks 481
Exercises 483

Chapter Eighteen MEASURES OF SECULAR TRENDS 490

18–1 Introduction to Time-Series Analysis 490
18–2 Types of Secular Trend and Methods of Trend Fitting 493
18–3 Arithmetic Straight-Line Trend 494
18–4 Second-Degree Polynomial Trends 499
18–5 The Exponential, or Geometric Straight-Line, Trend 502
18–6 Method of Moving Averages 506
18–7 Shifting the Trend Origin 509
18–8 Reducing Annual Trend Values to Monthly Values 510
18–9 Choice of Period and Mathematical Expressions for Trend 512
Exercises 514

Chapter Nineteen MEASURES OF SHORT-RUN FLUCTUATIONS 517

19–1 Introduction 517
19–2 Types of Seasonals 518
19–3 Calculation of the Typical Seasonal Index by the Method of Ratio to Moving Average 518
19–4 Alternative Methods of Measuring the Seasonal Index 528
19–5 Changing Seasonal 529
19–6 Measuring Cycles from Annual Data 531
19–7 Measuring Cycles from Monthly Data 533
19–8 Comments on the Estimate of *CI* 538
19–9 Time Series and Forecasting 539
Exercises 545

Chapter Twenty FURTHER TOPICS AND REFERENCES 547

20–1 The End of a Beginning 547
20–2 Bayesian Versus Classical Statistics 547
20–3 Mathematical Statistics 548
20–4 Methods of Sampling 549
20–5 Nonparametric Methods 550
20–6 Analysis of Variance and Experimental Design 556
20–7 Multiple Linear Regression Analysis 557
20–8 Nonlinear Regression Analysis 561
20–9 Index Numbers 564
20–10 Additional Reference Works 567

Appendix A SETS, PERMUTATIONS, COMBINATIONS, AND SUMMATIONS 569

A–1 The Notation of a Set and Ways of Specifying a Set 569
A–2 Universal Set and Subsets 570
A–3 Operations on Sets 573

A-4 Algebra of Sets 575
A-5 The Multiplicative Principle 577
A-6 Permutations 578
A-7 Combinations 580
A-8 The Σ Notation 582
A-9 Finite Double Sums 583
A-10 Theorems of Summation 584

Appendix B STATISTICAL TABLES 587

B-1 Squares and Square Roots 588
B-2 Common Logarithms 597
B-3 The Cumulative Binomial Distribution 599
B-4 The Hypergeometric Distribution 603
B-5 The Cumulative Poisson Distribution 605
B-6 Exponential Functions 607
B-7 Values of the Standard Normal Distribution Function 608
B-8 Unit Normal Loss Function 610
B-9 The Chi-Square Distribution 611
B-10 Student's t Distribution 612
B-11 F Distribution 613
B-12 Values of $z = \left(\frac{1}{2}\right) \ln \left(\frac{1+r}{1-r}\right)$ 616
B-13 Equidistributed Random Numbers 618

Index 619

Greek Alphabet

A	α	alpha	N	ν	nu
B	β	beta	Ξ	ξ	xi
Γ	γ	gamma	O	o	omicron
Δ	δ	delta	Π	π	pi
E	ϵ	epsilon	P	ρ	rho
Z	ζ	zeta	Σ	σ	sigma
H	η	eta	T	τ	tau
Θ	θ	theta	Υ	υ	upsilon
I	ι	iota	Φ	φ	phi
K	κ	kappa	X	χ	chi
Λ	λ	lambda	Ψ	ψ	psi
M	μ	mu	Ω	ω	omega

1

THE NATURE OF STATISTICS

1-1 THE SCOPE OF STATISTICAL INVESTIGATIONS

Traditionally, *statistics* is thought to deal with the collection, organization, presentation, and analysis of numerical data. This conception fails to stress the central function of statistical analysis.

In current usage, *statistics* is defined as a method of decision making under uncertainty. Its importance is obvious in a world where one is always uncertain to some degree about future events. In dealing with many physical, economic, social, industrial, business, or psychic phenomena, one is called upon to make decisions, or to take actions, in face of uncertainty as to the way things are—"the state of nature." Statistics becomes a valuable tool as a procedure of decision making whenever one is uncertain about the state of nature.

Problems in applying statistical methods in various areas are very similar. Our first step in this book will be to examine these common problems, which in fact define the scope of statistics. We shall consider them here in an orderly sequence, which one may follow in a statistical investigation.

Formulation or Definition of the Problem

Statistical analysis, as a manufacturing process, begins with raw materials, which are numerical, or categorical, data; the finished products are whatever useful information or valuable conclusions lie buried among the crude data. In manufacturing an output, the producer must first decide what is to be produced and then set up precise specifications of the output quality; otherwise he may not know what kind of raw materials and how much to feed into the process. In statistics, then, the first problem is to know exactly what is to be investigated: to formulate the problem or question as precisely as possible. Only then can the investigator decide what the relevant data for the problem are. If he fails in this formulation, data collected may be irrelevant or may even becloud the problem.

It is well to remember that the quality of statistical conclusions depends upon the appropriateness and accuracy of data, which in turn depend upon exactness in formulating the problem. Statistical techniques, no matter how refined and precise, cannot help one reach decisions if they are applied to inappropriate data. Furthermore, as we shall see, often decisions may have to be made with no data beyond the decision maker's personal judgment about the state of nature. In such situations a clear understanding of the problem itself will improve the "quality" of the decision.

In statistics, the totality of pertinent data that may be collected and observed in a given problem is called a *population*, or a *universe*. For instance, if the problem is to discover the percentage of unemployment in the labor force at a particular time in the United States, the population—or universe—will include all observations about the individual workers, as to whether they are employed or unemployed, in the United States at that time. Again, if we wish to determine the average life of a lot of 1000 automobile tires, the population for this problem consists of the different durabilities in terms of, say, number of miles traveled, of all the 1000 tires in the lot.

A statistical population, then, consists of a collection of observations of one kind or another, collected in accordance with the formulation of a particular statistical problem or with the definition of a specific statistical population.

Design of the Experiment

Once the problem has been precisely formulated, the investigator must decide whether to study the entire population or only a part of it. The study of a population is called a *complete enumeration*, or a *census*; the study with a sample is referred to as *sampling*. In practice, it is often costly, time-consuming, and even physically impossible to conduct a census, and we have to resort to sampling. This is what is done, for example, by the Bureau of the Census in its Current Population Survey each month. Among other things, this survey furnishes estimates of employment and unemployment based on data from a sample of 25,000 households distributed throughout the United States.

In sampling we may try to make an inference, on the basis of sample data, about the population from which the sample is drawn, or we may try to revise a prior (initial) judgment on the state of nature before making a decision. Hence, the sample must adequately represent the population. Obtaining a representative sample is fundamental in statistical theory. It involves such questions as these: What type of data should be collected? How should the data be collected? How large should the sample be? These questions belong to what is called *sample design* or the *experimental design*. Proper care must be taken in planning and designing an experiment; otherwise we may not be able to reach any valid conclusion. The design of the experiment is the second stage of a statistical investigation.

As we shall see, there are many types of sampling design, each with its merits, depending on the type of population and the aim of the survey. Usually, however, we shall deal with random samples. A *random* sample may be defined as one drawn under conditions such that each item in the population has an equal chance to be selected, when the drawing is made with replacement; or as one such that all samples of equal size are equally likely to be selected, when the drawing is made without replacement. (For a detailed discussion of these two types of sampling, see Section 2–7.)

Collection of Data

The third stage is to collect the data in accordance with the design of the experiment. This stage is mainly a matter of routine, even though it is often the most time-consuming and costly component of the whole statistical process. Formulating the problem and designing the sample require a good deal of judgment and a sound knowledge of the subject matter. Good results in collecting data, on the other hand, demand the suppression of interviewers' personal judgments.

Organization and Description of Data

When data are collected, they should be arranged in readable form. They may be classified in some systematic manner and presented in a table. Or, in order to convey the meaning of facts more simply or forcibly, data may be presented by graphs or diagrams. Descriptive measures, such as proportions, averages, or standard deviations, are then calculated. A measure derived from the sample data is called a *statistic*. If a measure is computed from observations of the entire population, however, it is referred to as a *parameter*. From a sample of 25,000 households, for instance, we may compute the number of unemployed as a percentage, which may turn out to be, say, 6 percent of all the observations comprising the sample. With this result, we might infer that 6 percent of all the labor force are unemployed.

This fourth stage of the process is called *descriptive statistics*.

Final Decision or Inference

When the sample includes the whole population—that is, when an investigation is made by a complete enumeration—a survey ends with descriptive statistics. By then the characteristics of the whole population can be described and revealed, and a decision of one sort or another on the problem can readily be made. However, when the sample consists of only a part of the universe, the study will not end with descriptive procedures. A sample is seldom examined for its own sake, but rather for the light it can throw on the true state of the population. In other words, the last stage consists of trying to answer, on the basis of the sample statistic, the original problem or question formulated, which is always concerned with the population parameters.

Here, in the last stage of a statistical investigation, the utilization of sample information depends upon whether one is concerned with decision problems or with inferential problems. Some statistical problems, especially in industry or business, are decision problems, in which the partial information about the states of nature is used as the basis for an immediate decision. Others, notably in scientific studies, have to do with the exploitation of data in order to advance our understanding of the laws of nature.

In a general decision problem, one is confronted with a set of possible actions that might be taken, and the task is the selection of one of these actions with the aid of sample information in coordination with the decision maker's prior beliefs about the states of nature. Inferential problems may be those of (1) *estimation*, in which the action set is identical with the set of parameter values that represent the family of possible states of nature, and sample statistics are employed as estimators of these parameter values, or (2) *hypothesis testing*, in which the action set often consists of two points, and sample data are used to guide the selection of one of these points. General decision problems fall into the scope of *statistical decision theory*, which is covered in Chapters 6, 7, 8, and 16; inferential problems belong to *inductive statistics*, or *classical decision procedures*, which are discussed in Chapters 12 through 15. In the remainder of this introductory chapter we shall not distinguish these two approaches to decision making, since we shall be concerned with a matter common to both: the definition of a statistical problem.

1–2 STATISTICAL DEFINITION OF THE DECISION PROBLEM

Even though sample information is the rule rather than the exception in statistical analysis, we must, before selecting the sample, ask many important questions: Is information needed at all? If so, what kind of information? How should we proceed to solve a problem if and when complete information can be made available? How much information is needed for a particular problem? How are decisions to be related to observations? Answers to all these questions are connected with the concept of *statistical definition of the decision problem*.

Whenever a decision problem exists, and the uncertainty attached to it can be reduced or removed by additional information, the very first step in the decision procedure is to translate the problem into statistical terms. Statistical definition of a problem involves two distinct but closely related steps: (1) defining the statistical population and (2) describing its characteristics. The latter step involves the use of statistical methods to summarize the pertinent facts about the population—that is, to compute the parameter or parameters necessary for the decision.

The concentration on statistical definition of a problem at the outset is very rewarding, even if the decision must or can be made with no additional information, or on the basis of a sample. First, a precise definition of the population is indispensable in the selection of the sample, preventing the frequent mistake of drawing the sample from the wrong population. Second, thinking in terms of complete information at the beginning will enable us to concentrate on designing a survey that goes beyond the mere collection or presentation of numerical or categorical facts—a survey that can lead directly to decisions. Third, too much initial stress on data gathering often converts a decision problem into a data-collection problem. Insufficient attention is given to the question of how to use the data; consequently, information gathered often proves useless for the decision problems in question.

In short, in stressing statistical definition of the decision problem we are stressing a priori planning. Modern statistics demands that *decisions be related to observations before observations are made*. The decision maker is called upon to specify a priori the possible decisions that are conditional upon the actual observations.

Now let us consider the various aspects of defining decision problems statistically.

1-3 POPULATIONS, ELEMENTARY UNITS, AND OBSERVATIONS

In statistics, the term *population* means an aggregate of individual items, whether people or things, that are to be observed in a given problem situation. The individual items in a population are called *elementary units*. To define a population is in a sense to limit the scope of the elementary units. Elementary units possess certain *characteristics*, sometimes referred to as *traits* or *properties*, which may be qualitative or quantitative in nature.

For instance, a decision problem may involve observing the effectiveness of a shipment of certain drugs. Here the population includes all the individual units of drugs—the elementary units—in the shipment, and the characteristics of the elementary units to be observed comprise its quality, its ability to cure some disease. The result of observing an elementary unit is called an *observation*. Since, during an investigation, an observation is made on each elementary unit, we may consider a *population* as the totality of all pertinent observations that may be made in a given decision problem.

Note that both the definition of a population and the characteristics of its elementary units to be observed depend upon the nature of the decision problem at hand. Reflective consideration of the following examples will reveal the importance of this concept.

EXAMPLE 1 (Male Adults' Hats)

1. Decision problem: To decide the appropriate quantities to produce of various sizes of hats for male adults.
2. Population: All American male adults.
3. Characteristic: Circumference of each male adult's head.

EXAMPLE 2 (Housing Supply)

1. Decision problem: To determine the effective supply of apartment units in New York City.
2. Population: All apartments in New York City.
3. Characteristic: Availability.

EXAMPLE 3 (Acceptance of an Order)

1. Decision problem: To decide whether to accept or reject a shipment of television tubes.
2. Population: All television tubes in the shipment.
3. Characteristic: Time to failure.

EXAMPLE 4 (Tax Policy)

1. Decision problem: (U.S. government) To determine a reasonable corporate profits tax.
2. Population: All corporations in America.
3. Characteristic: Net corporate incomes.

EXAMPLE 5 (Corporation Policy)

1. Decision problem: (A firm) To set up a pension plan.
2. Population: All employees in the firm.
3. Characteristic: Length of service of employees.

EXAMPLE 6 (Quality of Production Process)

1. Decision problem: To determine the quality of steel rods produced by a certain process.
2. Population: All steel rods that could have been produced by that process.
3. Characteristic: Tensile strength, or diameter, or weight, or all three traits.

EXAMPLE 7 (Sales Forecast)

1. Decision problem: To estimate future consumers' demand.
2. Population: All retail stores and mail-order houses.
3. Characteristic: Volume of sales.

EXAMPLE 8 (Testing Economic Theory)

1. Decision problem: To test whether marginal propensity to consume decreases with increase in income.
2. Population: All American families.
3. Characteristic: Family income and consumption expenditures.

EXAMPLE 9 (Personal Investment)

1. Decision problem: To decide on investment in common stocks.
2. Population: All common stocks listed on the New York Stock Exchange.
3. Characteristic: Dividend rates.

Considering these examples, we find that populations can be classified into two types: infinite and finite. An *infinite population* includes an infinitely large number of elementary units. Example 6 is such a population, since, theoretically, the population includes all the possible steel rods that could be produced if the manufacturing process continued to operate indefinitely under given operating conditions. Another infinite population comprises all possible outcomes when a die is thrown continuously and indefinitely. In the case of an infinite population, clearly, complete information cannot be secured, and so any decision must be based on a sample.

A population that is not indefinitely large or that contains only a finite number of items is said to be a *finite population*. There are a number of finite populations in the list above, and the student can easily think of many others.

Note, too, that an aggregate of elementary units considered to be the population in a given problem situation may be considered to be a sample in another. In Example 3, where the problem is to determine whether to accept or reject a shipment of television tubes, the population is defined to include all the tubes in the shipment. This very shipment, however, can also be considered as a sample drawn from an infinite population of television tubes, if we wish to make a decision on the process of production itself. In other words, the first problem is to determine the quality of the tubes in the shipment, and the second is to determine the quality of the production process.

For a given problem the population may be defined in different ways. For instance, to estimate future consumers' demand we may also define the population as all family units in the United States, and the characteristic to be observed becomes family income. For the same problem (Example 7) the population includes all retail stores and mail-order houses, and the characteristic to be observed is the volume of sales. The first definition is based on the theory that consumers' demand, among other things, may be considered as a function of income. The second is founded on the knowledge that past sales can be considered as a reliable estimate of future purchases. Clearly, both sets of data are relevant to the problem. The question of which definition to choose for better results is a matter for the investigator's personal judgment. In any statistical survey, then, the investigator must understand the field of investigation as well as statistical techniques. This is required of him not only in defining statistical population but at all other stages of the statistical procedure.

At this stage, the student may ask: Can the same population be used for decision making in different problem situations? A moment's reflection will reveal that the same population and the same characteristic of its elementary units—say, all American families and family incomes—can be used as a basis for decisions concerning (a) the future consumers' demand, (b) the standard of living of American people, (c) the pattern of distribution of income, (d) the formulation of an individual income-tax policy, and so forth. Actually, the possibility of using the same population to solve different problems makes it advisable, once the nature of the decision problem is known, to check whether data already exist (collected by others for other purposes) that we can use to advantage in solving our own problem.

1–4 MEASUREMENTS OF ELEMENTARY UNITS

Characteristics of the elementary units must be expressed in terms of numbers so that they can be fed to statistical methods for refinement and analysis. Traits, or characteristics, of elementary units are of two kinds: quantitative and qualitative. Quantitative traits can be transformed into numerical data simply by direct measurement in units, such as inches, pounds, dollars, calories. The results of measuring the quantities of the elementary units form the totality of observations, which are expressed numerically in terms of the units of measurement used. The values that the quantitative observations may assume form what is called a *variable*.

When the characteristics are qualitative, the elementary units can be classified only as having or not having a certain quality or property. Nevertheless, the qualitative observations of elementary units, which are called *attributes,* often may be expressed numerically. This we do by counting the numbers of the elementary units that have a certain attribute, such as males in a human population, or satisfactory or nondefective items in a population of some industrial products. In the process of enumerating, we score one for an element that has the attribute and zero for one that lacks it. We are in effect assigning the qualitative observations numerical values, 0 or 1. With the completion of accounting we would get the total by adding the ones together, obtaining a numerical result. This expression of qualitative data quantitatively opens up to statistical treatment many areas beyond the reach of direct measurement.

Often quantitative data can also be treated qualitatively if the nature of the problem demands. As an example, a collection of ages of all the persons in a community is a variable, but it can also be treated as attributes if we are interested in, say, estimating the eligible voters in that community. In this decision problem each person may be characterized by either of the two attributes: voting age or nonvoting age. To treat a variable as an attribute is quite common in statistical work.

In measuring elementary units we must define carefully both the elementary unit and its characteristic to be observed. Although in many cases it is simple and easy to identify the elementary unit in the population, in others, this proves quite difficult unless the elementary unit is precisely defined. For instance, in decision problems concerning a housing situation, we often need to know the number of rooms in a given area. But what is a room? Do we consider a bath, a kitchen, a terrace, or a foyer a room? Confusion in measurement often arises also when we lack a precise definition of the characteristic. The most obvious illustration is the characteristic of age. How are we to record a person's age? At his last birthday? At his approaching birthday? At his nearest birthday? Also, are we interested in age in years? Age to the nearest month? Age to the nearest day? The answers, of course, depend upon the purpose of the study. We should have these answers before making actual measurements.

Let us return for a moment to the concept of a variable. A variable is a set of individual values that, in turn, are the values of the quantitative observations. When the values of a variable are arranged in some order, from large to small or vice versa, a *statistical series* is formed. A series, or a variable, may be continuous or discrete. A *continuous* variable can assume any numerical value within a specific range. In such a series successive values may differ by infinitesimal amounts. In other words, a continuous series is one in which the units may be divided into fractions of any size, no matter how small, so that there is a continuous flow of measurements with graduations infinitely minute. Measurements of weight, height, length, time, or temperature represent continuous variables, since, though the scale in each case is graded in units, cases can be found coinciding with any point on the scale. Thus, a ton of steel can be divided into hundredweights, which, in turn, can be reduced to pounds, to ounces, or to millionths of an ounce, and so on without theoretical limits. Similarly, the diameters of steel rods may be measured to the nearest 0.001 inch, to the nearest 0.0001 inch, to the nearest 0.00001 inch, or as precisely as desired. Also, the height of a man may fall at 5 feet, $5\frac{1}{7}$ feet, $6\frac{13}{24}$ feet, $8\frac{2}{57}$ feet, or any other point on the scale between the shortest and the tallest height possible for a man. Clearly, all these examples are continuous variables, because in each case the small units of measurements touch each other in a continuous flow of individual values.

The nature of some type of data, on the other hand, is such that the unit of measurement is not divisible. That is, the unit of measurement can be defined only in terms of integers or fractional values, of which there is only a finite or countably infinite number. Observations of such data can assume only discontinuous values—that is, the values vary by definite amounts. For instance, the highest degree of precision in measuring family income is to the nearest cent. It cannot take any value ending in a fraction of a cent. A variable for which the individual values fall on the scale only with distinct gaps is called a *discrete* variable.

Although the theoretical distinction between continuous and discrete variation is clear and precise, in practical statistical work it is only an approximation. Even the most precise instruments of measurement can be used only to a finite number of places; thus, every theoretically continuous series can never be expected to flow continuously, one measurement touching another without any break in actual observations. This practical drawback should not invalidate the simplifying assumption of the continuity of quantitative observations in statistical theory. Moreover, as will be explained at various places in the text, in the application of statistical methods it is often helpful to treat discrete series as if they were continuous. The important point is that in interpreting and using statistical results, the investigator should pay due attention to the logical distinction between continuous and discrete variables.

1–5 SPECIFYING PARAMETERS

Our first step, whenever a problem can be solved with the aid of data, is to define the problem statistically—that is, define the statistical population and specify the parameters. We have discussed the aspects of defining a statistical population, and at this point we shall comment only briefly on the problem of specifying parameters.

A *parameter* is a single value obtained to describe in a summary fashion the pertinent characteristics or a state of nature about a population. A population may have many characteristics and therefore just as many parameters; it often has a minimum value, a maximum value, a range, a total value of all the individual values, and so forth. Clearly, not all parameters of a population are needed in a given problem situation; their selection depends upon the nature of the problem.

1-6 MATHEMATICAL PREREQUISITES

Knowing that statistics is a branch of applied mathematics, students often approach it with mixed feelings. On the one hand, they know they cannot go very far in their fields of specialized study without some knowledge of statistical methods. On the other hand, they look to a course in statistics with skepticism or even fear, owing to previous unfortunate encounters with mathematics.

Advanced mathematics is used in the development and evaluation of statistical theory and methodology, true, but the mathematics demanded for application of statistical methods to the study of various fields is very elementary. An applied statistician often is not interested in the derivation of formulas or the advancement of theoretical principles; he is concerned rather with applying the techniques perfected by mathematical statisticians to extract useful information from quantitative data and reach valid conclusions in his field. Thus, the success of an applied statistician or a decision maker depends upon his expertise in the field of application rather than his training in advanced mathematics.

This textbook is an introduction to elementary statistics as applied to decision making for business and economic problems. As preparation, a course in freshman mathematics is ample. The reader may not be entirely familiar, however, with two basic mathematical concepts that are used: the theory of sets and procedures of counting, and a convenient notation, the summation sign.

Set theory is required for convenient and clear treatments of probability theory. Efficient counting procedures, the concepts of permutations and combinations, will help us to compute complicated event probabilities and to discuss sampling methods. In statistics, the sums of large number of terms occur frequently and it is desirable to have a simplified notation for the summation of these terms. Appendix A at the end of this book will help the reader acquire or review some of the vocabulary and ideas of these three topics. The reader may find it rewarding to study this appendix before he ventures into the region of statistical analysis proper.

EXERCISES

1–1 As you know, in any scientific study it is of vital importance to define precisely the technical terms in the field. You should therefore try to understand every term thoroughly the first time you see it. What are the new terms you have learned in this chapter? What do they mean?

1–2 According to your present understanding of statistics, indicate the possible use of this method in the study of economics and business subjects.

2

PROBABILITY AND PROBABILITY CALCULUS

Statistics, as a method of decision making under uncertainty, has probability theory as its foundation, since probability is at once the language and the measure of uncertainty and the risks associated with it. We shall begin our study of statistics by investigating probability theory in this and the next three chapters.

The study of probability theory has two basic aspects: the interpretation of probability and the calculation of probability values. The first concerns the meaning of probability for an event; the second, the derivation of rules for computing probabilities for various types of events. These aspects will be considered in the present chapter.

We begin by introducing some basic concepts that we shall need later.

2–1 RANDOM EXPERIMENTS AND RELATED CONCEPTS

Basic to the discussion of probability theory are three concepts—sample space, sample points, and events—all of which are associated with what are called random experiments.

Suppose we are interested in finding out what is going to happen when a certain act is performed. We can proceed to perform the act one or more times under the same specified conditions and observe the results. Each performance of the act is called a *trial*, and all the trials conducted under the same conditions form an *experiment*. An experiment, of course, may consist of only a single trial. An *experiment* is said to be a *random*, or *chance*, or *stochastic, process* if the following conditions are fulfilled:

(1) The set of all possible outcomes of a trial is known before the trial is performed.
(2) No particular outcome can be predicted with certainty before the performance of the trial.
(3) Replication of the trial is possible or at least conceivable.

The set of all possible outcomes of a trial is called the *sample space*, denoted by S. The number of possible outcomes in S may be denoted by $N(S)$. Note that S is the universal set under discussion (which is usually called U in set theory).

The possible outcomes, or members, in S are called *sample points*, or *elements*, or *elementary events*. Sample points will be denoted by e_i or x_i.

Any subset defined on S is called a *composite event* or, simply, an *event*. We shall denote various events as $A, B, C, \ldots$, or as $E_1, E_2, E_3, \ldots$, and the number of sample points in each event set as $n(.)$, where the "." stands for any symbol that is designated to stand for an event in question.

The way the third condition is phrased stresses the fact that replication, which is defined as repetition under the same conditions, need not be realized. Replication may not be feasible on occasions, but we insist that it be conceivable. This insistence permits us to make sense of probabilistic statements for unique events. For example, the requirement of conceivable replication precludes the statement: "It is highly likely that Senator Robert F. Kennedy could have been the thirty-seventh President of the United States had he not been assassinated in 1968."

Now we note that the tossing of an ordinary die is a random experiment, since it satisfies all the three conditions. Before the die is tossed, we know that

$$S = \{1, 2, 3, 4, 5, 6\}.$$

However, before any toss of the die, we cannot be sure which of the six possible outcomes is going to occur. Furthermore, it is obviously possible to toss the die in the same fashion any number of times.

Again, the selection of an item blindly from a shipment of merchandise is also a random experiment. Before it is drawn, we know that the selection may produce either a defective or a nondefective item, even though we don't know which will emerge. Such a draw can be repeated any number of times under the same set of conditions.

The sample points in S are said to be mutually exclusive and exhaustive. They are *mutually exclusive* in the sense that no two sample points can occur at the same time from a single trial. They are *exhaustive* because no elementary events other than those listed in S can occur.

2-2 AXIOMS OF PROBABILITY FOR FINITE SAMPLE SPACES

Sample spaces, like universal sets, can be finite or infinite, depending upon whether the number of sample points is finite or infinite. Although the present chapter is concerned mainly with finite sample space, infinite sample space is discussed briefly at the end. The whole field of probability theory for finite sample spaces is based upon three simple axioms, or postulates.

Axiom I: The probability of an event is restricted to the range from zero to unity inclusive. That is, if we denote the probability of an event as $P(E)$, then $0 \leq P(E) \leq 1$.

Axiom II: The probability of the entire sample space is 1; i.e., $P(S) = 1$. It is the convention in probability theory that an event is said to have occurred if at least one of the elements in the event set has occurred. When an experiment is performed, we are certain that at least one of the sample points in S must occur and, therefore, the probability associated with S is unity.

Axiom III: The sample points in S are mutually exlusive and each is a simple event with a probability. Thus, if A and B are disjoint events—i.e., $A \cap B = \varnothing$—then $P(A \cup B) = P(A) + P(B)$.

The importance of this theorem is clear. If, by some procedure, a weight (probability) can be assigned to each of the sample points in S, the probability of any event defined on S can be obtained merely by summing the separate weights of all the sample points that are members of the event set under consideration.

To see the value of Axiom III, consider the following theorem:

> Let A be an event in a finite sample space S. If A is the empty set, then $P(A) = 0$. If A is nonempty, then $P(A)$ is the sum of the probabilities of the elementary events whose union is A.

For proof of the foregoing theorem, we first suppose $A = \varnothing$. In Axiom III, take $A = \varnothing$ and $B = S$, the entire sample space. Clearly A and B are mutually exclusive, since $\varnothing$ is empty. Thus

$$P(\varnothing \cup S) = P(\varnothing) + P(S). \tag{1}$$

Furthermore, since S is the entire sample space, we have $\varnothing \cup S = S$; hence,

$$P(\varnothing \cup S) = P(S). \tag{2}$$

Subtracting (2) from (1), we obtain

$$0 = P(\varnothing),$$

and thus $P(A) = 0$ if $A = \varnothing$.

Next, suppose A is nonempty and is the union of k distinct elementary events, $E_1, E_2, \ldots, E_k$, where $E_i = \{e_i\}$, $i = 1, 2, \ldots, k$. For convenience, we assume the sample points have been labeled in such a way that the k points in A are the first k elements in S. Now, if $k = 1$, $A = E_1$ and $P(A) = P(E_1)$. If $k = 2$, $A = E_1 \cup E_2$ is the union of the two mutually exclusive events, since E_1 and E_2 are distinct elementary events. Axiom III then gives this result:

$$(3) \qquad P(A) = P(E_1) + P(E_2).$$

If $k = 3$, then $A = (E_1 \cup E_2 \cup E_3)$ and, again by Axiom III,

$$P(A) = P(E_1 \cup E_2) + P(E_3).$$

By (3) we have

$$P(A) = P(E_1) + P(E_2) + P(E_3).$$

Finally, the extension of values of $k > 3$ is readily apparent by mathematical induction. Assuming the theorem is true for $k - 1$ elementary events and considering A as the union of E_k, we apply Axiom III to obtain

$$P(A) = P(E_1 \cup E_2 \cup \cdots \cup E_{k-1}) + P(E_k),$$

which is the desired extension.

We have just seen that the probability of an event is a number between 0 and 1; but how is such a number assigned? This question refers to the basic interpretation of probability. As often occurs with basic concepts in any scientific discipline, leading authorities disagree on the definition of probability. There are three major schools of thought on this matter: the classical, the relative-frequency, and the subjective theories of probability.

2-3 THE CLASSICAL THEORY

The classical theory of probability is the oldest, having originated in games of chance. Mathematicians in eighteenth-century France were confronted with questions raised by gamblers: How can the uncertainty of the outcomes of a game, such as coins, dice, or decks of cards, be measured? What constitutes a fair bet? Are the odds favorable? Is it possible to figure out the chances of winning at craps or to evaluate different systems of betting at a roulette wheel? Attempts to answer these questions led to the birth of probability theory.

The classical theory is based on the simple assumption of equally likely outcomes of a random experiment. Under this assumption, we should have no reason to prefer any one of the possible outcomes to any other. This argument was called the *principle of insufficient reason* by a Swiss mathematician, Jacob Bernoulli (1654–1705). By this principle, all the possible outcomes in S should have the same weights as their probabilities of occurring. Thus, if a fair die is tossed, each of the six numbers should be considered equally likely to occur; as a consequence, the probability that any number (say, a 3) will occur is $\frac{1}{6}$.

In general, if there are N possible equally likely outcomes of a random experiment, then the probability for each sample point should be $1/N$.

The foregoing remarks lead to the simple, straightforward classical interpretation of probability. If the sample space of an experiment has $N(S)$ mutually exclusive and equally likely outcomes, and if an event, as defined on this sample space, has $n(E)$ elements, then the probability of this event, denoted as $P(E)$, is the ratio of $n(E)$ to $N(S)$. That is,

$$P(E) = \frac{n(E)}{N(S)}. \tag{2–1}$$

For example, a die is tossed. Let A be the event that an odd number will occur, B the event that the number will be greater than 4, and C the event that either a 1 or a 2 will occur. Then the respective probabilities of these events, by (2–1), are

$$P(A) = \frac{n(E)}{N(S)} = \frac{3}{6} = \frac{1}{2},$$

$$P(B) = \frac{n(B)}{N(S)} = \frac{2}{6} = \frac{1}{3},$$

and

$$P(C) = \frac{n(C)}{N(S)} = \frac{2}{6} = \frac{1}{3}.$$

We should point out that the classical theory, under the assumption of equally likely outcomes, depends upon logical reasoning. Thus, we meet no difficulty if we are concerned with a well-balanced coin, an unbiased die, an honest roulette wheel, or any other experiment whose outcomes are symmetrical and can be deduced by logic. But what about a coin that is unbalanced? A loaded die? A crooked roulette wheel? In each of these cases, the classical approach of assigning equal probabilities would offer us nothing but confusion. Fortunately, when a priori reasoning fails owing to the lack of symmetry, we can consult the relative-frequency theory of probability.

2-4 THE RELATIVE-FREQUENCY THEORY

The relative-frequency theoreticians claim that the only valid procedure for determining event probabilities is through repetitive experiments. For example, when a coin is tossed, what is the probability that the coin will turn up heads? The relative-frequency theorist would approach the problem by actually tossing the coin, say, 100 times under the same conditions, and then calculating the proportion of times the coin fell heads. Suppose the coin falls heads 45 times out of 100; then the ratio $45/100$ is used as the estimate of the probability of heads, $P(H)$, of this coin. A moment's reflection will show that even if the coin were unbiased, we might not have exactly 50 heads out of 100 tosses. In other words, we cannot expect to obtain the true probability from repeated experiments. However, if the coin were well balanced, the estimate would approach the true ratio (probability) of $1/2$ as the number of trials increased.

This discussion leads us to the following interpretation of probability in terms of relative frequency: If an experiment is performed n times under the same conditions and there are x outcomes, $x \leq n$, favoring an event, then an estimate of the probability of that event is the ratio x/n. Furthermore, the estimate of the probability of an event x/n approaches a limit, the true probability of the event, when n approaches infinity; that is,

$$P(E) = \lim_{n \to \infty} \left(\frac{x}{n} \right).$$

Clearly, in reality we can never obtain the probability of an event as given by this limit. In practice, we can only seek a close estimate of $P(E)$ based on a large n. For convenience we shall treat the estimate of $P(E)$ as if it were actually $P(E)$, writing the working relative-frequency definition of probability as

$$P(E) = \frac{x}{n}. \tag{2-2}$$

Defining $P(E)$ as a limit as n approaches infinity, however, does emphasize that probability involves a long-run concept. This means that when we toss a balanced die six times, it is almost impossible for each of the six numbers to appear once. If, however, we toss the die over and over again, for a large number of times, we can expect, in the long run, or on the average, each of the six faces of the die to appear about one-sixth of the time. It is exactly in this sense that we say the probability of getting any one of the numbers on a die in a random toss is $1/6$.

Both the classical and relative-frequency theories are called the *objective* definitions of probability. The classical definition is *objective* in the sense that it is based on deduction from a set of assumptions. The relative-frequency definition is *objective* because the probability of an event is determined by repeated empirical observations, or by induction.

2–5 THE PERSONALISTIC THEORY

An objectivist is quite at home in talking about probability in connection with the toss of a coin or the manufacture of a mass-produced output. He can readily think of the number of automobile tires produced vis-à-vis the probability of one defective tire as the long-run ratio of the number of defective tires to the total number of tires produced. He might, however, be helpless with respect to the problem of unique events—events that occur just once or that cannot be subjected to repetitive experiments. Thus, he would not care to talk about the probability of Columbus' discovering America, or of Bacon's writing the works of Shakespeare. As a result, a large class of problems is beyond the reach of the objectivist. This limitation of both the relative-frequency theory and the classical assumption of symmetry has prompted the birth of the personalistic view of probability. While the relative-frequency theory is still the most popular, the personalistic position is steadily gaining strength.

The *personalistic*, or *subjective*, theorist regards probability as a measure of personal confidence in a particular proposition, such as a belief that Red China will be admitted to the United Nations no later than the mid-1970's. A subjectivist assigns a weight between zero and one to an event, according to his degree of belief in its possible occurrence. For example, if he is twice as confident of the occurrence of event A as he is of event B, and if A and B are the only two possible propositions in a situation, he then assigns the values $P(A) = 2/3$ and $P(B) = 1/3$. In general, the probabilities he assigns to various events for a given proposition are weighted averages, adding up to unity.

The subjective point of view grants that different reasonable individuals may differ in their degrees of confidence, even when offered the same evidence. Consequently, personal probabilities for the same event may differ in the eyes of different decision makers. Furthermore, the subjectivist can apply probabilities to all problems that an objectivist studies, and to many more. When the condition of symmetry is met, the personalist will obviously acknowledge a priori reasoning, and when the amount of data is large, the subjectivist will usually get the same answers as the relative-frequency theoretician. But, in addition, a personalist would take in his stride such propositions as those presented above concerning Columbus, Bacon, and Red China.

Despite these three different positions on the interpretation of probability, there is hardly any disagreement at the mathematical level. The basic axioms are valid for each of these three concepts. Thus, once weights are assigned to sample points, or the probabilities of events are determined, by any one of the three procedures, the same probability calculus can be applied. Since each approach has its own merits, we shall use whichever is convenient and appropriate for the problem under consideration.

2-6 PROBABILITIES VERSUS ODDS

Our consideration of the meaning of probability is incomplete unless we mention the concept of "odds," which is a common, everyday expression of event probabilities. *Odds* relate the chances for an event to the chance against it. When one says, for example, that the odds are 2 to 1 that A will win the next election, he means that there are two chances in three that A will win the election. To convert odds to probabilities, we note that odds of 2 to 1 can also be written as 2:1 (e.g., two successes to one failure). This can be converted into a probability of $\frac{2}{3}$, since $2/(2 + 1) = \frac{2}{3}$.

In general, if the odds are $x:y$ in favor of an event E, then $P(E) = x/(x + y)$. Furthermore, to say that the odds are $x:y$ in favor of an event is the same as to say that the odds are $y:x$ against the event. The probability of nonoccurrence of the event is $y/(x + y)$.

2-7 SIMPLE RANDOM SAMPLING

In discussing probability calculus, we shall use such phrases as "a random selection" or "n items are selected at random from a certain population." These phrases are employed in connection with a sampling process called *random sampling*.

A simple random sample is a "probabilistic" sample and it can be selected with or without replacement. In the case of drawing *with replacement*, the resulting sample is said to be an *unconditional simple random sample* if the selection process is such that, at each draw, each and every item in the population has the same chance of being included in the sample, and samples of equal sizes are equally likely to be drawn. For this process each individual item drawn must be measured, recorded, and returned to the population before another draw is made; thus the same item may be included in the sample more than once.

Sampling with replacement is clearly impossible if observation is destructive, since an item drawn, observed, and thereby destroyed cannot be returned to the population. In that case, a sample may be selected without replacement. That is, items selected need not be returned to the population before additional draws are made, or else a sample of n items may be selected simultaneously. When sampling is made *without replacement*, the resulting sample is said to be a *conditional simple random sample* if the process of selection insures that each item at each draw has the same chance of being included in the sample and that all the possible samples of the same size are equally likely to be selected. The second condition means that, if a sample of 3 is selected from a population of 10 items, then there are ${}_{10}C_3$ possible samples of $n = 3$, and each must have a probability of $1/({}_{10}C_3)$ of occurring.

A simple random sample is drawn by chance selection, which must be differentiated from selection in a "haphazard," or "hit-and-miss," manner. Unless this distinction is observed, random methods may be misapplied, and a sample may be assumed to be random that is not random at all. The simplest method of obtaining randomness is the "goldfish bowl" procedure. All elements in the universe are numbered and the numbers written on chips, balls, cards, or other articles that are physically homogeneous. These articles are put into a bowl and mixed thoroughly. Then the specific number of items to be included in the sample are chosen either with or without replacement. With replacement, the probability of each element's being drawn at each trial is $1/N$; without replacement, the probability of selecting the subsequent items increases successively. Without replacement, for example, the probability of drawing each item at the first selection is $1/N$, at the second selection, $1/(N - 1)$, at the third, $1/(N - 2)$, and so on.

The "goldfish bowl" procedure can be shortened, money and time saved, and reliability in sampling increased by use of a table of *random numbers*, such as Table B–13 in Appendix B. Such a table is so constructed that positive integers from 0 to 9 appear with approximately the same frequency. When numbers are combined in pairs, numbers from 00 to 99 are obtained. Combining three at a time gives numbers from 000 to 999.

To illustrate, suppose you wish to draw a sample of 500 items from a population containing 9000 items. First, prepare a list of the individual items in the population, such as names of employees, retail prices, households, or the like, and assign each item on the list a number from 1 to 9000. Then obtain your sample in the following fashion.

1. Enter the table in a random manner. Open the table and put it in front of you. Close your eyes. Point your finger to the table. The figure under your finger becomes the starting point. Then read the table vertically, horizontally, or diagonally. The table of random numbers can be used in any systematic way once the random start is made.
2. Read all the four-digit numbers in the table (since our population consists of 9000 members). If the number is 9000 or smaller, record it. If the number is greater than 9000, skip it. (If the sample is to be taken without replacement, also skip a number when it appears a second time.)

Proceed in this fashion until you obtain 500 numbers. (This procedure is quite general. If N is between 100 and 999, use the first three columns of the table; if N is between 1000 and 9999, use the first four columns; if N is between 10,000 and 99,999, use the first five columns; and so on.)

3. Take from the list those five hundred items that have the corresponding recorded numbers, for they are the members of the desired sample.

2-8 MUTUALLY EXCLUSIVE EVENTS

Axiom III is sometimes called the *additive rule* for disjoint events. We shall restate it here for convenience.

Two events, A and B, are said to be *mutually exclusive*, or *disjoint*, if they do not have any sample points in common—that is, $A \cap B = \emptyset$. Thus, disjoint events cannot occur simultaneously and, by Axiom III, the probability of occurrence of either A or B is the sum of their separate probabilities:

$$\begin{aligned} P(A \text{ or } B) &= P(A \cup B) \\ &= P(A) + P(B). \end{aligned} \tag{2-3}$$

If a card is drawn at random from a bridge deck, what is the probability that it will be (1) either an ace or a queen? (2) a heart or a club? (3) an honor card or a 3?

To solve this problem, note that there are 52 cards (that is, 52 equally likely outcomes) in the deck, made up of four suits: spades, hearts, diamonds, and clubs, each with thirteen cards; the honor cards are aces, kings, queens, jacks, and tens. Thus,

(1) $P(\text{an ace or a queen}) = 4/52 + 4/52 = 2/13,$
(2) $P(\text{a heart or a club}) = 13/52 + 13/52 = 1/2,$
(3) $P(\text{an honor or a 3}) = 20/52 + 4/52 = 6/13.$

The law of addition for disjoint events can be generalized to any number of events. For example, n events are said to be mutually exclusive if no two of them have any sample points in common. Thus, if $E_1, E_2, \ldots, E_n$ are n disjoint events, then

$$P(E_1 \cup E_2 \cup \cdots \cup E_n) = P(E_1) + P(E_2) + \cdots + P(E_n). \tag{2-4}$$

Furthermore, if the union of these n disjoint events form a partition of S, we have

$$P(E_1 \cup E_2 \cup \cdots \cup E_n) = P(S) = 1.$$

A coin and a die are tossed. Let A be the event "$H5$ or $H6$," B the event "$T4$, $T5$, or $T6$," and C the event "$H3$.". The probability of occurrence of any one of these events is then

$$P(A \cup B \cup C) = P(A) + P(B) + P(C)$$
$$= 2/12 + 3/12 + 1/12 = 1/2.$$

(Note that there are twelve sample points in this experiment and each point is assigned a weight of $1/12$.)

2–9 JOINT EVENTS

Two events A and B are said to be *joint* if they have sample points in common. Those sample points that belong to both A and B form a subset called the *intersection* of A and B, denoted by $A \cap B$ or by AB (Figure 2–1). We know that, in the case of disjoint events, $P(A \cup B)$ is the sum of probabilities of the points in the set $A \cup B$. That is, $P(A) + P(B)$ is the sum of probabilities of the points in A and the sum of probabilities of the points in B. Now, for the overlapping events, the quantity $P(A) + P(B)$ includes the probabilities of the sample points in the intersection set twice. Therefore, if we subtract $P(A \cap B)$—the probability of the intersection—once, we will have the sum of probabilities of all points in AB, each taken once. The logic of this procedure is shown by Figure 2–1. We conclude finally that, if A and B are joint events, then the probability of obtaining either A or B is

$$P(A \cup B) = P(A) + P(B) - P(A \cap B). \tag{2–5}$$

For example, if a card is drawn at random from a bridge deck, what is the probability that it will be either a spade or a king? Clearly a spade and a king are not mutually exclusive events, since one of the four kings is a spade. It is also clear that there is only one sample point in the intersection set, the

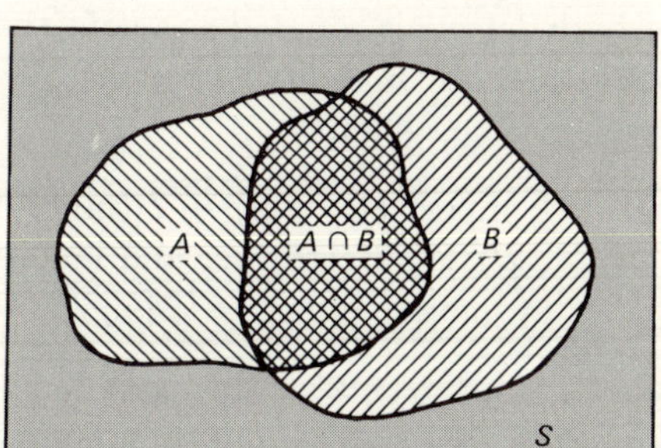

Figure 2–1 Venn Diagram for Two Joint Events

king of spades. Thus, by (2–5), we have

$$P(\text{a spade or a king}) = P(\text{spade}) + P(\text{king}) - P(\text{spade and king})$$
$$= {}^{13}\!/_{52} + {}^{4}\!/_{52} - {}^{1}\!/_{52} = {}^{4}\!/_{13}.$$

A coin and a die are thrown. Let A be the event that the coin turns up tails and B the event that a 3 or a 4 shows up on the die. What is the probability that either A or B will appear? In this example, as shown by Figure 2–2, there are two sample points, $T3$ and $T4$, in the intersection.

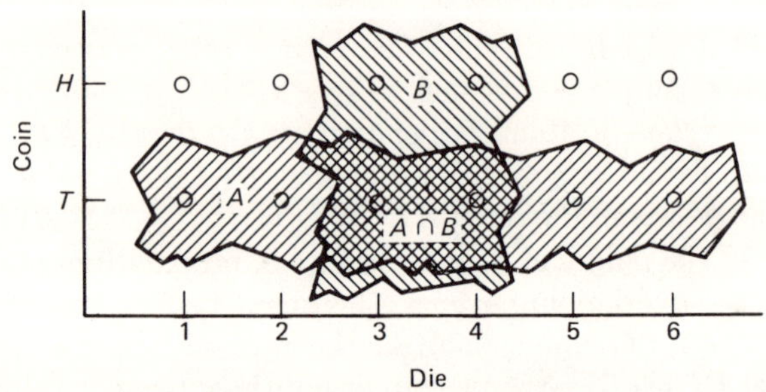

Figure 2–2 The Event *A* That the Coin Turns Up Tails and the Event *B* That the Die Shows 3 or 4 in the Coin-Die Experiment

Thus,

$$P(A \cup B) = {}^{1}\!/_{2} + {}^{2}\!/_{6} - {}^{2}\!/_{12} = {}^{2}\!/_{3}.$$

Let us consider another application of the additive rule. In a lot of 1200 golf balls, 40 have imperfect covers, 32 cannot bounce, and 12 have both defects. If a ball is selected at random from the lot, what is the probability that the ball is defective?

The probability of a ball's having an imperfect cover is ${}^{40}\!/_{1200}$, and the probability of its not bouncing is ${}^{32}\!/_{1200}$. Here we have two joint events, and a ball that has both defects is counted as a defective twice—once for its imperfect cover and once for its inability to bounce. We are told that twelve balls have both defects. Consequently,

$$P(\text{defective}) = P(\text{imperfect cover}) + P(\text{inability to bounce})$$
$$- P(\text{imperfect cover and inability to bounce})$$
$$= {}^{40}\!/_{1200} + {}^{32}\!/_{1200} - {}^{12}\!/_{1200} = 0.05.$$

The additive rule for joint events can be extended to cover any number of events. In particular, if A, B, and C are three joint events, as suggested by Figure 2–3, the probability that any one of these events will occur is

$$P(A \cup B \cup C) = P(A) + P(B) + P(C) - P(A \cap B) - P(A \cap C) - P(B \cap C) + P(A \cap B \cap C). \quad (2\text{–}6)$$

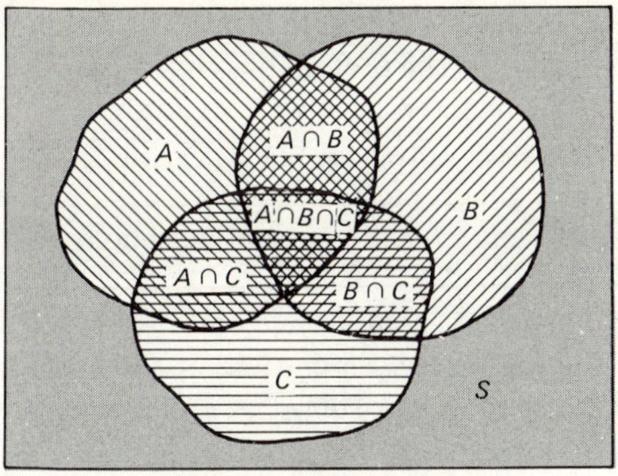

Figure 2-3 Venn Diagram for Three Joint Events

2-10 COMPLEMENTARY EVENTS

Two events, E and $\bar{E}$, are said to be *complementary events* if the second is a subset that contains all the elementary events in the sample space that are not in the first; that is, complementary events are mutually exclusive and their union is the sample space S. It follows from Axioms II and III of probability that

$$P(\bar{E}) = P(S) - P(E) = 1 - P(E). \tag{2-7}$$

What is the probability of obtaining a 1, 2, 3, 4, or 5 when an ordinary die is tossed? In a problem of this sort it is much more convenient and compact to obtain $P(E)$ by first obtaining $P(\bar{E})$, which in this case is $P(6)$, and then applying the law of *complementation*. Hence,

$$P(1, 2, 3, 4, \text{ or } 5) = 1 - P(6) = 1 - 1/6 = 5/6.$$

(What is the alternative way to calculate this probability?)

Consider one more example. If two ordinary dice are tossed, what is the probability that the sum of the points on the dice will not be 7? This experiment has 36 possible outcomes, of which six correspond to the event that the sum is 7. Assuming equally likely outcomes and denoting the event "7" by E, then we we have

$$P(\bar{E}) = 1 - P(E) = 1 - 6/36 = 5/6.$$

2-11 CONDITIONAL PROBABILITY

Often when we are concerned with event probabilities we already have some information with respect to an experiment. The availability of such information in effect reduces the original sample space to one of its subsets.

That is, we are dealing with probabilities of a section rather than of the whole sample space. Clearly, the probability of an event about which we had some information would be different from a situation in which we had no information. For example, the probability that a card drawn at random from a bridge deck will be an ace is greater if we know that the card is an honor card. Also, we can easily appreciate that the probability that a household selected at random from greater Los Angeles will have an annual income of over $25,000 is different from that for a household chosen randomly from, say, Beverly Hills.

In each of the foregoing examples, attention is focused on the probability of an event in a subset, or subpopulation, of the original sample space; the probability of an event in the subset is greater than it is in the original sample space. Each such subset is a reduced sample space and is specified by new conditions (information) beyond the initial conditions of the original sample space. Probabilities associated with events defined on the subpopulations are called *conditional probabilities*. We shall introduce the law governing conditional probabilities with an example.

Consider the experiment of tossing a coin and a die together. Given that the coin falls heads, find the probability that the die falls even.

We know that the original sample space for this experiment consists of twelve equally likely outcomes. Among them, the coin falls heads six times. Ignoring the other six outcomes associated with tails, we have a reduced sample space S' defined as

$$S' = \{(H1), (H2), (H3), (H4), (H5), (H6)\}.$$

The elementary events in S' are equally likely, so equal weights may be assigned as their probabilities, making them add up to unity. Thus, we have for each sample point a probability of $1/6$. Next, we observe that the event "the die falls even" consists of three points in this reduced sample space—that is, $(H2)$, $(H4)$, and $(H6)$. Consequently, the probability of this event's occurring in S' is $3/6 = 1/2$; or, the number $1/2$ is the conditional probability that the die falls even, given that the coin falls heads.

Let us study the concept of conditional probability further by considering the events "die falls even," denoted by A, and "coin falls heads," denoted by B, with reference to the whole sample space of the coin-die experiment. Clearly

$$A = \{(H2), (H4), (H6), (T2), (T4), (T6)\}$$

and

$$B = \{(H1), (H2), (H3), (H4), (H5), (H6)\}.$$

A moment's reflection will show that if we want to determine the probability of A given B, we must know the elements that belong to A and B—that is, the set $A \cap B$. From sets A and B given above, we see that

$$A \cap B = \{(H2), (H4), (H6)\}.$$

We know that there are twelve equally likely sample points in the original sample space; therefore,

$$P(A) = 6/12, \quad P(B) = 6/12, \quad \text{and} \quad P(A \cap B) = 3/12.$$

Since we wish to find the probability of A given B, the event set B becomes the reduced sample space, and $A \cap B$ becomes the event set in question. Thus, for this example, the probability that the die falls even given that the coin falls heads is

$$\frac{P(A \cap B)}{P(B)} = \frac{3/12}{6/12} = \frac{1}{2},$$

as obtained before.

We may now state the *law of conditional probability* as follows: The probability of A given B, denoted as $P(A \mid B)$, where the vertical line is read "given," is defined by the expression

$$P(A \mid B) = \frac{P(A \cap B)}{P(B)}, \qquad \text{if } P(B) > 0. \tag{2-8}$$

It can be shown, using the same line of reasoning, that the probability of B given A is

$$P(B \mid A) = \frac{P(B \cap A)}{P(A)}, \qquad \text{if } P(A) > 0. \tag{2-9}$$

For instance, in the coin-die experiment above, the probability that the coin falls heads, B, given that the die falls even, A, is

$$P(B \mid A) = \frac{P(B \cap A)}{P(A)} = \frac{3/12}{6/12} = \frac{1}{2},$$

which turns out to be the same as the probability that the die falls even given that the coin falls heads. But are $P(A \mid B)$ and $P(B \mid A)$ always equal to each other?

An urn contains four white balls, numbered 1, 2, 3, and 4, and six black balls, numbered 5, 6, 7, 8, 9, and 10. A ball, selected at random from the urn, is white; what is the probability that the ball chosen is numbered 1? We have

$$P(\text{white ball}) = 4/10 = 0.4,$$
$$P(1 \text{ and white}) = 1/10 = 0.1.$$

Therefore,

$$P(\text{numbered } 1 \mid \text{white ball}) = \frac{0.1}{0.4} = 0.25.$$

In a certain town, blacks and whites each form 50 percent of the population. It is known that 20 percent of the blacks and 5 percent of the whites are unemployed. An economist studying the employment situation selects an unemployed person at random. What is the probability that the person so selected is (a) black, (b) white?

The statement of the problem gives us probabilities as shown in Table 2–1. For example, 20 percent of 50 percent of the population, or 10 percent, are both black and unemployed, so 40 percent are black and employed. Similarly,

Table 2–1 Unemployment Data

	Unemployed	*Employed*	*Row sums*
Black	0.100	0.400	0.500
White	0.025	0.475	0.500
Column sums	0.125	0.875	1.000

5 percent of 50 percent, or 2.5 percent, are white and unemployed. From probabilities in this table, we have the conditional probability of the event "person selected is black," B, given the event "person is unemployed," U, as

$$P(B \mid U) = \frac{P(B \cap U)}{P(U)}$$

$$= \frac{0.100}{0.125} = 0.8,$$

and the probability of white, W, given unemployed, as

$$P(W \mid U) = \frac{P(W \cap U)}{P(U)}$$

$$= \frac{0.025}{0.125} = 0.2.$$

Several observations are of interest in connection with the definition of conditional probability.

First, all event probabilities are associated with some sample space. In effect, $P(A) = P(A \mid S)$, where S is omitted and understood. The conditional sign is employed only when there is some subset of S, S', that contains all outcomes of an experiment. In such a case we have, in particular, $P(A \mid A) = 1$.

Second, the probabilities in the ratio on the right side of (2–8) are event probabilities in the original sample space. The same result can be obtained if we first convert to the reduced sample space, which is B. Then we have increased the probability in B to 1; and, if we compute the probability on the right side of (2–8) in sample space B, the denominator is 1 whereas the numerator is the probability of $A \cap B$ in B. Similar interpretations can be given to (2–9). This is why the conditional probability of an event is greater than in the original sample space. We can see that additional information is a valuable thing. Figure 2–4 illustrates this point.

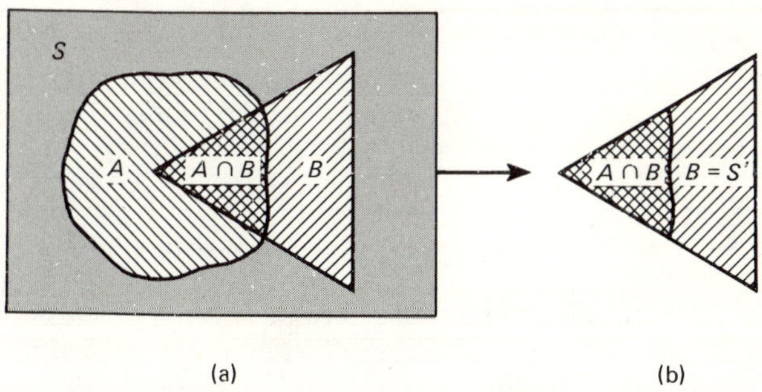

Figure 2–4 Reducing the Sample Space for Evaluating the Conditional Probability $P(A \mid B)$

Third, the probability of the reduced sample is greater than zero because otherwise the conditional probability is not defined.

2–12 DEPENDENT EVENTS

The discussion of conditional probability leads to the differentiation of statistical dependence from independence. When we are interested in the probability of two or more events' occurring, we naturally wish to know if the events affect each other.

Two events are said to be *dependent* if the occurrence or nonoccurrence of one event in any trial affects the probability of other events in other trials. For example, two cards are drawn at random from a bridge deck without replacement. The probability that the first selection will be a red card is $26/52$; the probability that the second card will be red, however, depends upon the outcome of the first draw. If the first is a red card, the probability of drawing a red card the second time is $25/51$. However, if the first card drawn is black, then the probability of drawing a red card on the second selection is $26/51$. Thus, in the case of dependent events, the probability of any event is conditional on, or depends upon, the occurrence or nonoccurrence of other events. From definitions of conditional probabilities, we can see that if A and B are dependent events, then,

$$P(A \cap B) = P(B)P(A \mid B) \tag{2-10}$$

and

$$P(B \cap A) = P(A)P(B \mid A). \tag{2-11}$$

Order is of no significance in the intersection set, since $A \cap B = B \cap A$. This property of intersection yields the following results:

$$P(A \cap B) = P(B)P(A \mid B) = P(B \cap A) = P(A)P(B \mid A). \tag{2-12}$$

The manager of a firm estimates that if he puts an ad in a certain magazine, the ad will be read by half its subscribers, and 1 percent of those who read the ad will buy the product advertised. Assuming these estimates are accurate, then we can say,

$$\begin{aligned} P(\text{read ad and buy product}) &= P(\text{read ad})P(\text{buy product} \mid \text{read ad}) \\ &= (0.5)(0.01) = 0.005. \end{aligned}$$

In a lot of 100 manufactured items, 15 are defective. Suppose 2 items are drawn at random without replacement; what is the probability that both are defective? Letting D_1 and D_2 denote the defectives, we have

$$P(D_1 \cap D_2) = P(D_1)P(D_1 \mid D_2) = (15/100)(14/99) = 21/990.$$

An urn contains three white and seven black balls; two balls are drawn, without replacement, at random. We desire to set up the sample space of this experiment and assign probabilities to the sample points. Evidently, the sample space of this experiment is

$$S = \{(W_1W_2), (W_1B_2), (B_1W_2), (B_1B_2)\},$$

where, for instance, (B_1W_2) means the first ball is black and the second ball is white. Random selection here ensures that all balls on any drawing are equally likely to be chosen. Thus, we have the following results:

$$\begin{aligned} P(W_1W_2) &= P(W_1)P(W_2 \mid W_1) = (3/10)(2/9) = 6/90; \\ P(W_1B_2) &= P(W_1)P(B_2 \mid W_1) = (3/10)(7/9) = 21/90; \\ P(B_1W_2) &= P(B_1)P(W_2 \mid B_1) = (7/10)(3/9) = 21/90; \\ P(B_1B_2) &= P(B_1)P(B_2 \mid B_1) = (7/10)(6/9) = 42/90. \end{aligned}$$

This example illustrates the application of the multiplicative rule to assign weights in a sample space. The sum of weights assigned in any sample space, as in this illustration, must add up to unity.

The multiplicative law for dependent events can be extended to cover more than two events. For example, the probability of the simultaneous, or joint, occurrence of three dependent events, A, B, and C, is

$$P(A \cap B \cap C) = P(A)P(B \mid A)P(C \mid A \cap B). \tag{2-13}$$

If three cards are drawn from a deck of bridge cards in succession without replacement, what is the probability that all are aces? Denote aces by A_i; then by (2–13) we obtain

$$P(A_1A_2A_3) = P(A_1)P(A_2 \mid A_1)P(A_3 \mid A_1A_2)$$
$$= (4/52)(3/51)(2/50).$$

Again, if five random selections are made simultaneously from a deck of bridge cards, what is the probability that all five cards are hearts? The answer, by analogy with (2–13), is

$$P(5 \text{ hearts}) = (13/52)(12/51)(11/50)(10/49)(9/48).$$

2–13 INDEPENDENT EVENTS

Two events A and B are independent if and only if

$$P(A \cap B) = P(A)P(B), \tag{2–14}$$

which implies that the independence between A and B requires that

(1) $$P(A \mid B) = P(A)$$

and

(2) $$P(B \mid A) = P(B).$$

These two expressions imply each other. We obtain (1) and (2) from (2–14) by recalling that

$$P(A \cap B) = P(A \mid B)P(B) = P(B \mid A)P(A).$$

If each of these expressions is set to equal $P(A)P(B)$, then all consequences of (1) and (2) follow (2–14), provided that $P(A) \neq 0$, $P(B) \neq 0$.

As an example, a white die, w, and a blue die, b, are tossed. What is the probability that $w \geq 5$ and $b \leq 4$?

We note that the sample space consists of the 36 possible combinations of the white and the blue die, each with six possible outcomes, and that the event under consideration requires that the two conditions be satisfied simultaneously. If A is the event that $w \geq 5$ and B is the event that $b \leq 4$, then we need to know the number of sample points that both sets have in common—that is, the intersection, $A \cap B$. In Figure 2–5 we can see that there are 8 points in the intersection set. We have then $P(A \cap B) = 8/36$. By counting, we find that there are 12 sample points in A and 24 sample points in B. As a result, $P(A) = 12/36$ and $P(B) = 24/36$. Thus

$$P(A \cap B) = (12/36)(24/36) = 8/36,$$

which verifies that the multiplicative rule gives the correct answer.

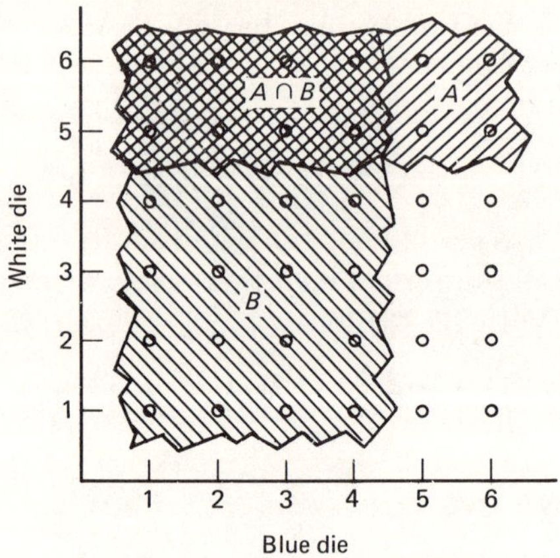

Figure 2–5 The Intersection of A: $w \geq 5$ and B: $b \leq 4$

This example reveals two important aspects of the multiplicative rule for independent events. First, if two events are independent, they must satisfy (2–14). Second, if A and B are two independent events with nonzero probabilities, then there must be at least one common elementary event in the intersection set; that is, $A \cap B \neq \emptyset$.

A coin and a die are tossed. Show that the event "head on the coin" and the event "the die shows even" are independent.

We may let A be the first event defined on the sample space for outcomes of the coin, S_1, and B be the second event defined on the sample space for outcomes of the die, S_2, and then proceed to find the probability of the joint outcomes of A and B to see if the multiplicative law for independent events is satisfied. From the statement of the problem, we have

$$P(A \cap B) = \left[\frac{n(A)}{N(S_1)}\right]\left[\frac{n(B)}{N(S_2)}\right].$$

We know that $P(A) = 1/2$, since

$$A = \{H\} \quad \text{and} \quad S_1 = \{H, T\}.$$

Similarly, $P(B) = 3/6 = 1/2$, since

$$B = \{2, 4, 6\} \quad \text{and} \quad S_2 = \{1, 2, 3, 4, 5, 6\}.$$

Now, we note that

$$A \cap B = \{(H2), (H4), (H6)\}$$

and

$$S = \{(H1), (H2), \ldots, (H6), (T1), \ldots, (T6)\}.$$

Therefore,

$$P(A \cap B) = 3/12 = 1/4 = P(A)P(B) = (1/2)(1/2).$$

This result verifies that A and B are independent. It is interesting to observe that if two events are independent, their joint probability is the product of their separate probabilities defined on their respective sample spaces.

The extension of the definition of independence to three or more events is not immediate, and some changes are required. In general, n events $E_1, E_2, \ldots, E_n$ are completely independent if they are independent by pairs, by triples, by quadruples, and so forth. In particular, three events A, B, and C are completely independent if and only if

$$P(A \cap B) = P(A)P(B),$$
$$P(A \cap C) = P(A)P(C),$$
$$P(B \cap C) = P(B)P(C),$$

and

$$P(A \cap B \cap C) = P(A)P(B)P(C). \tag{2–15}$$

Hence, the events must be independent not only by pairs but also by triples.

Suppose, for example, a random sample of three is drawn with replacement for a shipment of merchandise, 10 percent of which is defective; what is the probability that all three units in the sample are nondefectives? If we let A, B, and C, be the three nondefective units drawn, then

$$P(A) = P(B) = P(C) = 0.9,$$
$$P(A \cap B) = P(A \cap C) = P(B \cap C) = (0.9)(0.9),$$

and

$$P(A \cap B \cap C) = (0.9)(0.9)(0.9) = 0.729.$$

The fact that A, B, and C are independent in this example is obvious, since random selection with replacement dictates that occurrence of a nondefective item on any draw cannot affect whether a nondefective item will occur on another draw.

Now, consider the repeated tossing of a balanced coin. Here the tosses are independent in the sense that the outcome of one toss does not affect the outcome of any other. The probability of getting four heads in four tosses is $(1/2)(1/2)(1/2)(1/2) = 1/16$. If there are 100 tosses and the results are 100 heads, $P(H)$ on the 101st toss is still $1/2$ because of independence. By this time, of course, you may very well question the assumption that the coin is well balanced, since $P(100H) = (1/2)^{100}$, which is almost zero. Our discussion on independence here, however, does bring out the fallacy of the gambler's maturity theory of chance, which entertains the false idea that a long run of failures heralds a success.

It is important to clarify now the difference between mutually exclusive events and independent events. We want to know whether events are mutually exclusive when we consider the union of events. We are interested in independence, which we consider the intersection of the events. Consider, for example, the two events A and B in Figure 2–6, where the fifteen outcomes in S are assumed to be equally likely. Here, clearly A and B are

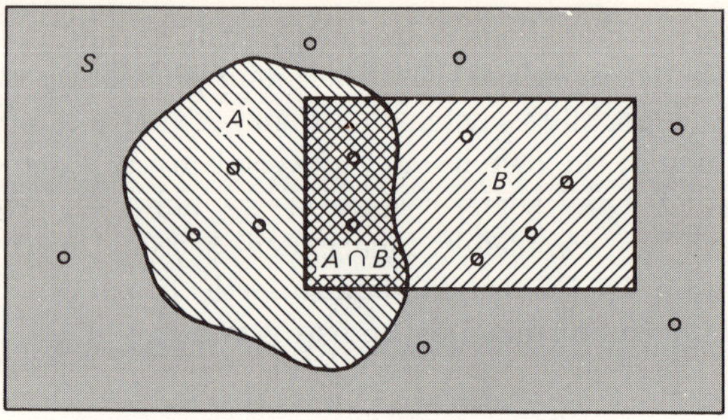

Figure 2–6 Venn Diagram Illustrating Independence

not mutually exclusive, since $A \cap B = 2 \neq 0$. They are, however, independent, since

$$\begin{aligned} P(A) &= P(A \mid B) &&= 5/15, \\ P(B) &= P(B \mid A) &&= 6/15, \\ P(A \cap B) &= P(A)P(B) &&= 2/15. \end{aligned}$$

It is interesting to note that a slight change in the definitions of A and B here—say, moving one element from $\bar{A}$ to A—would make A and B no longer independent.

Note carefully that, if A and B are mutually exclusive, then $P(A \cap B) = 0$. From the definition of conditional probability, we see that

$$P(A \cap B) = P(A)P(B \mid A) = P(B)P(A \mid B).$$

From this, we would have

$$P(B \mid A) = P(A \mid B) = 0, \qquad \text{if } P(A) \neq 0 \quad \text{and} \quad P(B) \neq 0,$$

respectively. Thus, unless either A or B is the null event $\varnothing$, A and B are not independent when they are mutually exclusive. Alternatively, if A and B are mutually exclusive, the occurrence of B must depend on A, since if A occurs, then B can never do so. If either A or B is equal to $\varnothing$, then A and $\varnothing$, or B and $\varnothing$, are independent.

To conclude our discourse on independence, let us provide the following general results (which the reader may wish to verify or explain as an exercise):

1. For any event E in S
   (a) E and the null event $\varnothing$ are independent,
   (b) E and S are independent.
2. If E_1 and E_2 in the same sample space S are independent, then so are
   (a) $\bar{E}_1$ and $\bar{E}_2$,
   (b) E_1 and $\bar{E}_2$, and
   (c) $\bar{E}_1$ and E_2.

2–14 THE THEOREM OF TOTAL PROBABILITY

Often we are interested in finding the probability of an *ultimate* event that is dependent on the occurrence or nonoccurrence of events in the intermediate stages of an experiment. The rule that enables us to determine such a probability is called the *theorem of total probability*. This theorem turns out to be a combination of the additive and multiplicative rules. Let us develop it by way of an example.

According to the judgment of a political scientist, the probability of the United States' having a Democratic administration in the 1980's is 0.55, and of its having a Republican adminstration is 0.45. Moreover, under a Democratic administration the probability for legislation of socialized medicine is 0.6, and under a Republican administration it is only 0.1. What we would like to know at the present is the probability of having socialized medicine in the 1980's.

We shall try to determine the probability of the ultimate event—socialized medicine—with the aid of the tree diagram in Figure 2–7, where

D = Democratic administration,
R = Republican administration,
S = socialized medicine legislation, and
$\bar{S}$ = no socialized medicine legislation.

We can see that the probability of having a Democratic administration and socialized medicine, and the probability of having a Democratic administration and no socialized medicine, are, respectively,

$$P(DS) = P(D)P(S \mid D) = (0.55)(0.60) = 0.330,$$
$$P(D\bar{S}) = P(D)P(\bar{S} \mid D) = (0.55)(0.40) = 0.220.$$

Likewise,

$$P(RS) = P(R)P(S \mid R) = 0.045,$$
$$P(R\bar{S}) = P(R)P(\bar{S} \mid R) = 0.405.$$

The foregoing four outcomes are mutually exclusive, and the first and the third represent the event of socialized medicine legislation in the 1980's.

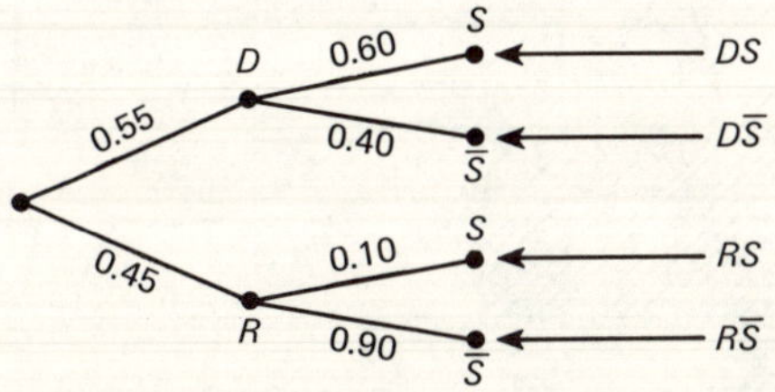

Figure 2–7 Tree Diagram Showing the Probability of Socialized Medicine

Thus, either the first or the third outcome includes the possibility of socialized medicine. Clearly, the desired probability is the sum of the separate probabilities of these two favorable outcomes. That is,

$$P(S) = P(DS) + P(RS) = 0.330 + 0.045 = 0.375.$$

This result indicates that if the political scientist's estimates are accurate, the prospects for socialized medicine in the United States in the 1980's are not very bright.

We shall now formally generalize the *theorem of total probability*.

First, we consider a partition of S. Let $H_1, H_2, \ldots, H_n$ be subsets of S. They constitute a partition of S if they are mutually exclusive and exhaustive. Now let E be a subset of S that may intersect with all, none, or some of the subsets of H_i. Then an expression for $P(E)$, as suggested by Figure 2–8, is

$$\begin{aligned} P(E) &= P(H_1 \cap E) + P(H_2 \cap E) + \cdots + P(H_n \cap E) \\ &= P(H_1)P(E \mid H_1) + P(H_2)P(E \mid H_2) + \cdots + P(H_n)P(E \mid H_n) \\ &= \sum P(H_i)P(E \mid H_i). \end{aligned} \tag{2–16}$$

This refers to a situation where E is an arbitrary event generated from the sample space, and where $P(E \mid H_i)$ is the probability of E given the appearance of H_i.

An ordinary die is tossed once. If a 1 appears, a ball is drawn from Urn I. If a 2 or a 3 appears, a ball is drawn from Urn II; otherwise, a ball is drawn from Urn III. Urn I contains five white, three green, and two red balls. Urn II contains one white, six green, and three red balls. Urn III contains three white, one green, and six red balls. We wish to determine (a) the probability that a red ball is chosen, and (b) the probability that Urn II is selected, given that the ball is red.

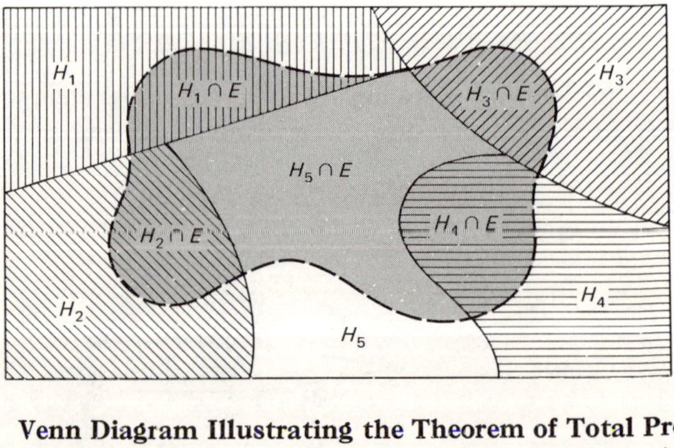

Figure 2–8 Venn Diagram Illustrating the Theorem of Total Probability, where the Rectangle is *S* and the Central Area Enclosed by the Broken Line is *E*

For this experimental design, the following probabilities are implied:

$$\begin{aligned} P(\text{I}) &= 1/6, & P(W \mid \text{I}) &= 5/10, \\ P(G \mid \text{I}) &= 3/10, & P(R \mid \text{I}) &= 2/10; \\ P(\text{II}) &= 2/6, & P(W \mid \text{II}) &= 1/10, \\ P(G \mid \text{II}) &= 6/10, & P(R \mid \text{II}) &= 3/10; \\ P(\text{III}) &= 3/6, & P(W \mid \text{III}) &= 3/10, \\ P(G \mid \text{III}) &= 1/10, & P(R \mid \text{III}) &= 6/10. \end{aligned}$$

From these probabilities and by the theorem of total probability, we have

$$\begin{aligned} \text{(a)} \quad P(R) &= P(\text{I})P(R \mid \text{I}) + P(\text{II})P(R \mid \text{II}) + P(\text{III})P(R \mid \text{III}) \\ &= (1/6)(2/10) + (2/6)(3/10) + (3/6)(6/10) \\ &= 13/30, \end{aligned}$$

and by the rule of conditional probability we have

$$\text{(b)} \qquad P(\text{II} \mid R) = \frac{P(\text{II} \cap R)}{P(R)} = \frac{P(\text{II})P(R \mid \text{II})}{P(R)}$$

$$= \frac{(2/6)(6/10)}{13/30} = \frac{3}{13}.$$

Can the reader explain why the probability of selecting Urn II, given a red ball, is much less than the probability of choosing a red ball? Note also that result in (b) follows Bayes' theorem, which is discussed in detail in Chapter 8.

2-15 SAMPLE SPACES WITH MANY OUTCOMES

When the number of possible outcomes of an experiment is large, the method of combinations can be used to calculate probabilities for the sample space that is generated by sampling without replacement and whose elements are equally likely.

As an example, a box contains 15 electronic tubes, of which 5 are defective. If two tubes are drawn together, at random, from the box, what is the probability that (a) both are good? (b) one is good and one is bad? (c) at least one is good?

We see that for all three cases the sample space is the same; it contains ${}_{15}C_2$ equally likely sample points, with the same probability of $1/{}_{15}C_2$ for each outcome. Among these elementary events in the sample space there are ${}_{10}C_2$ elements with exactly two good tubes. Thus,

$$\text{(a)} \qquad P(\text{both tubes are good}) = \frac{\binom{10}{2}}{\binom{15}{2}} = \frac{9}{21}.$$

Next, among the ${}_{15}C_2$ combinations, some satisfy the description "one is good and one is bad." There are ${}_{10}C_1$ ways to select one good and ${}_5C_1$ ways to select one bad tube. Consequently,

(b) $$P(\text{one good and one bad tube}) = \frac{\binom{10}{1}\binom{5}{1}}{\binom{15}{2}} = \frac{10}{21}.$$

Finally, the event "at least one is good" is the complement of the event "none is good." Thus,

(c) $$P(\text{at least one is good}) = 1 - P(\text{none is good})$$

$$= 1 - \frac{\binom{5}{2}}{\binom{15}{2}} = \frac{19}{21}.$$

There are 100 instructors at a certain college, 70 men and 30 women. Five are selected at random to evaluate the college's admission procedures. What is the probability that the sample consists of (a) all men? (b) exactly three men?

For this example the number of equally likely samples, assuming selections are made without replacement, is

$$\binom{100}{5} = 75{,}287{,}520.$$

A sample consisting of men only can be selected in ${}_{70}C_5 = 12{,}103{,}014$ ways. A sample consisting of exactly three men and two women can be drawn in $({}_{70}C_3)({}_{30}C_2) = 23{,}811{,}900$ ways. With these values, the desired probabilities become

(a) $$P(5\text{ men}) = \frac{\binom{70}{5}}{\binom{100}{5}} = \frac{12{,}103{,}014}{75{,}287{,}520}$$

$$= 0.1608;$$

(b) $$P(\text{3 men and 2 women}) = \frac{\binom{70}{3}\binom{30}{2}}{\binom{100}{5}}$$

$$= 0.3163.$$

2–16 INFINITE SAMPLE SPACES

Throughout this chapter we have been assuming that a random experiment can have only a finite number of outcomes. On occasion the random phenomenon of interest may generate a sample space that is infinite. Infinity, furthermore, may be countably infinite or uncountably infinite.

Countably Infinite Sample Space

A sample space is *countably infinite*, or simply *countable*, if it can be put into a one-to-one correspondence with the set of positive integers so it can be enumerated as 1, 2, 3, ... That is, when the sample space is countable, we can denote it as $S = \{e_1, e_2, e_3, \ldots\}$. Then what we need to do is to assign a number $P(e_i)$ to each outcome e_i such that

$$0 \leq P(e_i) \leq 1, \qquad \sum_{i=1}^{\infty} P(e_i) = 1.$$

Furthermore, if $E = \{e_i : i \in I_E\}$, it will be assigned a probability

$$P(E) = \sum_{i \in I_E} P(e_i).$$

Here I_E may be infinite, since S is infinite, but the sum in $P(E)$ is well defined, since $P(e_i)$ is finite.

As an example, let us consider the experiment where a well-balanced coin is tossed independently and a tally is kept of the number of trials until the first head occurs. The sample space clearly consists of integers 0, 1, 2, ... We get the outcome, i, if the sequence

$$\underbrace{TTT\ldots T}_{i}H$$

emerges. It is easy to see that the probability of such a sequence is $(\frac{1}{2})^{i+1}$. That is,

S	0	1	2	3	. . .
Probability	$\frac{1}{2}$	$(\frac{1}{2})^2$	$(\frac{1}{2})^3$	$(\frac{1}{2})^4$	. . .

It can also be verified that

$$\frac{1}{2} + \frac{1}{2^2} + \frac{1}{2^3} + \cdots = 1.$$

We see that the countable case is exactly the same as the finite case. As a result, all the definitions and consequences obtained before for the finite case can be carried over to the countable case. For this reason we sometimes lump the finite and countable cases together under the name *discrete sample space*.

Uncountably Infinite Sample Space

A sample space is uncountably infinite or continuous if it is not countable. Assigning probabilities to individual elementary outcomes, as in the discrete cases, does not work in the case of continuous sample space. In most uncountably infinite cases the probabilities of each individual outcome must be zero, and so it would not be possible to define the probability for any individual outcome. The way out of this difficulty is to assign probabilities to composite events rather than to individual outcomes.

When continuous, the sample space may be a set of ordinary real numbers, such as the interval (0, 1) or the whole real line $(-\infty, \infty)$, or it may be a set of points in a plane. In the former cases, events may be defined in terms of intervals; in the latter, in terms of areas. Once events can be clearly defined in either case, event probabilities can be assigned, and rules for discrete cases can then be applied.

Let us consider first the case of assigning probabilities to events that are intervals. Let the sample space be the set of real numbers on the interval from 0 to 1. It seems appropriate to say that $P(S) = 1$ here as for any other type of sample space. Intuitively, it also seems reasonable to say that a random selection from this sample space will produce a number less than 0.5 with a probability of 0.5. Similarly, the probability that the number selected at random will fall between 0.75 and 0.85 is 0.10.

Now let us define three interval events on this sample space as follows:

$$A = \{x\colon 0.6 < x < 0.9\},$$

$$B = \{x\colon x > 0.7\},$$

$$C = \{x\colon x < 0.5\}.$$

If we assign to each event a probability that is the proportion of its length to the total length of the sample space, we have

$$P(A) = 0.3, \qquad P(B) = 0.3, \quad \text{and} \quad P(C) = 0.5.$$

From the definitions of these events, we may derive other events on the same sample space via set operations. Thus, for example,

$$A \cup B = \{x: 0.6 < x < 1.0\},$$

for which

$$P(A \cup B) = 0.4.$$

Alternatively, we note that

$$A \cap B = \{x: 0.7 < x < 0.9\}$$

and

$$P(A \cap B) = 0.2;$$

hence,

$$\begin{aligned} P(A \cup B) &= P(A) + P(B) - P(A \cap B) \\ &= 0.3 + 0.3 - 0.2 \\ &= 0.4. \end{aligned}$$

Intuitively, we also see that the following results hold:

$$\begin{aligned} \bar{C} &= \{x: x \geq 0.5\}, \\ P(\bar{C}) &= 0.5 \\ &= 1 - P(C) \\ &= 1 - 0.5; \\ B \cup C &= \{x: x < 0.5 \text{ or } x > 0.7\}, \\ P(B \cup C) &= P(B) + P(C) \\ &= 0.5 + 0.3 \\ &= 0.8; \\ B \cap C &= \varnothing, \\ P(B \cap C) &= 0 \neq P(B)P(C). \end{aligned}$$

The last two results indicate that A and B are mutually exclusive but they are not independent. While $A \cap B = \varnothing$, neither A nor B is the null set, and independence requires that one of them must be $\varnothing$.

From our example of "selecting a number at random" we may take notice of two important properties. First, the set whose probabilities are computable are unions of intervals. Given two interval events A and B, we see that $\bar{A}$, $\bar{B}$, $A \cup B$, and $A \cap B$ are again unions of intervals. Second, the assignment of probability is additive. That is, if A and B are any two sets consisting of unions of intervals and if they are disjoint, then the sum of the lengths of the intervals in $A \cup B$ is just the length of A plus the length of B. These are clearly consistent with the axioms given earlier for the finite sample space.

Turning our attention to area events and their probabilities on an uncountably infinite sample space, let us consider a stopwatch with a minute hand and with numerals on the circumference from 1 to 10, indicating minutes. If the watch is started and then stopped at your blindfold command, what is the chance that the minute hand will stop exactly at the numeral 4? By exactly, we mean 4.000 . . . minutes with zeros carried on forever. Clearly, there is no chance at all. Yet the watch must stop somewhere, and the probability that it stops is 1. Or we may ask: "If the watch stops at random time, what is the probability that the minute hand stops between the numbers 2 and 4?" Many people would say the probability is 0.2, since the space between the numerals 2 and 4 is $2/10$ of the circumference.

To express the foregoing probability as an area, we may construct what is called an area probability graph. The total area under the graph is the total probability of the continuous sample space and must equal 1. For our example, the base should extend from 0 to 10 along the time axis, T. (Note that on the circumference of the watch, 0 and 10 coincide.) Since the watch can stop at random at any point between 0 and 10, we may assign equal probabilities to two time intervals with equal widths. If their probabilities are to be measured by areas, the altitudes of the corresponding rectangles should have equal heights. Hence, because the total area under it must equal 1, the area graph is bounded above by a straight line parallel to the T axis at a distance $1/10$ above the axis. Figure 2–9 gives us this area

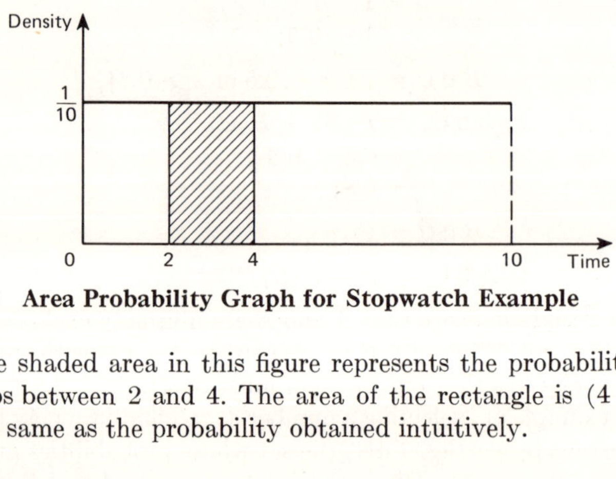

Figure 2–9 Area Probability Graph for Stopwatch Example

graph. The shaded area in this figure represents the probability that the watch stops between 2 and 4. The area of the rectangle is $(4 - 2)(1/10) = 0.2$, the same as the probability obtained intuitively.

EXERCISES

2–1 Give the sample space of each of the following random experiments in the form of set notation, a Venn diagram, a table, or a tree diagram, whichever is the most convenient in a given situation:
- (a) A perfect coin is tossed.
- (b) A coin and a die are thrown together.
- (c) Two ordinary dice are tossed and the sample points are given as sums of the numbers of the two faces turning up.
- (d) A bag contains five identical chips, numbered 1, 2, 3, 4, and 5. The chips are stirred, two chips are drawn blindly together, and the sample points are given in terms of the numbers on the chips.

2–2 Specify in a convenient form the event set for the following:
(a) Let A be the event of a head in tossing a coin.
(b) Let B be the event that the sum is 7 in tossing two dice.
(c) Let C be the event that the coin turns up heads and the die shows a 3 or less in tossing a coin and a die together.
(d) Let E be the event that the sum of the numbers is equal to or greater than 9 defined on the sample space in (d) of the preceding problem.

2–3 Determine the probabilities for events A, B, C, and E in the preceding problem in accordance with the classical theory.

2–4 Determine the probability for each of the following events by repeating the random trial 100 times:
(a) The probability of tails in tossing a coin.
(b) The probability of 6 in tossing a die.
(c) The probability of obtaining a sum of 7 in tossing two dice.

2–5 If the odds are 6 to 4 that an agreement between a labor union and management will be reached, what are the respective probabilities of reaching and of not reaching an agreement?

2–6 Three contractors are bidding for a contract to construct a school building. It is believed that A is twice as likely to obtain the contract as B, who in turn is twice as likely to obtain the contract as C. What are the respective probabilities for each to obtain the contract?

2–7 What are the odds in favor of rolling a total of seven on a throw of two dice?

2–8 In the game of "craps" the roller wins if he rolls a seven or eleven on the first roll, and he loses if he rolls two, three, or twelve. Any other number rolled is called a "point," and the roller wins if he rerolls "point" before he rolls a seven. What are the odds of winning the game if the "point" is five?

2–9 One card is drawn at random from a standard deck of cards. Show that
(a) $P(\text{ace or king}) = 2/13$.
(b) $P(\text{a heart or a diamond}) = 1/2$.

2–10 An ordinary and fair die is tossed. Show that the probability of obtaining a number divisible by 3 is $1/3$.

2–11 Two dice are tossed. Verify that the probability of getting a 7 or 11 is $2/9$.

2–12 A die is tossed. Show that the probability of getting 1, or 2, or 3 is $1/2$.

2–13 "The probability of not getting a double (both dice of the same number) when two perfect dice are tossed is $5/6$." Is this correct?

2–14 If the probability that the price of a stock will rise is 0.7, what is the probability that it will fall or remain unchanged?

2–15 Two coins are tossed. A is the event "two tails," and B is the event "two heads." Are A and B mutually exclusive? Are they complementary? Evaluate $P(A \cup B)$.

2–16 Items produced by a certain process each may have one or both of two types of defects, A and B. It is known that 10 percent of the items have type A defects and 5 percent have type B defects. Furthermore, 3 percent are known to have both types of defects. What is the probability that a randomly selected item will be defective?

2–17 At a party male and female guests are divided evenly. It has been observed that 5 percent of males and 10 percent of females are blondes. One guest pointed out at random is a blonde. What is the probability that the guest we selected is (a) male, (b) female?

2–18 An urn contains three bright white, four bright green, five dull white, and six dull green balls. If a ball is drawn at random, what is the probability that
(a) It is white?
(b) It is bright green, given it is green?
(c) It is dull white, given it is white?
(d) It is green or bright?

2–19 A coin is tossed three times. What is the probability of having three heads, given that the first toss turned up heads?

2–20 If five cards are randomly drawn, one at a time, without replacement, what is the probability of having five honor cards(A, K, Q, J, 10)?

2–21 An urn contains two black and two white balls. If two balls are drawn without replacement, show that the probability that one is white and the other is black is $2/3$.

2–22 An urn contains four red and six green balls. Two balls are drawn randomly without replacement.
(a) Set up a sample space for the possible outcomes of the experiment.
(b) Determine the probability for each sample point.

2–23 Repeat the preceding problem except that sampling is made with replacement.

2–24 Find the probability that a 5 will turn up at least once in two tosses of a fair die.

2–25 Two coins are tossed. Show that the event "head on the first coin" and the event "coins fall alike" are independent.

2–26 Two dice are tossed. Is it true that the probability that the first die shows even and the second die shows odd is $1/4$?

2–27 A bin of parts contains 10 percent defectives. A random sample of three is drawn from the bin with replacement. Is it true that the probability of at least one defective is 0.271?

2–28 Four cards are drawn at random with replacement. What is the probability that all are kings?

2–29 Recall that three events, A, B, and C, are said to be completely independent if and only if the following two conditions are satisfied:

(1) $P(A \cap B \cap C) = P(A)P(B)P(C)$, (2) All events are stochastically independent in pairs.

(a) Let a fair die be tossed and on this sample space let A be the even number, B a number equal to or greater than 3, and C a number divisible by either 3 or 5 without remainder. Show that in this case, while $P(A \cap B \cap C) = P(A)P(B)P(C)$, A, B, and C are not completely independent.

(b) Two coins are tossed. If A is the event "head on the first coin," B the event "head on the second coin," and C the event "the coins match," show that A, B, and C are pairwise independent events, but they are not completely independent.

2–30 A blue and a clear die are tossed.

(a) If $b + c = 11$, what is the probability that the clear die shows 4?

(b) Given that the blue die shows 4, what is the probability that the clear die shows a number greater than 4?

2–31 Prove that if A and B are independent, then $P(A \mid B) = P(A)$ and $P(B \mid A) = P(B)$.

2–32 A coin is tossed until a tail appears, or until it has been tossed three times. Given that the tail does not appear in the first toss, what is the probability that the coin is tossed three times?

2–33 Two slates, A and B, are competing for the control of a corporation. The probabilities that these two slates will win are 0.7 and 0.3, respectively. If A wins, the probability of introducing a new product is 0.8; if B wins, the corresponding probability is 0.4. What is the probability that the new product will be introduced?

2–34 A, B, and C are bidding on a contract for the construction of a bridge in Boston. The probabilities that A, B, and C will get it are 0.5, 0.3, and 0.2, respectively. If A gets it, he will select E as subcontractor with probability 0.8. If B or C gets it, E will be chosen with probabilities 0.4 and 0.1, respectively. What is the probability that E will eventually get the subcontract?

2–35 A committee of five is to be chosen at random without replacement from among six men and four women. What is the probability that it will be composed of (a) two women and three men? (b) four women and one man?

2–36 In a lot of 20 TV picture tubes, 5 are known to be defective. What is the probability that in a conditional random sample of 3 tubes, there are (a) 0 defectives? (b) 1 defective? (c) 2 defectives? (d) 3 defectives?

2–37 Five cards are drawn at random without replacement from a bridge deck. Check whether the following probabilities are correct.

(a) $P(\text{4 kings}) = 1/54{,}145$.

(b) $P(4 \text{ kings and } 1 \text{ ace}) = 1/649{,}740$.
(c) $P(3 \text{ queens and } 2 \text{ jacks}) = 1/108{,}290$.
(d) $P(\text{ace, king, queen, jack, and } 10 \text{ in any order}) = 64/162{,}435$.
(e) $P(3 \text{ of any suit and } 2 \text{ of another}) = 429/4165$.

2–38 In a lot of 100 radios, 10 are defective. The following acceptance procedure is adopted: take a conditional random sample of 10, and accept the shipment if there is no more than 1 defective in the lot. What is the probability that the shipment will be rejected? (Just write out the expression without evaluation.)

2–39 Let the sample space be $S = \{x: 0 < x < 12\}$. Given the following events:

$$A = \{x: x \leq 8\},$$

$$B = \{x: 6 < x < 9\},$$

$$C = \{x: x > 7\}.$$

Determine $P(A)$, $P(B)$, $P(C)$, $P(\bar{A})$, $P(A \cup B)$, $P(B \cup C)$, $P(A \cup C)$, $P(A \cap B)$, and $P(B \cap C)$.

2–40 Consider the stopwatch example in the text. What are the following probabilities?
(a) $P(5 < T < 10)$.
(b) $P(0 < T < 2 \text{ or } 7 < T < 9)$.
(c) $P(T > 8)$.
(d) $P(0 < T < 4 \text{ and } 2 < T < 5)$.

3

RANDOM VARIABLES

The preceding chapter introduced basic concepts of probability theory in terms of qualitative outcomes such as "defectives or nondefectives," "heads or tails," and finite sequences of symbols. In the application of probability theory to practical situations, it is usually more convenient to work with numbers than with qualitative outcomes, since real numbers are more adaptable to mathematical description and analysis than objects. For this reason we are interested in "random variables" and "probability functions." In this chapter we shall discuss these notions and some techniques of their manipulation. An understanding of these concepts will also greatly simplify certain problems in connection with continuous sample spaces.

3–1 RANDOM VARIABLES AND THEIR PROBABILITY FUNCTIONS

The notion of a random variable is a simple one. Imagine that a simple number could be assigned to each and every elementary event in a sample space. Then various sample points could be paired with various values of a variable. We would then have what is called a random variable. Thus, the idea of a random variable is related to that of a function.

A function, f, of a numerical variable, X, is defined to be any rule that assigns a unique value $f(x)$ (read as f at x) in some prescribed domain of numbers. This definition can be extended to random experiments. For example, the number of heads, X, occurring when two coins are tossed simultaneously can be viewed as a function on the sample space consisting of the following four points:

$$x_1 = (T, T), \quad x_2 = (T, H), \quad x_3 = (H, T), \quad x_4 = (T, T)$$

Clearly, a numerical value of 0 can be assigned to x_1; of 1, to x_2 and x_3; and of 2, to x_4. Thus we have a random variable X—the number of heads in tossing two coins—with values of 0, 1, and 2.

From the foregoing discussion we see that a *random variable* is a real-valued function defined on a sample space. We recall that a probability of an event, $P(E)$, is also defined on a sample space. The difference is that an event probability is defined on the set of subsets of the sample space, while a random variable is defined on the individual sample points only. The range of an event probability is the subset of real numbers between 0 and 1 inclusive, while the range of a random variable is not quite so restricted.

Notice that any value of a random variable is an event, since there is always some elementary event, or a set of elementary events, assigned that value. A value of a random variable, being an event, must have a probability. Thus, if X is a random variable that can assume values $x_1, x_2, \ldots, x_k$ with associated probabilities $p_1, p_2, \ldots, p_k$, then the set whose elements are the ordered pairs (x_i, p_i), $i = 1, 2, \ldots, k$ is called the *probability function*, or *distribution*, of X. In order to determine more readily the probabilities of the values of a random variable, we can state the probability function of a random variable in a more obvious and practical manner: denote the probability that the random variable X will assume the value of x_i by $f(x)$; then the probability function of X can be written as

$$f(x) = P(X = x_i). \tag{3–1}$$

Let us illustrate the notions of a random variable and its corresponding probability function by a detailed example. Consider the tossing of four well-balanced coins, when we are interested in the number of heads. This number may be 0, 1, 2, 3, or 4. But what is the distribution of the total probability of 1 over the range of these numbers? To answer this question, let us observe Table 3–1. The appropriate sample points for the sample space are given in the first column; the second column gives the number of heads corresponding to the sample points; and the third column contains the probabilities for the sample points. From the information in the last two columns we can compute the probability for each possible number of heads resulting from this experiment. For instance, the probability of exactly 1 head is the sum of the probabilities for the sample points TTTH, TTHT, THTT, HTTT; the same holds true for other probabilities. Now, if we let X represent the number of heads, the values that X can assume and their associated probabilities must be as shown in Table 3–2.

Table 3–2 contains the probability function, or probability distribution, of X, because it is a set of ordered pairs, each of the following form: (number of heads, probability of that number.)

A probability function may be represented by a table or by an equation. To illustrate how a probability function can be expressed in terms of an

Table 3–1. Derivation of the Number of Heads from the Sample Space of Four Coins

Sample points	*Number of heads*	*Probability*
TTTT	0	1/16
TTTH	1	1/16
TTHT	1	1/16
THTT	1	1/16
HTTT	1	1/16
HHTT	2	1/16
THHT	2	1/16
TTHH	2	1/16
HTTH	2	1/16
HTHT	2	1/16
THTH	2	1/16
HHHT	3	1/16
HHTH	3	1/16
HTHH	3	1/16
THHH	3	1/16
HHHH	4	1/16

equation, let us consider the throw of an ordinary die. Let Z represent the variable number of each of the six faces of the die; we can then write

$$h(z) = P(Z = z_i) = 1/6, \quad i = 1, 2, \ldots, 6.$$

Next, for a less obvious example, let us take up the experiment of the four coins discussed before. The values of $f(x) = P(X = x_i)$ are

$$1(1/16),\quad 4(1/16),\quad 6(1/16),\quad 4(1/16),\quad 1(1/16).$$

The numbers 1, 4, 6, 4, and 1 are called *binomial coefficients* and can be obtained from $\binom{4}{x}$ for $x = 0, 1, 2, 3, 4$. As a result, as we shall explain in detail in a later chapter, the general mathematical statement for all values

Table 3–2. Probability Distribution for the Number of Heads from Tossing 4 Coins

Values of X: x_i	0	1	2	3	4
Probability: $P(X = x_i)$	1/16	4/16	6/16	4/16	1/16

of this probability function becomes

$$f(x) = P(X = x_i) = \binom{4}{x} (1/2)^x (1/2)^{4-x},$$

$$i = 0, 1, 2, 3, 4.$$

For example,

$$P(X = 0) = \binom{4}{0} (1/2)^0 (1/2)^4 = 1/16.$$

The graphic representation of a random variable is constructed by putting the values of the random variable on the horizontal scale; and the vertical axis shows the corresponding probabilities. In effect, the graph of a random variable turns out to be the graph of its probability function; see, for example, Figure 3–1. For this reason many authors define a random variable by explicitly stating its probability function—for instance: a *random variable* X is a numerical-valued function on the outcome of a random experiment, each of the number of events having a probability $f(x) = P(X = x_i)$.

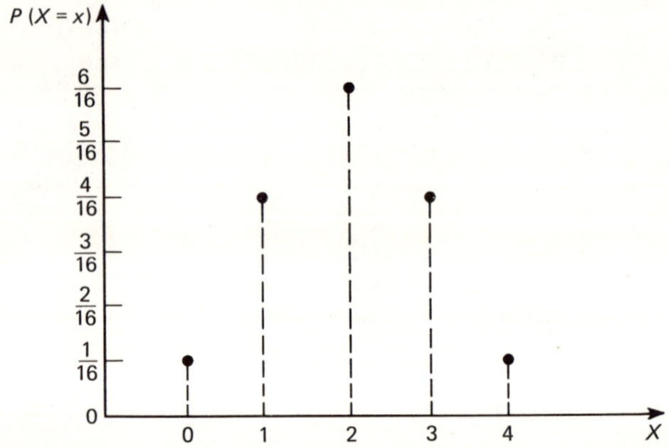

Figure 3–1 Probability Distribution of the Number of Heads in the Four-Coin Experiment

Let us summarize the steps in constructing a probability function:

(1) List all possible sample points with their corresponding probabilities for an experiment.
(2) List the value of the random variable under consideration corresponding to each sample point.
(3) List all possible values of the random variable, $x_1, x_2, \ldots, x_k$ with their associated probabilities, $f(x_1), f(x_2), \ldots, f(x_k)$.

(4) $P(X = x_i) = f(x_i)$ is the sum of the probabilities of all sample points corresponding to x_i.

(5) Thus, the set of ordered pairs $[x_i, f(x_i)]$ is the probability function of a random variable; it is usually given by a table or by a general mathematical expression.

3-2 CUMULATIVE DISTRIBUTION FUNCTIONS

Quite often, in connection with random variables, we need to ask questions such as: what is the probability that a random variable assumes a value less than or equal to a particular prescribed number? Answers are given by the cumulative distribution function, CDF, defined as follows: if X is a random variable and x is a real number, then the *cumulative distribution function*, denoted as $F(x)$, showing the probability that X assumes values less than or equal to x_i, is

$$F(x) = P(X \leq x_i). \tag{3-2}$$

Thus, in defining the CDF of a random variable, probabilities are assigned to all events that are intervals from $-\infty$ to x, such as the event $X \leq x$, where x could be any real number. A distribution function is immediately defined from the corresponding probability function. For example, Table 3-3 shows the possible numbers of heads when four coins are tossed, together with their probabilities, $f(x)$, and the cumulative probabilities, $F(x)$, as given by equation (3-2).

Table 3-3. Number of Heads, Probability, Function, and Distribution Function for Four Coins

X	$f(x)$	$F(x)$
0	$1/16$	$1/16$
1	$4/16$	$5/16$
2	$6/16$	$11/16$
3	$4/16$	$15/16$
4	$1/16$	$16/16$

The CDF has a value for every real number x, not just those listed in the table. For instance,

$$F(1.5) = P(X \leq 1.5) = f(0) + f(1) = F(1) = 5/16.$$

In general if x is any number greater than or equal to a but less than b and if a and b are two adjacent values, then $F(b) = F(a)$. This reveals that the

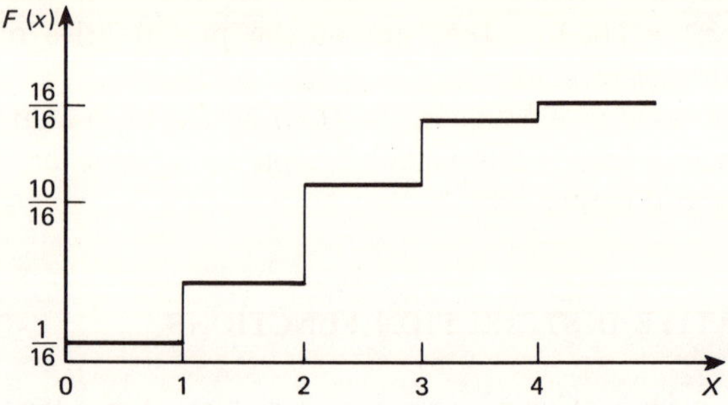

Figure 3–2 The Cumulative Distribution Function of the Number of Heads in Tossing Four Coins

cumulative probability between two adjacent numbers a and b is the same as that of its lower bound in a discrete distribution. For example, if a die is tossed and we let $a = 2$ and $b = 3$, there will be no point between a and b that has a positive probability. The graph of the CDF of four coins is shown by Figure 3–2. We see that jumps occur at $X = 0, 1, 2, 3,$ and 4.

The CDF, as a consequence of being a probability, must satisfy certain conditions. First, because probability is a number between 0 and 1, we must require that $0 \leq F(x) \leq 1$. Second, since any interval is an event and an event cannot have negative probability, we must have $F(b) > F(a)$, if $a < b$. This condition implies that $F(x)$ is a nondecreasing function, as we move from left to right on the horizontal scale of the graph for a CDF. Third, assuming that $-\infty \leq X \leq +\infty$, we must have $F(\infty) = P(X \leq \infty) = 1$ and $F(-\infty) = P(X \leq -\infty) = 0$. Finally, we must also note that

$$P(X > x_i) = 1 - P(X \leq x_i). \tag{3–3}$$

3–3 EXPECTATION

To facilitate analysis and application, it is desirable to have summary measures of important properties of probability distributions. Such measures are called parameters, since a probability distribution refers to a population—all possible outcomes of a stochastic process. The most important parameters of a random variable are the mean and variance. The *mean* measures the central location of the distribution of values, and the *variance* measures the dispersion of the values over the domain of the variable. As we shall see, there are other measures of central tendency and variability, but owing to their well-defined mathematical properties the mean and variance are the most versatile and, therefore, the most frequently employed measures of these basic characteristics.

The mean is called by many other names, such as the *mathematical expectation*, the *expected value*, or simply the *expectation*. The mathematical expectation of a random variable is the sum of the products that are obtained by multiplying all possible values of the random variable by their corresponding probabilities. Thus, let X be a random variable with probability function as follows:

Values of X: x_i	x_1	x_2	$\ldots$	x_k
Probability: $f(x_i)$	$f(x_1)$	$f(x_2)$	$\ldots$	(fx_k)

The *mathematical expectation* of X, denoted by $E(X)$, or by μ_X, is defined to be

$$\begin{aligned} E(X) &= x_1 f(x_1) + x_2 f(x_2) + \cdots + x_k f(x_k) \\ &= \sum_i x_i f(x_i). \end{aligned} \tag{3-4}$$

If a perfect die is tossed, what is the expectation as to the number of dots on the top face? To answer this question, we may recall that this random variable has possible values of 1, 2, 3, 4, 5, and 6, each with a probability of $1/6$. Call this variable Z, and by equation (3-4) we have

$$\begin{aligned} E(Z) &= 1(1/6) + 2(1/6) + 3(1/6) + 4(1/6) + 5(1/6) + 6(1/6) \\ &= 3.5. \end{aligned}$$

This result shows that the expected value need not be a possible value of the random variable.

Four coins are tossed; what is the expectation of the number of heads? The answer for this problem is obtained with the aid of Table 3-4. Thus, if we toss four coins together each time, over and over again, on the average we shall obtain two heads per toss.

Table 3-4. Computation of Expectation of the Number of Heads with Four Coins

x_i	$f(x_i)$	$x_i f(x_i)$
0	$1/16$	0
1	$4/16$	$4/16$
2	$6/16$	$12/16$
3	$4/16$	$12/16$
4	$1/16$	$4/16$
	Sum = $E(X)$ =	$32/16$

The probability that a house of a certain type will be burned down by fire in any twelve-month period is 0.005. An insurance company offers to sell the owner of such a house a $20,000 one-year-term fire insurance policy for a premium of $150. What is the company's expected gain?

The "gain," G, for the company is a random variable with possible values of $150, if the house does not have a fire accident, and −$19,850, if the house is burned down during the year covered by the policy. The probability function of G is then

Values of G: g_i	150	$-19{,}850$
Probability: $f(g_i)$	0.995	0.005

With the information above we see that

$$E(G) = (150)(0.995) + (-19{,}850)(0.005) = \$50.$$

From the viewpoint of the insurance company, the expected gain for an insurance policy must be positive in order to enable the company to pay for administrative costs and to build up reserves for paying its beneficiaries and policyholders. From the buyer's viewpoint, however, insurance, like any game of chance that is run for profit, has a negative expected value.

For example, a lottery sells 10,000 tickets at $1 per ticket; a prize of $5000 will be given to the winner of the first draw. Suppose you have bought a ticket; how much should you expect to win?

Here, the random variable "win," W, has two possible values: −$1 and $4999. Their respective probabilities are 9999/10,000 and 1/10,000. Thus,

$$E(W) = (-1)\left(\frac{9999}{10{,}000}\right) + (4999)\left(\frac{1}{10{,}000}\right) = -\$0.50.$$

This amount, a *minus* 50 cents, is the amount you expect to win (lose) each time, if you play this game over and over again. Thus, you will lose an average of $0.50 each time you play.

3-4 PROPERTIES OF EXPECTATION

If X is a random variable whose probability distribution is known, we are often interested in other random variables that are defined in terms of X. Such *functions* of the *random variable* X have expectations that we can compute from X without first finding the probability function of the related variable. A few laws governing properties of expectation make this possible. We shall introduce them now via a simple random variable X with the following probability function:

Values of X: x_i	-1	0	2
Probability: $f(x_i)$	0.3	0.1	0.6

For X, we have

$$E(X) = (-1)(0.3) + (0)(0.1) + (2)(0.6) = 0.9.$$

If we add a constant, c, to a random variable X, then the expectation of the new random variable $(X + c)$ has the same expectation as X plus the constant. That is, $E(X + c) = E(X) + c$. For instance, if we add 3 to our illustrative random variable, X, we have

$$\begin{aligned} E(X+3) &= (-1+3)(0.3) + (0+3)(0.1) + (2+3)(0.6) \\ &= 3.9 = E(X) + 3. \end{aligned}$$

This relation is just what we would expect, since if X is increased by c for all its values, it will be increased by c on the average.

Next, multiplication by a constant gives $E(cX) = cE(X)$. That is, if we multiply each value of X by a constant c, its average value will be multiplied c times. Consider, for instance, the expectation of the variable $(3X)$ for our illustrative problem. Clearly,

$$\begin{aligned} E(3X) &= (-3)(0.3) + (0)(0.1) + (6)(0.6) \\ &= 2.7 = 3E(X). \end{aligned}$$

Finally, the first two properties indicate that in multiplying X by a constant c and adding a constant d to the results, we multiply the mean of X by c and add d to the result. For example,

$$\begin{aligned} E(3X+4) &= (-3+4)(0.3) + (0+4)(0.1) + (6+4)(0.6) \\ &= 6.7 = 3E(X) + 4. \end{aligned}$$

To formalize the foregoing discussion, we introduce an intuitively obvious theorem: if X is a random variable, then for any numerical constant a and b,

$$E(aX + b) = aE(X) + b. \tag{3–5a}$$

This property can be stated in another way. Let X be a random variable and let $Z = aX + b$; then

$$E(Z) = E(aX + b) = aE(X) + b. \tag{3–5b}$$

From equation (3–5a) we can see that if X is a random variable with mean $E(X) = \mu$, then $E(X - \mu) = 0$. That is, if we deduct from each value of X the mean of X, then the mean of the random variable, $(X - \mu)$, is zero.

Also, a random variable that can assume only a single value, say c, is called a *constant random variable*. The expectation for such a random variable is c. You can easily prove this yourself.

3–5 VARIANCE

An expectation is a parameter that describes the central tendency of a random variable, telling us where the center of mass of the probability distribution of a random variable is located. However, two distributions with the same expectation may still be quite different. For instance, if X takes on values of -1, 0, and 1, with probabilities $\frac{1}{5}$, $\frac{3}{5}$, $\frac{1}{5}$, respectively, then $E(X) = 0$. If another variable, Y, takes on values of -10, 0, and 10, with probabilities $\frac{1}{5}$, $\frac{3}{5}$, $\frac{1}{5}$, then $E(Y) = 0$. But X and Y are obviously different, since Y's values have a much greater spread or variability from one trial of an experiment to another trial (see Figure 3–3). Thus, expectation furnishes us a quick picture of the long-run average when an experiment is performed over and over; for a more complete description of a probability distribution we need to know the extent to which outcomes of an experiment may deviate from this average value.

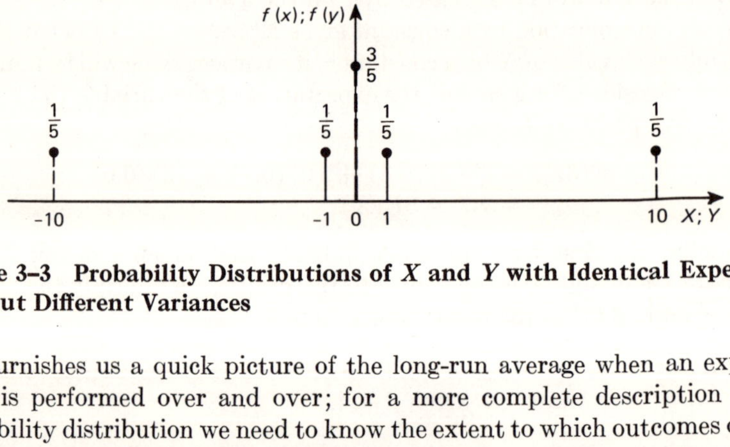

Figure 3–3 Probability Distributions of X and Y with Identical Expectation but Different Variances

The most important measure of variability is the variance. If X is a random variable with expectation $E(X) = \mu$, the *variance* of X, denoted by $V(X)$, or $\sigma_X{}^2$, is defined as follows:

$$V(X) = E[(X - \mu)^2] = \sum_i (x_i - \mu)^2 f(x_i). \tag{3–6}$$

Thus, variance of X is the mean of squared deviations of X from its mean. Note that the variance is a weighted mean of all of the squared deviations, each weighted by its probability.

Formula (3–6) is the definitional equation of variance. For actual computation it is more convenient to use its equivalent

$$V(X) = E(X^2) - [E(X)]^2 = E(X^2) - \mu^2, \qquad (3\text{–}7)$$

which dispenses with the difference $X - \mu$.

Consider again, for example, the toss of an ordinary die. The expectation of the number of spots showing on the die, Z, calculated before, is 3.5, or $7/2$. Now, we wish to determine the variance of Z. First we find

$$E(Z^2) = 1^2(1/6) + 2^2(1/6) + 3^2(1/6) + 4^2(1/6) + 5^2(1/6) + 6^2(1/6) = 91/6.$$

Next, by equation (3–7), we have

$$V(Z) = E(Z^2) - \mu^2 = 91/6 - (7/2)^2 = 35/12.$$

For random variables, X and Y, as represented by Figure 3–3, we have $E(X) = E(Y) = 0$, and by (3–7)

$$V(X) = [-1^2(1/5) + 0^2(3/5) + 1^2(1/5)] - 0^2 = 0.4,$$

$$V(Y) = [-10^2(1/5) + 0^2(3/5) + 10^2(1/5)] - 0^2 = 40.$$

Note that $V(Y)$ is many times greater than $V(X)$. The total probability of X is concentrated within a narrow range, from -1 to 1, while the total probability of Y is distributed over a wide range, from -10 to 10.

3–6 PROPERTIES OF VARIANCE

The computation of variance is quite time-consuming if the values of the random variable are at all large. To save time and effort, we rely upon a procedure that involves shifting the origin of the domain of the values of a variable—changing the scale, so to speak. We note, first of all, that if a random variable is always equal to a constant, c, then its expectation will be c. Furthermore, such a variable will never differ from its expectation, and so the variation about this expectation will be zero. Second, if we add a constant to a random variable, X, then $E(X)$ will be increased by the same constant. However, the difference $X - E(X)$ remains unchanged, and therefore $V(X)$ is unaffected by addition of the constant to X. Third, if we multiply each value of a random variable by a positive number, say a, then we also multiply its expectation by a, so that the quantity $X - E(X)$ will be a times as large as before. The square $[X - E(X)]^2$ will evidently be a^2

times as large as before; that is, if $Z = aX$, then

$$\begin{aligned}[Z - E(Z)]^2 &= [aX - aE(X)]^2 \\ &= [a(X - E(X)]^2 \\ &= a^2[X - E(X)]^2.\end{aligned}$$

Consequently, if a random variable is multiplied by a constant, a, the variance of the variable is multiplied by a^2. In particular, if a random variable is multiplied by $c = -1$, we have $V(-X) = V(X)$. Summarizing: if X is a random variable with variance σ_X^2, and if a and b are numbers, then

$$V(aX + b) = \sigma_{aX+b}^2 = a^2\sigma_X^2. \tag{3-8}$$

Let the probability distribution of X be

x_i	5050	5100	5150
$P(X = x_i)$	0.1	0.4	0.5

Determine $V(X)$. To simplify the solution, we introduce a new random variable, Y, defined as

$$Y = \frac{X - 5100}{50},$$

whose probability function is, clearly,

y_i	-1	0	1
$P(Y = y_i)$	0.1	0.4	0.5

For Y, we have

$$E(Y) = (-1)(0.1) + (0)(0.4) + (1)(0.5) = 0.4,$$

$$E(Y^2) = (-1^2)(0.1) + (0^2)(0.4) + (1^2)(0.5) = 0.6,$$

and, therefore, the variance of Y is

$$\sigma_Y^2 = E(Y^2) - \mu_Y^2 = 0.6 - (0.4)^2 = 0.44.$$

Now, we may find σ_X^2 on the basis of σ_Y^2 by recalling that, from the definition of Y, we must have

$$X = 50Y + 5100.$$

Therefore, by (3-8),

$$\begin{aligned}\sigma_X^2 &= \sigma_{50Y+5100}^2 = \sigma_{50Y}^2 \\ &= (50)^2\sigma_Y^2 = 2500\sigma_Y^2 \\ &= 2500(0.44) = 1100.\end{aligned}$$

The reader should go through the example above, recalling various properties of variance at every step.

For applications, it is more convenient to have a measure of variability in the original units instead of their square. This leads to the introduction of the *standard deviation*, which is simply the positive square root of the variance. If X is a random variable with $V(X) = \sigma_X^2$, then the standard deviation of X, denoted by σ_X, is

$$\sigma_X = \sqrt{V(X)} = \sqrt{\sigma_X^2}. \tag{3-9}$$

Corresponding to the properties of variance, we see that the following relations hold true for standard deviation:

$$\sigma_{cX} = |c|\,\sigma_X \tag{3-10}$$

and

$$\sigma_{X+c} = \sigma_X. \tag{3-11}$$

Can you give a verbal interpretation of each of these two expressions?

3-7 STANDARD RANDOM VARIABLES

As we shall see, it is much more compact and convenient to deal with a standard random variable than with one that is not. A *standard random variable* is one with zero expectation and unity variance. Any random variable can be standardized by subtracting from it its expectation and dividing the result by its standard deviation. That is, if we let X be a random variable with mean μ_X and standard deviation σ_X, the corresponding standardized random variable, denoted as Z, is

$$Z = \frac{X - \mu_X}{\sigma_X}. \tag{3-12}$$

The expectation of a standardized variable is always 0, since

$$\begin{aligned}E(Z) &= E\left(\frac{X - \mu_X}{\sigma_X}\right) \\ &= \frac{E(X) - E(X)}{\sigma_X} \\ &= 0,\end{aligned}$$

where $E(X)$ and σ_X are constants over all the possible values that X can assume.

The variance of a standardized variable is always unity, since

$$\begin{aligned} V(Z) &= E(Z^2) - [E(Z)]^2 \\ &= E\left[\frac{X - E(X)}{\sigma_X}\right]^2 \\ &= \frac{E(X - \mu_X)^2}{\sigma_X^2} \\ &= \frac{\sigma_X^2}{\sigma_X^2} = 1, \end{aligned}$$

which indicates, of course, that σ_Z is also unity.

Clearly, from our previous discussions on functions of random variables, the Z transformation does not in any way change the pattern of the original probability function. The probability of any Z value is simply the probability of the corresponding value of the original random variable. We note also that each X value corresponds to one, and only one, Z value. The Z transformation has the effects of changing the scale of the original variable and of converting the individual deviations of the original variable from the mean to standard-deviation units.

3–8 CHEBYSHEV'S THEOREM

We have discussed the expectation, the variance, and the standard deviation for probability distributions. Let us now consider how the standard deviation can be used to provide information about the pattern of probability accumulation in intervals centered on their expectation, as the intervals grow wider. Intuitively, we feel that when the standard deviation is small the probability masses up close to the expectation, and when the standard deviation is large the probability disperses out more. But it seems desirable that we should be able to answer questions such as these: (1) What percent of the total probability lies in a given interval centered at the expected value? (2) How wide an interval about the expected value is required to guarantee that, say, 95 percent of the total probability is included in that interval?

A simple but quite remarkable inequality called *Chebyshev's theorem* can help us give approximate answers to these questions. This theorem states: if X is a random variable with mean μ and finite variance σ^2 and if $h > 0$, then

$$P(|X - \mu| \geq h\sigma) \leq \frac{1}{h^2} \tag{3–13}$$

or

$$P(|X - \mu| \leq h\sigma) \geq 1 - \frac{1}{h^2}. \tag{3–14}$$

Form (3–13) of this inequality shows that the probability is never more than $1/h^2$ that X assumes a value outside the closed interval from $\mu - h\sigma$ to $\mu + h\sigma$. The alternative form, (3–14), says that at least the fraction $1 - (1/h^2)$ of the total probability of X lies within h standard deviations of the mean. Clearly, these two expressions are complementary to each other. This theorem states, for example, that at most one-fourth of the total probability falls outside the interval of $\mu \pm 2\sigma$ for any random variable. Alternatively, this also means that at least $1 - \frac{1}{4}$, or $\frac{3}{4}$, of the total probability is within the interval $\mu \pm 2\sigma$ for any random variable.

The significance of Chebyshev's theorem is due to its complete universality: it can be applied to any random variable. However, it does not provide a practical method of estimating probabilities, since the right side of (3–13), for example, is usually much larger than the left side. Indeed, no new information is obtained from it when $h \leq 1$. The reader is advised to apply Chebyshev's theorem to some of the random variables introduced in this chaper in order to get a feeling of the relative sizes of the right and left sides of both (3–13) and (3–14).

3–9 CONTINUOUS RANDOM VARIABLES

So far our discussion has been concentrated on *discrete random variables*—random variables that can take on only a finite number of values. For a discrete random variable, the cumulative distribution function $F(x)$ is a step function—a function that increases only in finite jumps and that is constant between jumps. The distribution of probability is thus made in discrete jumps at the points where the jumps occur, in amounts equal to the sizes of the jumps.

When a random variable is defined on a continuous sample space, such as time, length, or weight, it is called a *continuous random variable*. The probability distribution of a continuous random variable is also continuous, and so its *CDF*, $F(x)$, has no jumps and no lumps of probability at any value of X. See, for example, Figures 3–6 and 3–8.

A discrete probability distribution is analogous to a discrete mass distribution, which is a collection of "point masses" at certain points along an axis. A continuous probability distribution is analogous to a continuous mass distribution, as in a long rope of varying thickness. In this case we say that there is no mass at any given point, but there is mass in an interval.

A continuous probability distribution is called the *probability density function*, or the *density function*, or the *distribution density*. It is so called because we think of the density function as the rate at which probability would accumulate if we were to sweep from the left to the right on the axis of the values of the continuous random variable. Over the intervals of the axis where probability is uniformly distributed, for example, this rate (density) would be constant. Thus, the density function $f(x)$ is a measure of the concentration of probability within an interval. This probability can be interpreted as an area under the graph of $f(x)$, called the *density curve*, between the ordinates at the endpoints of an interval. More precisely, we state that the probability density function $f(x)$ is the function or curve that, for any interval (x_1, x_2) on the X axis, gives the probability that X lies in (x_1, x_2) as the area under $f(x)$ between x_1 and x_2. This is shown graphically by Figure 3–4.

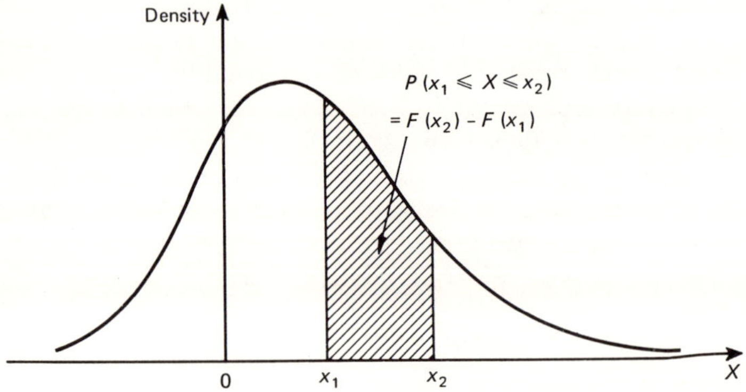

Figure 3–4 A Density Curve Showing the Relationship between $f(x)$ and $F(x)$

The distribution function of a continuous random variable, as in the discrete case, gives the probability that the random variable will assume a value less than or equal to a particular value of X. That is,

$$\begin{aligned} F(x) &= P(X \leq x_i) \\ &= P(-\infty < X \leq x_i). \end{aligned} \tag{3–15}$$

This definition is shown by Figure 3–5 in connection with the density function of X.

Now, we note that $f(x)$ may be interpreted as the rate of change of $F(x)$. Since the cumulative distribution function $F(x)$ is nondecreasing, it follows that its rate of change (or slope) is nonnegative—i.e., $f(x) \geq 0$. Also, we

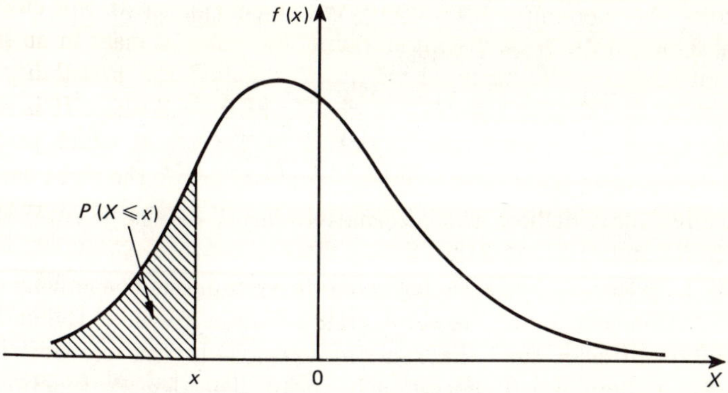

Figure 3–5 A Density Curve Illustrating the Definition of CDF of a Continuous Random Variable

see that the probability between any two values, x_1 and x_2, of a continuous random variable can be evaluated in terms of $F(x)$ as

$$P(x_1 < X < x_2) = F(x_2) - F(x_1). \tag{3–16}$$

This is illustrated by Figure 3–4. In particular, the total area under the density curve is $F(+\infty) = 1$:

$$\begin{aligned} P(-\infty < X < +\infty) &= F(+\infty) - F(-\infty) \\ &= 1 - 0. \end{aligned} \tag{3–17}$$

Since events may be defined as intervals, the reader should be familiar with some frequently encountered intervals expressed in terms of inequalities. First, consider a finite interval of the form $X \leq x_2$. This event is equivalent to the compound event ($X \leq x_1$ or $x_1 < X \leq x_2$). Thus, we must have

$$P(X \leq x_2) = P(X \leq x_1) + P(x_1 < X \leq x_2),$$

or, transposing,

$$P(x_1 < X \leq x_2) = F(x_2) - F(x_1).$$

Next, for example, the event $|X - 2| < 5$ is simply the interval from -3 to 7, because this inequality is merely an alternative way of writing

$$-5 < X - 2 < +5.$$

Adding 2 to each member of this extended inequality gives $-3 < X < 7$. As a result,

$$\begin{aligned} P(|X - 2| < 5) &= P(-2 \leq X \leq 7) \\ &= F(7) - F(-2). \end{aligned}$$

Finally, the inequality $|X - 2| > 5$ defines the set of all values of X farther than 5 units from 2, and

$$\begin{aligned} P(|X-2|>5) &= P(X \geq 7 \text{ or } X \leq -3) \\ &= P(X > 7) + P(X \leq -3) \\ &= [1 - F(7)] + F(-3). \end{aligned}$$

The reader may deduce the alternative forms of the events, such as $(|X| > 5)$ and $(|X| \leq 3)$.

Let us now turn to furnish some concrete examples of continuous random variables. Consider again the stopwatch example given at the end of the last chapter. Recall that the random variable in this case is time, T, ranging in value from 0 continuously to 10. The density function for T gives the ordinate of the area probability graph, which is uniform throughout the range 0–10. Thus, we have

$$\begin{aligned} f(t) &= 1/10, && \text{if } 0 \leq T \leq 10, \\ &= 0, && \text{otherwise.} \end{aligned}$$

This density curve is reproduced as Figure 3–6(a).

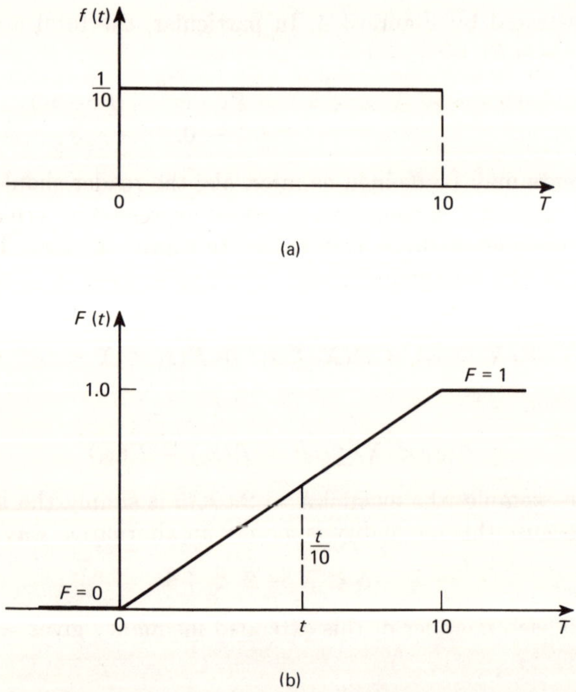

Figure 3–6 The Density and Cumulative Distributions of the Stopwatch Example

With $f(t)$ specified, probability for the random stop of the minute hand within any subinterval, say from t_1 to t_2, is simply $(\frac{1}{10})(t_2 - t_1)$. Thus,

$$P(2 \leq T \leq 4) = \tfrac{1}{10}(4 - 2) = 0.20,$$

$$P(T \leq 4) = P(0 \leq T \leq 4) = \tfrac{1}{10}(4 - 0) = 0.40,$$

$$P(T > 7) = P(7 \leq T \leq 10) = \tfrac{1}{10}(10 - 7) = 0.30,$$

and so on.

It is often easier and more convenient to evaluate continuous random variables in terms of their cumulative distribution functions rather than their density functions. For the stopwatch example, the distribution function is $F(t) = P(T \leq t)$. The probability given by this function is zero for $T \leq 0$ and 1 for $T \geq 10$. For T between 0 and 10, the probability is given by the area of a rectangle of base from 0 to t and of altitude $\frac{1}{10}$. With these conclusions, F in this case can be described as follows:

$$F(t) = \begin{cases} 0, & \text{if } T \leq 0, \\ \dfrac{t}{10}, & \text{if } 0 < T < 10, \\ 1, & \text{if } T \geq 10. \end{cases}$$

The graph of $F(t)$ is given by Figure 3-6(b). To find the three probabilities obtained before with $F(t)$, we have

$$\begin{aligned} P(2 \leq T \leq 4) &= F(4) - F(2) \\ &= \tfrac{4}{10} + \tfrac{2}{10} = 0.20, \end{aligned}$$

$$P(T \leq 4) = F(4) = \tfrac{4}{10} = 0.40,$$

and

$$\begin{aligned} P(T > 7) &= 1 - P(T \leq 7) \\ &= 1 - F(7) \\ &= 1 - \tfrac{7}{10} = 0.30. \end{aligned}$$

Let us consider the second example involving continuous random variables. A process is adjusted to fill one pound of instant coffee into each jar. Because of random variability, the weight of coffee filled into each jar actually ranges from 0.97 to 1.03 pound. The relative frequencies of various weights suggest that the density curve for the random variable W, weight, can be approximated very well by an isosceles triangle.

For the sake of simplifying the analysis, let $W = 0$, corresponding to 1 pound, the midpoint of W; $W = -3$, corresponding to the minimum value, 0.97, of W; and $W = 3$, corresponding to the maximum value, 1.03, of W. That is, the density curve for W would now be an isosceles triangle with base ranging from $W = -3$ to $W = 3$. To satisfy the condition that the area under this density must equal 1, the vertex above $W = 0$ must be at a distance h such that

$$\tfrac{1}{2}(h)6 = 1;$$
$$h = \tfrac{1}{3}.$$

The density curve for W is shown by Figure 3–7.

To derive the cumulative distribution function $F(w)$ in our example, we note that $P(W \leq -3) = 0$ and that, for $-3 < W < 0$, the probability is given by the area of the triangle like the shaded one in Figure 3–7. From

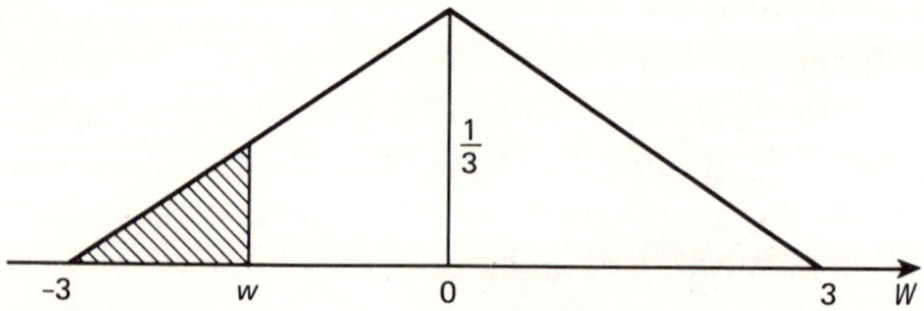

Figure 3–7 The Density Curve for the Coffee-Jar Example

the properties for similar triangles, we know that

$$\frac{h}{w+3} = \frac{\tfrac{1}{3}}{3} = \frac{1}{9}.$$

Hence, we obtain a general expression for h for the range from -3 to 0:

$$h = \frac{w+3}{9}.$$

To verify that this result is reaasonable, we see that if $W = -3$, $h = 0$; if $W = 0$, $h = \tfrac{1}{3}$. These certainly agree with what we have already learned. Since the foregoing expression for h is linear, it is also correct for $-3 < W < 0$. Therefore, the shaded area in Figure 3–7 can be defined as

$$F(w) = \frac{\tfrac{1}{2}(w+3)(w+3)}{9}$$
$$= \frac{(w+3)^2}{18}, \qquad \text{if } -3 < W < 0.$$

For $0 < W < 3$, we may compute the area of the small triangle to the right of $W = 0$. Again, making use of similar triangles, we may obtain the following result:

$$F(w) = 1 - \frac{(3-w)^2}{18}, \qquad \text{if } 0 < W < 3.$$

For $W \geq 3$, $F(w) = 1$. Combining the several pieces of information, we arrive at the following description:

$$F(w) = P(W \le w_i) = \begin{cases} 0, & \text{if } W \le -3, \\ \dfrac{(w+3)^2}{18}, & \text{if } -3 < W < 0, \\ 1 - \dfrac{(3-w)^2}{18}, & \text{if } 0 < W < +3, \\ 1, & \text{if } W \ge 3. \end{cases}$$

The graph for this cumulative distribution function is shown by Figure 3–8.

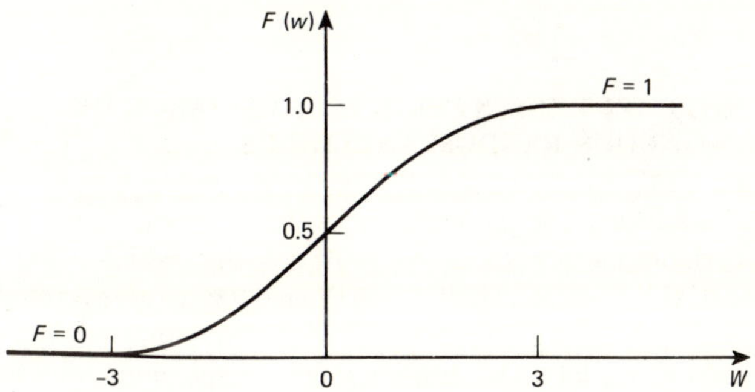

Figure 3–8 The Cumulative Distribution Function for the Coffee-Jar Example

Now, suppose a jar is selected at random from a day's production and the net weight of instant coffee in it is measured. We have

(1) $$F(-1) = \frac{(-1+3)^2}{18} = 0.222,$$

which would be the probability that the net weight in the jar is less than 0.99 pound;

(2) $$F(1.5) = 1 - \frac{(3-1.5)^2}{18} = 0.875,$$

which would be the probability that the net weight in the jar is $W \le +1.5$— i.e., 1.015 pound;

(3) $$\begin{aligned} P(|W| < 1.5) &= P(-1.5 \le W \le +1.5) \\ &= F(1.5) - F(-1.5) \\ &= 0.750, \end{aligned}$$

which would be the probability that the net weight in the jar falls within the interval from 0.985 to 1.015 pound;

$$
\begin{aligned}
(4)\quad P(|W| \geq 1.5) &= P(W > 1.5 \text{ or } W < -1.5) \\
&= [1 - F(1.5)] + F(-1.5) \\
&= \left[1 - \left(1 - \frac{(3 - 1.5)^2}{18}\right)\right] + \frac{(-1.5 + 3)^2}{18} \\
&= 0.250,
\end{aligned}
$$

which would be the probability that the net weight of the jar falls outside the interval $-1.5 \leq W \leq +1.5$, or outside the interval from 0.985 to 1.015 pound. Note that the results in (3) and (4) add up, as they should, to unity.

3–10 A NOTE ON EXPECTATION AND VARIANCE OF CONTINUOUS RANDOM VARIABLES

Precise definitions for the expectation and the variance of a continuous random variable require a knowledge of calculus, which is not assumed for the reader of this text. We shall, nevertheless, try to give a rough evaluation of these concepts by arguing in terms of discrete approximations.

For convenience, let us assume we have a continuous random variable X whose probability is distributed over a finite interval from $X = a$ to $X = b$. Let this interval be subdivided into n equal parts, of which $\Delta x = (b - a)/n$. Denote the left endpoint of the ith subinterval as x_i (for $i = 1, 2, 3, \ldots, n$). Round off all values in the ith subinterval to x_i. These rounded-off values of X may be considered as a new random variable, X', which is discrete since it can assume only the endpoint values, $x_1, x_2, \ldots, x_n$. Now, the probability that X' can take on any of these possible values, say x_5, is the probability that the original variable X takes on a value that would be rounded off to x_5. This occurs if X takes on a value in the fifth subinterval—that is,

$$P(X' = x_5) = P(x_5 < X \leq x_5 + \Delta x) \doteq f(x_5)\, \Delta x.$$

This approximation becomes more and more exact as n, the number of subintervals, increases.

Now, by the formula for the expectation of a discrete random variable, we have

$$
\begin{aligned}
E(X') &= \sum_{i=1}^{n} x_i P(X' = x_i) \\
&\doteq \sum_{i=1}^{n} x_i f(x_i)\, \Delta x.
\end{aligned}
$$

In this expression, as $n \to \infty$, then $\Delta x \to 0$. Hence, we expect the discrete variable X' to approach the continuous variable X. That is, the "limit" of the foregoing expression for $E(X')$ should be taken as the expectation of X, $E(X)$.

Similarly,

$$\begin{aligned} V(X') &= E[X' - E(X')]^2 \\ &= \sum_{i=1}^{n} (x_i - \mu')^2 P(X = x_i) \\ &\doteq \sum_{i=1}^{n} (x_i - \mu)^2 f(x)\, \Delta x. \end{aligned}$$

Thus, $V(X')$ should be taken as $V(X)$ as $n \to \infty$ and $\Delta x \to 0$.

Considering a continuous probability distribution as a continuous mass distribution, as suggested previously, we may interpret the expectation and variance as the center of gravity and radius of gyration, respectively.

3–11 ADDITIONAL PROPERTIES OF RANDOM VARIABLES

In this chapter we have introduced only the most important descriptive measures of random variables—expectation, variance, and standard deviation. Needless to say, other properties of random variables are also of interest and practical importance. We shall mention some of these properties briefly here, mainly for future reference.

First we should notice that, in addition to the mean, two other measures, the median and the mode, are frequently employed to measure the central tendency of a random variable. Median is a concept related to the notion of "percentiles." Consider the distribution function of a random variable, $F(x)$, whose values range from 0 to 1. If this range of values of X is divided into 100 equal parts, the corresponding values of X are called the *percentiles*. For instance, the value of X corresponding to, say, $F(x) = 0.35$ is called the 35th percentile and may be denoted by $X_{.35}$.

Percentiles that have been given special names are *quartiles*. The first quartile is the 25th, the second quartile, the 50th, and the third quartile, the 75th percentiles, respectively. The second quartile is also called the *median*. A definition of the median, which can be applied to both discrete and continuous random variables, is this: the median, denoted by $X_{0.50}$, is that value of the random variable X that satisfies

$$P(X \leq m) \geq ½ \quad \text{and} \quad P(X \geq m) \geq ½,$$

where $m = X_{.5}$—the value of a random variable that divides its probability (or density) function into two equal parts. Thus, for the four-coin example, $X_{.5} = 2$; for the coffee-jar example, $X_{.5} = 0$ (which is the coded value of 1 pound).

The *mode* for a discrete variable, if it exists, may be defined as the value of the random variable that has the highest probability of occurrence. Denote the measure as X_m; for the four-coin example, $X_m = 2$. In the continuous case the maximum density function, when the density has a single maximum, is the mode of the variable. Again, for the coffee-jar example, $X_m = 0$ with a maximum ordinate of $\frac{1}{3}$. For the stopwatch ex-

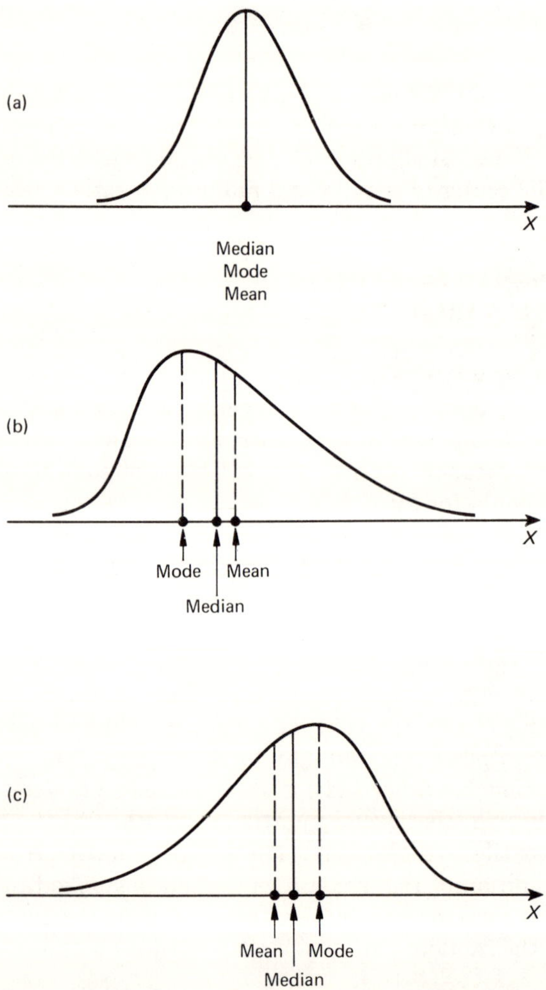

Figure 3–9 Illustrations of (a) Symmetry, (b) Positive Skewness, and (c) Negative Skewness

ample, however, there is no mode, since the density function is uniform and has no single maximum.

A random variable may have no mode or it may have more than one. In particular, when a random variable has two modal values, we say it is *bimodal* (in contrast to the *unimodal* case where it has only one).

A simple but quite useful measure of variability is the *range*, which is simply the difference between the maximum and minimum values of a random variable. For example, the ranges for random variables represented by Figure 3–3 are 2 for X and 20 for Y, respectively. Here the respective ranges seem quite sufficient to indicate the difference in variability between the two variables, since $E(x) = E(y)$.

Besides central tendency and variability, a random variable may also be characterized by "symmetry" and "asymmetry." For unimodal distributions, Figure 3–9 shows three broad classes of symmetry and asymmetry. As panel (a) indicates, if the distribution is symmetric about the mean, the mean, the median, and the mode coincide. For asymmetric distributions, as panels (b) and (c) show, these measures vary, with μ and $X_{.5}$ being pulled away from X_m in the direction of the outlines in the long tail. Since μ is sensitive to the precise location of these more remote (that is, outlying) values while $X_{.5}$ is not, μ is pulled further away from X_m than is $X_{.5}$. The sensitivity to outlying values differs because the mean is a weighted average while the median is a position average.

Note that the departure from symmetry is called *skewness*, which is described in terms of the direction of the distribution's long tail. In Figure 3–9, panel (b) illustrates *right*, or *positive skewness*; panel (c) *left*, or *negative skewness*. The direction of skewness can thus be easily determined geometrically, but its degree must be measured algebraically (the algebraic measure will be given in Chapter 10).

EXERCISES

3–1 Three coins are tossed. Find and graph the probability function of the number of heads, X.

3–2 Show that the number of tails, Y, has the same probability distribution as the number of heads when three coins are tossed.

3–3 Three identical marbles, numbered 1, 2, and 3, are placed in a bag. If two marbles are drawn at random with replacement, what is the probability function of the random variable Z, where Z is the sum of the numbers of the marbles?

3–4 Let X be the random variable showing the number of girls in families with four children. What is the probability distribution of X if the births of boys and girls are equally likely?

3–5 Four balls are drawn at random without replacement from an urn that contains six white and four black balls. What are the probability and distribution functions of the number of black balls?

3–6 Sketch the graphs of $f(x)$ and $F(x)$ in the previous problem.

3–7 From a lot of twelve radios containing five defectives, a random sample of three is drawn without replacement. Find the probability and distribution functions of the number of defectives in the sample.

3–8 A box of articles, nine good and one defective, is tested, one article at a time. If X is the number of the test in which the defective article is located, what are the possible values of X and the corresponding probability function?

3–9 Give a general expression of the probability function in Exercise 3–5.

3–10 Five coins are tossed. Let X = number of heads and Y = number of tails. Find the distribution of (a) $X - Y$, (b) $X + Y$, and (c) XY.

3–11 Three balls are drawn without replacement from a bag containing five white and two black balls. Check whether the expected number of black balls is $6/7$.

3–12 In a given business venture a man can make a profit of \$1000 or suffer a loss of \$500. The probability of a profit is 0.6. What is the expected profit (or loss) in that venture?

3–13 The probability is 0.99 that a 30-year-old man will survive one year. An insurance company offers to sell such a man a \$10,000 one-year term life insurance policy at a premium of \$110. What is the company's expected gain?

3–14 In a lottery, 1000 tickets are sold at 25 cents each. There are five cash prizes of \$25, \$20, \$10, \$5, and \$1, respectively. What is the expected net gain for a purchaser of two tickets?

3–15 A random sample of three conferees is selected without replacement from a group of four men and three women. Show that the expected number of men in the sample is $12/7$.

3–16 A roulette wheel has 38 equally spaced openings with numbers 00, 0, 1, 2, ..., 36. A gambler may bet, say, \$1 on any number. The roulette wheel is spun and a small ball is dropped onto it while it is spinning. If the ball comes to rest on the number on which the gambler has bet, he receives \$35 in addition to his bet of \$1; otherwise he loses his \$1. Show that the expected gain is $-\$2/38$.

3–17 A gambler who has \$700 plays the single-die game with the following system: At the first toss of the die, he bets \$100 on even numbers and quits if he wins. If he loses, he bets \$200 on even numbers at the second toss and quits if he wins. If he loses again, he bets his final \$400 on even numbers at the third toss. Is the game fair? [*Hint:* A game is fair if its expectation is 0.]

3–18 The probability that a man aged 55 will live another year is 0.98. What premium should the insurance company charge him for a \$10,000 term life insurance policy for a year if the premium should include a profit and administrative cost of \$50?

3–19 For the following distribution, show that (a) $E(Y) = 7$, (b) $E(Y^2) = 590$, (c) $E[Y - E(Y)]^2 = 541$, (d) $E(Y^3) = 10{,}900$.

y_i	-20	-10	30
$P(Y = y_i)$	$3/10$	$2/10$	$5/10$

3–20 A random variable X has the following probability function:

x_i	-1	0	1
$f(x_i)$	$1/5$	$3/10$	$1/2$

Find probability distributions for $2X$, $X + 1$, $2X + 1$, X^2, and $(X - 0.3)^2$, and then compute their expectations.

3–21 Let X be the number of tails in a toss of three coins. Compute $E(X)$, $V(X)$, and σ_X.

3–22 If the variance of a random variable is 0.8, what is the variance of the random variable $5X$? $2X$? $X/2$?

3–23 Let X be a random variable with the following probability function:

x_i	10,025	10,050	10,075
$f(x_i)$	0.2	0.3	0.5

Show that $V(X) = 381.25$.

3–24 A random variable X takes on the value 1 with probability p and the value 0 with probability $1 - p$. Show that $E(X) = p$ and $V(X) = p(1 - p)$.

3–25 Assuming that $E(X) = 4$ and $V(X) = 2$, answer the following:
(a) At least how much of the probability lies within the range 0 and 8?
(b) What is the minimum value of $P(-2 \leq X \leq 10)$?
(c) What is the minimum value of $P(|X - 4| \geq \sqrt{2h})$?
(d) What value of h guarantees $P(|X - 4| \leq \sqrt{2h}) \geq 0.96$?

3–26 Using Chebyshev's theorem, determine what value of h guarantees that at least 90 and 99 percent of probabilities, respectively, are within $h\sigma$ of the mean.

3–27 An urn contains four identical chips each with two numbers, one on each side: (1, 1), (2, −1), (3, 1), and (4, −1). Let the chip selected be (X, Y). Find $E(X)$, $E(Y)$, $E(X + Y)$, and $E(X - Y)$.

3–28 If a clock is stopped at random time, what is the probability that the hour hand will stop at (a) exactly numeral 2? (b) between 2 and 6?

3–29 Buses in a certain town run every half hour between 10 P.M. and 6 A.M. Verify that the probability that a man arriving at a bus stop at random time will have to wait 20 minutes or less is ⅔.

3–30 A random variable X has the following CDF:

$$F(x) = \begin{cases} 0, & \text{if } X \leq 0, \\ x^2, & \text{if } 0 < X < 1, \\ 1, & \text{if } X \geq 1. \end{cases}$$

For this random variable, show that
(a) $P(X \leq 0.6) = 0.36$.
(b) $P(0.1 \leq X \leq 0.6) = 0.35$.
(c) $P(X > 0.5) = 0.75$.
(d) $P(-1.5 < X < 2.7) = 1$.
(e) $P(X \leq 0.5 \text{ or } X > 0.8) = 0.61$.

3–31 Consider the random variable Y with the following density function:

$$f(y) = \begin{cases} \frac{1}{2} - bY, & \text{if } 0 \leq Y \leq 4 \\ 0, & \text{if } Y < 0 \quad \text{or} \quad Y > 4 \end{cases}$$

where b is a constant.
(a) What should the value of b be? Draw the graph for $f(y)$.
(b) Find (1) $P(Y \leq 2)$, (2) $P(Y > 3)$, and (3) $P(1 \leq Y \leq 2)$.

4

SPECIAL DISCRETE PROBABILITY MODELS

One of the basic tasks of a decision maker or a scientific investigator is to find a probability model to describe data on hand. In a probabilistic sense, the notion "model" usually refers to a mathematical expression or formula used in studying the results of a random experiment. This definition is not inconsistent with the idea that a model is a set of assumptions necessary to derive a formula.

Probability models are important to the extent that they are reasonable reflections of real processes. Such models can help us predict the behavior of future repetitions of an experiment. They are also of practical interest, since if we could have precise mathematical formulations for various distributions frequently encountered in statistical analysis, computations of probabilities involved could be reduced to routine operations, and mathematical analysis could be used in the development of statistical methods in connection with these models.

In this chapter we shall study distributions of a few special discrete random variables that can be employed to approximate a large number of real-world situations. In the next chapter we shall introduce the basic models for a few important continuous random variables. We shall actually derive the mathematical expressions for each model from a set of assumptions whenever the required mathematical reasoning is not beyond the reader's reach. We shall also rely extensively on prepared tables so that computational labor can be minimized. The reader should take note at this stage, however, that it is more important for him to identify a problem with the correct model than to memorize the detailed mathematical definitions.

4–1 THE BERNOULLI MODEL

We begin our study of probability models with the Bernoulli model—not only because it is the simplest but also because it furnishes a basis for the derivation of the binomial probability model, which is associated with a number of frequently occurring real processes.

The *Bernoulli* model (named in honor of Jacob Bernoulli, who lived during the latter half of the seventeenth century) applies to a random variable that can assume only two values. For simplicity, let the two values be 1 and 0, with p and $q = 1 - p$ as their respective probabilities; then the probability function of a Bernoulli variable is simply

x_i,	$f(x_i)$
1	p
0	q
Sum:	1

For the distribution above we note that

$$\sum_i x_i f(x_i) = p$$

and

$$\sum_i x_i^2 f(x_i) = E(X^2) = p.$$

Consequently, we have for the Bernoulli variable

$$E(X) = \mu = p \tag{4–1}$$

and

$$\begin{aligned} V(X) &= \sigma^2 \\ &= E(X^2) - \mu^2 \\ &= p - p^2 \\ &= p(1 - p) \\ &= pq. \end{aligned} \tag{4–2}$$

The Bernoulli model is appropriate whenever one is interested in an experiment that would result in an event E or its opposite $\bar{E}$, such as success or failure, yes or no propositions, male or female subjects, defective or nondefective items, and so forth.

4–2 THE BINOMIAL MODEL

A binomial process may be considered as the sum of n independent Bernoulli variables. More precisely, a *binomial variable* is generated under the following three postulates:

(1) There must be a fixed number of repeated trials that are stochastically independent.

(2) Each trial must result in a Bernoulli variable—that is, a "success" or a "failure."

(3) All trials must have identical probabilities of success, p, with the probability of failure for each trial remaining at a constant value of q that is equal to $1 - p$.

These conditions are satisfied if we toss a coin, say, ten times, and we define success in terms of the number of heads that turn up; if we select five items at random with replacement from a lot, say, of one hundred units of incoming merchandise and we define success as the number of defective items in the sample; and so on.

A random variable generated under the foregoing conditions is called a *binomial variable*; it is *discrete* and has $n + 1$ possible values. For example, if a coin is tossed twice, then the possible numbers of heads are 0, 1, and 2; there are $(2 + 1)$ terms.

This model is useful for answering questions such as: If we conduct an experiment under the stated conditions n times, what is the probability of obtaining exactly x successes? More specifically, suppose six dice are tossed together, or one die is tossed six times; what is the probability of obtaining exactly two aces?

We note first that the outcome of each die is independent of those of all others. Furthermore, the outcome of each die may be either an ace or any one of the other five numbers. That is, this experiment satisfies all premises of a binomial model: we have $p = 1/6$, $q = 5/6$, and $n = 6$. Now, any sequence that contains two aces and four other numbers can be considered a successful outcome. The probability for any such sequence, by the product rule, is p^2q^4, or $(1/6)^2(5/6)^4$. For instance, the probability for a sequence in which the first and the last outcomes are aces, and the remaining four are not, is

$$(1/6)(5/6)(5/6)(5/6)(5/6)(1/6) = (1/6)^2(5/6)^4.$$

Furthermore, all sequences that contain exactly two aces each are mutually exclusive events; by the law of addition, therefore, the probability of success is the sum of the individual probabilities of all these sequences—that is,

$$(1/6)^2(5/6)^4 + (1/6)^2(5/6)^4 + \cdots,$$

where there are as many identical terms as there are ways of selecting two things at a time from six objects, $\binom{6}{2}$. These considerations lead us to conclude that the probability of obtaining exactly two aces in six tosses of a die, with the probability of success in each trial being $1/6$, is

$$P(2; n = 6, p = 1/6) = \binom{6}{2}\left(\frac{1}{6}\right)^2\left(\frac{5}{6}\right)^4 = \frac{3125}{15{,}656}.$$

To generalize: the *binomial probability function* is the probability of obtaining exactly x successes in n independent trials of an experiment, with p as the probability of success for each trial; it is given by the expression

$$b(x; n, p) = \binom{n}{x} p^x q^{n-x}. \tag{4–3}$$

The binomial model derives its name from the fact that (4–3) is a term of the "binomial expansion" of $(q + p)^n$.

In a lot of twelve television tubes three are defective. If a random sample of three is drawn from the lot with replacement, what is the probability that (1) exactly one is defective, and (2) none or one is defective?

Sampling with replacement ensures independence of successive selections, so the binomial model applies. As stated in the problem, $p = 3/12 = 0.25$. Thus

$$(1)\qquad b(1; 3, 0.25) = \binom{3}{1}(0.25)(0.75)^2 = 0.42;$$

$$(2)\qquad b(0; 3, 0.25) + b(1; 3, 0.25) = \binom{3}{0}(0.25)^0(0.75)^3 + \binom{3}{1}(0.25)(0.75)^2 = 0.84.$$

The *distribution function* of a *binomial variable*, like that of any discrete random variable, gives the probability of obtaining r successes or less in n trials, with $r \leq n$, and is obtained by adding the individual probabilities for all binomial values equal to or less than r. That is,

$$\begin{aligned} B(r; n, p) &= P(X \leq r) \\ &= b(0; n, p) + b(1; n, p) + \cdots + b(r; n, p) \\ &= \sum_{x=0}^{r} b(x; n, p). \end{aligned} \tag{4–4}$$

Extensive tables for the binomial distribution function are available. A shorter table of the binomial cumulative probabilities appears as Table B-3 in Appendix B. This table, as shown by the following examples, can be used to obtain cumulative as well as individual binomial probabilities.

A coin is tossed ten times; what is the probability of getting eight heads or more? Here, we have a binomial model with $n = 10$ and $p = 1/2 = 0.5$. The desired probability is that of obtaining 8, 9, or 10 heads. What we need then is

$$\sum_{x=8}^{10} b(x; 10, 0.5) = 1 - \sum_{x=0}^{7} b(x; 10, 0.5)$$

$$= 1 - 0.94531 \qquad \text{(from Table B-3)}$$
$$= 0.05469.$$

In a 20-question, 4-answer, multiple-choice examination, what is the probability of getting (1) exactly 7 correct answers, or (2) 7 or more correct answers, if a student answers randomly? To solve this problem, let us assume that all questions are answered, that all questions are independent, and that the answer is either right or wrong. Under these premises, we would then have a binomial model with $p = \frac{1}{4}$ and $n = 20$. Then

(1) $$b(7; 20, 0.25) = B(7; 20, 0.25) - B(6; 20, 0.25)$$
$$= 0.89819 - 0.78578 \qquad \text{(from Table B-3)}$$
$$= 0.11241;$$

(2) $$\sum_{x=7}^{20} b(x; 20, 0.25) = 1 - \sum_{x=0}^{6} b(x; 20, 0.25)$$

$$= 1 - 0.78578 \qquad \text{(from Table B-3)}$$
$$= 0.21422.$$

Output produced by a certain process is 90 percent perfect; the items that comprise the defective remainder cannot be detected without being destroyed through inspection. If a random sample of 15 is drawn from a certain day's production run—which is more than 100,000 units—what is the probability that the sample will contain 12 or fewer perfect units?

We now have a case where sampling with replacement is impossible because sampling is destructive. However, the population is extremely large by comparison with the size of the sample; as a result, the probability of getting a perfect item remains approximately at the same value of 0.90 at each selection, and so the binomial model can be employed as an approximation. Here, we have a binomial model with $n = 15$ and $p = 0.90$. We wish to evaluate

$$\sum_{x=0}^{12} b(x; 15, 0.90).$$

However, because our binomial table gives p's up to only 0.50, probabilities for p's greater than 0.50 cannot be read from it directly. To use the table when $p > 0.50$, we must make some transformation. That is, when $p > 0.50$, we can find the cumulative binominal probabilities from the table by interchanging the roles of p and q and, thereby, the roles of successes and failures.

For our present problem, to obtain 12 or fewer perfect items is the same as to get 3 or more defective ones. It follows that

$$\begin{aligned}\sum_{x=0}^{12} b(x; 15, 0.90) &= \sum_{x=3}^{15} b(x; 15, 0.10)\\ &= 1 - \sum_{x=0}^{2} b(x; 15, 0.10)\\ &= 1 - 0.81594 \qquad \text{(from Table B–3)}\\ &= 0.18406.\end{aligned}$$

A binomial variable, X, the number of successes in n independent trials, may be considered as a sum

$$X = X_1 + X_2 + \cdots + X_n,$$

where each X_i is a Bernoulli variable with expectation p. Thus, the expectation of the binomial variable may be thought of as the sum of expectations of the n Bernoulli variables. That is,

$$\begin{aligned}E(X) &= E(X_1) + E(X_2) + \cdots + E(X_n)\\ &= \underbrace{p + p + \cdots + p_n}_{n \text{ times}} = np. \end{aligned} \tag{4–5}$$

Furthermore, the Bernoulli variables are identical and independent. As such, their variances are also additive. That is, the binomial variable, as the sum of n Bernoulli variables, must have as its variance the sum of the variances of n Bernoulli variables:

$$\begin{aligned}V(X) &= V(X_1) + V(X_2) + \cdots + V(X_n)\\ &= \underbrace{pq + pq + \cdots + pq}_{n \text{ products}} = npq. \end{aligned} \tag{4–6}$$

Thus, for the previous example,

$$E(X) = np = 15(0.9) = 13.5 \text{ nondefective items},$$

and $$V(X) = npq = 15(0.9)(0.1) = 1.35.$$

Before concluding our discussion on the binomial probability model, we should note that the binomial distribution is symmetric when $p = 0.5$ and skewed when $p \neq 0.5$. When $p < 0.5$, it is skewed to the right; when $p > 0.5$, it is skewed to the left. The interchange of p and q in any binomial distribution would yield its mirror image. The skewness of a binomial distribution, regardless of the size of p, becomes less and less pronounced as n continues to increase. All of these properties are revealed by Figure 4–1.

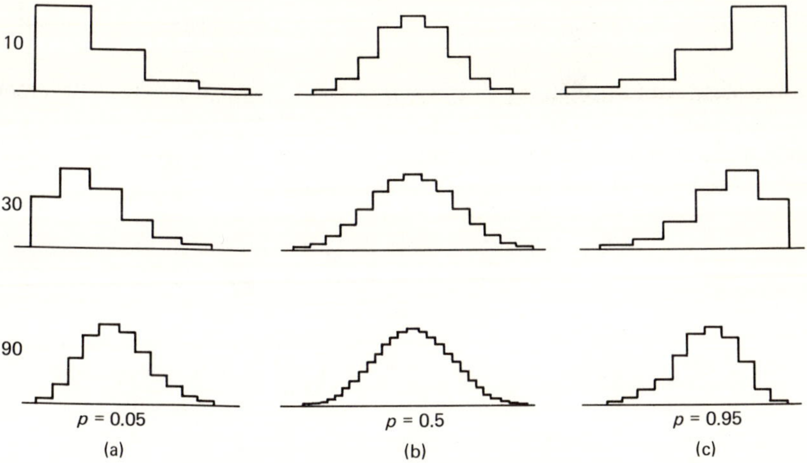

Figure 4–1 Histograms of Binomial Distributions

4–3 THE HYPERGEOMETRIC MODEL

The *hypergeometric probability model* applies to sampling without replacement from a finite population whose elements can be classified into two categories. Consider a population of N items, k of which possess a certain characteristic and $N - k$ of which do not possess such a characteristic. Obviously, if a random selection is made from such a population, the result must be either one of the k's (successes) or one of the $(N - k)$'s (failures). It is also clear that if n random selections are made without replacement from the population, each subsequent draw is dependent and the probability of success changes on each draw. Under these conditions, we may wish to find the probability of obtaining exactly x items that are of type k, or successes, in a random sample of size n. The number of successes in n dependent trials is called a *hypergeometric variable* and is generated with three fixed numbers: N, n, and k. Let us now proceed to derive the probability function for this random variable.

First, recall that the total number of ways to draw n items from N things is $\binom{N}{n}$. Next, note that to draw exactly x items of a certain characteristic is the same as to draw $n - x$ items that do not possess the characteristic from a total of $N - k$ items available. Thus, the total number of ways to draw x k's and $n - x$ non-k's is $\binom{k}{x}\binom{N-k}{n-x}$. Finally, if the process of sampling is random, then all items that are still available after any number of selections have been made are equally likely to be chosen. As a result, we can assign equal probabilities to each of the $\binom{N}{n}$ points. These arguments lead us to conclude that the *probability function* of a *hypergeometric variable*, the probability that x of the n items are of type k, is

$$h(N, n, k, x) = \frac{\binom{k}{x}\binom{N-k}{n-x}}{\binom{N}{n}}, \tag{4–7}$$

from which we obtain the *distribution function* of X as follows:

$$\begin{aligned} H(N, n, k, r) &= P(X \leq r) \\ &= \sum_{x=\max[0,n-(N-k)]}^{r} h(N, n, k, x). \end{aligned} \tag{4–8}$$

Table B-4 in Appendix B gives values for both $h(x)$ and $H(x)$ for $N = 10$. (An extensive table for these values is *Table of the Hypergeometric Probability Distribution* by G. J. Lieberman and D. B. Owens, published by Stanford University Press in 1961.)

To derive the expectation of a hypergeometric distribution, note that the hypergeometric variable X may also be considered as a sum, just as in the binomial case, of n variables X_i, except that in the hypergeometric model, $X_1, \ldots, X_n$ are dependent. But, because the additive property of expectation does not require independence of X_i, it is still true that $E(X) = E(X_1) + \cdots + E(X_n)$, where each $E(X_i)$ is the probability of X at the ith trial (k/N), if it is not known what has happened at preceding or subsequent trials. Thus, the expectation of the hypergeometric variable X becomes

$$E(X) = n\left(\frac{k}{N}\right). \tag{4–9}$$

The variance is not additive for dependent variables. It can be shown, however, that in the hypergeometric case, the variance is given by the expression below:

$$V(X) = \left(\frac{N-n}{N-1}\right) n \left(\frac{k}{N}\right)\left(\frac{N-k}{N}\right). \tag{4–10}$$

We note that in the binomial case, where sampling is made with replacement, $V(X) = npq = n(k/N)[(N-k)/N]$. When sampling without replacement is assumed for the hypergeometric model, a factor $(N - n)/(N - 1)$ is introduced.

In a lot of 50 baby chickens, 6 are females. If a random sample of 3 is drawn without replacement, then X, the number of female chickens in the sample, is a hypergeometric variable and its probability function can be evaluated by the general term

$$h(50, 3, 6, x) = \frac{\binom{6}{x}\binom{44}{3-x}}{\binom{50}{3}}.$$

Table 4-1 contains calculations of these terms, together with their cumulative sums.

Table 4-1. Probability and Distribution Functions of the Number of Female Chickens with $N = 50$, $n = 3$, and $k = 6$

x	$3 - x$	$\binom{6}{x}$	$\binom{44}{3-x}$	$\binom{6}{x}\binom{44}{3-x}$	$h(x)$	$H(x)$
0	3	1	13,244	13,244	0.6757	0.6757
1	2	6	946	5,676	0.2896	0.9653
2	1	15	44	660	0.0337	0.9990
3	0	20	1	20	0.0010	1.0000
—	—	—	—	19,600*	1.0000	—

* $\binom{50}{3} = 19{,}600.$

For this model, we have

$$E(X) = n\left(\frac{k}{N}\right) = 3\left(\frac{6}{50}\right) = 0.36 \text{ female chickens,}$$

and

$$V(X) = \left(\frac{N-n}{N-1}\right) n \left(\frac{k}{N}\right)\left(\frac{N-k}{N}\right)$$

$$= \left(\frac{50-3}{50-1}\right) 3 \left(\frac{6}{50}\right)\left(\frac{50-6}{50}\right) = 0.3039.$$

Whenever sampling is destructive, the hypergeometric model should be adopted rather than the binomial model. Fortunately, as mentioned earlier, a comprehensive hypergeometric probability table has been made available. In using such a table, one should observe a very interesting property of the hypergeometric distribution: when k and n are interchanged, there will be no effect on the probabilities. That is,

$$h(N, n, k, x) = h(N, k, n, x) \tag{4–11}$$

and

$$H(N, n, k, r) = H(N, k, n, r). \tag{4–12}$$

Consider this example: if 4 chips are drawn together at random from an urn containing 6 black and 4 white chips, what is the probability that 2 or fewer black chips are drawn?

We have here a hypergeometric model with $N = 10$, $k = 6$, and $n = 4$. The desired probability is

$$\begin{aligned} P(X \leq 2) &= H(10, 4, 6, 2) \\ &= H(10, 6, 4, 2) && \text{[by (4–12)]} \\ &= 0.547619 && \text{(from Table B-4).} \end{aligned}$$

Now if 5 chips are drawn without replacement from the same urn, what is the probability that (1) all are black? (2) 4 or more are black?

Answer:

$$\begin{aligned} \text{(1)} \quad P(\text{All five are black}) &= h(10, 5, 6, 5) \\ &= h(10, 6, 5, 5) && \text{[by (4–11)]} \\ &= 0.023810 && \text{(from Table B-4);} \end{aligned}$$

$$\begin{aligned} \text{(2)} \quad P(X \geq 4) &= 1 - P(X \leq 3) \\ &= 1 - H(10, 5, 6, 3) \\ &= 1 - H(10, 6, 5, 3) && \text{[by (4–12)]} \\ &= 1 - 0.738095 && \text{(from Table B-4)} \\ &= 0.261905. \end{aligned}$$

4–4 THE DISCRETE UNIFORM MODEL

We have encountered the discrete uniform probability model before, though it was not given such a name. For instance, the probability distribution of the number of spots on a die generated by tossing the die is uniform because the probability associated with each and every outcome is a constant $1/6$. The probability function of this model can be written as

$$f(x; 6) = 1/6, \qquad x = 1, 2, \ldots, 6.$$

In general, the uniform distribution applies to an experiment that can terminate in n mutually exclusive and equally likely ways. Consider, for example, n numbers, one each on n identical chips, are placed in a box and mixed thoroughly, and then one chip is drawn from the box. The mathematical statement for this experiment is clearly

$$u(x; n) = \frac{1}{n}, \qquad x = 1, 2, \ldots, n, \tag{4–13}$$

which gives the probability that x assumes for any specific value as $1/n$; it is therefore called the *discrete uniform*, or *rectangular, probability model.*

One important aspect of the uniform distribution is that it furnishes a practical means of drawing a random sample from a finite population. Recall that sampling from an infinite population is random if all observations in the sample are made from the same population and if these observations are drawn independently. In the case of a finite population, a sample is said to be random if the sampling procedure is such that all possible samples of the same size have the same probability of being chosen. This distinction exists, of course, if sampling is made without replacement. When replacement is employed, a finite population can be in effect treated as an infinite one, because then we can never exhaust observations, and each observation is made independent of all the rest. It is interesting to note that both types of samples—one drawn under the condition that all the items in the population have equal probability to be included in the sample, an unconditional random sample, and one drawn under the condition that all samples of the same size have the same probability to be selected, a conditional random sample—are called *simple random samples,* and they are the only kinds of samples we shall consider when we discuss the theory of sampling in a later chapter. Now, let us explain how to use the uniform probability model for selecting a random sample from a finite population.

Consider a population of ten elements designated as 1, 2, . . . , 10. We wish to draw from it a sample of size three. To do this, we first write the ten numbers on ten chips and put them in a box. Next, we select one chip at a time without replacement until three chips have been drawn. By this procedure, we see that the probabilities associated with all selections are uniform. For the first choice, we have $u(x, 10) = 1/10$. For the second choice, we have $u(x, 9) = 1/9$. For the third choice, we have $u(x, 8) = 1/8$. Finally, let us evaluate the probability of drawing a sample that contains any three items, say, 1, 3, and 7. Such a sample can be obtained by drawing first any one of the three numbers with a probability of $3/10$, by drawing second either of the two remaining numbers with a probability of $2/9$, and by drawing third the remaining number with a probability of $1/8$. By the product rule for dependent events, we have the probability of getting this sample as follows:

$$(3/10)(2/9)(1/8) = 3!7!/10! = \frac{1}{10!/3!7!} = \frac{1}{\binom{10}{3}},$$

where $\binom{10}{3}$ is the number of all possible samples of size three from a population of ten items. From similar reasoning, each and every one of the possible samples has the same probability of $1\Big/\binom{10}{3}$ of being selected. As a result, the method has yielded a random sample.

As a matter of fact, the preceding arguments constitute a verification of the statement that all combinations are equally likely. We may now generalize that if a sample of size n is drawn from a population with N observations, then there are $\binom{N}{n}$ possible samples, each of which has a probability of $1\Big/\binom{N}{n}$ of occurring. This means that the distribution of samples itself is uniform, with a probability function

$$u(x; r) = \frac{1}{\binom{N}{n}},$$

where $r = \binom{N}{n}$, the number of possible samples of size n, and x stands for the r possible samples. Here x is a variable of n dimensions, since each sample contains n items. It seems that what we need is an "N-sided" die to generate a uniform population of N items. Such a die is clearly impractical to construct. Fortunately, instead, we can use a table of random digits for the same purpose.

A table of random digits is assumed to represent a large random sample whose values are those of a random variable with sample space

$$S = \{0, 1, 2, 3, 4, 5, 6, 7, 8, 9\}$$

and uniform distribution. That is,

$$u(x; 10) = 1/10, \qquad x = 0, 1, 2, \ldots, 9.$$

4–5 THE POISSON MODEL

Some events occur not as outcomes of a definite number of trials of an experiment but rather at random points of time, space, or volume. Look at Figure 4–2, where each point represents an occurrence of the event. For events of this kind we are interested only in the number of occurrences of the event, never in its nonoccurrences. We may wish to know the number of trains arriving at a certain station during a certain hour, but obviously it makes no sense to investigate the number of trains that will not arrive at the station during that hour. Or, we may need to know the number of printer's errors on each page of a book, but there is no need to determine the number of correct words on each page.

Figure 4–2 Random Points in Time or Space

There is a large class of events of this variety. The event may be the number of occurrences of accidents, errors, breakdowns, or other calamities that appear randomly and independently per stated unit. The event may be the demand for service per unit of time upon a cashier or a saleswoman of a department store, upon a clerk at a factory, upon a tollbooth of a bridge or tunnel, upon a cargo-handling facility of a port, upon the trunk lines of a telephone exchange, or upon a maintenance man at a machine shop. The event may be the number of defects on the surface of a table, or on an item of glassware, or on a length of fabric. The event may be the emission of a radioactive particle—and so on.

For each such random variable, the following postulates seem to be appropriate:

(1) The number of occurrences of the event is independent from one stated unit to another.
(2) The probability of a single occurrence of the event in an infinitesimal unit is proportional to the size of the stated unit.
(3) The probability of more than one occurrence of the event in an infinitesimal unit is negligible in comparison to the probability of a single occurrence.

We are concerned, under these conditions, with what is called the *Poisson* model, attributed to the French mathematician S. D. Poisson (1781–1840). The Poisson probability distribution gives the probability of the number of occurrences per stated unit and is defined completely by the mean number of occurrences per stated unit. Given the mean number of occurrences per stated unit of a Poisson variable, μ, the probability function is

$$p(x; \mu) = e^{-\mu}\left(\frac{\mu^x}{x!}\right), \tag{4–14}$$

where p stands for the functional notation of a Poisson distribution and e is a constant that is approximately equal to 2.71828.

It is interesting to note that (4–14) is the general term of the Poisson series,

$$e^{-\mu} + \mu e^{-\mu} + \frac{\mu^2 e^{-\mu}}{(1)(2)} + \frac{\mu^3 e^{-\mu}}{(1)(2)(3)} + \cdots,$$

which says that $P(0) = e^{-\mu}$, the probability of one occurrence is μ times the probability of none, the probability of two occurrences is $\mu/2$ times the probability of one occurrence, the probability of three occurrences is $\mu/3$ times the probability of two occurrences, and so on.

Suppose that calls go through a telephone exchange at random at a mean rate of four per minute. The probability that a call will occur at any instant is infinitely small, but the number of calls during a period of time, say, an hour, is extremely large. Under these conditions, the number of calls per minute is a Poisson variable with $\mu = 4$ and the following probability model:

$$p(x; 4) = e^{-4}\left(\frac{4^x}{x!}\right).$$

The individual terms of this model can be evaluated in terms of the Poisson series and are recorded in Table 4–2.

Table 4–2. Probability Distribution of the Number of Calls with $\mu = 4$ per Minute

x	$P(X = x)$	x	$P(X = x)$
0	$P(0) = e^{-4} = 0.0183$*	8	$P(8) = \frac{4}{8}P(7) = 0.0298$
1	$P(1) = \frac{4}{1}P(0) = 0.0733$	9	$P(9) = \frac{4}{9}P(8) = 0.0132$
2	$P(2) = \frac{4}{2}P(1) = 0.1465$	10	$P(10) = \frac{4}{10}P(9) = 0.0053$
3	$P(3) = \frac{4}{3}P(2) = 0.1954$	11	$P(11) = \frac{4}{11}P(10) = 0.0019$
4	$P(4) = \frac{4}{4}P(3) = 0.1954$	12	$P(12) = \frac{4}{12}P(11) = 0.0006$
5	$P(5) = \frac{4}{5}P(4) = 0.1563$		.
6	$P(6) = \frac{4}{6}P(5) = 0.1042$		.
7	$P(7) = \frac{4}{7}P(6) = 0.0595$		.

* $P(0) = e^{-4} = 0.0183$ by Table B-6 in Appendix B.

The distribution function of the Poisson variable, as in any discrete case, is obtained by adding the proper number of Poisson probabilities, which gives the probability of r or less occurrences. The mathematical statement is

$$P(r;\mu) = \sum_{x=0}^{r} p(x;\mu). \tag{4-15}$$

It can be shown that, for the Poisson variable, the expectation and variance are identical. That is,

$$E(X) = V(X) = \mu. \tag{4-16}$$

Table B-5 in Appendix B gives cumulative Poisson probabilities for selected values of μ. A fairly extensive table, *Poisson's Exponential Binomial Limits,* tabulated by E. C. Molina, has been published by the Van Nostrand Reinhold Company.

In evaluating Poisson probabilities, it is important to note that the Poisson model is completely defined by its mean μ, which is dependent on the size of the stated unit, and that μ changes proportionately whenever the stated unit is changed. For example, doubling the stated unit also doubles the expected number of occurrences per new stated unit.

To illustrate the preceding observations, let us use the telephone-call example mentioned before and ask: What is the probability that no more than two calls will occur during the next two minutes?

We note that the expected number was given as four calls per minute. Since, for a Poisson variable, expectation is proportional to the size of the stated unit, we have now $\mu = 4(2) = 8$ calls per two minutes. Thus, the desired probability is

$$P(2;8) = 0.01375 \qquad \text{(from Table B-5)}.$$

From the same distribution, what is the probability that no calls will occur in any interval of 30 seconds?

Now, $\mu = 4/2$ calls per 30 seconds and

$$p(0;2) = e^{-2} = 0.13534 \qquad \text{(from Table B-5)}.$$

A manufacturer of woolen piece goods claims that the average number of flaws in his products is one per two square yards. A sample square yard of his product, selected at random, shows three flaws. What is the probability of obtaining three or more flaws in any one square yard if the manufacturer's claim is valid?

Here, we have a Poisson variable with $\mu = 0.5$ per square yard, and the desired probability is

$$\begin{aligned}\sum_{x=3}^{\infty} p(x;0.5) &= 1 - \sum_{x=0}^{2} p(x;0.5)\\ &= 1 - P(2;0.5)\\ &= 1 - 0.98561 \qquad \text{(from Table B-5)}\\ &= 0.01439.\end{aligned}$$

In view of this result, what can we conclude about the manufacturer's claim?

A certain state in the United States has, on the average, twelve traffic deaths every three months. What is the probability that there are (a) more than four traffic deaths, and (b) exactly four deaths, in any one month?

Note that $\mu = 4$ per month. Thus

$$\begin{aligned}
\text{(a)} \quad \sum_{x=5}^{\infty} p(x;4) &= 1 - \sum_{x=0}^{4} p(x;4) \\
&= 1 - P(4;4) \\
&= 1 - 0.62884 \qquad \text{(from Table B-5)} \\
&= 0.37116; \\
\text{(b)} \quad p(4;4) &= P(4;4) - P(3;4) \\
&= 0.62884 - 0.43347 \qquad \text{(from Table B-5)} \\
&= 0.19537.
\end{aligned}$$

The Poisson distribution is positively skewed, but its skewness decreases as the size of its expectation increases (see Figure 4–3). Thus, a Poisson distribution is completely defined by μ, which, however, cannot be determined theoretically. Its value must be estimated empirically. The method of estimating μ involves, first, the selection of an appropriate fixed stated unit (time or space). Next, this fixed unit is subdivided into equal subunits

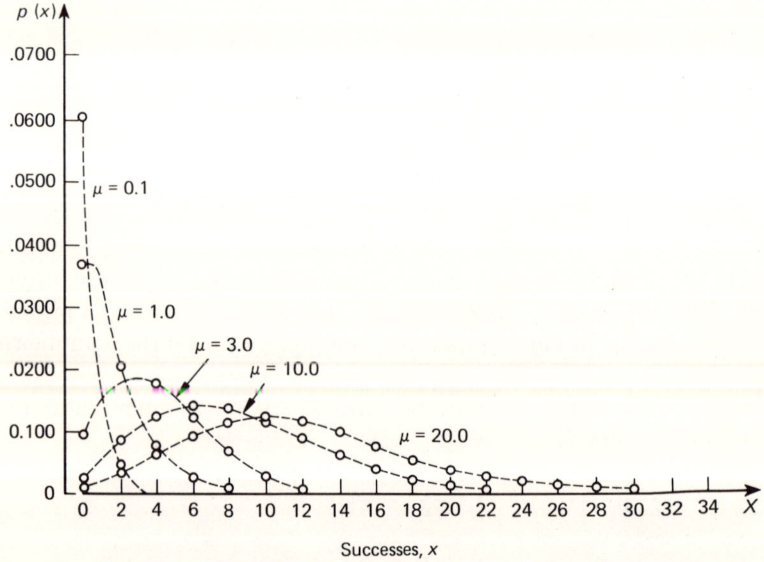

Figure 4–3 Poisson Distributions for μ = 0.1, 1.0, 3.0, 10.0, and 20.0

that generate a Poisson process. Then, we count the number of occurrences of the event in each subunit and denote this number by N_x, $x = 0, 1, 2, \ldots, k$. For example, if there are four subunits without any occurrence of the event, we write $N_0 = 4$; if there are three subunits each with one occurrence of the event, we write $N_1 = 3$; and so on. The total number of these equal subunits is

$$N_0 + N_1 + N_2 + \cdots + N_k = N,$$

and the total number of occurrences of the event in the fixed stated unit is

$$0(N_0) + 1(N_1) + 2(N_2) + \cdots + k(N_k) = T.$$

Finally, the ratio T/N is used as an estimate of μ, that is,

$$\hat{\mu} = \text{estimate of } \mu = \frac{T}{N}. \tag{4-17}$$

Let us use, as an example of estimating the Poisson expectation, data on the vacancies in the U.S. Supreme Court, either by death or by resignation of members, reported by W. A. Wallis in "The Poisson Distribution and the Supreme Court," *Journal of the American Statistical Association*, vol. 31 (1936), pp. 376–380. These data are given in the first two columns of Table 4–3.

Table 4–3

x: *Number of vacancies in the year*	N_x: *Observed number of years with* x *vacancies*	p(x; 0.5): *Probability of* x *vacancies*	96p(x; 0.5): *Expected number of years with* x *vacancies*
0	59	0.6065	58.22
1	27	0.3033	29.12
2	9	0.0758	7.28
3	1	0.0126	1.21
over 3	0	0.0018	0.17
	96	1.0000	96.00

For the data in Table 4–3 there are 96 fixed intervals of one year each, which can be thought of as 96 repetitions of a random experiment. Thus, we have $N = 96$. In these 96 intervals, the total number of vacancies is

$$\begin{aligned} T &= 0(N_0) + N_1 + 2N_2 + 3N_3 \\ &= 0 + 27 + 2(9) + 3(1) = 48. \end{aligned}$$

With the values of N and T, we have the estimate of μ for this Poisson variable as

$$\hat{\mu} = \frac{T}{N} = \frac{48}{96} = 0.5.$$

For the model under consideration, the probability function becomes

$$p(x; 0.5) = e^{-0.5}\left[\frac{(0.5)^x}{x!}\right].$$

The individual terms of this function are given in column three of Table 4–3. The last column of this table contains what may be considered the theoretical values of the number of years in which x vacancies occur. Observe how close the expected numbers are to the actual numbers in this case.

4–6 RELATIONSHIPS AMONG DIFFERENT PROBABILITY MODELS

So far we have been concerned with some special discrete probability models. Before taking up continuous models, we should realize that all the discrete models introduced are closely related to each other. We have already shown that the binomial variable may actually be considered as the sum of n identical Bernoulli variables. We shall proceed in this section to observe the relations among the binomial, the hypergeometric, and the Poisson models.

First, let us establish the relationship between the hypergeometric and the binomial distributions. Suppose that 30,000 of the 100,000 members of a union are in favor of a strike; if 10 members are selected at random, without replacement, then the probability that exactly x members in the sample are in favor of a strike is given by the hypergeometric law

$$h(100{,}000; 10, 30{,}000, x) = \frac{\binom{30{,}000}{x}\binom{100{,}000 - 30{,}000}{10 - x}}{\binom{100{,}000}{10}}.$$

If, however, the sample of 10 is drawn at random with replacement, and the probability of each independent trial remains constant at 30,000/100,000, or 0.3, then the binomial model is appropriate for the number of members in favor of a strike in the sample, and the probability function becomes

$$b(x; 10, 0.3) = \binom{10}{x}(0.3)^x(0.7)^{10-x}.$$

Now, the ratio of those in favor of a strike to the total number of union members will change only very slightly when a sample as small as 10 is drawn without replacement from a population as large as 100,000. In other words, when n is very small and N is very large, the probability of success will not change significantly from one draw to another, and successive draws can in effect be considered as independent. Thus, we would expect the binomial probability to approximate very closely the hypergeometric probability. So, in our numerical example, we would expect $b(x; 10, 0.3)$ to yield very close approximations to $h(100{,}000, 10, 30{,}000, x)$. In general, it can be shown that if n, and therefore x, is very small compared to both k and $N - k$, then

$$\frac{\binom{k}{x}\binom{N-k}{n-x}}{\binom{N}{n}} \doteq \binom{n}{x}\left(\frac{k}{N}\right)^x\left(\frac{N-k}{N}\right)^{n-x}, \tag{4–18}$$

which says: if a very small sample is taken from a very large population, then nonreplacement (the hypergeometric model) and replacement (the binomial model) give approximately identical results.

The practical importance of (4–18) is that, when conditions for its validity are met, we can obtain a good approximation of a hypergeometric probability by the more easily computed binomial probability. This approximation improves, given n, as N increases. When N becomes infinite, the hypergeometric and binomial probabilities become identical. Because of this, we say that the binomial model is the limit of the hypergeometric model as N approaches infinity.

While the binomial can be used to approximate the hypergeometric probabilities, the binomial, in turn, can be approximated by the Poisson distribution. It can be shown that if n is large and p is small, then the binomial and Poisson models are related by the following equation:

$$\binom{n}{x}p^x q^{n-x} \doteq e^{-np}\left[\frac{(np)^x}{x!}\right], \tag{4–19}$$

where the individual binomial terms are replaced by corresponding Poisson terms with $\mu = np$.

When we find it convenient to approximate the binomial by the Poisson, we must be sure to know what is meant by large n and small p. As a rule of thumb, if $n > 100$ and $p < 0.01$, the binomial and Poisson probabilities agree approximately (to three decimal places) for every value of the random variable. Furthermore, for those x values for which $[(x - np)^2/n] < 0.01$, the two models agree to within 1 percent. Finally, if p is sufficiently small, satisfactory approximations of the binomial by the Poisson can be made even when n is as small as 10. Given a small p, the larger the n, the better the results. Given n, the smaller the p is, the more satisfactory the approximations.

Table 4–4 gives the binomial probabilities for $x = 0, 1, 2, 3, 4$, and 5 with $n = 10$ and $p = 0.1$, and with $n = 20$ and $p = 0.05$, respectively. In each case the Poisson approximations are the same, since $\mu = 10(0.1) = 20(0.05) = 1$.

Table 4–4. Binomial Distributions and Their Poisson Approximations

X	0	1	2	3	4	5
$b(x; 10, 0.10)$	0.349	0.387	0.194	0.057	0.011	0.0015
$b(x; 20, 0.05)$	0.358	0.377	0.187	0.060	0.013	0.0022
$p(x; 1)$	0.368	0.368	0.184	0.061	0.015	0.0031

The numerical illustrations reveal that Poisson approximations of the binomial probabilities are quite good even when n is as small as 10 or 20, and that for $n = 20$ the approximations are better than for $n = 10$. However, in practice, it would make no sense to approximate binomial probabilities by the Poisson values when the former can be read directly from tables. Speed is gained without sacrificing accuracy only when n is large and p is sufficiently small, and when the binomial probabilities cannot be obtained from published tables.

Suppose that the probability that a newborn baby will die of a certain disease is 0.00002; what is the probability that out of 100,000 newborn babies (a) 4 or more, (b) exactly 4, will die of this disease?

The probabilities required are clearly binomial.

(a) The probability that 4 or more will die of this disease is

$$\sum_{x=4}^{100,000} \binom{100,000}{x} (0.00002)^x (0.99998)^{100,000-x}.$$

Rather than trying to evaluate this seemingly formidable expression, we would be well advised to approximate it by the Poisson law, with

$$\mu = np = 100,000(0.00002) = 2.$$

So,

$$\sum_{x=4}^{100,000} p(x;2) = 1 - \sum_{x=0}^{3} p(x;2)$$

$$= 1 - P(3;2)$$

$$= 1 - 0.85712 \qquad \text{(from Table B-5)}$$

$$= 0.14288.$$

(b) The probability of exactly 4 deaths is

$$P(4;2) - P(3;2) = 0.94735 - 0.85712 \qquad \text{(from Table B-5)}$$

$$= 0.09023.$$

In concluding this section, let us observe that just as the hypergeometric model takes the binomial model as its limit, so the binomial model takes the Poisson model as its limit. Thus the Poisson can also be employed to approximate the hypergeometric probability via the binomial distribution. We shall see, however, that all of these probabilities can be approximated by a single continuous probability model—the normal probability law—under appropriate conditions.

EXERCISES

4–1 A random sample of size 10 is taken with replacement from a production process that produces 5 percent defective items. What is the probability that the sample will include exactly one defective item?

4–2 Forty percent of high-school graduates will go on to college next year. Verify that in a random sample of two high-school graduates, the probabilities that 0, 1, or 2 will go to college are 0.36, 0.48, or 0.16, respectively.

4–3 A coin is tossed five times. Would you agree that the probability of obtaining no heads and the probability of obtaining five heads are identical? If so, what is this identical probability value?

4–4 Five ordinary dice are tossed. Do you expect that the probability of obtaining zero ace is the same as the probability of obtaining five aces? If not, what are the exact respective probabilities?

4–5 The output of a certain manufacturing process is 1 percent defective.
(a) In a random sample of four items, taken with replacement, what is the probability that none is defective? that at most, one is defective?
(b) What is the mean number of defectives in a lot of 50 items? the variance? the standard deviation?

4–6 The probability that A will make a profit on any business deal is 0.7. What is the probability that he will make a profit exactly seven times in ten successive deals?

4–7 A certain make of computer has a probability of 0.05 of giving wrong results owing to electrical or mechanical failure. To improve the accuracy of calculations, the operator feeds the same data into five such computers and accepts as accurate the answer that the majority of the computers give. How does this procedure reduce the probability of an error due to a faulty machine?

4–8 Mr. Smith has purchased five electronic tubes produced by a certain manufacturer. This product is known to be defective 2 percent of the time.

(a) What is the probability that all the five tubes purchased by Smith are perfect?

(b) Upon making the purchase, Mr. Smith adopted the following decision rule: If more than one tube in the purchase is defective, the tubes will be returned and the order canceled. If there is exactly one defective tube, the defective tube will be returned for replacement. If the replacement is also defective, the order will be canceled. Assuming the replacement sent to Mr. Smith is a random selection, verify that the probability of canceling the order, by this rule, is 0.0057.

4–9 A production process is shut down for adjustment whenever a sample of five items yields two successive defectives. What is the probability that the process will be shut down after an inspection if it is producing

(a) 20 percent defectives?

(b) 10 percent defectives?

(c) 5 percent defectives?

4–10 A certain process is considered in statistical control if the proportion of defective items is 0.1 or less. To determine whether the process is out of control, two plans have been suggested.

I. The process is judged out of control if five or more defectives appear in a sample of 25.

II. The process is judged out of control if two or more defectives appear in a sample of 10.

Evaluate the efficiency of the two plans with respect to (a) concluding falsely that a good process is out of control; (b) detecting a process that is producing 20 percent defectives.

4–11 In a lot of ten manufactured parts, two are known to be defective. If a random sample of two parts is taken without replacement, what is the probability that there will be no defectives in the sample? one defective? two defectives?

4–12 A bag contains eight oranges and four apples. A random sample of four is drawn from the bag without replacement.

(a) What is the probability model for the number of oranges in the sample?

(b) How should the probability function be stated?

4–13 Defective guidance mechanisms have been mistakenly mounted on three of a group of nine missiles. It is not known which missiles have the defective mechanisms. If a random sample of three is taken without replacement, what is the probability that the sample will contain no defective mechanism? one defective mechanism? more than one defective mechanism?

4–14 A certain make of automobile tire has a flat due to external causes, on the average, once every 2500 miles. Assuming that the occurrences follow the Poisson law, determine the probability that
(a) More than one flat will occur in a 500-mile run.
(b) No flat will occur in a trip of 5000 miles.

4–15 If a Poisson variable has a variance of 2.5, what is its probability function?

4–16 What probability model is appropriate to describe a situation where 100 misprints are distributed randomly throughout the 100 pages of a book? For this model, what is the probability that a page observed at random will contain at least two misprints?

4–17. Automobiles arrive at a tollbooth at random at a rate of 300 per hour. Calculate the probability that
(a) One car arrives during a given one-minute period.
(b) At least two cars arrive during a given one-minute period.
(c) No car arrives during a period of t minutes.

4–18 Flaws in a certain kind of woolen piece goods occur at random with a mean number of one flaw per 100 square feet. What is the probability that a roll measuring 50 by 10 feet will have no flaws? at most one flaw? [*Hint:* $\mu = 50(10)/100 = 5$.]

4–19 A taxi company has, on the average, 8 flat tires per week. During the past week, however, 16 flat tires occurred. What conclusion can you draw in view of this event?

4–20 One thousand television sets are inspected as they come off the production line and the number of defects per set is recorded below. Estimate the average number of defects per set for the production process.

No. of defects, x	*No. of sets,* N_x
0	955
1	42
2	2
3	1
4	0
	1000

4–21 A sample of 100 is taken without replacement from an incoming shipment of 10,000 light bulbs. It is found that there are three defective bulbs in the sample. What is the approximate probability for this sample result if it is known that there are 500 defective bulbs in the shipment?

4–22 A life insurance company has established as 0.00001 the probability that a person in the 50–60 age group will die from a certain rare disease during a period of one year. If the company has 100,000 policyholders in this age group, what is the approximate probability that the company must pay off more than four claims because of death from this disease?

4–23 An intercontinental ballistic missile has 10,000 parts that work independently of each other. The probability that any part will succeed is 0.99998 in any flight. The failure of any part will make the flight a failure. What is the approximate probability of a successful flight?

4–24 A company will accept a shipment of 500 articles if, in a random sample of 50 drawn without replacement, there is at most one defective item. If the shipment is 10 percent defective, what is the approximate probability of acceptance?

4–25 What is the Poisson approximation for the preceding probability? Do you think that the Poisson is better than the binomial approximation in this case? Explain.

5

SPECIAL CONTINUOUS PROBABILITY MODELS

In Chapter 3 we said that a random variable is continuous if its cumulative distribution is continuous—that is, if it leaves no finite jumps at any point. Recall also that the density function of a continuous random variable has been interpreted as the rate of change of the distribution function. In calculus terminology, $f(x)$ is said to be the first derivative of $F(x)$. It can be shown that given $F(x)$, the corresponding $f(x)$ is found by "differentiating" $F(x)$, and that given $f(x)$, the corresponding $F(x)$ is found by "integrating" $f(x)$. These observations are made merely for future reference; no knowledge of calculus is required for an understanding of our discussion on continuous probability models. As a matter of fact, the evaluation of areas under the special density functions we shall introduce is difficult even with the benefit of calculus methods. Thus, tables of the areas under these density functions have been provided in the Appendix at the end of this text.

There are two main reasons for our interest in continuous probability models. First, many real-life random variables are continuous. Second, it is easier and more convenient to develop statistical theories with continuous variables than with discrete ones. Too, the study of continuous random variables has a practical advantage in that they can be employed to approximate discrete probabilities. Much theoretical development in statistical methods has been prompted by the need for approximating procedures.

In this chapter we shall introduce three basic and widely used continuous probability models: the rectangular, the exponential, and the normal distributions. Others will be taken up in later chapters.

5-1 THE CONTINUOUS UNIFORM DISTRIBUTION

We begin our discussion of continuous models with a simple and familiar case. A random variable whose value can lie only in a certain finite interval for finite numbers a and b—for example, $X = \{\text{real number } x: a \leq x \leq b\}$—has a *uniform,* or *rectangular, distribution* if its probability density function is constant over the interval from a to b. Thus, for a continuous uniform model, the probability of any event that is a subinterval of (a, b) is just the length of that interval in (a, b); but the probability of an event that has no points in common with (a, b) is zero. The density function of a continuous random variable, X, can therefore be written as

$$r(x) = \frac{1}{b - a}, \quad \text{if } a \leq x \leq b,$$
$$= 0, \quad \text{otherwise.} \tag{5-1}$$

It can be shown that by integrating (5-1) we obtain the distribution function of the uniform model as

$$R(x) = 0, \quad \text{if } x \leq a,$$
$$= \frac{x - a}{b - a}, \quad \text{if } a < x < b,$$
$$= 1, \quad \text{if } x \geq b. \tag{5-2}$$

Figure 5-1 gives the density function of the continuous rectangular distribution.

The rectangular model has a and b as its parameters. Its expectation and variance, respectively, are

$$E(X) = \frac{(a + b)}{2} \tag{5-3}$$

and

$$V(X) = \frac{(b - a)^2}{12}. \tag{5-4}$$

Note that when a random variable is uniformly distributed over the interval from $X = a$ to $X = b$, its density is symmetrical about the center of the interval $(a + b)/2$, and thus this value is both the mean and the median of the distribution. Furthermore, a uniform model is quite analogous to a uniform distribution of mass along a line, such as a uniform thread.

Subway trains on a certain line run every half hour between midnight and six in the morning. What is the probability that a man entering the

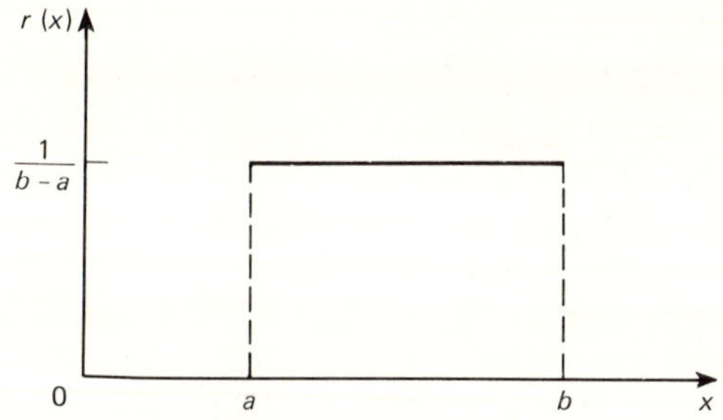

Figure 5–1 Density of a Uniform Distribution

station at a random time during this period will have to wait at least 20 minutes?

The random variable T = time to the next train, under the assumption of random time of the man's arrival, is uniformly distributed on $0 \leq T \leq 30$. The probability that he has to wait at least 20 minutes is

$$\begin{aligned} P(T > 20) &= 1 - R(t) \\ &= 1 - \frac{20 - 0}{30 - 0} \\ &= \tfrac{1}{3}. \end{aligned}$$

The expected wait is, clearly, 15 minutes.

5–2 THE EXPONENTIAL MODEL

The exponential model can be derived in a number of ways. One way that can be readily understood is to think of the exponential model as generated from the Poisson process. For example, if the arrivals of automobiles at a tollbooth follow the Poisson law, then the distribution of the length of time between successive arrivals is an exponential variable. The exponential model arises, therefore, in response to the question: "If a sequence of events occurs in time according to the Poisson law, at the rate of μ events per unit of time, how long do we have to wait in order to observe the first occurrence of an event?"

The answer to the foregoing question suggests a method of constructing the exponential probability model from the Poisson process as follows: Let us denote T as the variable time between events and proceed to determine the distribution function of T by evaluating the event $T > t$ for any specific time interval t. The event $T > t$ is an event that has not yet occurred in the time interval $(0, t)$. By the Poisson law, this probability is seen to be

$$\begin{aligned} P(T > t) &= P[\text{no occurrence in } (0, t)] \\ &= e^{-\mu t}\left[\frac{(\mu t)^0}{0!}\right] \\ &= e^{-\mu t} \\ &= e^{-t/\beta}. \end{aligned} \tag{5-5}$$

As a result, the *CDF* of T—the cumulative distribution function of the exponential variable—becomes

$$\begin{aligned} F(t) &= P(T \leq t) \\ &= 1 - e^{-t/\beta}. \end{aligned} \tag{5-6}$$

The density function of the exponential variable, as we can show by differentiating $F(t)$, is

$$f(t) = \frac{1}{\beta} e^{-t/\beta}. \tag{5-7}$$

In the previous three equations, t is the time interval between occurrences of the Poisson variable, e is the base of natural logarithms, which is approximately equal to 2.71828, μ is the average number of occurrences per time unit of the Poisson variable, and β is the expectation of the exponential variable T. Notice that $\beta = 1/\mu$. This is a natural result, since the exponential variable refers to time between successive Poisson occurrences. In this connection we see that the mean of the exponential distribution can be interpreted as the average time interval between Poisson occurrences, or the expected time until the first occurrence of Poisson event.

It can be shown that the variance of the exponential distribution is β^2. Thus, in the Poisson case the expectation and the variance are equal, $\mu = E(X) = V(X)$; in the exponential case, the expectation and standard deviation are equal, $E(T) = \sqrt{V(T)} = \beta$.

The distribution of lengths of time during which a certain make of computer operates effectively—that is, hours of effective operation before the first breakdown—is exponential with $\beta = 360$ hours. What is the probability that the computer will operate effectively for 180 hours or less? More than 720 hours?

As stated in the problem, the density function of T, the exponential variable lengths of time, is

$$f(t) = \tfrac{1}{360} e^{-t/360} \qquad (0 \leq T \leq \infty),$$

whose graph is shown by Figure 5–2.

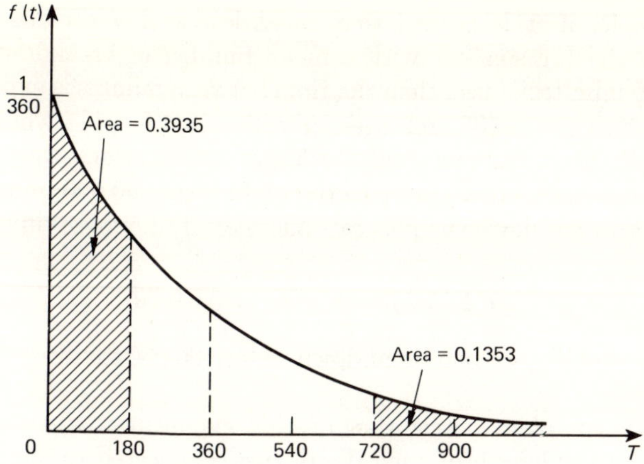

Figure 5–2 Exponential Distribution with $\beta = 360$

The exponential distribution function in this case is

$$F(t) = 1 - e^{-t/360}.$$

Now, at points of 180 and 720 (see Figure 5–2), we have

$$\begin{aligned} F(180) &= 1 - e^{-180/360} \\ &= 1 - e^{-0.5} \\ &= 0.3935 \qquad \text{(from Table B-6)} \end{aligned}$$

and

$$\begin{aligned} F(720) &= 1 - e^{-720/360} \\ &= 1 - e^{-2} \\ &= 0.8647 \qquad \text{(from Table B-6)}. \end{aligned}$$

Thus, the probability that the computer will operate effectively less than 180 hours is

$$P(T \leq 180) = F(180) = 0.3935,$$

and the probability that it will operate effectively more than 720 hours is

$$\begin{aligned} P(T > 720) &= 1 - P(T \leq 720) \\ &= 1 - F(720) \\ &= 1 - 0.8647 \\ &= 0.1353. \end{aligned}$$

Sometimes we may find it more convenient to compute exponential probabilities with (5–5), which gives the probability that the event will not occur in t intervals. That is,

$$\begin{aligned} P(\text{no success in a given interval } t) &= P(\text{success after the stated interval } t) \\ &= P(T > t) \\ &= e^{-t/\beta}. \end{aligned}$$

For example, if it is known that breakdowns of a certain electronic device obey the Poisson law with a mean number of breakdowns of two every 80,000 minutes of use, then the time between failures is exponentially distributed with $\beta = 40{,}000$ minutes, or with $\beta = 1/\mu = \frac{1}{2}$ when a trial is viewed as a 40,000-minute interval. With these designations, the probability that the device will not fail prior to the end of, say, 60,000 minutes is the probability that the device will operate successfully 1.5 more time intervals. That is,

$$P(T > 1.5) = e^{-1.5/0.5} = 0.04979.$$

Thus, the probability of one breakdown before 60,000 minutes are elapsed is $1 - 0.04979$.

To have a complete description of this exponential distribution, the distribution of the length of time the device is expected to operate before a breakdown occurs, we may sketch its density curve by determining the ordinates of the exponential density at various time intervals with expression (5–7). For example, at $T = 1.5$ we have

$$f(1.5) = \frac{1}{0.5}\,(e^{-1.5/0.5}) = 0.09958.$$

Other ordinates for a few selected values of T are computed in the same way and the results summarized in Table 5–1. These ordinates are employed to sketch the density curve for the illustrative example shown by Figure 5–3. The shaded area in this figure gives $P(T > 1.5)$, calculated before.

The exponential model is of great importance in applied probability theory because it can be used to describe a variety of random situations. In management science, one often encounters such operational problems as

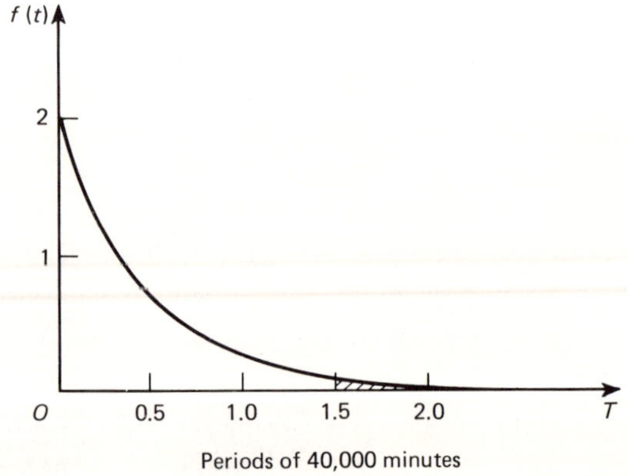

Figure 5–3 Exponential Distribution of Time between Failures of an Electronic Device

Table 5-1

Total elapsed time (min.)	*Interval (periods of 40,000 min.)*	*Ordinate:* $f(t) = 2e^{-t/0.5}$
0	0.0	2.00000
20,000	0.5	0.73580
40,000	1.0	0.27060
60,000	1.5	0.09958
80,000	2.0	0.03664

determining the number of tollgates required at highway entrances, the number of telphone lines to service a given area, the number of helpers at a cafeteria, and so forth. In each of these situations there is a common feature—the existence of a variable demand over time for service that can be satisfied by a specific number of servers. This variable demand over time is found to obey the exponential law. Furthermore, recent studies also reveal that the exponential model describes well the lengths of waiting time, as well as such numerically valued random phenomena as the life of an electronic tube, the time intervals between breakdowns of electrical mechanisms or between accidents, and so on. We shall have another opportunity to consider the applications of this model when we take up waiting-line problems in Chapter 7.

5-3 THE GENERAL NORMAL MODEL

The most frequently used probability model in economic and business decisions is the *normal distribution*, which may be stated in the general and standard forms. A random variable is said to have a general normal distribution if it is continuous, if the constants μ (with a value between $-\infty$ and ∞) and σ (with a value greater than zero) exist, and if its density function is given by the following expression:

$$n(x) = n(\mu, \sigma) = \frac{1}{\sigma\sqrt{2\pi}} e^{-(x-\mu)^2/2\sigma^2}, \tag{5-8}$$

where μ and σ are, respectively, the mean and the standard deviation of the normal variable, $e \doteq 2.71828$, and $\pi \doteq 3.14159$. The reader should not worry about his inability to comprehend this seemingly formidable formula, since he can easily calculate normal probabilities from published tables. He is, however, advised to learn about the geometric properties of the

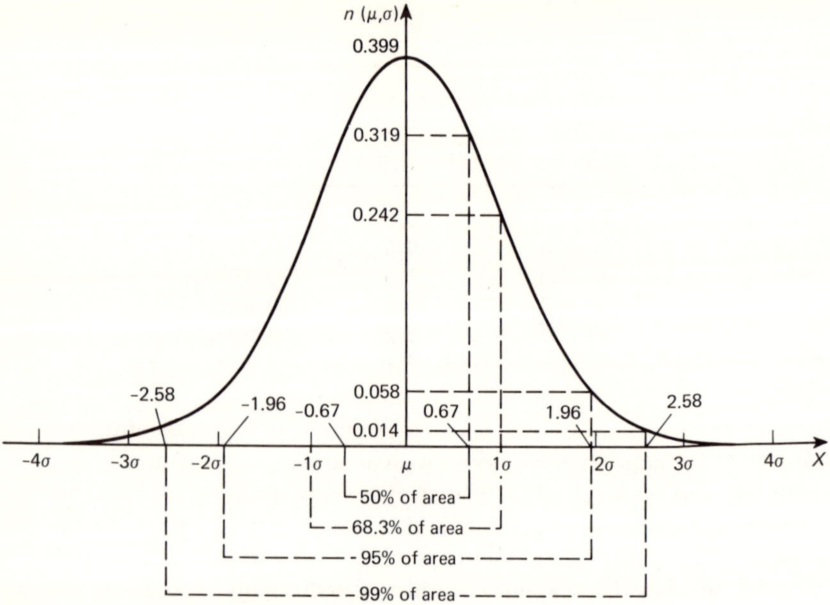

Figure 5–4 Graph of the General Normal Density Function

graph of the normal density function as shown by Figure 5–4 and as discussed in the following paragraphs.

First, a normal distribution has as its parameters μ and σ (or σ^2) in the sense that the area under its density curve (normal probability), given the magnitude of σ, is completely defined by the distance between μ and a specified value of σ. As such, the working part of the normal density is the exponent $-(x - \mu)^2/2\sigma^2$ because it contains a particular value of the normal variable, X, and the parameters μ and σ^2 of the distribution. The greater the deviation of a particular x from μ, the greater is the numerator of this exponent. However, this deviation is a squared quantity; and so two different values of X showing the same absolute deviation from μ have the same probability. This reflects the fact that the normal distribution is symmetrical with respect to μ. That is, $n(\mu + x) = n(\mu - x)$. For example,

$$\begin{aligned} P(\mu - 0.67\sigma < X < \mu) &= P(\mu < X < \mu + 0.67\sigma) \\ &\doteq 0.2500, \text{ or} \\ P(\mu \pm 0.67\sigma) &\doteq 0.5000; \end{aligned}$$

$$\begin{aligned} P(\mu - \sigma < X < \mu) &= P(\mu < X < \mu + \sigma) \\ &\doteq 0.3415, \text{ or} \\ P(\mu \pm \sigma) &\doteq 0.6830; \end{aligned}$$

$$P(\mu - 1.96\sigma < X < \mu) = P(\mu < X < \mu + 1.96\sigma)$$
$$\doteq 4750, \text{ or}$$
$$P(\mu \pm 1.96\sigma) \doteq 0.9500;$$

$$P(\mu - 2.58\sigma < X < \mu) = P(\mu < X < \mu + 2.58\sigma)$$
$$\doteq 0.4950, \text{ or}$$
$$P(\mu \pm 2.58\sigma) \doteq 0.9900.$$

Second, the fact that the exponent $-(x - \mu)^2/2\sigma^2$ is negative indicates that the greater the deviation of X from μ, the smaller will be the probability density of X. That is, both tails of the normal distribution experience decreasing density, since the further X is from μ, the lower the height of the density function curve becomes. In this connection we see that, when the value of X is identical with μ, the exponent is zero, and so the density is reduced to $1/\sigma\sqrt{2\pi}$, the largest value of the normal density. Thus, the normal distribution is unimodal, with the modal value at $X = \mu$.

Third, the curve for $n(\mu, \sigma)$ has two points of inflection, located at a distance of σ from either side of the mean μ.

Fourth, the normal distribution has an infinite range, so that its density curve never touches the X axis. As a consequence, any interval of numbers will have nonnegative probability. However, the probability of large deviations from μ is negligibly small, as can be seen from the fact that the density curve falls off quickly, more than 99 percent of its area being enclosed by $\mu \pm 3\sigma$. This property enables us to use the normal distribution to approximate other distributions for which the true range is finite.

Fifth, a change in the value of μ displaces the whole normal distribution, while a change in the value of σ merely alters its relative position with reference to a fixed scale. These facts indicate that the normal distribution is really a family of distributions. (See Figure 5–5.) This is true, of course,

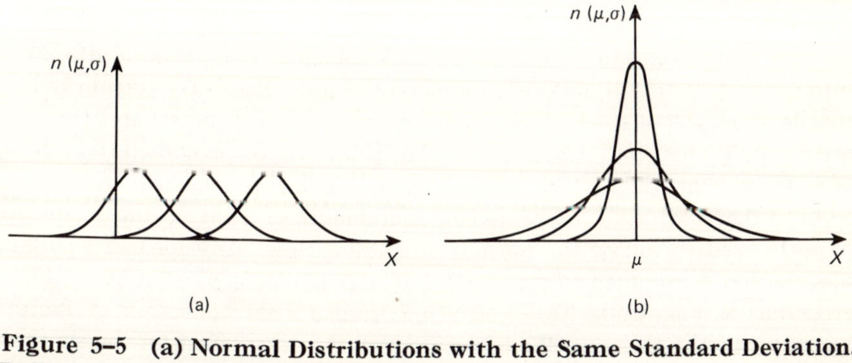

Figure 5–5 (a) Normal Distributions with the Same Standard Deviation, Different Means; (b) Normal Distributions with the Same Mean, Different Standard Deviations

for many other distributions. It has been mentioned here because the student often thinks of *a* normal distribution as *the* normal distribution.

Sixth, a linear change of scale for a normal distribution results in a new normal distribution. That is, if X is a normal variable, then $Y = a + bX$ for $a \neq 0$ is also a normal variable.

Seventh, if $X_1, X_2, \ldots, X_n$ are independent (or merely uncorrelated) normal variables, then their sum, S, is also a normal variable. Furthermore, because of independence, the additivity property holds for both the expectation and the variance here. That is, the expectation of S is the sum of the expectations of the n normal variables. Also, the variance of S is the sum of the variances of the n normal variables.

Finally, if X is normally distributed with μ and σ, then $Z = (X - \mu)/\sigma$ is also normally distributed. This transformation from X to Z, as it may be recalled, has the effect of reducing X to units in terms of standard deviations. In other words, given a value of X, the corresponding Z value tells us how far away and in what direction X is from its mean, μ, in terms of its standard deviation, σ. For example, $Z = 1.5$ means that the particular value of X is 1.5σ to the right of μ. Similarly, $Z = -2$ means that the particular X value is 2σ to the left of μ. This property of a normal variable, as we shall see immediately, enables us to compute normal probabilities, irrespective of values of μ and σ, from a single probability table for the standard normal distribution.

The normal distribution was first discovered by De Moivre as the limiting form of the binomial model in 1733. It was also known to Laplace no later than 1774, but through a historical error it has been credited to Gauss, who first made reference to it in 1809. Throughout the eighteenth and nineteenth centuries various efforts were made to establish the normal model as the underlying law ruling all continuous random variables: thus the name "normal." These efforts failed because of false premises. The normal model has, nevertheless, become the most important probability model in statistical analysis, for a number of reasons.

First, many continuous random variables, such as heights of adult males, intelligences of school children, diameters of auto tires of a certain make, tensile strengths of steel wires produced by a certain process, and the like, are normally distributed because of what is known, or presumed to be true, of measurements themselves. In this connection, it is particularly interesting to note that "errors" of repeated measurements of a given dimension are hypothesized to follow the normal probability law. Any measurement or observation is assumed to represent a true magnitude *plus* an error. Each error has a magnitude itself, resulting from a vast collection of factors operative at the moment. Each factor has but a tiny effect on the size and direction of the error; furthermore, errors of measurement work independently with equal force to push the observed measurement up or down, and therefore cancel out in the long run. Thus, we think of errors of measurements as reflections of chance variations that are normally distributed with zero expectation.

Second, as we shall see, the normal distribution serves as a good approximation for many discrete distributions, such as the binomial or the Poisson model, whenever the exact discrete probability is laborious to obtain or impossible to calculate accurately.

Third, in theoretical statistics, many problems can be solved only under the assumption of a normal population. In applied work, we often find that methods developed under the normal probability law yield satisfactory results, even when the assumption of a normal population is not fully met, despite the fact that the problem can have a formal solution only if such a premise is hypothesized.

Finally, and most important of all, as we shall find in Chapter 11, distributions of many sample statistics computed from large samples approach the normal distribution as a limit. As a result, our work on statistical inferences is made easier.

Let us now turn to study the standard normal distribution, which enables us to compute probabilities for any general normal distribution simply by looking at published normal probability tables.

5-4 THE STANDARD NORMAL MODEL

As we discussed in Chapter 3, it is often easier to work with standardized probability distributions. This is especially true when we have a density function, such as that for the normal variable, involving indefinite integrals that cannot be reduced to elementary functions. For speed and efficiency in practical work, the concept of the standard normal distribution is indeed indispensable.

A *normal distribution* is said to be of the *standard form* if its mean is zero and its variance, therefore the *standard deviation*, is unity. From the density function of the general normal distribution, we can see that the probability density of the standard normal variable, Z, must be

$$n(z) = n(0, 1) = \frac{1}{\sqrt{2\pi}} e^{-z^2/2}. \tag{5-9}$$

The cumulative distribution function corresponding to the standard density, $n(0, 1)$, as does any other CDF, gives the probability that the standard normal variable assumes a value equal to or less than z, namely,

$$\begin{aligned} N(z) &= N(0, 1) \\ &= P(Z \leq z). \end{aligned} \tag{5-10}$$

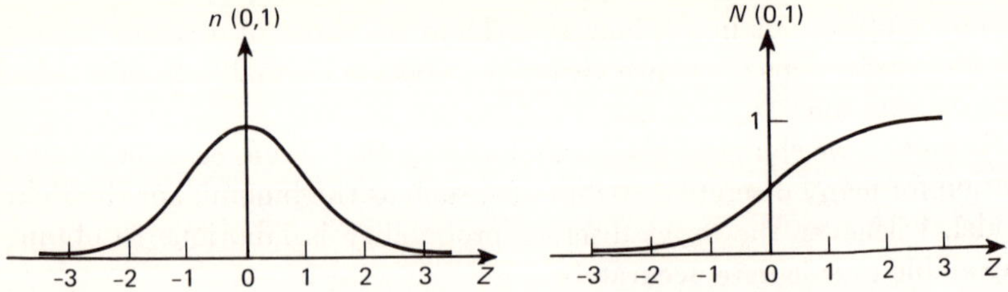

Figure 5–6 Standard Normal Density and Distribution Functions

The standard normal density and distribution functions are graphed in Figure 5–6.

Any general normal distribution, $n(\mu, \sigma)$, can be converted into the standard normal distribution, $n(0, 1)$, by the Z transformation; that is, if X is any normal variable, then its distribution function can be expressed as follows:

$$\begin{aligned} N(\mu, \sigma) &= P(X \leq x) \\ &= P(\mu + \sigma Z \leq x) \\ &= P\left(Z \leq \frac{x - \mu}{\sigma}\right) \\ &= N\left(\frac{x - \mu}{\sigma}\right). \end{aligned} \qquad (5\text{–}11)$$

Thus, given $n(\mu, \sigma)$ for any real numbers a and b, with $a < b$, we would have

$$\begin{aligned} P(a \leq X \leq b) &= N(b) - N(a) \\ &= N\left(\frac{b - \mu}{\sigma}\right) - N\left(\frac{a - \mu}{\sigma}\right). \end{aligned} \qquad (5\text{–}12)$$

As a result, we can compute probabilities for any normal model if we have available an extensive table for $n(0, 1)$ or for $N(0, 1)$. A table of $N(0, 1)$ is given as Table B-7 in Appendix B. We shall now see how to compute normal probabilities by employing expression (5–12) and Table B–7.

An aluminum factory produces, among other things, a certain type of aluminum-alloy channel. According to past experience, it is known that the stiffness, measured in effective psi, is normally distributed with $\mu = 2{,}425$ psi and $\sigma = 115$ psi. This distribution is portrated by Figure 5–7(a).

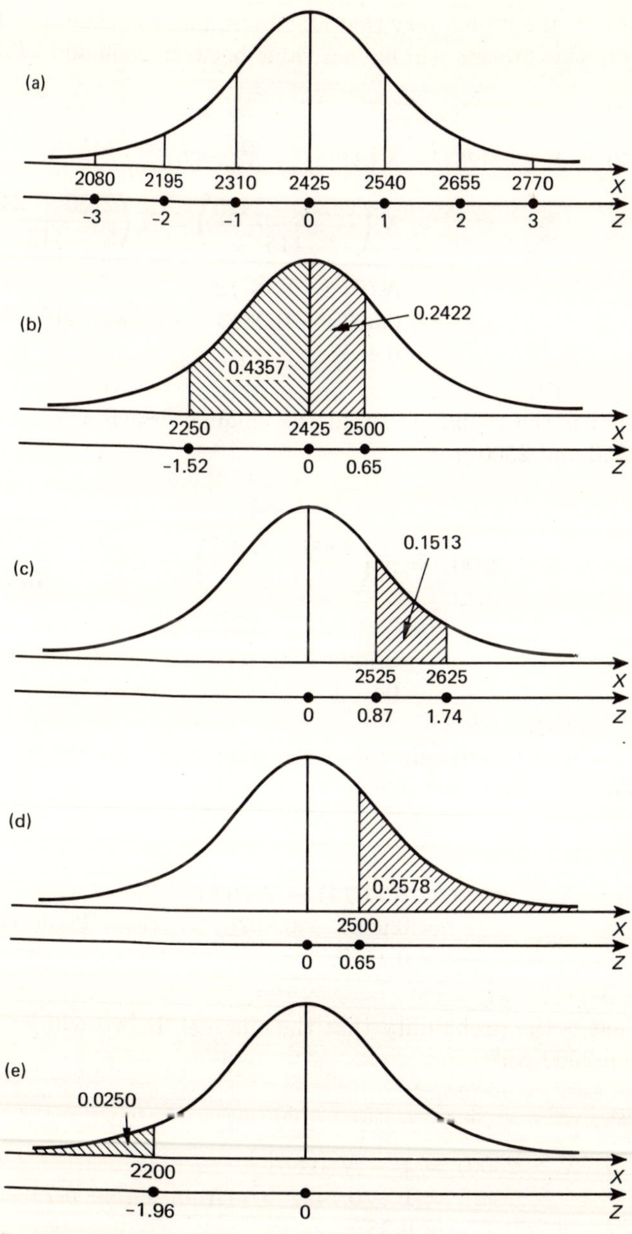

Figure 5-7 Normal Area Probabilities for the Aluminum-Alloy Example

(1) What is the probability that an aluminum-alloy channel selected at random from this process will have a value between 2250 and 2425 psi?

ANSWER:

$$\begin{aligned}
P(2250 \leq X \leq 2425) &= N(2425) - N(2250) \\
&= N\left(\frac{2425 - 2425}{115}\right) - N\left(\frac{2250 - 2425}{115}\right) \\
&= N(0) - N(-1.52) \\
&= 0.5000 - 0.0643 \qquad \text{(from Table B-7)} \\
&= 0.4357.
\end{aligned}$$

See Figure 5–7(b).

(2) What is the probability that the channel drawn will have a value between 2250 and 2500 psi?

ANSWER:

$$\begin{aligned}
P(2250 \leq X \leq 2500) &= N\left(\frac{2500 - 2425}{115}\right) - N\left(\frac{2250 - 2425}{115}\right) \\
&= N(0.65) - N(-1.52) \\
&= 0.7422 - 0.0643 \qquad \text{(from Table B-7)} \\
&= 0.6779.
\end{aligned}$$

See Figure 5–7(b).

(3) What is the probability that the channel drawn will have a value between 2525 and 2625 psi?

ANSWER:

$$\begin{aligned}
P(2525 \leq X \leq 2625) &= N(1.74) - N(0.87) \\
&= 0.9591 - 0.8078 \qquad \text{(from Table B-7)} \\
&= 0.1513.
\end{aligned}$$

See Figure 5–7(c).

(4) What is the probability that the channel drawn will have a value greater than 2500 psi?

ANSWER:

$$\begin{aligned}
P(X > 2500) &= 1 - N(0.65) \\
&= 1 - 0.7422 \qquad \text{(from Table B-7)} \\
&= 0.2578.
\end{aligned}$$

See Figure 5–7(d).

(5) What is the probability that the channel drawn will have a value less than 2200 psi?

ANSWER:

$$
\begin{aligned}
P(X \le 2200) &= N\left(\frac{2200 - 2425}{115}\right) \\
&= N(-1.96) \\
&= 0.0250 \qquad \text{(from Table B-7).}
\end{aligned}
$$

See Figure 5–7(e)

If X is a normal variable with $\mu = 80$ and $\sigma = 30$, then

(1)
$$
\begin{aligned}
P(X \le 100) &= N\left(\frac{100 - 80}{30}\right) \\
&= N(0.67) \\
&= 0.7486 \qquad \text{(from Table B-7);}
\end{aligned}
$$

(2)
$$
\begin{aligned}
P(X > 100) &= 1 - P(X \le 100) \\
&= 1 - N(0.67) \\
&= 1 - 0.7486 \qquad \text{(from Table B-7)} \\
&= 0.2514;
\end{aligned}
$$

(3)
$$
\begin{aligned}
P(100 \le X \le 150) &= N\left(\frac{150 - 80}{30}\right) - N\left(\frac{100 - 80}{30}\right) \\
&= N(2.33) - N(0.67) \\
&= 0.2415 \qquad \text{(from Table B-7);}
\end{aligned}
$$

(4)
$$
\begin{aligned}
P(|X - 100| \ge 25) &= 1 - P(|X - 100| < 25) \\
&= 1 - P(-25 < X - 100 < 25) \\
&= 1 - P(75 < X < 125) \\
&= 1 - [P(X < 125) - P(X < 75)] \\
&= 1 - [N(1.5) - N(-0.17)] \\
&= 0.4993 \qquad \text{(from Table B-7).}
\end{aligned}
$$

A product consist of four components that are produced independently. The overall length of this product is the sum, S, of the lengths of the four components, X_1, X_2, X_3, and X_4, all normal random variables with means and variances as follows:

$$
\begin{aligned}
\mu_1 &= 2 \text{ in.}, & \sigma_1^2 &= 0.010; \\
\mu_2 &= 1 \text{ in.}, & \sigma_2^2 &= 0.006; \\
\mu_3 &= 0.5 \text{ in.}, & \sigma_3^2 &= 0.004; \\
\mu_4 &= 1.5 \text{ in.}, & \sigma_4^2 &= 0.011.
\end{aligned}
$$

What is the probability that the overall length for S will meet the specification of 5 ± 0.10 inches?

Here, the desired probability is $P(4.9 < S < 5.1)$. Since lengths of the four components are independent and normally distributed, S must be normally distributed also and we must have

$$\mu_s = \mu_1 + \mu_2 + \mu_3 + \mu_4 = 5 \text{ in.}$$

and

$$\sigma_s = \sqrt{\sigma_1^2 + \sigma_2^2 + \sigma_3^2 + \sigma_4^2} = \sqrt{0.031} = 0.176 \text{ in.}$$

Consequently,

$$\begin{aligned} P(4.9 < S < 5.1) &= N(0.57) - N(-0.57) \\ &= 0.4314 \qquad \text{(from Table B-7)}. \end{aligned}$$

This result indicates that, on the average, only 43 percent of the items produced can be expected to meet the predesignated specifications. This is a very low percentage for problems of this kind; the management should either increase the precision of producing the components (reducing variability in production) or relax the standards if relaxation does not impair the products usefulness.

It is estimated that a certain machine, on the average, can replace 10,000 man-hours, and chances are 50:50 that the saving in terms of mean time could be more than 10,500 hours or less than 9500 hours per week. Now, assuming that a specific normal distribution will fit these estimates, what is the probability that the machine will actually replace less than 9000 man-hours per week?

To solve this problem, we must first find the value of the standard deviation for this distribution. To do so, we note that the central 50 percent of the area of a normal distribution is enclosed by $\mu \pm 0.67\sigma$. For the problem, we have

$$\frac{10{,}500 - 9500}{2} = 0.67\sigma.$$

That is, if we shift in either the positive or negative direction by 500 hours from the mean, we shift 0.67σ away from the mean. So

$$500 = 0.67\sigma$$

and

$$\sigma = \frac{500}{0.67} = 746 \text{ hours.}$$

With this, the desired probability is seen to be

$$\begin{aligned} P(X \leq 9000) &= N\left(\frac{9000 - 10{,}000}{746}\right) \\ &= N(-1.31) \\ &= 0.0951 \qquad \text{(from Table B-7)}. \end{aligned}$$

Obviously, the probability that the machine will replace more than 9000 man-hours per week is greater than 0.90.

5-5 SUMMARY AND CONCLUSIONS

Discussion of probability models is an essential step in statistical decision and inference, since we are required to select a model to describe the behavior of the data at hand. We have in this and the last chapter investigated some important special probability models that are known to characterize a large class of random experiments. While in most cases we have tried to derive formulas under different sets of assumptions, we have emphasized throughout that the student should learn to associate the correct model with the problem at hand instead of trying to memorize mathematical statements. As the reader can now agree, calculations of probabilities for various models are quite simple because extensive tables are available.

We began with what is called the Bernoulli model with reference to a random variable that can assume only two values. Then we introduced the binomial distribution as the sum of n independent Bernoulli trials, with the probability of success for each trial remaining as a constant, p. In the binomial case, the two numbers n and p determine the probabilities and may be called parameters of the distribution. The binomial model is appropriate when we think of sampling from an infinite population or sampling with replacement from a finite population, where the elements of the population, infinite or finite, can be classified into two mutually exclusive categories.

In general, the hypergeometric model is appropriate when a random sample of size n is drawn without replacement from a population of N items, k of which have a certain characteristic. In contrast to the binomial model, the hypergeometric probability of success changes from one selection to another because successive draws are dependent. It is intuitively clear, however, that if N is large relative to n, the change in k/N could be small enough to be of no consequence. To an extreme, as N becomes infinite, k/N would become a fixed ratio and the hypergeometric probabilities would approach the binomial probabilities with $p = k/N$. In practice, we are satisfied to approximate hypergeometric probabilities by the binomial values when the sample size is small in relation to the size of the population.

The discrete uniform model refers to a random variable generated by an experiment that can terminate in n mutually exclusive and equally likely ways. One important application of this model arises when we seek a simple random sample from a finite population. When a sample of size n is drawn from a population with N items, on each selection the distribution is uniform and the probability of drawing any specified set of n elements is

$1 \Big/ \binom{N}{n}$. In practice, requirements of simple random samples, unconditional and conditional, are satisfied by using a table of random numbers constructed with an underlying uniform distribution of the ten digits from 0 to 9.

The Poisson model is useful in certain situations in which some kind of event occurs repeatedly at random during a time interval or an area of space. A Poisson variable may be considered as the number of occurrences of an event in a time interval or an area of space with possible values 0, 1, 2, ... The Poisson model can also be considered as the limit of the binomial when n is large and p is small. To approximate the binomial probabilities by the Poisson values, we use $\mu = np$. In this connection, situations may arise in which the hypergeometric is approximated by the binomial, which, in turn, is approximated by the Poisson probability.

As the first continuous model, we introduced the continuous uniform, or rectangular, distribution. If the density function of a random variable is constant over some region of values and zero elsewhere, the variable is said to be uniformly distributed over that region. The density of a uniform model is symmetrical about the center of the interval on which it is defined; hence the central number is both the mean and median of the distribution.

Next, we took up the exponential model, which is also called the negative exponential or Laplace distribution. This model has its origin in the Poisson process. For example, if the number of incoming telephone calls per unit of time obeys the Poisson law, then the distribution of lengths of time between successive telephone calls is exponential. The distribution function of the exponential variable gives us the probability of the length of time one has to wait to observe the first occurrence of a Poisson event. In the case of Poisson, the mean and the variance are equal; in the case of exponential, the mean and the standard deviation are equal. The exponential model is especially useful in describing the queuing phenomena, to be discussed in Chapter 7.

The last probability model we discussed is the normal distribution, which has been given in both the general and standard forms. This model is by far the most important among all the special probability distributions we have so far investigated. It is important for a number of reasons. First, many random phenomena are known to follow the normal probability law. Second, this model gives a reasonable enough approximation to many other distributions. Third, many other continuous distributions, such as the t, the chi-square, and the F distributions (to be discussed in the future) are derived from the normal distribution. Finally, as we shall see in Chapter 11, the distribution of many a sample statistic approaches the normal model as a limit when the size of the sample becomes sufficiently large. Thus, as we proceed with our analytical work, we should have the normal probability model in mind much of the time.

EXERCISES

5–1 The circumference of a roulette wheel is divided into 37 arcs of equal length, numbered from 0 to 36. A ball is placed in the wheel, and the wheel is spun. After the wheel comes to rest, the position of the ball is observed. What is the probability that the ball will be in the arc with a number
(a) From 10 to 20 inclusive?
(b) That is even?
(c) One?

5–2 The traffic light at a certain intersection is 35 seconds green, 5 seconds yellow, and 10 seconds red. Assuming a uniform distribution, what is the probability that a motorist, arriving at the intersection at random time, will go through without stopping? will have to stop and wait?

5–3 Let X be uniformly distributed over the interval $0 \leq X \leq 4$.
(a) Determine and sketch $r(x)$.
(b) Compute $P(|X| > 4/3)$.

5–4 The life of a certain make of radio tube is exponentially distributed with a mean life of 1000 hours. What is the probability that a tube will last less than 1000 hours? more than 1200 hours?

5–5 A certain digital computer, which operates 24 hours a day, suffers breakdowns at a rate of 0.1 per hour. Suppose the computer has operated satisfactory for ten hours; what is the probability that it will continue to operate satisfactorily during the next 10 hours? [*Hint*: An interesting property of the exponential model is that the distribution of future life remains the same as the initial distribution. For such computers, an old one is just as good as a new one.]

5–6 Workers join a waiting line in front of a workshop to receive tools with an average time between arrivals of two minutes. What is the probability that
(a) Four minutes will elapse in which no worker arrives?
(b) In a four-minute interval at least two workers arrive?

5–7 Customers arriving at a certain shop follow the Poisson law, with a mean rate of one customer per minute. What is the probability that
(a) Five or more minutes have elapsed since the last customer arrived?
(b) Five or more minutes have elapsed since the next to the last customer arrived?

5–8 Ships arrive in a certain port according to the Poisson law, with an average time of two hours between arrivals. What is the probability that
(a) Five hours will elapse with no ships arriving?
(b) During a five-hour interval at most, three ships arrive?
(c) Sketch the density curve for this exponential distribution.

5–9 The random variable X is normally distributed with $\mu = 10$ and $\sigma^2 = .4$ Determine $P(|X - \mu| > 3)$.

5–10 Let X be a normal variable with mean 100 and variance 49. Verify the following:
(a) $P(X = 120) = 0$.
(b) $P(X > 100) = 0.5$.
(c) $P(|X - 95| > 5) = 0.5764$.
(d) $P(|X - 100| < 10) = 0.8472$.

5–11 If Y is a normal variable with $\mu = 65$ and $\sigma = 6$, then what is
(a) $P(Y = 36)$?
(b) $P(Y > 60)$?
(c) $P(Y < 55)$?
(d) $P(|Y - 65| < 9)$?

5–12 Given that X is a normal variable with $\mu = 10$ and $P(X > 12) = 0.1587$, what is the probability that X will be included in the interval $(9, 11)$?

5–13 Inside diameters of steel pipes produced by a certain process are normally distributed with $\mu = 10$ inches and $\sigma = 0.1$ inch. Pipes with diameters beyond the range of 10.05 ± 0.12 inches are considered as defectives.
(a) Verify that the probability of a defective item is 0.2866.
(b) If the process is adjusted so that the inside diameters will have $\mu = 10.10$ and $\sigma = 0.10$ inch, would the probability of a defective item be the same as before?
(c) If the process is adjusted so that the inside diameters will have $\mu = 10.05$ and $\sigma = 0.06$ inch, then 95.44 percent of the pipes produced will meet the specifications. Verify this.

5–14 If W is a normal variable and if $P(W < 10) = 0.8413$, what would be $E(W)$ and $V(W)$, respectively?

5–15 Suppose that heights of American adult males are normally distributed with $\mu = 70$ inches and $\sigma^2 = 9$ inches². How long should mattresses be in order to accommodate at least 99 percent of them?

5–16 There are two procedures for getting fighter planes ready to take off. Procedure A requires a mean time of 27 minutes with a standard deviation of 5 minutes. For procedure B, $\mu = 30$ minutes and $\sigma = 2$ minutes. Both A and B are normally distributed. Which procedure should be used if the available time is 30 minutes? 34 minutes?

5–17 Given that

$$Y = \frac{X_1 + X_2}{2} - X_3,$$

where X_i are normal variables with $\mu_1 = 20$, $\mu_2 = 16$, $\mu_3 = 25$, $\sigma_1^2 = 5$, $\sigma_2^2 = 9$, and $\sigma_3^2 = 7$.
(a) How is Y distributed?
(b) What are $E(Y)$ and $V(Y)$?

5–18 If $Y = X_1 - X_2$, where both X_1 and X_2 are normal variables with $\mu_1 = 27$, $\mu_2 = 25$, $\sigma_1 = 3$, and $\sigma_2 = 4$, what is $P(Y > 5)$?

5–19 A large flashlight is powered by five batteries. Suppose the life of a battery is normally distributed with $\mu = 120$ hours and $\sigma = 10$ hours. The flashlight will cease functioning if one or more of its batteries go dead. Assuming the lives of the batteries are independent, what is the probability that the flashlight will operate more than 100 hours?

5–20 We know that if a fair coin is tossed 10,000 times, the probability of obtaining 5000 heads or less is a binomial probability. But can you try to approximate this probability by the normal law?

6

INTRODUCTION TO STATISTICAL DECISION THEORY

It has been pointed out that statistics is a method of decision making under uncertainty and that uncertainty is measured in terms of probability theory. Having learned all the fundamental theorems of probability in the previous chapters, we are now prepared to see how our knowledge can be used for decision making.

Let us note at the very outset that there are two main approaches to statistical decisions. One is modern statistical decision theory, or Bayesian statistics; the other is classical, or traditional, decision procedure. As the reader goes through the rest of this text, he will learn about the differences between these two schools of thought. For the time being, it is sufficient to observe that both approaches have probability theory as their foundation, that classical statistics embraces inferential and decision procedures that are based on sample information alone, and that Bayesian statistics contains inferential and decision procedures that can be applied with or without empirical evidence.

In this chaper we shall bring out some of the basic aspects of modern decision theory under the assumption that no sample information is available. In the chapter that follows we shall take up some decision problems that can be solved with special probability models. In Chapter 8 we shall discuss the mechanism of combining available information with sample evidence for decision making. Then, we shall turn our attention to the theory of sampling and classical analysis. Finally, in Chapter 16, we shall return to consider the Bayesian inferential techniques.

6-1 THE STRUCTURE OF DECISIONS

The term "decision" has several fundamental implications that must be clearly understood. To begin with, a decision is needed in a problem situation where two or more alternative courses of action are available, and where only one can be taken. Indeed, a *decision* may be defined as the selection by the decision maker of an act, judged to be the best according to some predesignated standards, among all the available courses of action in a given problem situation.

Second, decision making is predictive: to make a decision the decision maker must predict the outcome or outcomes for each alternative course of action. Future outcomes are called events, or, more frequently, states of nature. A *state of nature* is any combination of factors resulting from outside forces, such as sociopolitical conditions, weather, consumers' preferences, labor strikes, technological changes, general economic environment, and so forth. States of nature are usually beyond the decision maker's control, and a decision is often independent of the states of nature.

Decision making under uncertainty refers to the fact that the actual state of nature is not known with certainty. If one knows the actual state of nature related to a decision that will prevail, then a decision is said to be made under the condition of certainty. For example, if you wanted to buy a Grand Prix automobile from one of the, say, four dealers in town, and if you found that those dealers' quotations on such a car with identical accessories ranged from $4950 to $5200, your decision to buy from the dealer submitting the quotation of $4950 would be made under conditions of certainty because all the facts would be known. Suppose, next, you decided to buy an apartment building instead of a commercial building in a certain city, reasoning that the former would appreciate in value more than the latter. If you were not absolutely sure that this would occur (and how could you be?), the decision would be made under uncertainty.

Thus, depending upon our knowledge of the states of nature, decisions can be classified as decision making under certainty, as in the case of linear programming, and decision making under uncertainty, with which statistical decision theory is concerned.

In the face of uncertainty, we can only make predictions about the possible states of nature in terms of probabilities. Such probabilities may be purely subjective, or they may be estimates from additional information obtained, say, from sampling. In this chapter we shall use state probabilities of the first kind, assuming that the decision maker is not free to defer his choice of action until he obtains further information. This restriction will be relaxed in Chapter 8.

A third implication of the term "decision" is that, under conditions of uncertainty, for each course of action, there are numerous results called *consequences, outcomes,* or *payoffs*. To evaluate the consequence of each and every action, we need a measure of these results. In business and economic decision problems we have available two standards, or units, by which to measure consequences: money and utility.

Money can be used to measure not only the direct monetary gains and costs of an outcome, but also intangibles, such as convenience, goodwill, self-satisfaction, and peace of mind. One way to measure an intangible variable is to assign a monetary value to it by evaluating the relevant abstract factors that might contribute to its value. Thus, taking into consideration a firm's reputation, the loyality of its customers, the advantageous competitive position of its products, and some other abstract factors, an accountant may assign, say, a value of $100,000 to the firm's goodwill.

Another way of converting an intangible into monetary values is to ask an introspective question, such as: Would I be willing to stand in a line ten blocks long to buy a ticket at Radio City Music Hall, or would I prefer to give up a quarter, fifty cents, a dollar, or two dollars to avoid standing in the line? In answering a question of this sort, one can eventually determine the monetary value of the act of standing in the line.

It is also possible to convert time into monetary values, since each individual consciously or unconsciously values his time at so much money per hour or per day. For example, suppose it costs you $20 to have your apartment cleaned by a maid, or it costs you four hours to clean it yourself, and suppose you decide to hire a maid; then you must value your time at more than $5 per hour. Of course, in this example we have ignored many factors that may enter the decision situation; for example, you may actually enjoy cleaning, or you may consider cleaning as a sort of exercise, and so on.

For a given action, there is a payoff for each and every state of nature. A table (such as Table 6–1) that contains all the payoffs for all courses of action corresponding to all states of nature in a given decision situation is called the *payoff table*, or *payoff matrix*. Clearly, a payoff stands for the outcome determined jointly by a specific act and a specific state of nature.

Payoffs expressed in terms of money can also be transformed, as we shall see, into units of utility. The mechanics of making decisions, however, are the same whether payoffs are given in terms of money or of utility.

Finally, given the payoff matrix, we must now establish a *strategy*, a *decision criterion*, or a *decision rule* for selecting a course of action among the alternatives. The purpose of decision making is the eventual selection of an act that will accomplish some predesignated objective. Different decision makers, when confronted with the same decision situation, may pursue different objectives, owing to their differences in personality and/or circumstances. Thus we would expect to identify different strategies and to judge each by the degree to which it achieves the objective of a given decision maker. Indeed, a large body of current literature on decision theory is concerned with the formulation of decision criteria for various assumed objectives. We shall now introduce some of the strategies most often used.

6–2 SOME DECISION STRATEGIES

The selection of an action when the states of nature are unknown is similar to a two-person game, in which the opponent is nature. Carrying the terminology and concepts of game theory over to decision theory, we have, for example, defined the choice of an action as a strategy, and thus a state of nature can be considered as a strategy of the opponent (though it is doubtful that nature is really trying to defeat the decision maker). Also, many strategies employed in decision theory originated in game theory.

A concrete example will help introduce some decision strategies. Suppose that the manager of a manufacturing firm is considering whether or not to expand the plant capacity by purchasing one of two machines, one of which has twice the capacity of the other and costs over 50 percent more. The firm's existing plant capacity is not quite sufficient to meet the current demand for its output. The firm's sales, according to past experience, have a high correlation with general economic conditions. If economic conditions improve in the near future, then the high-capacity machine will be required. If conditions remain stable, then the low-capacity machine will be adequate. If a recession is in the making, then the current capacity will do. Thus, at the moment, three alternative courses of action are available to the manager:

a_1: purchase the high-capacity machine,
a_2: purchase the low-capacity machine,
a_3: purchase no machine.

In general, the *action set* may be denoted as

$$A = \{a_i\}, \qquad i = 1, 2, \ldots .$$

Corresponding to each act, there are three possible states of nature for this comparatively simple decision situation:

θ_1: prosperity,
θ_2: stability,
θ_3: recession.

The *state set* may be denoted as

$$S = \{\theta_j\}, \qquad j = 1, 2, \ldots .$$

With action and state sets specified, the decision maker can now estimate, or determine, the set of payoffs associated with each act on the basis of available information on possible gains and losses. Note that the payoff is a random variable, X, whose values are real numbers (money or utility) assigned by the decision maker; these values may be denoted in a number of ways. For example, the payoff for the outcome when the decision maker selects a_i, given that the state θ_j prevails, may be denoted by

$$X(\theta_{ij}), \qquad X(a_i, \theta_j), \qquad X_{ij}, \quad \text{or} \quad X(a, \theta).$$

Determination of payoffs may sometimes require some algebraic reasoning. But for the time being, let us suppose that the manager, in view of the possible revenues and costs under various combinations of acts and states, has estimated payoffs (in millions of dollars) as shown in Table 6–1.

Table 6–1. Payoff Table

	State of nature		
Act	θ_1	θ_2	θ_3
a_1	2.5	1.5	−1.0
a_2	1.6	2.0	0.0
a_3	1.2	1.2	0.8

Note that a payoff table is an efficient format for organizing and displaying the outcomes produced jointly by the alternative acts and states of nature. It allows payoffs to be compared readily, and it assures that the alternatives are being evaluated on equivalent basis. For example, in Table 6–1 the relative merits of the three courses of action relative to the various states of nature are clearly evident. Thus, suppose a_1 (purchasing the high-capacity machine) is taken; then if θ_1 (prosperity) prevails, the firm will make a net profit of \$2.5 million; if θ_2 (stability) emerges, the firm will make a net profit of \$1.5 million; and if θ_3 (recession) occurs, the firm will suffer a loss of \$1 million. Other rows of the table can be interpreted in exactly the same fashion.

The payoff matrix also enables us to detect if all acts in a decision situation are admissible. An act is *admissible* if it is not dominated by any other act. An act a_k is said to *dominate* another act a_r if, for each possible state of nature, a_k leads to a payoff which is at least as high as that for a_r, and, for at least one state of nature, a_k leads to a higher payoff than a_r. Our illustrative payoff matrix here clearly indicates that all three acts are admissible since none dominates, or is dominated by, any other act. However, if $X(a_3, \theta_3) \leq 0$ instead of 0.8 in Table 6–1, then a_3 would be dominated by a_1 and a_2, and a_3 should be eliminated from further consideration.

Finally, given the payoff table, the manager (the decision maker) may follow any one of the strategies to be introduced in the next few pages, depending on the objective he wishes to achieve.

The Maximin Criterion

This strategy, suggested by Wald, attempts to maximize the minimum payoffs of all acts. That is, the decision maker determines the worst possible outcome of each act and then picks the act whose most unattractive contingency is the least disastrous. According to this rule, the manager should select a_3 as the optimal act, since $X(a_3, \theta_3) = \$0.8$ million is the maximum of all minimum payoffs contained in Table 6–1.

Maximin should clearly appeal to a confirmed pessimist whose highest aspiration is to get the best among the worst. Pessimism is not necessarily irrational; there is much to be said for it, especially in situations where self-preservation seems more important than windfall gains. One drawback of this rule is that it disregards most of the information in the payoff matrix, since it considers only the worst outcomes for each act. This criticism, however, applies to many other decision criteria.

The Maximax Criterion

In direct opposition to the maximin criterion, the maximax rule calls for the selection of the act that would maximize the maximum payoffs. In terms of this rule, the decision maker should employ a_1, whose highest payoff of \$2.5 million exceeds the highest payoffs of all other acts.

The maximax rule, well suited to a confirmed optimist who always looks through rose-colored glasses, considers only the most glittering possible reward and ignores all other contingencies. This rule is quite appropriate where the decision maker can well sustain the worst penalties. The maximax rule, like the maximin, ignores all intermediate payoffs, even though some may be only negligibly lower.

The Coefficient-of-Optimism Criterion

Instead of being completely pessimistic or completely optimistic, a decision maker may have an intermediate attitude. To accommodate this possibility, Hurwitz proposes a measure called the *coefficient of optimism*, c, ranging in value from 0 for a complete pessimist to 1 for a complete optimist. If one's $c = 0$, his rule coincides with the maximin rule; if 1, with the maximax rule. If one is partially optimistic, he should assign proper weights to the maximum and minimum payoffs of each act; in other words, he should compute a quantity called the *Hurwitz value* for each act and select as the optimal act the one that has the highest Hurwitz value. This quantity, denoted by $H(a_i)$, is defined as

$$H(a_i) = (\max X_{ij})c + (\min X_{ij})(1 - c). \qquad (6\text{–}1)$$

Now, suppose the decision maker in our example is a rather cautious and pessimistic person, and he assigns his coefficient of optimism a value of

$c = 0.25$. Then the optimal act would be a_3—to buy no machine—since, as shown below, $H(a_3)$ is the highest.

$$H(a_1) = 2.5(0.25) + (-1.0)(0.75) = -\$0.125 \text{ million},$$
$$H(a_2) = 2.0(0.25) + (0.0)(0.75) = \$0.500 \text{ million},$$
$$H(a_3) = 1.2(0.25) + (0.8)(0.75) = \$0.900 \text{ million}.$$

The Hurwitz criterion is, in effect, a weighted average of the maximin and maximax rules, in which the weights are judged to reflect the decision maker's degree of optimism. Like the other two rules, this criterion ignores less extreme payoffs in its calculations.

The Minimax Regret, or Opportunity-Loss, Criterion

The three decision rules introduced so far are based on absolute values. A strategy may be established on the basis of relative merits in terms of *regrets*, or *opportunity losses*. Precisely, the regret, or opportunity loss, for any outcome is the absolute difference between the cost incurred (gain realized) and that which would have been incurred (realized) if the optimal act had been selected for the state of nature that was obtained. Symbolically, the opportunity loss for a particular act i and state j, denoted by $L(a_i, \theta_j)$, is given as

$$L(a_i, \theta_j) = |X(a_i, \theta_j) - X(a^*, \theta_j)|, \tag{6-2}$$

where a^* is the optimal act for state θ_j under consideration and $X(a^*, \theta_j)$ is the value of the payoff function for the optimal act for the state θ_j.

Thus, the measure of regret is the difference between the payoff actually received and the payoff that could have been received by the decision maker had he known in advance which state of nature would prevail. For example, with reference to Table 6–1, if the manager knows that θ_3 is to prevail, then his optimal act will be $a^* = a_3$, since then his payoff will be 0.8 with a regret of $0 = |0.8 - 0.8|$. However, he might not know this and might select, say, a_1 with a payoff of -1.0; then his regret would be $1.8 = |-1.0 - 0.8|$, representing the amount of payoff foregone by failing to take the optimal act a_3 given the occurrence of θ_3. Table 6–2 gives the regrets corresponding to the payoffs in Table 6–1.

Table 6–2. Regret, or Opportunity Loss, Table

	State of Nature		
Act	θ_1	θ_2	θ_3
a_1	0.0	0.5	1.8
a_2	0.9	0.0	0.8
a_3	1.3	0.8	0.0

The strategy applied to regrets is called the *minimax principle*, by which one should expect the worst and be prepared for it. Namely, for each act, determine the maximum possible loss or regret that might be incurred under the various states of nature and select the one with the smallest (minimum) maximum loss as the optimal act. For our illustrative example, the minimax principle leads to the selection of a_2, with a minimum regret of 0.9 among all the maximum regrets—1.8, 0.9, and 1.3.

Strategies involving regrets concentrate on the opportunity costs of incorrect action. They are designed to protect the decision maker against excessive cost of mistakes. This rule also has the limitation of considering the extreme figures in each row and ignoring other information contained in the regret table.

The Expected-Value Criterion

In all the decision criteria introduced so far, the decision maker is assumed to know nothing about the probability distribution over the state variable. Furthermore, each of these rules takes into consideration only the extreme payoffs. Provided that the probability of each of the states of nature can be assigned, there exists a strategy, called the *expected-value*, or *Bayesian, criterion*, that actually utilizes all information furnished by a payoff matrix.

The decision maker is advised first to assign *prior*, or *a priori*, probabilities to the various possible states of nature, using past data, other available information, or merely his subjective judgement. In shear desperation, when the likelihoods for the states cannot even be subjectively evaluated, all the states may be considered as equally likely to occur. Next, the expected payoff, also called the *Bayesian measure*, for each act is computed. Denoting this measure by $EP(a)$, we have

$$EP(a) = E[X(a, \theta)] = \sum_j X_{ij} f(\theta_j). \qquad (6\text{-}3)$$

Finally, the expected-value criterion would direct the decision maker to select as the optimal act the one with the highest $EP(a)$.

Now, suppose the decision maker in our illustrative example believes, with whatever information is at his disposal, that the state of nature represented by θ_1 (prosperity) is twice as likely to occur as state θ_2 (stability), and that state θ_2, in turn, is twice as likely to occur as state θ_3 (recession). That is, he implies a particular probability distribution for S such that if $P(\theta_3) = p$, then $P(\theta_2) = 2p$ and $P(\theta_1) = 4p$. Since the sum of these probabilities must equal unity, the value of p can be obtained by solving the following equation:

$$\begin{aligned} p + 2p + 4p &= 1.00, \\ 7p &= 1.00, \\ p &= 1/7. \end{aligned}$$

With p determined, the probability function for the state variable becomes:

$$f(\theta_j) = \begin{cases} 4/7, & \text{for } \theta_j = \theta_1, \\ 2/7, & \text{for } \theta_j = \theta_2, \\ 1/7, & \text{for } \theta_j = \theta_3. \end{cases}$$

Given the foregoing probabilities, we can now compute the expected payoff for each act by (6–3). For our example, we have

$$EP(a_1) = 2.5(4/7) + 1.5(2/7) + (-1.0)(1/7) = 1.7143,$$

$$EP(a_2) = 1.6(4/7) + 2.0(2/7) + 0.0(1/7) = 1.4857,$$

$$EP(a_3) = 1.2(4/7) + 1.2(2/7) + 0.8(1/7) = 1.1429.$$

Thus, the Bayesian optimal act is a_1, with a maximum expected payoff of \$1.7143 million.

The expected-value criterion can also be applied to regrets in the same fashion. In terms of regrets, the Bayesian measure is defined as

$$EL(a) = E[L(a, \theta)] = \sum_j L(a_i, \theta_j)f(\theta_j) \tag{6–4}$$

and the optimal Bayesian act becomes that act for which $EL(a)$ is a minimum. Whether expectation is applied to payoffs or regrets, the Bayesian strategy will lead to the same optimal act, since an act that has the maximum expected payoff must also have the minimum expected regret. It can easily be verified that, referring to Table 6–2, the Bayesian optimal act is still a_1 with a minimum expected loss of

$$EL(a_1) = 0.0(4/7) + 0.5(2/7) + 1.8(1/7) = 0.4.$$

The Bayesian strategy is the most frequently employed decision criterion in statistical decision theory. There are many reasons for its popularity. One is that it characterizes the attitudes of many a decision maker in a large class of decision situations where the decision maker's main objective is either to maximize the expected gain or to minimize the expected loss in the long run. Another is that, compared to other strategies enumerated earlier, the Bayesian criterion is the only one that takes into account all the information in a payoff, or regret, table.

The application of the Bayesian strategy in the no-data case, however, requires a subjective evaluation of the state probabilities. There is the question as to how accurately the decision maker in a given situation can determine the appropriate weights for the states of nature. Some people reject the Bayesian strategy entirely for the reason that a prior probability law is required for it. However, if one accepts the necessity and validity of subjective probability, then systematic procedures for developing the values of the prior probability law over the states can be established to represent the beliefs held by the decision maker. One such procedure, as we shall see, is the gamble (or lottery) technique, for which prior probabilities can be found from answers of the decision maker to a series of yes-no questions phrased in terms of simple odds. Furthermore, if time and resource are available in a decision situation, prior probabilities can be revised with sample or experimental information before a final decision is made. For the time being, the reader may also be assured that very often the selection of the optimal act by the Bayesian criterion may not be sensitive to moderate changes in the probabilities assigned to the states of nature. That is, even when different subjective state probabilities are proposed, the same optimal act may still be obtained.

We shall turn now to apply the Bayesian strategy to a number of simple but quite important decision models.

6–3 A DECISION MODEL FOR PERISHABLE PRODUCTS

Some firms deal with merchandise that has practically no value if it is not sold within a very short time. Products of this variety include printed programs for special events, newspapers, fresh flowers, fresh vegetables, and other perishable commodities. Such items usually have a high markup, the large difference between purchasing (or production) cost and selling price representing the great risk a vendor (or producer) faces in stocking the commodity.

With perishable products, the seller cannot forecast exactly what his demand will be during a given day or period. He must balance the expected loss from unsold items owing to insufficient demand against the expected profits foregone owing to insufficient supply. If he has records of past sales and believes that they represent ordinary demand conditions, he may determine the optimal size of stock or output that would enable him to maximize profits or to minimize losses in the long run.

Consider, for the example, the case of a baker who produces a certain type of fancy cake at a total average cost of \$10 and sells it for \$15. The cakes are made during the weekend and are sold during the following week. Such cakes produced but not sold during the week are worthless and must be thrown away. The baker has been in business for 100 weeks, during which the weekly demand has fluctuated between 10 and 15. Thus, in this problem, the states of nature, D, represent the various levels of demand ranging from 10 consecutively to 15. The action set consists of the different number of cakes the baker may decide to produce. Clearly, this set contains six elements: producing 10, 11, . . . , or 15 cakes.

The baker makes \$5 profit on each unit sold and loses \$10 on each unit produced but not sold. Thus, the payoff function here can be stated as:

$$X(a_i, d_j) = \begin{cases} 5d_i - 10(a^*_i - d_j), & \text{if } d_j < a^*_i, \\ 5a^*_i, & \text{if } d_j \geq a^*_i. \end{cases}$$

Here, a^*_i is the number of cakes produced corresponding to the action, a_i, selected. For example, if a_2 is selected, then $a^*_2 = 11$. With the payoff function determined, its various values can now be computed easily (see Table 6–3).

Table 6–3. Payoff Matrix for the Baker Example

Act: units to be produced	*State: demand*					
	d_1: 10	d_2: 11	d_3: 12	d_4: 13	d_5: 14	d_6: 15
a_1: 10	$50	$50	$50	$50	$50	$50
a_2: 11	40	55	55	55	55	55
a_3: 12	30	45	60	60	60	60
a_4: 13	20	35	50	65	65	65
a_5: 14	10	25	40	55	70	70
a_6: 15	0	15	30	45	60	75

The regret or loss function of this decision situation can be defined as

$$L(a_i, d_j) = \begin{cases} 10(a^*_i - d_j), & \text{if } d_j \leq a^*_i, \\ 5(d_j - a^*_i), & \text{if } d_j > a^*_i. \end{cases}$$

The values of $L(a_i, d_j)$ computed with this loss function are tabulated in Table 6–4. Note these quantities also can be obtained directly from the payoff matrix.

Table 6–4. Loss Matrix for the Baker Example

Act	*State*					
	10	11	12	13	14	15
10	0	5	10	15	20	25
11	10	0	5	10	15	20
12	20	10	0	5	10	15
13	30	20	10	0	5	10
14	40	30	20	10	0	5
15	50	40	30	20	10	0

Now, to apply the Bayesian strategy, the baker must assign probabilities to the various states or demand levels. Suppose his sales records of the past 100 weeks are as shown in Table 6–5. If the baker thinks the relative

Table 6–5. The Baker's Sales Record

	No. of cakes sold	*Frequency: no. of weeks*	*Relative frequency*
	10	10	0.10
	11	15	0.15
	12	25	0.25
	13	35	0.35
	14	10	0.10
	15	5	0.05
Total		100	1.00

frequencies reasonably represent future probabilities, then he can use them as estimates and determine the optimal number of cakes to produce each week by computing and comparing the expected values of each act in terms of payoffs or regrets. Furthermore, he can modify these probabilities as he thinks reasonable and necessary during holdiday seasons such as Easter and Christmas. In any event, given these state probabilities, the expected payoff or expected loss for each act can now be computed. For example,

$$E[X(a_1, d_j)] = 50(0.10) + 50(0.15) + \cdots + 50(0.05)$$
$$= \$50.00$$

and

$$E[L(a_1, d_j)] = 0(0.10) + 5(0.15) + \cdots + 25(0.05)$$
$$= \$11.75.$$

Bayesian measures for other acts in terms of payoffs and regrets are computed in similar fashion (Table 6–6).

Table 6–6. Expected Payoffs and Losses for the Baker Example

Act	$E[X(a_i, d_j)]$	$E[L(a_i, d_j)]$
a_1: 10	\$50.00	\$11.75
a_2: 11	53.50	8.25
a_3: 12	54.75	7.00
a_4: 13	52.25	9.50
a_5: 14	44.50	17.25
a_6: 15	35.25	26.50

We see that the Bayesian decision criterion leads to the selection of a_3 (producing 12 cakes) as the optimal act, having the maximum expected payoff of \$54.75 or the minimum expected loss of \$7.00. Note that, in practice, we need determine only the maximum expected payoff or the minimum expected loss, not both. Both are considered here merely to show that the Bayesian criterion leads to the same optimal act whether applied to payoffs or opportunity losses.

6-4 EXPECTED VALUE OF PERFECT INFORMATION

Suppose the payoffs or losses and probabilities given in the preceding decision situation represents a frequently occurring problem. What is the expected value of one occurrence if the decision maker, the baker, knows each time which of the six possible states is going to occur? How much is such added information worth?

To answer the foregoing question, we note that if the baker could obtain perfect information on demand, he would produce exactly the amount demanded. Namely, if he knew the demand to be 10, he would take a_1 and produce 10 cakes. This would result in a profit of \$50. If he knew the demand to be 11, he would take a_2 and produce 11 cakes to obtain a profit of \$55. But the demand is 10 with a probability of 0.10, 11 with a probability of 0.15, and so on. Continuing this line of reasoning, we come to the general conclusion that the *expected payoff with perfect information*, denoted as EPPI, for a given decision situation should be

$$\text{EPPI} = \sum_j X(a^*_i, \theta_j)f(\theta_j), \qquad (6\text{–}5)$$

where a^*_i is the optimal act for state θ_j. Thus, for the baker's case, we have

$$\begin{aligned}\text{EPPI} &= 50(0.10) + 55(0.15) + 60(0.25) + 65(0.35) + 70(0.10) + 75(0.05)\\ &= \$61.75.\end{aligned}$$

With EPPI obtained, the *expected value of perfect information*, denoted as EVPI, is simply the difference between EPPI and the expected value of the Bayesian optimal act under uncertainty. That is,

$$\text{EVPI} = \text{EPPI} - \text{optimal } EP(a). \qquad (6\text{–}6)$$

For the baker's problem,

$$\text{EVPI} = 61.75 - 54.75 = \$7.00.$$

This value represents the maximum amount of money that a decision maker could spend to obtain additional information, such as through sampling. In other words, EVPI provides an upper limit to the expected value of sampling information—a value, as we shall see in Chapters 8 and 16, that is alwavs less than the expected value of perfect information.

It is also interesting to note that the expected value of perfect information is identical with minimum expected regret or opportunity loss. Indeed, we can define the expected value of perfect information directly as the expected opportunity loss for the Bayesian optimal act. This is obvious, since the minimum expected opportunity loss is the smallest expected loss the decision maker will sustain if he acts in the optimal way without perfect information, and he has no reason then to spend more than this amount to obtain such information.

6–5 AN INVENTORY MODEL

One important decision situation in business is to determine the "best" level of inventory. Decision making in connection with inventory levels depends on the type of inventory model on hand. We are interested here in probabilistic models, in which management must decide the optimal number of inventory units to stock under the condition that the number of units that can be sold, or is demanded, is a random variable. As a matter of fact, the baker's problem just analyzed is but an inventory problem for perishable goods. We shall introduce here an inventory model for semi-durable goods.

Consider an example. The manager of a men's clothing store has to decide how many suits of a particular size and type to stock for the coming season. Suppose suits of a given size and a given style made by a certain manufacturer can be purchased in lots of one dozen at a price of $600 per lot. The retail price for suits of this type is $800 per lot during the season. Suits unsold at the end of the season can be cleared at a price of only $450 per lot. For the sake of simplicity, we shall assume that unsatisfied demand due to understock will not affect future sales, and we shall also ignore fixed costs in handling. Sales records indicate that the probability distribution of demand (states of nature) for this particular type of suits is as follows:

Demand (dozens)	*Probability*
1	0.05
2	0.10
3	0.30
4	0.40
5	0.10
6	0.05
	1.00

Table 6–7. Payoff Matrix

	State: demand					
Act Stock	1	2	3	4	5	6
1	$200	$200	$200	$200	$200	$200
2	50	400	400	400	400	400
3	−100	250	600	600	600	600
4	−250	100	450	800	800	800
5	−400	−50	300	650	1000	1000
6	−550	−200	150	500	850	1200

The payoff function can be determined easily as

$$X(a_i, d_j) = \begin{cases} 200d_j - 150(a^*_i - d_j), & \text{for } d_j < a^*_i, \\ 200d_j, & \text{for } d_j \geq a^*_i. \end{cases}$$

Here, a^*_i is the number of lots stocked corresponding to action a_i. The values of $X(a_j, d_j)$ are given in Table 6–7. From these, the Bayesian measures, or expected profits, for alternative stock actions can be computed in the usual fashion (Table 6–8).

Table 6–8

Stock	*Expected profits*
1	$200.00
2	382.50
3	530.00
4	572.50
5	475.00
6	342.50

The optimal act in this decision situation is to stock 4 dozen, since the expected profit for it is the highest possible.

6–6 SPECIAL FORMS OF BAYESIAN STRATEGY AND LOSS FUNCTIONS

The application of the Bayesian strategy to decision problems is quite time-consuming and at times rather complicated. Under certain circumstances, fortunately, the Bayesian decision criterion can be simplified to some equivalent forms, and computations can be considerably reduced. The first of these situations involves *equivalent state-action* problems, where there is a different optimal act for each state. The second involves *nonequivalent state-action* problems, where one action may be optimal over more than one state. In this section we shall take up situations of the first type, deferring our discussion of the second until Chapter 7. The special forms of the Bayesian decision criterion are conditional upon the types of loss functions. We shall pay special attention here to the piecewise linear loss function and consider briefly two other frequently encountered loss functions under the names of quadratic and all-or-nothing loss functions.

For both the perishable-product and the inventory models introduced before, as in many other problems of this kind, the payoff function for each act is piecewise linear in two segments (Figure 6–1). As a result the cor-

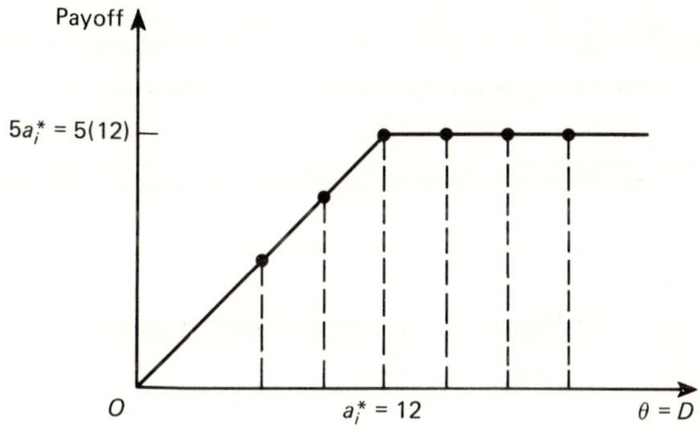

Figure 6–1 Piecewise Payoff Function for a_3: the Baker Example

responding loss functions are also piecewise linear in two segments (Figure 6–2). (Note if all actions were plotted on a graph, they would constitute a series of identical and nonsymmetric V-shaped figures, each with the bottom of the V resting on the state level corresponding to a^*_i determined by the optimal state action. The nonsymmetric shape of the loss function in Figure 6–2 results from the fact that the loss from overstocking is not the same as the loss from understocking.) In general, a piecewise linear loss function in two segments is expressed as

$$L(a_i, \theta_j) = \begin{cases} b_o(a^*_i - \theta_j), & \text{if } \theta_j \leq a^*_i, \\ b_u(\theta_j - a^*_i), & \text{if } \theta_j > a^*_i. \end{cases} \tag{6–7}$$

In this expression

a^*_i = the value needed to compute the opportunity loss for the action a_i,
b_o = the unit loss from overestimating the state by one unit,
b_u = the unit loss from underestimating the state by one unit.

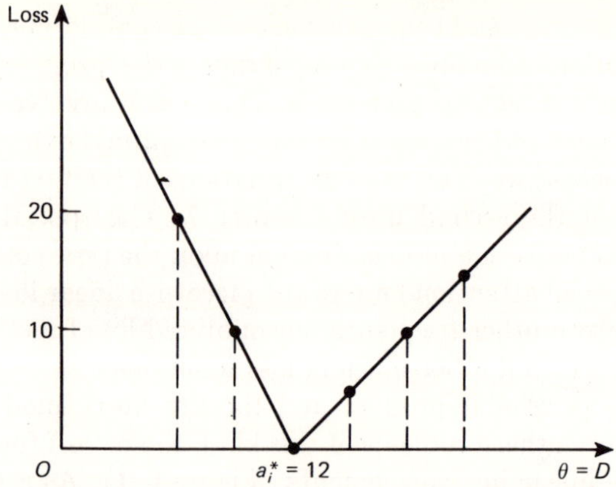

Figure 6–2 Piecewise Linear Loss Function for a_3: the Baker Example

Percentiles, quartiles, and the median were discussed in Section 3–11. These are but a few of a family of summary measures called fractiles. In general, if

$$F(a) = P(X \leq a) \geq f \quad \text{and} \quad P(X \geq a) \geq 1 - f, \tag{6–8}$$

then a is said to be an f *fractile* of the distribution of X.

Now, it can be shown that for a piecewise linear loss function as defined by (6–7), the optimal Bayesian measure is merely the $b_u/(b_o + b_u)$ fractile of the distribution of θ. In terms of (6–8), this would be a value of $a_i{}^*$ such that

$$P(\theta \leq a^*{}_j) \geq \frac{b_u}{b_o + b_u} \quad \text{and} \quad P(\theta \geq a^*{}_i) \geq 1 - \frac{b_u}{b_o + b_u}. \tag{6–9}$$

In other words, this special form of the Bayesian criterion directs the decision maker to accumulate the state probabilities and select as the optimal act the one for which the cumulated probability first exceeds the ratio $b_u/(b_o + b_u)$. This special form simplifies to a great extent the computational effort required. Instead of calculating the expected losses for all the acts, the decision maker is required now, given b_o, b_u, and the distribution function of θ, to find a single fractile of the distribution.

For the baker's problem, recall that the loss function was given as:

$$L(a, \theta) = \begin{cases} 10(a^*{}_i - d_j), & \text{for } d_j \leq a_i{}^*, \\ 5(d_j - a^*{}_i), & \text{for } d_j > a^*{}_i. \end{cases}$$

From this we have

$$\frac{b_u}{b_o + b_u} = \frac{5}{10 + 5} = 0.333.$$

Now, from the first two columns of Table 6–9, we find for a_3, $F(\theta) = 0.50$, which just exceeds 0.333. Therefore, as before, a_3 (to produce 12 cakes) is the Bayesian optimal act.

Similarly, for the men's clothing store illustration, the loss function can be given as

$$L(a_i, d_j) = \begin{cases} 150(a^*_i - d_j), & \text{for } d_j \leq a_i^*, \\ 200(d_j - a^*_i), & \text{for } d_j > a^*_i. \end{cases}$$

For this,

$$\frac{b_u}{b_o + b_u} = \frac{200}{200 + 150} = 0.57.$$

This leads to the selection of a_4, as before, as the optimal act, since for this act $F(\theta) = 0.85$, which just exceeds 0.57 as shown in the last two columns of Table 6–9.

Table 6–9. Cumulative Probability Function

Baker		*Men's store*	
State and act	$F(\theta)$	*State and act*	$F(\theta)$
10	0.10	1	0.05
11	0.25	2	0.15
12	0.50	3	0.45
13	0.85	4	0.85
14	0.95	5	0.95
15	1.00	6	1.00

This special form of the Bayesian strategy can also be applied to a continuous state random variable, provided that the loss function is defined as (6–7). For a continuous case, (6–9) is reduced to

$$P(\theta \leq a^*_i) = \frac{b_u}{b_o + b_u}. \tag{6–10}$$

Suppose a grocer has to decide how many cartons of milk to order for a particular week. The milk costs the grocer 24 cents per carton and he sells it for 30 cents per carton. Any carton of milk that remains unsold at the end of the week will spoil and thus becomes worthless. If he orders less than demanded, he suffers an opportunity loss of 6 cents per carton. If he orders more than demanded, he suffers a spoilage loss of 24 cents per carton. His loss function is therefore piecewise linear with $b_u = 6$ and $b_o = 24$. Furthermore, suppose past experience indicates that the weekly demand for milk is approximately normal with $\mu = 120$ and $\sigma = 30$. Under these conditions, what is the optimal order for milk?

We note first that $b_u/(b_o + b_u) = 6/(6 + 24) = 1/5$ fractile of the distribution of the demand variable $F(\theta)$. Next, from the table of values of the standard normal distribution function, we find that the $1/5 = 0.2$ fractile is approximately -0.84. This means that the 0.2 fractile of $n(\mu, \sigma)$ is 0.84 standard deviation to the left of the mean; viz., $\mu - 0.84\sigma$. Hence, the 0.2 fractile of the grocer's distribution is

$$120 - 0.84(30) \doteq 95 \text{ cartons.}$$

To minimize expected loss, the grocer should order 95 cartons of milk per week.

Quadratic and All-or-Nothing Loss Functions

When there is a different optimal act for each state, the Bayesian strategy also takes on special forms under two other types of loss function. The first is the *quadratic loss function*, where the decision maker views the loss from either overestimating or underestimating the state value as a function of the squared error. It is expressed as

$$L(a_i, \theta_j) = c(a^*_i - \theta)^2, \tag{6–11}$$

where $c > 0$ and may stand for the amount of money the decision maker is willing to pay for perfect information on the state—i.e., knowledge that the true state is only one unit above or below a given value of θ, say θ^*. The graph for a quadratic loss function is shown by Figure 6–3.

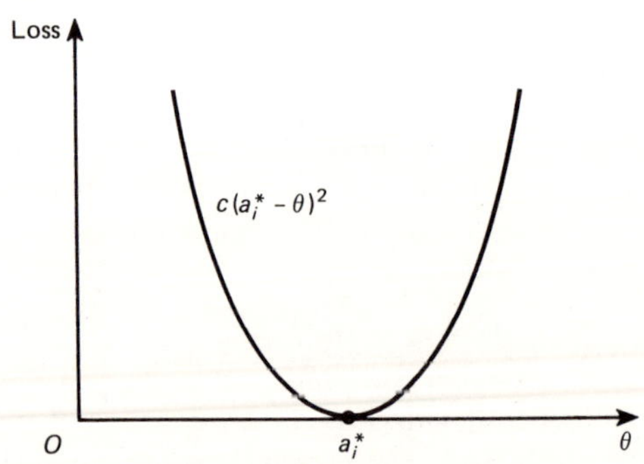

Figure 6–3 Quadratic Loss for a_i

The Bayesian strategy specifies that the optimal act is the one that minimizes the expected loss. It can be shown mathematically that if the state variable is continuous, the expectation for the right side of (6–11) is a minimum at

$$a^*_i = E(\theta). \tag{6–12}$$

This form of the Bayesian strategy states that the decision maker should select as the optimal act the one whose numerical representation is given by the mathematical expectation of the random variable whose value represent the states. When the state variable is discrete, intuitively, one should select the act whose numerical representation is the closest to $E(\theta)$ according to this special form of the Bayesian criterion.

The second, the *all-or-nothing loss function*, is given as

$$L(a_i, \theta_j) = \begin{cases} c, & \text{for } a^*_i \neq 0, \\ 0, & \text{for } a^*_i = \theta. \end{cases} \qquad (6\text{–}13)$$

The graph of this function is given by Figure 6–4. This function indicates that the loss, however large the error, is a constant c.

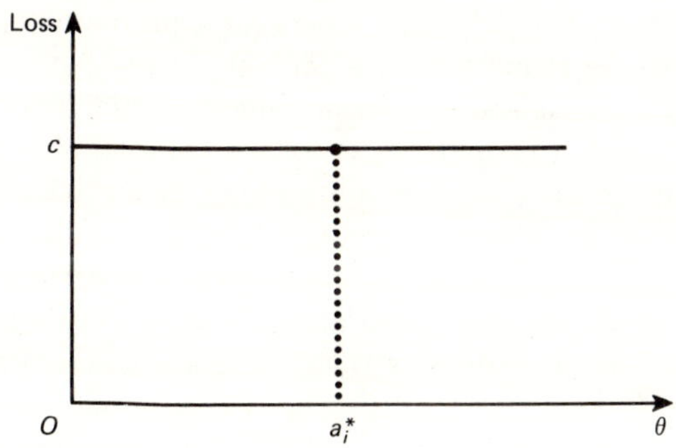

Figure 6–4 All-or-Nothing Loss for a_i

Let $f(\theta)$ be the probability function for a discrete state variable; then the expected loss for any action is

$$E[L(a_i, \theta_j)] = c \sum_{\theta \neq a^*_i} f(\theta),$$

where the sum is over all the values of the state variable except for $\theta = a^*_i$. Thus,

$$\sum_{\theta \neq a^*_i} f(\theta) = 1 - f(\theta = a^*_i)$$

$$= 1 - f(a^*_i)$$

and

$$E[L(a_i, \theta_j)] = c[1 - f(a^*_i)].$$

The left side of this equation is a minimum when $f(a^*_i)$ on the right side is a maximum. Thus, this form of the Bayesian strategy instructs the decision maker to select the action a_i for which

$$\max f(a^*_i) \qquad (6\text{–}14)$$

occurs.

Form (6–14) is valid for discrete state variables only. The special Bayesian strategy here is to select the act a_i for which $f(a^*_i)$ is a maximum. If the value of the state variable that maximizes $f(a^*_i)$ does not correspond to any act in A, then this form of the Bayesian strategy is to select the act for which the state corresponding to its numerical representation maximizes $f(a^*_i)$.

We shall have opportunity to employ these two loss functions in Chapter 16.

6–7 A SEQUENTIAL DECISION MODEL

So far we have been concerned with what are called *single-stage* decision problems, where the decision maker selects one action at a single point of time. Occasions may arise, however, where multiple decisions, or a sequence of decisions, must be made before a problem is resolved—i.e., where the entire set of decisions at different time points should be evaluated at the time the initial decision is made.

Let us introduce sequential decision making by an example. Confronted with the decision of introducing a new product, a manufacturer is considering whether a small or large plant should be built. Construction costs for the small and large plants are estimated at one and two million dollars respectively. The appropriate plant size depends upon whether there will be a weak or a strong demand for the proposed product. According to the manufacturer's marketing research staff the net profit, or payoff, (as the difference between gross revenues and construction costs, in millions of dollars) for the first five years, conditional to a weak or a strong demand are as shown in Table 6–10.

Table 6–10. Payoff Matrix

	State of nature	
Act	*Weak demand*	*Strong demand*
Small plant	1	3
Large plant	0	4

For this decision situation, the research and production staff suggests the possibility of building a small plant at the beginning and expanding it later when it is known for sure whether a weak or a strong demand is going to emerge. The cost of expansion will be \$1.2 million. The expanded facilities will equal that of the large plant built initially and, therefore, will not alter the estimated gross revenues but will, of course, affect the payoffs.

Thus, the manager may either make a single decision to construct a small plant, a_1, or a large plant, a_2, now; or he may make a sequence of decisions to build a small plant for the present and to expand, e, or not to expand, $\bar{e}$, later as future conditions render appropriate.

The adoption of either process requires that probabilities for weak demand, θ_1, and strong demand, θ_2, be assigned if the Bayesian strategy is to be applied. Suppose that after careful consideration of future economic conditions, the relative merits of the product under consideration, and so on, the manufacturer concludes that the odds are about 3:2 in favor of a strong demand. Thus, the decision maker believes that $P(\theta_1) = 0.4$ and $P(\theta_2) = 0.6$.

At this point, if the manufacturer adopts the first option, the optimal act is to build a large plant now, since

$$EP(a_1) = 1(0.4) + 3(0.6) = \$2.2 \text{ million}$$

and

$$EP(a_2) = 0(0.4) + 4(0.6) = \$2.4 \text{ million.}$$

As a matter of fact, as the reader can verify easily, as long as $P(\theta_2) > 0.5$, a_2 is the optimal Bayesian act.

Turning to the second process, the manufacturer would make a sequence of decisions: the first is to decide on a small or a large plant; the second, at a later date, is to decide whether or not to expand the small plant.

An instructive and efficient method of evaluating a sequence of decisions is to express the alternative courses of action and their respective consequences by a *decision-flow diagram*, or a *decision tree*. Such a diagram for our current example is shown by Figure 6–5. In a decision tree, squares are referred to as *decision forks* and circles as *chance forks*. At a decision fork, the decision maker is in the driver's seat and decides which action he wants

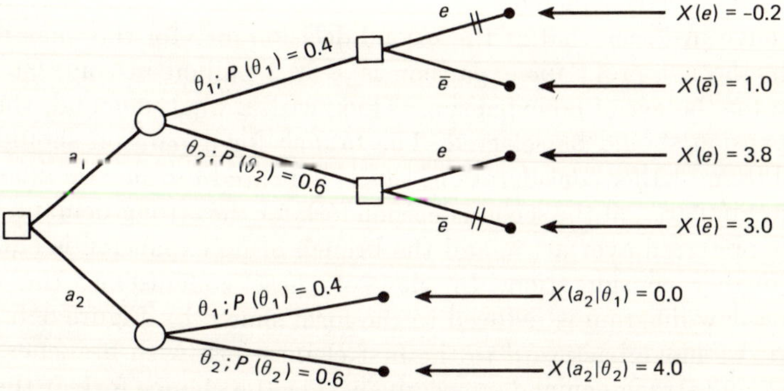

Figure 6–5 Initial Decision Tree for the New-Plant Example

to take. At a chance fork, the decision maker is no longer the master of destiny and chance is in the driver's seat.

Starting from the first decision fork at the extreme left of Figure 6–5, the decision maker is confronted with two alternative courses of action: to build a small or a large plant. At the end of each act he comes to the chance fork, which may result either in a weak demand with a probability of 0.4 or a strong demand with a probability of 0.6.

If the manufacturer travels down the "build a large plant" branch, the tree ends with the event branches and yields payoffs of $0 and $4 million for a weak and a strong demand, respectively.

If the manufacturer travels down the "build a small plant" branch, he arrives at another decision fork after each of the event branches: "weak demand" and "strong demand." Here, he is in the driver's seat once again and he can decide either to expand or not to expand the small plant, using additional information obtained between the two decision forks. Given a weak demand, the payoff for expansion is the difference between the estimated net revenue of $1 million and the cost of expansion, $1.2 million. This difference is −$0.2 million. The payoff for no expansion remains the same at $1 million as shown in Table 6–10. Similarly, payoffs for expansion or no expansion given a strong demand are $3.8 and $3.0 million, respectively. Observe that the payoff of $3.8 million for expansion given strong demand is derived from the fact that the net payoff for a large plant given a strong demand is $4 million. This is the same as saying the corresponding gross revenue should be $6 million. Now, if a small plant is built first and expanded later, the total cost will be $1 + 1.2 =$ $2.2 million and thus the payoff becomes $6 - 2.2 =$ $3.8 million.

With information given by the decision-flow diagram, we can proceed to find the best decision or sequence of decisions for the problem at hand. This we achieve by working backward from the end branches of the diagram to the first decision fork. In our example, the second decision fork is the first to be considered.

We have just seen that at the second decision fork for the weak-demand branch, the net profit for expansion is −$0.2 million in contrast to $1.0 million for the act of no expansion. Thus, with a weak demand, the act of no expansion should be selected. The branch for expansion should be removed from further consideration; this is indicated by the two lines drawn upon it. Similarly, at the second decision fork for the strong-demand branch, act e is preferred over act $\bar{e}$, and the branch of no expansion is eliminated from further consideration. In view of these conclusions, the original decision-flow diagram is reduced to the form shown by Figure 6–6.

Next, we move backward to the first chance fork with branches labeled as weak and strong demand, respectively. For the chance fork at the end of

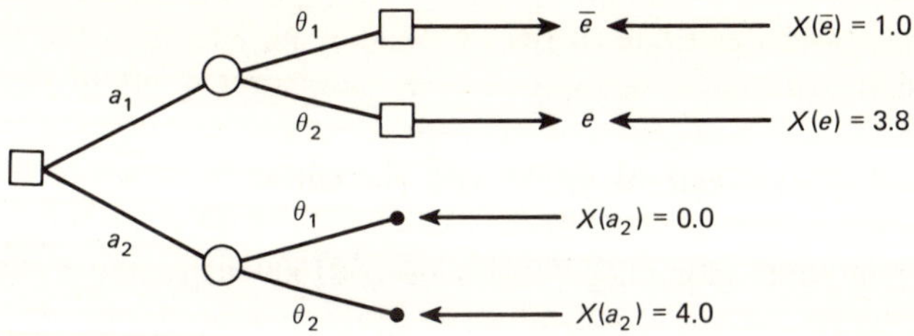

Figure 6–6 Intermediate Decision Tree for the New-Plant Example

the "build a small plant" branch, we have

$$EP(a_1) = 1.00(0.4) + 3.8(0.6) = \$2.68 \text{ million.}$$

For the chance fork at the end of "build a large plant" branch, we have

$$EP(a_2) = 0(0.4) + 4(0.6) = \$2.40 \text{ million.}$$

Finally, by substituting the Bayesian measures for the chance forks, we reduce the decision-flow diagram to its ultimate form (Figure 6–7).

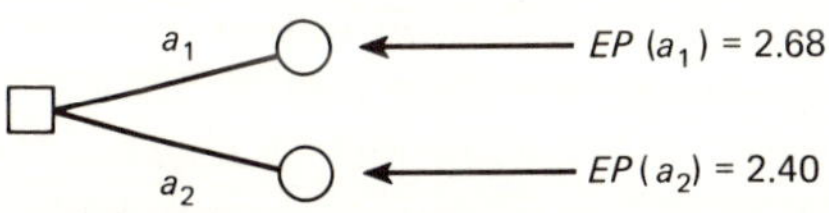

Figure 6–7 Final Decision Tree for the New-Plant Example

Thus, the best decision in this case is build a small plant initially and to decide upon expansion later when it is known whether the demand is going to be weak or strong. Note that the single-decision procedure yields a different conclusion: to build a large plant initially.

When either a single-decision or a sequential-decision procedure can be applied to a given situation, the latter often turns out to be better or more profitable. Furthermore, the sequential-decision procedure is especially productive in situations where subjective probabilities must be employed, since in between the several decisions the event probabilities can be revised to reflect new information (the method for doing so will be discussed in Chapter 8).

6–8 A NOTE ON UTILITY THEORY

The use of money as a measure of a consequence of an act can be questioned on the ground that different decision makers usually place different valuations on the same amount of money, owing to their differing financial circumstances and reactions to risk. Consider, for example, the following three decision situations:

I: a_1 = having a gain of \$1000 with certainty,
a_2 = having a gain of \$3200 or a loss of \$1000 with equal probabilities;

II: a_1 = having a loss of \$100 with certainty,
a_2 = having a gain of zero with probability of 0.995 or a loss of \$10,000 with probability of 0.005;

III: a_1 = having a gain of \$100,000 with certainty,
a_2 = having a gain of \$2,000,000 if the toss of a die comes up with a number 5 or less and a loss of \$8,000,000 if the die turns up 6.

In each of these three situations, if the expected monetary value is used' then a_2 should be preferred over a_1—yet, many people would select a_1. Why?

In case I, if one can gain \$1000 for sure, why should he risk losing \$1000 to gain only \$100 more in expected value (the expected value of a_2 is \$1100)?

In case II, while a sure loss of \$100 is painful, it may be a reasonable price to pay to avoid risking the huge loss of \$10,000, however slight the probability. This is indeed why any prudent man would buy fire insurance on his house, even though the expected monetary value of such an act is negative.

In case III, $EP(a_1)$ is less than one-third of $EP(a_2)$; if the decision maker is financially strong and ambitious, he may prefer a_2 over a_1. However, to many others, the sure gain of \$100,000 may be more attractive than the possible windfall gain of \$2,000,000 at the risk of a fantastic loss of \$8,000,000, which they can never afford. This illustrates clearly the famous St. Petersburg paradox, where rational men often refrain from games with favorable odds when stakes are too high.

The St. Petersburg paradox led Daniel Bernoulli, more than a century ago, to investigate utility rather than money as a basis of rational choice among alternatives. Utility theory was fully developed and applied by Von Neumann and Morgenstern in the 1940's.

In current usage, a *utility index* is a real number that gives a subjective preference measure of the decision maker to an outcome of an action. The utility index of an outcome O_i may be denoted as $U(O_i)$. Furthermore, if an act will result in n possible consequences, or payoffs, each with a number representing its utility, the utility of such an action is called its *expected utility*, defined as the weighted average of the utilities of the respective consequences with corresponding probabilities as weights. Let O_i be the ith outcome of an action a, $U(O_i)$ be the utility of O_i, which occurs with probability α_i for $i = 1, 2, \ldots, n$ and $\sum \alpha_i = 1$; then the expected utility of a becomes

$$E[U(a)] = U(O_1)\alpha_1 + U(O_2)\alpha_2 + \cdots + U(O_n)\alpha_n. \qquad (6\text{–}15)$$

This is interpreted as the expected payoff in terms of utilities for an act.

Before turning to consider the procedure of constructing utility indices, let us observe some of the important assumptions underlying modern utility theory.

TRANSITIVITY: If the decision maker is indifferent between two outcomes O_1 and O_2, and if he is also indifferent between O_2 and O_3, then he will be indifferent between O_1 and O_3. That is, if $U(O_1) = U(O_2)$ and $U(O_2) = U(O_3)$, then $U(O_1) = U(O_3)$. This is called the *transitivity of indifference*.

Also, if the decision maker is confronted with three consequences, O_1, O_2, and O_3, and if he prefers O_1 to O_2 and, in turn, O_2 to O_3, then he also prefers O_1 to O_3. That is, if $U(O_1) > U(O_2)$ and $U(O_2) > U(O_3)$, then $U(O_1) > U(O_3)$. This is called the *transitivity of preference*.

CONTINUITY OF PREFERENCE: If the decision maker prefers O_1 to O_2 and O_2 to O_3, there exists a probability α, $0 < \alpha < 1$, such that he will be indifferent between O_2 and $O_1(\alpha) + O_3(1 - \alpha)$. That is, given $U(O_1) > U(O_2)$ and $U(O_2) > U(O_3)$, there exists a probability α that can make $U(O_2) = U(O_1)\alpha + U(O_3)(1 - \alpha)$ hold.

For example, if $U(O_1) = 1.0$, $U(O_2) = 0.6$, and $U(O_3) = 0.2$, then for $\alpha = 0.5$ we would have

$$\begin{aligned} U(O_1)\alpha + U(O_3)(1 - \alpha) &= 1.0(0.5) + 0.2(0.5) \\ &= 0.6 = U(O_2). \end{aligned}$$

INDEPENDENCE: If the decision maker is indifferent between O_1 and O_2' and if he is also indifferent between O_3 and O_4, then for any probability α he will be indifferent between $O_1(\alpha) + O_3(1 - \alpha)$ and $O_2(\alpha) + O_4(1 - \alpha)$.

Thus, if $U(O_1) = U(O_2) = 0.7$, and $U(O_3) = U(O_4) = 0.5$, then for any probability such as $\alpha = 0.25$, we have

$$0.7(0.25) + 0.5(1 - 0.25) = 0.7(0.25) + 0.5(1 - 0.25).$$

PREFERENCE FOR HIGHER PROBABILITY OF SUCCESS: If the decision maker prefers O_1 to O_2 and there are two probabilities α and β such that $\alpha > \beta$, then he will also prefer $O_1(\alpha) + O_2(1 - \alpha)$ to $O_1(\beta) + O_2(1 - \beta)$.

To see that this is a reasonable assumption, let $U(O_1) = 0.8$, $U(O_2) = 0.6$, $\alpha = 0.5$, and $\beta = 0.4$; we have

$$0.8(0.5) + 0.6(0.5) = 0.70$$

which is greater than

$$0.8(0.4) + 0.6(0.6) = 0.68.$$

Compound Probability: If there are three probabilities, α, β, and λ, and there are two outcomes, O_1 and O_2, then the decision maker will be indifferent between

$$\alpha[O_1\beta + O_2(1 - \beta)] + (1 - \alpha)[O_1\lambda + O_2(1 - \lambda)]$$

and $$O_1[\alpha\beta + (1 - \alpha)\lambda] + O_2[1 - (\alpha\beta + (1 - \alpha)\lambda)].$$

Note that this property holds irrespective of the probability values and the utility indices. For example, let us arbitrarily set $\alpha = 0.2$, $\beta = 0.3$, $\lambda = 0.4$, $U(O_1) = 0.6$, and $U(O_2) = 0.5$; we then have

$$0.2[0.6(0.3) + 0.5(1 - 0.3)] + (1 - 0.2)[0.6(0.4) + 0.5(1 - 0.4)] = 0.538$$

which is equal to

$$0.6[0.2(0.3) + (1 - 0.2)0.4] + 0.5[1 - (0.2)(0.3) + (1 - 0.2)(0.4)] = 0.538.$$

With these assumptions stated, we now move on to the practical problem of determining a decision maker's utility function for money. It is important to note that a utility function is defined over the set of distinct actions, or outcomes of actions, available to the decision maker. Furthermore, a utility function is not unique, since it is a subjective measure of an individual's preference. Not only may different individuals have different utility functions, but also the same individual may have different utility functions under different sets of circumstances.

There are several ways of determining an individual's utility function. One method is based on what is called the *reference contract*, a device for taking into account the unique value that individual decision makers place on different alternatives. To develop the reference contract one must determine the amount of money a decision maker would demand or be willing to pay to be relieved of the obligation stated in a proposition, or gamble. Let us illustrate this via a payoff matrix.

For the payoff functions shown in Table 6–11 we note that $EP(a_1) =$ \$450 and $EP(a_2) = EP(a_3) =$ \$500. Thus, if expected-monetary-value criterion is used, any decision maker will be indifferent between a_2 and a_3, and either is preferred over a_1. However, if we take into consideration a

Table 6–11. Payoff Matrix

Act	θ_1 $f(\theta_i) = 0.3$	θ_2 $f(\theta_2) = 0.4$	θ_3 $f(\theta_3) = 0.2$	θ_4 $f(\theta_4) = 0.1$
a_1	\$600	\$600	\$300	−\$300
a_2	750	900	−600	350
a_3	1000	1000	−1000	0

decision maker's reaction to risk as a function of the degree of that risk to the importance of the gain or loss—i.e., if utility indices are employed—then a differential optimal act may result for a different decision maker. (This, of course, can also happen with different sets of subjective prior probabilities.) Consider, for example, two decision makers. X, who is of strong financial resources and is inclined to be a risk-taker, and Y, who is a struggling business man and is inclined to be risk-averse, are both facing the decision situation represented by Table 6–11. How can we go about to convert these monetary payoffs into utility indices?

Consider, first, X, who feels that he will not be hurt by a big loss but will receive disproportionately great satisfaction from a success. Now, by asking him to associate cash values with gains or losses entailed by various alternatives, we wish to establish the cash equivalence of his preferences. We will consider a range of monetary values from a gain of $1000 (the most desirable payoff) to a loss of $1000 (the most undesirable payoff) as indicated by Table 6–11, although we could use any other scale. We start by asking: "What amount of cash would you be willing to accept in lieu of a contract that assures you a gain of $1000 dollars with a probability of 1?" Any decision maker, irrespective of his attitude toward risk, would ask for at least $1000. We may next ask him: "What amount of cash would you be willing to pay in order to be relieved from a contract that assures you a gain of $1000 with a probability of 0 and a loss of $1000 with a probability of 1?" Again, any decision maker would pay up to $1000 to be freed from such a contract. Next we may ask what amount of money he would accept in place of a contract that gave him a gain of $1000 with a probability of 0.9 and a loss of $1000 with a probability of 0.1. Suppose, X, being a risk-taker, says that he would take a minimum of $850 to forego the opportunity of a 90 percent chance of winning $1000 and a 10 percent chance of losing $1000. In other words, he is indifferent between the value of $850 and the value of the gamble as stated in the contract. By following this procedure for a few more discrete probabilities within the range from 0 to 1, we would obtain the reference-contract table of his utility function, as depicted in the first two columns of Table 6–12.

Columns 1 and 3 of the same table give the utility function, derived in the same fashion, for the more cautious decision maker, Y. This utility function clearly shows Y's inclination to avoid risk. For example, he is willing to accept only $500 in lieu of the contract that would give him a 90 percent chance of winning $1000 and a 10 percent chance of losing $1000. Also, he would pay as much as $750 in order to avoid the gamble of winning $1000 with a probability of 0.2 and losing $1000 with a probability of 0.8.

Note that we have adopted the convention of using the equivalent probabilities as the utility indices. This procedure is entirely arbitrary, since any other scale can also be used. However, this method is very

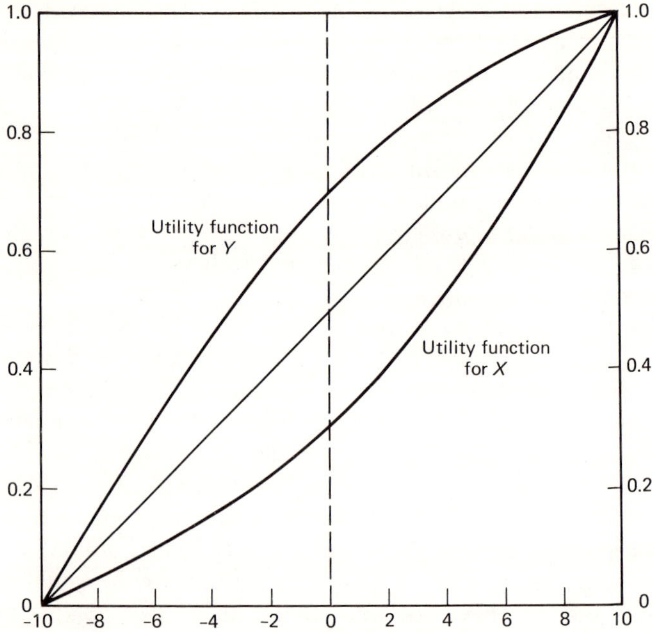

Figure 6–8 Utility Functions for the Illustrative Example in Table 6–12

convenient and is quite consistent with more involved procedures of establishing utility index. When a reference-contract schedule is established, it may be plotted on a graph paper, and a smooth line drawn through the points may be considered an approximation of the continuous utility function for the decision maker. (See, for example, Figure 6–8 for our current illustration.) It can then be used to convert monetary payoffs into utility payoffs by interpolation.

Using Table 6–12 or Figure 6–8, we can convert monetary payoffs in Table 6–11 into utilities for decision makers X and Y as shown in Table 6–13.

Now, if each decision maker is to apply the Bayesian strategy to utility payoffs, then for X we have

$$E[U(a_1)] = (0.66(0.3) + 0.66(0.4) + 0.46(0.2) + 0.19(0.1)$$
$$= 0.573 \text{ utils,}$$

$$E[U(a_2)] = 0.80(0.3) + \cdots + 0.50(0.1)$$
$$= 0.678 \text{ utils,}$$

$$E[U(a_3)] = 1.00(0.3) + \cdots + 0.30(0.1)$$
$$= 0.730 \text{ utils.}$$

Table 6–12. Reference-Contract Table

Probability of $1000 *Gain*	*Reference contract* *Mr. X*	*Mr. Y*
1.0	$1000	$1000
0.9	850	500
0.8	750	250
0.7	650	0
0.6	500	−200
0.5	350	−325
0.4	200	−450
0.3	0	−600
0.2	−250	−750
0.1	−600	−875
0.0	−1000	−1000

For Y we have

$$E[U(a_1)] = 0.92(0.3) + \cdots + 0.65(0.1)$$
$$= 0.873 \text{ utils},$$
$$E[U(a_2)] = 0.789 \text{ utils},$$
$$E[U(a_3)] = 0.770 \text{ utils}.$$

Thus, in terms of utility indices, a_3 is the optimal act for X and a_1 is the optimal act for Y, since each has the highest expected utility for the respective decision maker.

We can gain additional insights into the properties of utility functions by studying their graphs, such as Figure 6–8. We see in this figure that the shape of the curve for the risk-taker shows an increasing marginal utility for money. It means that when his cash position increases, he places more value on each additional dollar to be gained. The curve for the risk-averse Y, however, shows just the reverse—a decreasing marginal utility of money. If one's utility function is linear over the whole range, it appears as a straight line such as the diagonal line in Figure 6–8. Such a function shows

Table 6–13. Utility Payoff Matrix for X and Y

	X				Y			
Act	θ_1	θ_2	θ_3	θ_4	θ_1	θ_2	θ_3	θ_4
a_1	0.66	0.66	0.46	0.19	0.92	0.92	0.82	0.65
a_2	0.80	0.92	0.10	0.50	0.95	0.98	0.30	0.52
a_3	1.00	1.00	0.00	0.30	1.00	1.00	0.00	0.70

the utility in proportion to the monetary value; in this case, expected value computed directly from monetary payoffs would indicate the expected utility of each alternative.

In fact none of these three types of utility functions is very realistic. Recent literature suggests that, typically, the utility function of an individual or organization may follow a pattern such as shown by Figure 6–9. Here, the utility function has an inflection point associated with a cash asset position, C_0, called the *level of aspiration*. With a lower cash position, a decision maker will search for opportunities to increase his income and is willing to be a risk-taker. If fair gambles are available to him in this range, he will take them even if he has to pay some premium. With cash position equal to or greater than C_0, however, he tends to become risk-averse and would be willing to sacrifice some income to be relieved of the obligations of some risk alternatives.

Utility indices assigned to payoffs, or outcomes, are such, then, that the rank order of the expected utility reflects the decision maker's action preferences. Utility indices are attractive and easy to manipulate but very difficult and time-consuming to construct. Another problem is that individuals are often inconsistent in stating their preferences for specified combinations of risk and associated consequences. Thus, it may be questioned at times how accurately the resulting utility functions represent the decision maker's true preferences. Perhaps, too, different responses would have been obtained at different times because of changes in the decision maker's moods or outlook. It is equally apparent that a person's utility function may change from one decision situation to another. Who would contend that the same individual would have the same reaction to risk in

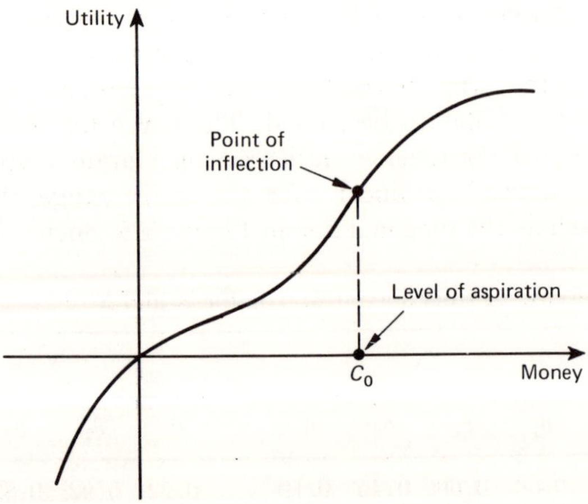

Figure 6–9 A Typical Utility Function for Money

the stock market as at the poker table? When we attempt to determine a utility function representing organizational policy, the difficulty may be compounded by the multitude of organization objectives. Thus, utility indices should always be used with some caution.

We observe, too, that when the range of payoffs is relatively small, the expected monetary value can be used directly without introducing serious errors. However, the question, "How small is small?" is a subjective matter. A range of $1000 may be quite large for this author, for example, but actually trivial for a large business firm.

EXERCISES

6–1 Consider the following payoff matrix:

Act	State θ_1	θ_2	θ_3	θ_4
a_1	1	−3	7	6
a_2	2	4	0	8
a_3	−2	5	7	7
a_4	9	6	−1	−1

Determine the optimal act in accordance with each decision strategy introduced in Section 6–2. [Make your own assumption when $f(\theta)$ is required.]

6–2 A certain output is manufactured at $5 and sold at $10 per unit. This product, if produced but not sold during a week's time, becomes worthless. The weekly sales records in the past are as follows:

Demand per week	*Number of weeks each sales level was recorded*
10	5
11	15
12	20
13	25
14	20
15	10
16	5

(a) Determine the payoff and opportunity-loss functions for this problem.
(b) Prepare the payoff and regret matrices.
(c) What are the expected payoff and expected opportunity loss for the Bayesian optimal act?

6–3 Using the data in Exercise 6–2, (a) compute the expected payoff under the assumption of certainty (that is, a perfect predicting device is available); (b) compute the expected value of perfect information.

6–4 Each day a grocery shop purchases for \$2 a product that it sells for \$4. Every unit sold costs the shop \$0.20 for wrapping. Because the product is perishable, units remaining unsold at the end of the day are returned, without wrapping, to the supplier for reprocessing, and a refund of \$1 per unit is received. The probability distribution of the daily demand is estimated as below:

Number of units sold daily D	*Probability* $f(d_i)$
0	0.10
1	0.40
2	0.30
3	0.10
4	0.10
	1.00

According to the Bayesian strategy, show that the optimum standing order the grocery shop should place each day is 2.

6–5 If in Exercise 6–4 we assume that there are no wrapping costs and no refunds, verify that the optimal act now is to order either 1 or 2 each day.

6–6 A newsstand at a certain subway station sells for 15 cents a copy of a daily newspaper for which it pays 10 cents. Unsold newspapers are returned for a refund of 5 cents a copy. The daily sales and corresponding probabilities are as follows:

Daily sales (no. of copies)	*Probability*
1000	0.4
2000	0.3
3000	0.2
4000	0.1

(a) Do the same as for Exercise 6–2.
(b) If unsold copies cannot be returned (that is, unsold copies are useless), what is the optimal order each day now?

6–7 Dr. Smith, the director of research for a laundry-products manufacturer, must decide whether to terminate a project for the development of a new intensified laundry detergent, or to request additional funds for its continuation. If the project is continued and turns out to be successful, there will be a net profit of $2 million to Dr. Smith's firm; if unsuccessful, there will be a loss of $1 million. If the project is discontinued, but another firm successfully develops the detergent, Dr. Smith's firm will sustain a loss of $0.4 million owing to competitive disadvantage; however, this action will also entail a gain of $0.7 million, accounted for by funds that would have been spent had the project been continued. Dr. Smith's research staff estimated that the odds are 3:2 against the success for this project.

(a) Present all the information above in a payoff table.
(b) What probabilities should be assigned, respectively, to the success and failure of the project?
(c) What is the optimal act according to data obtained in (a) and (b)?

6–8 Suppose our decision maker operates a hamburger stand in a large amusement park, and he must decide what price he should charge. Suppose he finally has narrowed his choice down to three possible prices: 50 cents, 60 cents, and 70 cents. By past experience he judges that the daily sales have a probability distribution at each of these prices as shown in Table 6–P1.

Table 6–P1

Price = $0.50		*Price* = $0.60		*Price* = $0.70	
D	$f(d_i)$	D	$f(d_i)$	D	$f(d_i)$
50	0.05	50	0.10	50	0.25
60	0.10	60	0.15	60	0.35
70	0.10	70	0.20	70	0.25
80	0.20	80	0.25	80	0.10
90	0.35	90	0.20	90	0.05
100	0.15	100	0.10	100	0.00
110	0.05	110	0.00	110	0.00
	1.00		1.00		1.00

(a) What observations can be made on the probability distributions in Table 6–P1 relative to price and quantity sold? Are these observations consistent with the economic theory of demand?
(b) If the total average cost is a constant quantity of $0.25, what price should the operator charge in view of his estimated sales and their corresponding probabilities? [*Hint*: Compute the Bayesian measure for each price and select.]

6–9 The constancy assumption of total average cost in the preceding exercise is quite unrealistic. In economic theory we have learned the average cost function is typically U-shaped, owing to the operation of the law of diminishing returns. Now, suppose the total-average-cost function of the hamburger-stand operator is as follows:

Quantity	*Total average cost*
50	$0.32
60	0.28
70	0.22
80	0.20
90	0.19
100	0.24
110	0.33

In view of this information, what is the optimal price decision for the operator in Exercise 6–8?

6–10 The service department of a hotel rents tuxedos to its guests at $15 per weekend. The hotel does not own its own tuxedos but leases them from a clothing firm at $10 each per weekend. The hotel must order on Monday the tuxedos it intends to lease the next weekend. According to past records, the number of tuxedos requested is a random variable with the relative frequencies shown in Table 6–P2. What is the optimal number of tuxedos to order each Monday?

Table 6–P2

Number of tuxedos requested	*Relative frequency*
5	0.10
6	0.10
7	0.15
8	0.20
9	0.25
10	0.15
11	0.05
	1.00

6–11 If you applied in Exercise 6–10 the special Bayesian decision form for a piecewise linear loss function, would you reach the same conclusion?

6–12 A toy-store manager must decide how many units of a certain make of electric train to stock for the coming Christmas season. The probability distribution for demand levels is estimated as follows:

Demand	*Probability*
10	0.05
15	0.05
20	0.15
25	0.20
30	0.25
35	0.20
40	0.10
	1.00

Each train set costs the store $30, and the retail price during the season is $50. Those not sold during the season must be marked down to $20 per set.

(a) Select the optimal act by using the special form of Bayesian strategy with a piecewise linear loss function.

(b) Determine the expected payoff for the optimal act.

6–13 A contractor wishes to submit a competitive bid for the modernization of an office building. Having studied the specifications, he estimates that the cost of the project can be represented by the following probability distribution:

Cost (c_i)	*Probability of cost* [$P(c_i)$]
$25,000	0.25
30,000	0.65
35,000	0.10

The contractor is now considering three possible bids. The probability of his getting the contract depends upon his bid, and according to his judgment the following data seem reasonable:

Bid (a_i)	*Probability of being awarded the contract* [$P(a_i)$]
$33,000	0.60
36,000	0.35
40,000	0.05

If the contractor is willing to use expected monetary returns as his decision rule, which of these three bids should he submit? [*Hint*: Whatever bid the contractor submits, he may or may not be awarded the contract. If he does not get it, his expected payoff will be zero. If he does get it, his conditional profits depend upon the difference between his bid amount and the possible construction costs. Therefore, three possible payoffs are associated with each bid. For instance, if he takes a_1 and the actual cost turns out to be c_1, his payoff will be $33,000 - 25,000 = \$8000$. The probability for this conditional profit to occur is the joint probability

$$P[X(a_1, c_1)] = P(a_1)P(c_1) = 0.60(0.25) = 0.15.$$

In the same manner, all the possible conditional profits and their associated probabilities can be calculated for each possible amount bid, and then the Bayesian measure for each bid can be determined.]

6–14 The management of a large canning factory is considering the possibility of modernizing, a_1, or not modernizing, a_2, its obsolescing though still operative canning process. This decision depends upon the uncertain events of future demand for its output. According to the estimates of the company's research staff, the payoffs (in millions per annum) net of modernization costs, of modernization or no modernization, relative to a high (θ_1) or a low (θ_2) demand, are as follows:

	State	
Act	θ_1	θ_2
a_1	3.0	0.5
a_2	2.0	1.8

Furthermore, it is the judgment of the research staff and sales personnel of the company that the odds are 11:9 in favor of a high demand. With these data, the management can already decide whether or not to modernize now. However, the sales manager suggests it is also possible to continue production with the existing process for another year and then to decide on modernization, when it will be known whether a high or a low demand will actually result. The drawback of this postponement is that modernization will then cost \$0.25 million more—i.e., net profits for modernization will be reduced by this amount from previous estimates.

(a) If management is to decide on modernization now, what is the optimal act?

(b) If management is to adopt a sequential decision procedure, how would it go about it?

(c) Which decision procedure is the best in terms of expected profits for this problem?

6–15 A French economist, M. Allais, posed the following pair of decision problems:

I:
- a_1: having a payoff of $1,000,000 for certain;
- a_2: having a payoff of $5,000,000 with a probability of 0.10, $1,000,000 with a probability of 0.89, and $0 with a probability of 0.01.

II:
- a_1: having payoffs of $5,000,000 with a probability of 0.10 and $0 with a probability of 0.90;
- a_2: having payoffs of $1,000,000 with a probability 0.11 and $0 with a probability of 0.89.

Can you explain why many people prefer a_1 over a_2 in each of these two situations?

6–16 A decision maker has a utility index of 0.1 for a loss of $1000 and 0.8 for a profit of $3000. He says he is indifferent between $400 for certain and a 0.6 chance at a $1000 loss and a 0.4 chance at $3000 profit. What is his utility index for $400? [The answer here is

$$\begin{aligned} U(400) &= U(-1000)(0.6) + U(3000)(0.4) \\ &= (0.1)(0.6) + (0.8)(0.4) \\ &= 0.36 \text{ utils.}] \end{aligned}$$

6–17 A manager has $U(\$18{,}000) = 10$ and $U(11{,}000) = 6$. He says he is indifferent between $15,000 for certain and a 0.6 chance at winning $18,000 and a 0.4 chance at winning $11,000. Explain why his $U(15{,}000) = 8.4$ utils.

6–18 Consider the following payoff matrix in a decision situation:

Table 6 P-3. Payoff Matrix

Act	θ_1 (0.4)	θ_2 (0.3)	θ_3 (0.3)
a_1	$5,000	$7,000	−$2,000
a_2	10,000	2,000	−3,000
a_3	8,500	8,500	−4,000

(a) Convert the payoffs into utility indices by using the utility function for Y in Figure 6–8 and assuming the horizontal scale is in thousands of dollars.

(b) What are the optimal Bayesian acts in terms of monetary payoffs and utility indices, respectively?

6–19 Assume that the utility function of a firm is $U(M) = 2M - 0.1(M^2)$, where M stands for millions of dollars. The firm is faced with two investment opportunities, a_1, and a_2, with the following payoff matrix:

Act	θ_1 (0.6)	θ_2 (0.4)
a_1	\$5M	\$3M
a_2	\$2M	\$8M

What is the expected utility for the optimal act?

6–20 For Exercise 6–19, what is the optimal act now if the firm's utility function is $U(M) = 5 + 0.2M$?

7

DECISION MAKING WITH SPECIAL PROBABILITY MODELS

We have noted that decision making under uncertainty requires a knowledge of the probability distribution of the state variable. Distributions of the decision variables often can be approximated by special probability models as those that have already been introduced. When this is the case, decision analysis becomes considerably simpler, because computations can then be made in accordance with the laws based upon appropriate probability models.

In this chapter we shall first discuss a sequential decision procedure called acceptance sampling, in which the binomial and Poisson probability laws are frequently employed. Next, we shall take up the special form of the Bayesian strategy for nonequivalent state-action problems under the assumptions of linear payoffs and normal states of nature. Then we shall introduce a waiting-line model by assuming that the decision variables follow the Poisson and exponential probability laws. Finally, we shall consider Monte Carlo simulation with reference to some waiting-line problems.

7–1 ACCEPTANCE SAMPLING: A SEQUENTIAL DECISION RULE

Often a purchaser must decide whether to accept or to reject an incoming lot of merchandise. One obvious scheme is to inspect the entire lot in order to determine whether, say, the proportion of defectives in the lot is at a level prescribed as satisfactory. This procedure, however, is generally rejected as impractical because it costs too much or takes too long. Instead, such decisions typically are made on the basis of sampling, and different acceptance sampling plans are evaluated in terms of *operating characteristic, OC, functions.* The OC function of an acceptance sampling plan provides probabilities of acceptance given various proportions of defective items in the lot. (For a more formal definition of an OC function, see Section 13–5.) Let us discuss acceptance sampling further via an example.

The U.S. Navy must decide whether or not to accept an incoming lot of 100,000 raincoats. The purchasing agent is considering to choose between the following two plans:

PLAN I Sample 50 with replacement and accept the lot if there is at most 1 defective raincoat in the sample.

PLAN II Sample 40 with replacement and accept the lot if there is no defective raincoat in the sample. If there are two or more defective raincoats in the sample, reject the lot. If there is only one defective raincoat in the sample, take a second sample of 20 with replacement and accept the lot if there is no defective raincoat in it; otherwise, reject the lot.

To evaluate these sampling plans, we note that the probability of acceptance follows the binomial law and it depends upon p, the proportion of defectives in the lot. Here, furthermore, p is expected to be quite small and n is rather large, so the binomial probability can be satisfactorily approximated by the Poisson law.

The operating characteristics for Plan I can be determined simply as follows:

If $p = 0$, then $\text{OC} = P(\text{acceptance}) = P(X \le 1) = 1$.

If $p = 0.005$, then $\mu = 50(0.005) = 0.25$ and

$$P(\text{acceptance}) \doteq P(1; 0.25) = 0.974.$$

If $p = 0.01$, then $\mu = 50(0.01) = 0.5$ and

$$P(\text{acceptance}) \doteq P(1; 0.5) = 0.910.$$

If $p = 0.02$, then $\mu = 50(0.02) = 1.0$ and

$$P(\text{acceptance}) \doteq P(1; 1.0) = 0.736.$$

If $p = 0.03$, then $\mu = 50(0.03) = 1.5$ and

$$P(\text{acceptance}) \doteq P(1; 1.5) = 0.558.$$

If $p = 0.04$, then $\mu = 50(0.04) = 2.0$ and

$$P(\text{acceptance}) \doteq P(1; 2.0) = 0.406.$$

If $p = 0.06$, then $\mu = 50(0.06) = 3.0$ and

$$P(\text{acceptance}) \doteq P(1; 3.0) = 0.199.$$

If $p = 0.08$, then $\mu = 50(0.08) = 4.0$ and

$$P(\text{acceptance}) \doteq P(1; 4.0) = 0.092.$$

If $p = 0.10$, then $\mu = 50(0.10) = 5$ and

$$P(\text{acceptance}) \doteq P(1; 5.0) = 0.04.$$

The graph for an OC function is called the OC curve. The OC curve for Plan I under consideration is sketched with the values obtained above and is labeled as OC(1) in Figure 7–1.

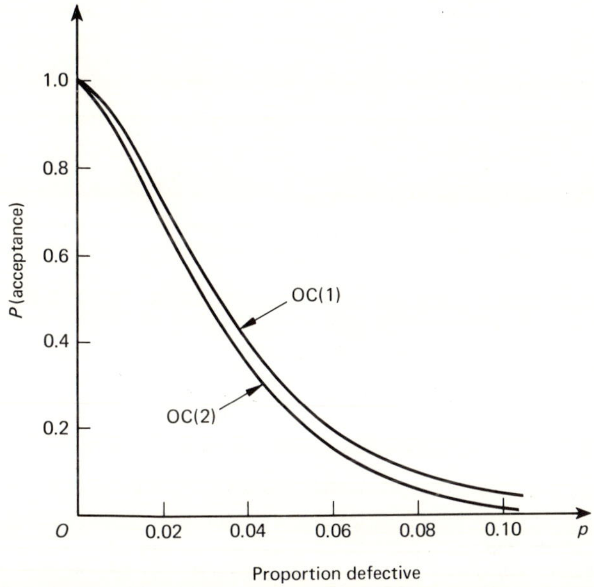

Figure 7–1 Operating Characteristics for Two Acceptance Plans

Turning now to Plan II, we note that the probability of acceptance is the sum of the probabilities for these two mutually exclusive events: (1) the event that there are no defective units in the first sample of 40 and (2) the compound event that there is only one defective unit in the first sample of 40 and that there are no defective units in the second sample of 20. Namely, for Plan II,

$$\begin{aligned} \text{OC} &= P(\text{acceptance}) \\ &= P(X = 0 \mid 40p) + P(X = 1 \mid 40p)P(X = 0 \mid 20p). \end{aligned}$$

As before, if $p = 0$, then $P(\text{acceptance}) = 1$.

If $p = 0.005$, then

$$\begin{aligned} P(\text{acceptance}) &\doteq p(0;0.2) + p(1;0.2)p(0;0.1) \\ &= 0.819 + (0.164)(0.905) \\ &= 0.967. \end{aligned}$$

If $p = 0.01$, then

$$\begin{aligned} P(\text{acceptance}) &\doteq p(0;0.4) + p(1;0.4)p(0;0.2) \\ &= 0.670 + (0.268)(0.819) \\ &= 0.889. \end{aligned}$$

If $p = 0.03$, then

$$\begin{aligned} P(\text{acceptance}) &\doteq p(0;1.2) + p(1;1.2)p(0;0.6) \\ &= 0.301 + (0.361)(0.549) \\ &= 0.499. \end{aligned}$$

If $p = 0.06$, then

$$\begin{aligned} P(\text{acceptance}) &\doteq p(0;2.4) + p(1;2.4)p(0;1.2) \\ &= 0.091 + (0.218)(0.301) \\ &= 0.157. \end{aligned}$$

If $p = 0.08$, then

$$\begin{aligned} P(\text{acceptance}) &\doteq p(0;3.2) + p(1;3.2)p(0;1.6) \\ &= 0.041 + (0.130)(0.202) \\ &= 0.067 \end{aligned}$$

If $p = 0.10$, then

$$\begin{aligned} P(\text{acceptance}) &\doteq p(0;4.0) + p(1;4.0)p(0;2.0) \\ &= 0.018 + (0.073)(0.135) \\ &= 0.028. \end{aligned}$$

The OC curve for Plan II is identified as OC(2) in Figure 7–1.

The quality of a given acceptance sampling plan is evaluated in terms of the predetermined tolerance limit together with the plan's OC curve. For our illustrative example, if the Navy is willing to tolerate, say, no more than 6 percent defective in the lot, then neither plan is entirely satisfactory. "Good" lots, those with $p < 0.06$, have high probabilities of being accepted for both plans. This is certainly desirable. However, even if a lot is as bad as 8 percent defective, there is still a probability of 0.092 of acceptance according to Plan I and a corresponding probability of 0.067 for Plan II. In other words, these two plans would rarely reject a good lot but would undoubtedly accept many bad lots. However, it is only necessary to sample 50 items out of 100,000 to obtain a probability as small as 0.199 for accepting a lot that is 6 percent defective, or a probability as high as 0.801 for rejecting a lot that is 6 percent defective. Also, it is only necessary to sample 60 out of 100,000 items for us to reject a lot with 6 percent defectives at a high probability of 0.843.

Next, it may be observed that while both sampling plans leave something to be desired, Plan II is evidently superior to Plan I. Both plans have about the same probabilities of accepting good lots but Plan II has lower probabilities of accepting bad lots. This is clearly indicated by the OC curves of these two plans. The main point of our illustration here is that a plan such as Plan II, known as *multiple-sampling plan*, is often at least as good as any *single-sampling plan* such as Plan I, but less expensive. With Plan I, we always have to sample 50 items to reach a decision; with Plan II, we can often reach a decision by sampling only 40 items. Of course, decisions with Plan II may occasionally require as many as 60 items; nonetheless, in the long run over a large number of lots, such a multiple-sampling plan permits us to reach decisions by sampling fewer items. These comments are true only if the fixed costs of sampling are about the same for the two kinds of plan or if variable costs are relatively more important than fixed costs, as is usually the case in acceptance sampling or quality control. Thus, typically, acceptance sampling is of the multiple variety.

It is quite common for a multiple-sampling plan to include three or more rather small samples of the same size, such as 10, 20, or 30 each.

Finally, it is important to note that the OC function is incapable of establishing the true state of nature—that is, the true proportion of defectives in a lot. It reveals, instead, the operating characteristics of a sampling plan, in the sense that if the lot contains a certain proportion of defectives, then the probability of acceptance would be at a given level. Thus, an ideal OC curve would remain close to unity for defective proportions that are equal to or less than the tolerance limit, and would fall rapidly beyond such proportions. The larger the sample size, or the combined sample size, the closer the OC curve approaches this ideal. This tendency is shown by Figure 7–2. This figure displays a set of operating characteristics for several sampling plans, all designed to have a probability of 0.5 of rejection when the proportion of defects is 0.07. As n increases, with $p = 0.07$, the operation characteristic becomes steeper at $p = 0.07$. When $n = 1000$, the OC curve differs but slightly from the ideal shape of an OC curve.

The curves in Figure 7–2 were made with the aid of large binomial tables, but, as we shall see in Chapter 11, the normal approximation could have been used for larger values of n. Plans such as those shown by the figure can be used not only to help a buyer decide whether the lot of material he purchases has the quality the seller claims, but also to help in the control of a production process.

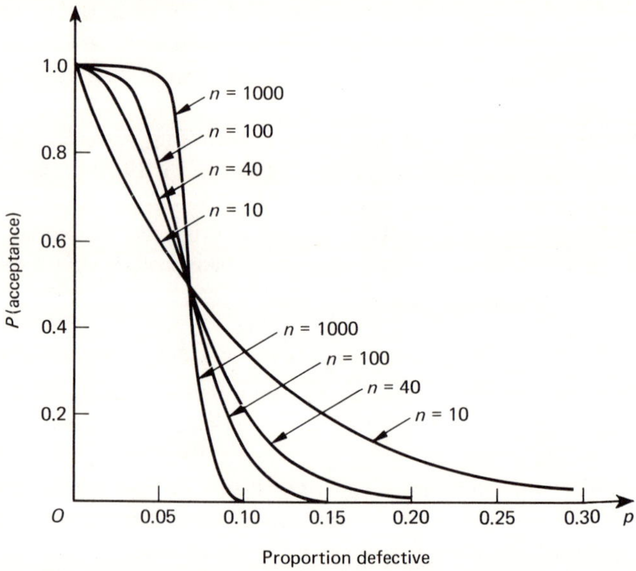

Figure 7–2 Operating Characteristics for Several Acceptance Plans

7–2 BAYESIAN STRATEGY FOR NONEQUIVALENT STATE-ACTION PROBLEMS

In Chapter 6 we noted that the application of the Bayesian decision criterion can be simplified under certain conditions, and we considered special forms of the Bayesian strategy for equivalent state-action problems with piecewise linear, quadratic, and constant functions. Now, we shall take up the case of applying the Bayesian strategy to nonequivalent state-action problems. Recall that in a nonequivalent state-action problem, one action may be optimal over more than one state. This happens especially when the state random variable is continuous. In general, for nonequivalent state-action problems, the expected losses for various actions must be evaluated separately. Fortunately, the Bayesian strategy takes a simple special form if the payoff functions over the state random variable are linear. The relatively simple but quite useful case is where there are only two actions and a continuous state variable. For this case, if the state variable can also be assumed to have a normal distribution, then the expected value of perfect information, or the expected opportunity loss, can also be easily determined by a simple expression. Let us now proceed to derive first the special form of Bayesian decision criterion under these assumptions.

Special Form of Bayesian Strategy for Nonequivalent State-Action Problems with Linear Payoffs

The special form of the Bayesian strategy for two-action and linear payoff problems involving a single density function over the states can be derived quite easily. In two-action problems, the linear payoff functions over the state variable X, with values of x, can be written in the following general form:

$$R(a_1, X) = R_1 = k_1 + b_1 x \qquad \text{for } a_1,$$

$$R(a_2, X) = R_2 = k_2 + b_2 x \qquad \text{for } a_2.$$

In these expressions, k_i and b_i are constants with $k_1 \leq k_2$ and with $b_1 \geq b_2$. These restrictions indicate that, as shown by Figure 7-3, the graph for R_1 crosses that for R_2 from below.

The break-even point, denoted by x_b, is the value of the state variable at which the decision maker would be indifferent between a_1 and a_2, and it is found by equating R_1 and R_2; that is,

$$k_1 + b_1 x_b = k_2 + b_2 x_b$$

or

$$x_b = \frac{k_2 - k_1}{b_1 - b_2}.$$

Now, as long as $E(X) > x_b$, the expected payoff of a_1 would exceed that for a_2. That is, if

$$E(X) > x_b,$$

then

$$E(X) > \frac{k_2 - k_1}{b_1 - b_2}.$$

Noting that $b_1 - b_2 > 0$, we must have

$$E(X)(b_1 - b_2) > (k_2 - k_1),$$

or

$$[b_1 E(X) + k_1] > [b_2 E(X) + k_2].$$

This last result implies that

$$E(b_1 X + k_1) > E(b_2 X + k_2)$$

and, therefore,

$$E(R_1) > E(R_2).$$

Here, R_1 and R_2 are the payoffs for a_1 and a_2, respectively. By following similar procedure, we could obviously conclude that if $E(X) < x_b$, then $E(R_1) < E(R_2)$.

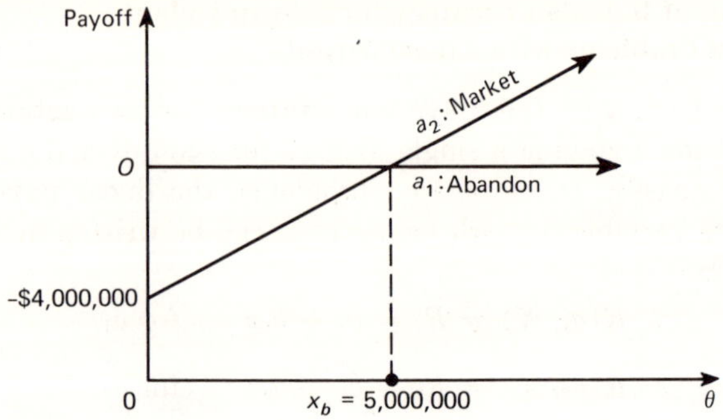

Figure 7–3 Payoff Functions: New-Product Example

The foregoing results lead to the special form of the Bayesian decision criterion for nonequivalent problems with two-action and linear payoffs: the optimal act is the one for which the payoff is a maximum for the expected value of the state variable. That is, the optimal act is the one for which

$$\max_a [R(a, E(X))]. \tag{7–1a}$$

Alternatively, the special form directs the decision maker to select the act for which the loss is a minimum for the expected value of the state variable—that is, the act for which

$$\min_a [L(a, E(X))]. \tag{7–1b}$$

Thus, in this case, the probability law can be replaced by its expectation. The application of this special form of the Bayesian strategy is very simple. All that we need to determine, given the payoff or loss functions, is the mean value of the state random variable. In other words, by this special form, the expectation of the state variable controls, and the particular pattern of the probability distribution of the state variable is of no significance.

Introduction of a New Product

As an illustration of the application of (7–1), let us consider an example. The research and development department of a manufacturing firm has just perfected a new product. The production and marketing of this new product entail a total fixed cost, for equipment and promotion, of $4 million. It has also been estimated that the average variable cost is $1.20 and that a price of $2 can be charged. The management thus faces a decision situation with the following alternative courses of action:

a_1: market the product,
a_2: abandon the product.

The payoff for a_2 is obviously 0. To determine the payoff function for a_1, we note that, according to the statement of the problem, both the revenue and cost functions are linear:

$$\text{total revenue} = \$2(x);$$

$$\text{total cost} = \$4{,}000{,}000 + \$1.20(x).$$

From these we obtain a payoff function for a_1:

$$\begin{aligned} R(a_1, X) &= \$2x - (\$4{,}000{,}000 + 1.2x) \\ &= -\$4{,}000{,}000 + \$0.8x. \end{aligned}$$

The break-even level of sales becomes

$$\begin{aligned} x_b &= \frac{0 - (-4{,}000{,}000)}{0.8 - 0} \\ &= 5{,}000{,}000. \end{aligned}$$

If the sales level is above 5,000,000, the product is profitable and should be marketed; otherwise, it should be abandoned. The payoff function for each action is shown by Figure 7–3. Each payoff is linear over all the values of the state random variable—sales volume.

Now, since the payoff functions are linear, the corresponding opportunity-loss functions must also be linear. In determining the loss functions, the value of x_b plays a key role. If it is decided to market the product in the case that $X > x_b$, then the optimal decision has been made and there is no loss. If, however, $X \leq x_b$, the opportunity loss becomes b for each unit of sales in short of x_b. Thus, for marketing the product, we have

$$\begin{aligned} L(a_1, X) &= \$0.8(x_b - x) && \text{for } X \leq x_b = 5{,}000{,}000, \\ &= 0 && \text{for } X > x_b = 5{,}000{,}000. \end{aligned}$$

Similarly, if it is decided to abandon the product, in the case that $X \leq x_b$ no loss will result, since a_2 is the optimal act; but in the case that $X > x_b$ a loss of b will be incurred for every unit of sales above x_b. Hence,

$$\begin{aligned} L(a_2, X) &= \$0.8(x - x_b) && \text{for } X > x_b = 5{,}000{,}000, \\ &= 0 && \text{for } X \leq x_b = 5{,}000{,}000. \end{aligned}$$

Note that loss functions must be expressed in equation forms because, for a continuous state random variable, there is an infinite number of possible states.

The opportunity-loss functions for our example are graphed in Figure 7–4. It is interesting to note that the opportunity losses are symmetric about $x_b = 5{,}000{,}000$. This means that the loss is just as serious if a_1 is selected but the sales level is c units below x_b as it is if a_2 is chosen but the sales level is c units above x_b. The reader may also be warned that these linear functions are different from the piecewise linear loss function introduced in Chapter 6, even though their graphs look alike—both are V-shaped. The former consist of two different loss functions, while the latter is a single loss function.

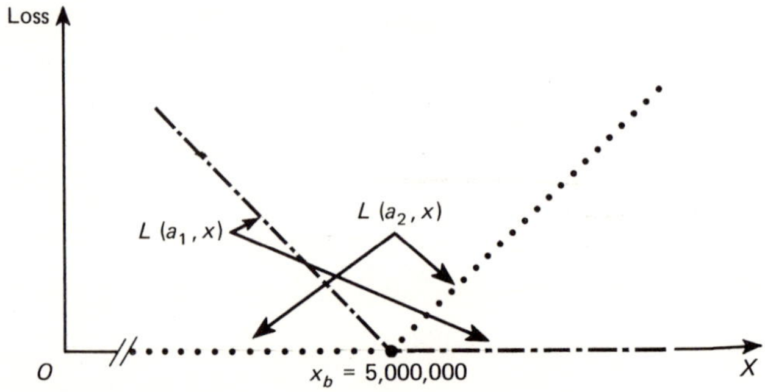

Figure 7–4 Loss Functions: New-Product Example

Now, if we wish to solve this problem by applying Bayesian strategy, all we need to know is the decision maker's subjective judgment on the expected value of the sales volume, since the specific form of the density of the state variable is immaterial here. Suppose the decision maker in question has reason to believe that the mean sales will be about 6,000,000 units; then we have $E(X) > x_b$. According to the special form of the Bayesian strategy, we conclude then that a_1 is the optimal act, and for this act

$$\begin{aligned} E[R(a_1, X)] &= -4{,}000{,}000 + 0.8(6{,}000{,}000) \\ &= \$800{,}000.00 \end{aligned}$$

in contrast to

$$E[R(a_2, X)] = 0.$$

Normally Distributed States and Expected Opportunity Loss

The particular shape of the density function for the state random variable is of no consequence, and we need only the expectation of this density, if our purpose is merely to determine the optimal act in a nonequivalent state-action problem with two actions and linear payoffs. However, if a special probability law could be assumed for the state random variable, then it would become much easier to evaluate the expected opportunity loss, or the expected value of perfect information, for the optimal act.

This is especially true if the state random variable can be assumed to be normal. The selection of a particular density function depends upon the decision maker's judgment and knowledge of the state variable. If the general shape of the probability law governing the state variable is judged to be of some special model, such as the uniform, exponential, or the normal distribution, then its parameter or parameters can usually be determined either from past data or by evaluating some relatively simple properties of the functional form of the probability law.

For instance, if the decision maker has reason to believe that state random variable is normally distributed, that its most probable value is, say, 100, and that the chances are 50:50 that the values of the state variable will fall within the range from, say, 75 to 125, then we may say that the state random variable is distributed normally with a mean of 100 and a standard deviation of approximately 37. We deduce that $\mu = 100$ because a normal distribution is symmetric and unimodal, and the most probable value is also the mean. We deduce that $\sigma \doteq 37$, since, for a normal distribution, approximately 50 percent of the area is contained within the range of $\mu \pm 0.67\sigma$. That is,

$$Z = 0.67 = \frac{125 - 100}{\sigma}$$

or

$$\sigma = \frac{25}{0.67} \doteq 37.$$

Returning to our illustrative example, assume that the decision maker is satisfied that the probabilities assigned to various states of sales obey the normal probability law with $\mu = 6{,}000{,}000$ and $\sigma = 2{,}000{,}000$. Then we could reproduce the information in Figure 7–4 in Figure 7–5 with the rele-

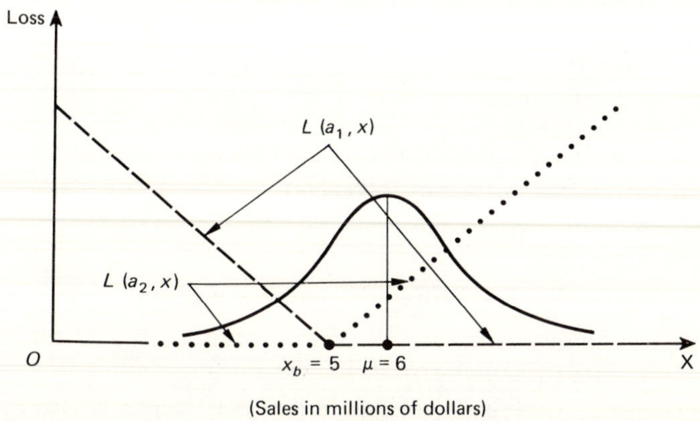

Figure 7–5 Loss Functions and Subjective Normal Probability Law for the State Variable X

vant density curve superimposed over the states. From this figure, the expected loss from marketing the product can be thought of as the loss for each value of the state variable less than x_b multiplied by the probability of that value, which is the area under the density curve to the left of $L(a_1, X)$. Similarly, the expected loss from abandoning the product is the product of $L(a_2, X)$ and the area under the density curve on the right of $L(a_2, X)$ for sales above x_b. Intuitively we see that the expected loss for a_1 is less than that for a_2; therefore, a_1 should be taken.

As can be shown, in a two-action problem involving linear payoffs and a normal subjective density function for the state variable, the expected opportunity loss from the optimal act—and therefore also the expected value of perfect information, EVPI—is given by the following simple expression:

$$\text{EVPI} = b\sigma L(D), \tag{7–2}$$

where

b = the absolute value of the slope of either segment of the loss function where the loss exceeds zero. (since these two segments are symmetrical, $|b| = |-b|$),

σ = the standard deviation of the subjective normal probability function for the state variable,

$L(D)$ = the normal loss integral, or the normal loss function, whose value depends upon the value of D. The value of D, in turn, indicates the relative position of the break-even value of the state variable in terms of σ; it is defined as

$$D = \left| \frac{x_b - \mu}{\sigma} \right|, \tag{7–3}$$

where μ is the mean of the subjective prior distribution. With the value of D determined, the value of $L(D)$ is found by using Table B–8 in Appendix B.

For our illustrative example we have

$$D = \left| \frac{5{,}000{,}000 - 6{,}000{,}000}{2{,}000{,}000} \right|$$
$$= 0.5$$

and $L(D) = 0.1978$ (from Table B–8).

Hence,

$$\begin{aligned} \text{EVPI} &= b\sigma L(D) \\ &= 0.8(2{,}000{,}000)(0.1978) \\ &= \$316{,}480.00. \end{aligned}$$

This is a rather large value. Thus, if the decision could be delayed until more information were available, the expected loss might be reduced. We must bear in mind, however, that additional information can be made available only at a cost, and, since even perfect information can not eliminate loss completely, the decision maker should never be willing to pay more than the expected value of perfect information. Since, as we shall see in future chapters, sampling or research seldom, if ever, yields perfect information, the amount spent for gathering information should be always less than EVPI.

7-3 A QUEUING MODEL

Basic Queuing Models

A. K. Erlang's effort to analyze telephone traffic congestion, with the aim of meeting the randomly arising demand for services of the Copenhagen automatic telephone system in 1909, produced a new theory that has become known as *queuing*, or *waiting-line, theory*. This theory is one of the most valuable tools in management science, because many managerial problems can be characterized as "arrival and departure" problems.

In queuing problems, we often speak about *customers*—such as people waiting for clear telephone lines, machines waiting to be repaired, and planes waiting to land—and *service stations*—such as tables at a restaurant, repairmen in a shop, runways at an airfield, and so on. Queuing problems often involve a variable rate of arrival of customers requiring some type of service, and a variable rate of completing the service at the service station.

When we talk about waiting lines we may refer to those created either by customers or by the service stations. Customers may wait in a line simply because existing facilities are inadequate to meet the demand for service; in this case, the waiting line tends to be explosive—that is, to become longer and longer as time goes on. Service stations may be waiting because the existing facilities are excessive relative to customer's demand; in this case, service stations could remain idle much of the time. Customers may be waiting temporarily, even when service facilities are adequate, because previously arrived customers are being attended to. Service stations may encounter temporary waiting when, although facilities are adequate in the long run, there is an occasional shortage of demand owing to happenstance. These last two cases typify a balanced situation constantly tending toward equilibrium, or a stable situation.

In queuing theory we usually call a group of physical units, integrated in such a way that they can operate in unison with a series of organized operations, a *system*. Queuing theory seeks a solution to the problem of waiting by first predicting the behavior of the system of waiting. A solution to a waiting problem, however, is concerned not only with minimizing the time the customers spend in the system but also with minimizing the costs of waiting time; that is, a waiting problem is generally solved in terms of reducing the total costs of those who request, and those who render, the service.

There is a great variety of queuing models; some of the basic ones are illustrated by Figure 7–6. In this section we shall present a simple queuing model, called the single-service-station model, as typified by Figure 7–6(a).

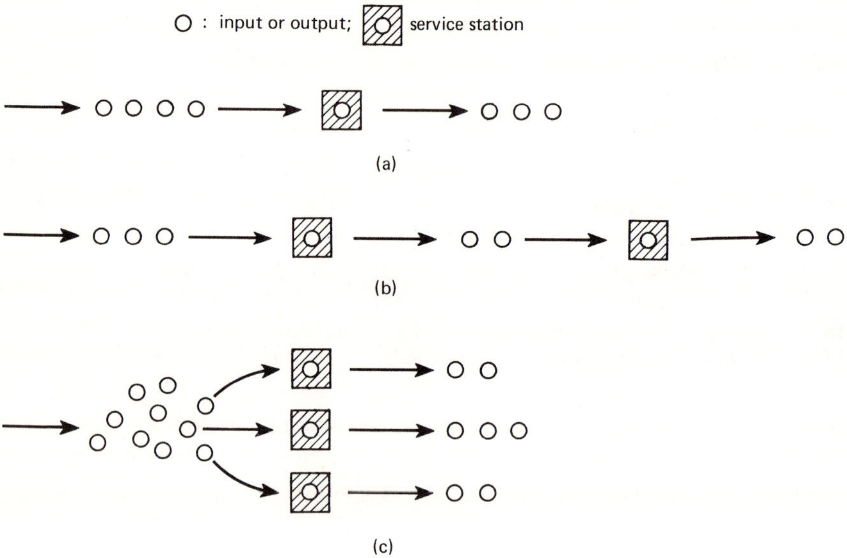

Figure 7–6 Some Queuing Models: (a) Single-Channel, (b) Multistage, (c) Multichannel

The Single-Service-Station Queuing Model

Analytically, this model is constructed with the following assumptions:

(1) ARRIVAL OF CUSTOMERS, OR INPUTS: We shall assume that the arrivals occur at random and that the probability of an arrival during any time interval of fixed length remains constant, regardless of what has happened previously and the length of the waiting line. The reader may recall that, under these assumptions, the arrivals would obey the Poisson probability law. We shall denote the *mean arrival rate*, or the *average number of arrivals*, per unit of time as μ, which is independent of time. Furthermore, the reciprocal, $1/\mu$, is the *mean length of time intervals between two successive arrivals*. In other words, we postulate a Poisson arrival pattern and an exponential pattern of interarrival time.

(2) QUEUE DISCIPLINE, or PRIORITY RULE: When a customer arrives at the system, he usually has to wait before he receives service. His departure is influenced, among other things, by the *queue discipline*, the established rule by which customers waiting in the line are served. We shall assume the time-honored rule (not always the best) of *first come, first served.* Our rule also includes the requirement that no customer in the system will depart before he receives service.

(3) OUTPUT: This criterion refers to the number of service stations available, the capacity of each service station, and the distribution of service time. In the discussion that follows, we shall assume that there is only one service station (one repairman, one tunnel, one information clerk, and so on).

(4) SERVICE TIME: Under the assumption that the arrival of customers follows the Poisson law, then the service time can be assumed to be an exponential variable. Recall that the Poisson and exponential variables always go together. The exponential postulate for service time holds for a majority of cases, even when the assumption that the probability that servicing will end at any instant is independent of what has happened previously is rather questionable. When the servicing time is exponentially distributed, the mean *service rate*, or the *expected number of services performed* per unit of time, may be denoted as β. Note also that the reciprocal $1/\beta$ is the *mean service time*, which is identical with the standard deviation of service time.

When the previous conditions are satisfied, we have a mathematical model for single-service-station, first-come first-served, waiting-line problems. Governing equations for this model can then be derived or established.

Equations Governing Queues for the Single-Service-Station Model

There are a number of important properties we must know in evaluating a queuing system. First, can the service station handle the customer's demand? Obviously, we can find out by comparing μ (the mean number of arrivals per unit of time) with β (the average number of services per unit of time). It is easy to see that if more customers arrive than can be serviced per unit of time, the waiting line will increase continuously and the system will eventually break down. When $\mu > \beta$, we are headed for trouble; drastic measures are called for without further analysis. Thus, our system must have $\mu < \beta$ in order to be workable. The ratio

$$U = \frac{\mu}{\beta} \tag{7–4}$$

is called the *utilization parameter*, since it measures the degree to which the capacity of the service station is utilized. For example, if customers of a repair shop arrive at a rate of 7 per hour, and they can be processed at an average rate of 10 per hour, then $U = 0.7$. It means that the repair shop is kept busy 70 percent of the time. The utilization parameter can also be interpreted as the *proportion of arrivals who must wait*. If the service station is kept busy 70 percent of the time when $U = 0.7$, then the probability that a customer must wait for service upon his random arrival must be 0.7.

Next we may ask: What is the probability that the service station remains idle? Clearly, if U is the proportion of time that the service station is kept busy, the *proportion of time it remains idle* must be

$$P_0 = 1 - U. \tag{7–5}$$

As P_0 is defined, it must also be the *probability that a customer*, upon his random arrival, *will not have to wait* at all for service. This definition also suggests that the *probability of having one customer in the waiting line* (excluding the one being serviced) is

$$P_1 = UP_0. \tag{7–6}$$

We may now extend (7–6) to establish the *probability of having a waiting line of length n*, for $n \geq 0$, by substituting for the P_i's the preceding P_{i-1} up to P_0. That is,

$$\begin{aligned} P_1 &= UP_{i-1} = P_0(U), \\ P_2 &= UP_1 = (U)(U)P_0 = P_0(U)^2, \\ P_3 &- UP_2 - P_0(U)^3, \end{aligned}$$

and so on. Generalizing,

$$\begin{aligned} P_n &= P_0(U)^n \\ &= (1 - U)(U)^n. \end{aligned} \tag{7–7}$$

In addition to $1/\mu$ and $1/\beta$, various expected values of a waiting line are of importance for solving queuing problems. For our present model, they are:

(1) *The mean number of customers in the system*, $E(n)$, both waiting and in service:

$$E(n) = \frac{U}{1 - U} = \frac{\mu/\beta}{1 - \mu/\beta} = \frac{\mu}{\beta - \mu}. \tag{7–8}$$

(2) *The mean length of the waiting line, $E(L)$, excluding the customer in service:*

$$E(L) = \frac{\mu^2}{\beta(\beta - \mu)}. \tag{7–9}$$

(3) *The average waiting time of an arrival in the queue, $E(W)$:*

$$E(W) = \frac{\mu}{\beta(\beta - \mu)}. \tag{7–10}$$

(4) *The average time an arrival spends in the system, $E(T)$:*

$$E(T) = \frac{1}{\beta - \mu}. \tag{7–11}$$

With these eight basic equations, we can answer many questions concerning waiting lines and obtain solutions to many queuing problems. Let us now turn to some illustrative applications of this simple queuing model.

Applications of the Single-Service-Station Queuing Model

First let us consider the problem of determining the number of phone booths. The owner of a restaurant observed that, in the long run, customers used his single phone booth every five minutes during rush hours, and the average length of conversation was four minutes. Suppose that the arrivals follow the Poisson pattern and that the service time is exponentially distributed; then we can outline the following relevant values of this system:

(1) The average number of arrivals and the average number of servicings, respectively, are

$$\mu = \tfrac{1}{5} = 0.20 \quad \text{arrival per minute,}$$

$$\beta = \tfrac{1}{4} = 0.25 \quad \text{servicing per minute.}$$

(2) With the values of μ and β determined, we see that the phone both is kept busy

$$U = \frac{0.20}{0.25} - 0.80$$

or 80 percent of the time.

(3) The probability of forming a queue of length n with $n = 0, 1, 2, \ldots$ becomes

$$P_0 = (1 - U)(U)^0 = (1 - 0.8)(0.8)^0 = 0.2,$$

$$P_1 = (1 - 0.8)(0.8) = 0.16,$$

$$P_2 = (1 - 0.8)(0.8)^2 = 0.128,$$

$$P_3 = (0.2)(0.8)^3 = 0.1024,$$

and so on. Thus, when a customer arrives at the booth he has a probability of 0.2 of making the phone call immediately, a probability of 0.16 of standing in the queue alone, a probability of 0.128 of being second in the waiting line, and so on. In general, the probability that a customer will be in the nth position decreases when n increases.

(4) The expected number of customers in the system is

$$E(n) = \frac{0.2}{0.25 - 0.2} = 4 \text{ persons.}$$

In general, $E(n)$ is linearly proportional to U, since, as can be seen, $E(n) = U/(1 - U)$. When U approaches unity, $E(n)$ grows rapidly and approaches infinity.

(5) The average length of the queue, excluding the one in service, is

$$E(L) = \frac{(0.2)^2}{0.25(0.25 - 0.2)} = 3.2 \text{ persons.}$$

Note that $E(n)$ differs from $E(L)$ in that the former includes, but the latter does not include, the customer in service. We note also that $E(L) = E(n)U$.

(6) The average time a customer spends in the system is

$$E(T) = \frac{1}{0.25 - 0.2} = 20 \text{ minutes.}$$

(7) The average waiting time a customer spends in the queue is

$$E(W) = \frac{0.2}{0.25(0.25 - 0.2)} = 16 \text{ minutes.}$$

It follows from the definitions that $E(W) = E(T)U$. We can also state, in view of our present illustration, that the average time a customer spends in the system is the sum of the average waiting time in the queue and the average time a customer spends in service.

What conclusions can we reach from the foregoing calculations? We observe that the value of U is rather high and, as a consequence, $E(n)$ and $E(T)$ are quite large. In order to reduce the traffic intensity and waiting time, the owner perhaps should consider the installation of a second telephone booth if space is available in the restaurant.

In our second illustration, a manager has to decide which of two repairmen, X and Y, to hire. The frequency of machine breakdown in his shop is known to follow the Poisson law, with an average of one breakdown per hour. The nonproductive cost of each idle machine is estimated as \$25 per hour. Repairman X asks for \$20, and Y for \$12, per hour. It is also known that the average repair rates of X and Y are 1.8 and 1.2 machines per hour, respectively. The manager would like to know if it is wise to hire the more expensive but faster X.

To solve this problem, we only need to compare the total expected daily cost for both repairmen. This cost is clearly the total wage per day plus the cost of those machines waiting to be, and being, repaired. The expected nonproductive time per hour is the mean repair time multiplied by the expected number of machines in the system (i.e., the repair facility), both those being repaired and those in the queue. We note also that the mean number of machine arrivals is $\mu = 1$ and that the service rate per hour is $\beta = 1.8$ for X, and $\beta = 1.2$ for Y. Thus, the expected numbers of machines in the system for X and Y, respectively, are

$$E(n) = \frac{1}{1.8 - 1} = 1.25 \qquad \text{for X,}$$

and

$$E(n) = \frac{1}{1.2 - 1} = 5.00 \qquad \text{for Y.}$$

Thus, if X is hired, the expected nonproductive machine time in an 8-hour day is $(1.25)(8) = 10$. The total daily cost in this case becomes

$$\text{labor cost} + \text{cost of nonproductive machine time} = (8)(\$20) + (10)(\$25) = \$410.$$

If Y is hired, the nonproductive machine time per day is $(5)(8) = 40$, and the total daily cost becomes

$$(8)(\$12) + (40)(\$25) = \$1096.$$

It is obvious that the cost of the cheaper repairman would be more than twice that of the more expensive repairman. The right decision is for the manager to hire X.

As our last example, let us consider an investment decision. A firm that manufactures a high-quality product has an average of 24 pieces of equipment to be repaired each week. The probability of the breakdown of a piece of equipment is approximately constant, so that the arrivals are distributed as under the Poisson law. The repair time is close to an exponential distribution. The opportunity cost of having an equipment breakdown is estimated as \$150 per day.

The firm has two repair facilities, A and B, from which to choose. Facility A requires \$20,000 to install, facility B, \$60,000. Operation of each facility entails an annual labor cost of \$8000. Furthermore, A can repair machines at a rate of 30 per week, and B at a rate of 60. Finally, both investments are estimated to have economic lives of four years. For simplicity we shall ignore interest charges and returns to investment. Under these circumstances, which facility should the firm purchase?

The solution can be obtained by comparing the total annual costs of the two facilities. For each the total annual cost consists of three parts: (1) annual capital-recovery cost, which is one-fourth of the total investment expenditure, (2) annual labor cost, and (3) annual cost of lost production-equipment time. Component (3) may be thought to be based on the sum of time waiting to be repaired and time being repaired. The former is to be computed on the basis of $E(W)$—the mean waiting time of an arrival—and the latter on the basis of the average service time per repair, which is the reciprocal of the mean service rate per unit of time, $1/\beta$.

Now, for facility A, we have

(1) Annual capital recovery cost = \$5000.
(2) Annual labor cost = \$8000.
(3) Cost of lost production-equipment time:
  (a) time waiting to be repaired:
    (i) average waiting time of an arrival:

$$E(W) = \frac{24}{30(30 - 24)} = 0.133333 \text{ week.}$$

    (ii) Total waiting time per year = (24)(52)(0.133333) = 166.4 weeks.
  (b) time being repaired:
    (i) mean time per repair = $1/\beta = 1/30$ week.
    (ii) total repair time per year = (24) (52) (1/30) = 41.6 weeks.
  (c) total lost time per year = 166.4 + 41.6 = 208 weeks.
  (d) total cost of lost production-equipment time = (5) (\$150) (208) = \$156,000 (where "5" refers to a 5-day week).
(4) Total annual cost for facility A = 5000 + 8000 + 156,000 = \$169,000.

For facility B, we have

(1) Annual capital recovery cost = \$15,000.
(2) Annual labor cost = \$8000.

(3) Cost of lost production-equipment time:

(a) time waiting to be repaired:

(i) $E(W) = \dfrac{24}{60(60 - 24)} = 0.011111$ week.

(ii) total waiting time per year = (24)(52)(0.011111) = 13.87 weeks.

(b) time being repaired:

(i) mean time per repair = $\frac{1}{60}$ week.

(ii) total repair time per year = (24) (52) ($\frac{1}{60}$) = 20.8 weeks.

(c) total lost time per year = 13.87 + 20.8 = 34.67 weeks.

(d) total cost of lost production-equipment time = (5)($150) (34.67) = $26,002.50.

(4) Total annual cost for facility B = 15,000 + 8000 + 26,002.50 = $49,002.50.

Thus, in this case the more expensive investment proves to be less costly in the long run. The firm should purchase facility B in order to minimize total cost incidental to repair of production equipment.

7-4 MONTE CARLO SIMULATION

We have now discovered the essence of queuing problems. When customers need services at arbitrary moments, two unpleasant situations may arise: either a customer has to wait to be served, or a service station is idle. Both cost money. The problem is to find the optimum size of service capacity, or the optimum number of service stations, that will minimize costs, given certain probabilities of arrivals and servicing time.

Solutions have been given to a number of waiting problems, mathematically, under the assumptions that arrivals follow the Poisson probability law and that the service time is exponentially distributed. With such a judiciously selected set of premises, we have presented computations that might have given the impression that queuing problems are quite simple. However, our assumptions were deliberately chosen to make calculations amenable to relatively simple algebraic manipulations. When models become more complex and realistic, the level of mathematics required becomes quite involved. A queuing problem may even be so complicated that it frustrates a competent mathematician. In such situations, the best tactic is to run a "simulation" study of the problem. We shall call such a study a *Monte Carlo simulation*, because simulation is a statistical method related to chance probabilities. Monte Carlo simulation can be easily adapted to many situations, provided that alternatives can be specified quantitatively and that data required can be predicted with acceptable confidence.

Monte Carlo is a process of solving a problem by simulating a process with random-number generators. Its application has only two basic requirements. First, we must have a *model* that represents a reality as we see it. The model here is nothing more than the probability distribution of the variable under consideration. The significant merit of simulation is that it can be applied even when the probability distributions cannot be expressed explicitly in any of the theoretical forms that have been introduced in this text. All that is required is a table or a graph of a cumulative probability distribution of a variable, which may be obtained either directly, by observation (such as an appropriate survey to determine the actual behavior of, say, arrivals), or indirectly by the use of past records.

The second requirement is a *mechanism* to simulate the model. The mechanism can be any random-number generator such as a pair of dice, a spinning pointer, a roulette wheel, a table of random digits, or a properly instructed high-speed computer. In short, the Monte Carlo method is to simulate, via random devices, real-world situations that are probabilistic in nature. Let us now consider two examples of this method.

The Optimal Number of Checkout Counters

The manager of the children's clothing department of a new department store wishes to determine the optimal number of checkout counters for his department. He wishes to do so via simulated sampling. After some thought, he decides to simulate 30 one-minute periods. With this designation, there can be at most one arrival at the counter or counters in any single minute. Thus, he has hypothesized a binomial arrival rate with $p = \frac{1}{2}$. These time periods are entered in the first column of Table 7–1 under the heading P_i. To simulate the arrival of customers he can use any random device that can generate the two numbers 0 and 1. He decides on tossing a coin, associating arrivals with heads and no arrivals with tails. The results are entered in the second column under the heading A_i, the simulated arrivals.

Next, according to his past experience, the time required to serve each customer ranges from one to six minutes. However, he is not quite certain about the probability distribution of service time, so he simply assumes that all six possible service times are equally likely. That is, he tentatively postulates a discrete uniform model for service time with $S = 1, 2, \ldots, 6$. Then, he simulates the service time for each arrival by tossing an ordinary die. The simulated service times, S_i, are recorded in the third column.

With these simulated data he proceeds to determine the waiting time for the ith customer in the queue, W_i, and the total time the ith customer spends in the system, T_i, under the assumption of one or two checkout counters by observing the time the ith customer begins to receive service, B_i, and the time the ith customer is checked out, E_i.

Table 7–1. Simulation Table for Thirty One-Minute Periods

			One counter				*Two counters*			
P_i	A_i	S_i	B_i	E_i	W_i	T_i	B_i	E_i	W_i	T_i
1	1	2	1	2	0	2	1	2	0	2
2	0	—	—	—	—	—	—	—	—	—
3	1	4	3	6	0	4	3	6	0	4
4	1	6	7	12	3	9	4	9	0	6
5	0	—	—	—	—	—	—	—	—	—
6	1	4	13	16	7	11	7	10	1	5
7	0	—	—	—	—	—	—	—	—	—
8	0	—	—	—	—	—	—	—	—	—
9	1	2	17	18	8	10	10	11	1	3
10	0	—	—	—	—	—	—	—	—	—
11	1	5	19	23	8	13	11	14	0	5
12	1	3	24	26	12	15	12	14	0	3
13	1	4	27	30	14	18	15	18	2	6
14	0	—	—	—	—	—	—	—	—	—
15	0	—	—	—	—	—	—	—	—	—
16	1	1	31	31	15	16	16	16	0	1
17	0	—	—	—	—	—	—	—	—	—
18	1	2	32	33	14	16	18	19	0	2
19	0	—	—	—	—	—	—	—	—	—
20	0	—	—	—	—	—	—	—	—	—
21	1	4	34	37	13	17	21	24	0	4
22	1	3	38	40	16	19	22	24	0	3
23	1	5	41	45	18	23	25	29	2	7
24	1	1	46	46	22	23	25	25	1	2
25	0	—	—	—	—	—	—	—	—	—
26	1	3	47	49	21	24	26	28	0	3
27	0	—	—	—	—	—	—	—	—	—
28	0	—	—	—	—	—	—	—	—	—
29	1	6	50	55	21	27	29	34	0	6
30	0	—	—	—	—	—	—	—	—	—
Total	16	55	—	—	192	247	—	—	7	62

For the single checkout counter, computations are made as follows:

(1) The ith customer begins to receive his service immediately upon his arrival, or $B_i = P_i$, if he is the only customer in the system at the time. If he arrives when there is already a queue, then he begins to receive service a minute after the ending time, E_i, of the customer immediately before him. For example, the customer who arrives at the fourth minute begins to

receive his service at the seventh minute, since the customer before him is checked out at the sixth minute.

(2) The time at which the ith customer is checked out, E_i, may be computed as the sum of the ending time of the previous customer and the time required to serve the ith customer. For example, the customer who arrives at the nineth minute needs two minutes to be served, and the previous customer is checked out at the sixteenth minute; therefore the checkout time for the current customer is at the eighteenth minute $(= 16 + 2)$.

(3) The waiting time of the ith customer is simply the difference between the time he begins to receive service and the time he arrives at the system; i.e., $W_i = B_i - P_i$.

(4) Finally, the total amount of time the ith customer spends in the whole system is the sum of his service time and waiting time. Thus, the customer who arrives at the eleventh minute, for example, who needs five minutes for service but who has to wait for eight minutes before he begins to receive service, spends a total of $5 + 8 = 13$ minutes in the system.

Computations for the two checkout counters are the same as for the single-counter case, except for the determination of B_i. Now, B_i is related to P_i and the preceding two ending times, E_i. If P_i is greater than both of the two preceding checkout times, then $B_i = P_i$. For example, the customer who arrives at the fourth minute begins to receive service at the fourth minute, since the preceding two E_i's are 2 and 6, and $P_i = 4 > 2$. If P_i is equal to or less than either of the two preceding checkout times, then B_i is the sum of the smaller of the two preceding E_i's and 1. Consider, for example, the customer arriving at the sixth minute. He begins to receive service at the seventh minute, because the smaller of the two preceding E_i's is 6, which is equal to P_i.

With computations obtained from simulated arrivals and service times, the manager may now estimate the expected waiting time and expected time spent in the whole system. For the single-checkout-counter case, we have

$$\text{estimated } E(W) = \frac{\text{total waiting time}}{\text{total arrivals}}$$

$$= \frac{192}{16} = 12.000 \text{ minutes,}$$

$$\text{estimated } E(T) = \frac{\text{total time spent in the system}}{\text{total arrivals}}$$

$$= \frac{247}{16} = 15.438 \text{ minutes.}$$

With two checkout counters, we have

$$\text{estimated } E(W) = \frac{7}{16} = 0.438 \text{ minutes,}$$

$$\text{estimated } E(T) = \frac{62}{16} = 3.875 \text{ minutes.}$$

From these estimates we can see that having only one checkout counter is quite unsatisfactory, since it is unthinkable to have customers waiting in the queue as long as 12 minutes and spending more than 15 minutes in the system on the average. Furthermore, as can be seen from the columns for both W_i and T_i, the values tend to increase progressively with time. Had we simulated for 60 or even more one-minute periods, as we should in practice, the waiting time and time spent in the system would become explosive. Clearly, this is a case where average number of arrivals is greater than the average number of servicings per unit of time and the system is unworkable.

But are two checkout counters adequate in this waiting-line problem? The answer seems to be affirmative. In the first place, the average waiting time is less than half a minute, and average total time spent in the system is about four minutes. These values seem to be reasonable in situations of this kind. Furthermore, neither W_i nor T_i seems to get progressively longer as time progresses. Instead, W_i fluctuates between 0 and 2 minutes, and T_i varies between 1 and 7 minutes in a random manner. In other words, they seem to have reached stable situations.

The foregoing conclusions must, however, be viewed with caution—for two main reasons. One is that the simulated sample is rather small. More reliable results require simulating perhaps 100 or more time periods. Another is that the assumption of a uniform distribution for service time may be questionable. This being the case, the probability distribution of service time can be estimated by actual observations for, say, a few hundred arrivals, and service time required may then be simulated with a table of random digits via the estimated relative cumulative frequency distribution, as illustrated in the next example.

The Optimal Number of Trucks

Owing to a rise in its sales, a manufacturing firm needs more factory space. One alternative being considered is the purchase of a warehouse for storing finished goods. This plan requires the moving of finished items from the plant to the warehouse, ten miles distant, at the end of each workday. Management must determine the optimal number of trucks to be hired or bought. In order to do this, assuming various numbers of trucks, estimates must be made of the average amount of finished goods that will remain unmoved each day and the daily average of truck capacity that will remain unused.

In this case, clearly, a knowledge of the weight of the output produced and the weight of the output that can be moved each day is essential. Daily production volumes in pounds, according to past data, are distributed as shown in Table 7–2.

Table 7–2 Distribution of Daily Production

Pounds produced daily	*Relative frequency*	*Cumulative frequency*
400,000–419,999	0.08	0.08
420,000–439,999	0.12	0.20
440,000–459,999	0.25	0.45
460,000–479,999	0.40	0.85
480,000–499,999	0.10	0.95
500,000–519,999	0.05	1.00

The type of truck that should be used for moving this output is known to have an average loading capacity of 4500 pounds. The number of trips that can be made each eight-hour day depends on loading and unloading time as well as on travel time. The distribution of the number of pounds of product a fully loaded truck can move from the plant to the warehouse during an eight-hour period is estimated as shown in Table 7–3.

Table 7–3. Distribution of Pounds of Product Moved per Eight-Hour Day per Truck

Pounds moved	*Relative frequency*	*Cumulative frequency*
38,000–39,999	0.15	0.15
40,000–41,999	0.25	0.40
42,000–43,999	0.35	0.75
44,000–45,999	0.15	0.90
46,000–47,999	0.06	0.96
48,000–49,999	0.04	1.00

From data given in Table 7–2, we find that the expected number of pounds of product produced per day is

$$\begin{aligned}410{,}000(0.08) &+ 430{,}000(0.12) + 450{,}000(0.25) + 470{,}000(0.40)\\ &+ 490{,}000(0.10) + 510{,}000(0.05)\\ &= 459{,}400 \text{ pounds}\end{aligned}$$

From data in Table 7–3, we find that the expected number of pounds of product moved per day per truck is

$$\begin{aligned}39{,}000(0.15) &+ 41{,}000(0.25) + 43{,}000(0.35) + 45{,}000(0.15)\\ &+ 47{,}000(0.06) + 49{,}000(0.04)\\ &= 42{,}680 \text{ pounds.}\end{aligned}$$

Suppose we wish to use ten identical trucks; then the expected total number of pounds to be moved per day would be 42,680(10) = 426,800 pounds, which is smaller than the expected number of pounds of product produced per day.

We can now simulate the process on any number of days we desire to sample. Of course, the more days we sample, the more reliable our results will be. However, merely for the purpose of illustration, we shall take a sample of only 20 days.

The first step in our simulation is to construct a cumulative probability graph for each of the two variables now under consideration. The graph for production data is given by Figure 7–7, and that for data on quantities

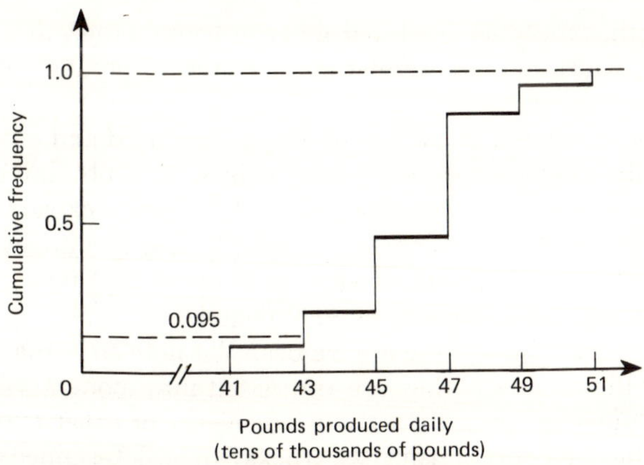

Figure 7–7 Cumulative Distribution of Daily Production

moved is given by Figure 7–8. We note that in both diagrams class marks are used on the horizontal scales. Also, the class marks on the horizontal scale for Figure 7–8 are ten times those in Table 7–3, since ten identical trucks are being simulated.

To simulate production, we enter the Table of Random Digits and select 20 three-digit numbers, each for a day's production. These numbers are given in column (1) of Table 7–4. Note that the first number selected is 095. We then enter the cumulative distribution in Figure 7–7 on the vertical axis at 0.095 and draw a horizontal line across until it meets the cumulative probability curve. Finally, we drop a vertical line from this point to the abscissa and read off the sample value, which is 430,000 pounds. This value

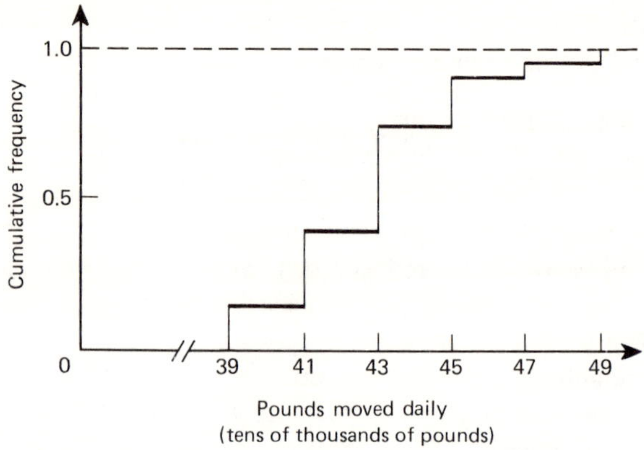

Figure 7–8 Cumulative Distribution of Product Moved

is our first predicted daily production figure. All the production figures in column (2) are so determined with the corresponding random numbers in column (1).

Following the procedure described above, we obtain the values in columns (3) and (4), which are our simulated quantities of products moved daily for 20 days.

Using these simulated quantities of output produced and output moved, we can readily compare the two sets of values, thus obtaining values for quantities produced that are unmoved, and for truck capacities that are unused. These are given in columns (5) and (6). The results reveal that the daily average unmoved finished product amounts to 29,000 pounds, and the daily average idle truck capacity is 8000 pounds.

How can we use these summary values? We note first that the figures take on new dimensions if they are translated into monetary values. The daily cost of unmoved items, due to inconvenience or other reasons, can be estimated. The daily cost of using ten trucks can be determined quite easily. The total daily combined costs of using ten trucks can then be calculated. Employing the same simulation procedure and calculations, we can estimate the total daily combined costs of using, say, 8, 9, 11, or 12 trucks. By comparing total costs per day at various levels of truck utilization, we can finally ascertain the optimal number of trucks to be hired or purchased.

The example just presented can be solved mathematically without much difficulty. Its purpose has been to make clear the basic methodology of Monte Carlo simulation. In actual practice, Monte Carlo models must be much more complicated to be realistic, and simulated samples must be much larger to be reliable. In our illustration, for example, more reliable results would be obtained if we had simulated 100 days, or even 1000 days. When large simulated samples are used, even a table of random digits

Table 7-4. Simulated Daily Production and Moved Quantity

Random number (1)	*Production* (2)	*Random number* (3)	*Quantity moved* (4)	*Quantity unmoved* (5)	*Idle truck capacity* (6)
095	430,000	601	430,000	—	—
973	510,000	889	450,000	60,000	—
410	450,000	882	450,000	—	—
176	430,000	245	410,000	20,000	—
030	410,000	230	410,000	—	—
164	430,000	842	450,000	—	20,000
676	470,000	839	450,000	20,000	—
879	490,000	443	430,000	60,000	—
954	510,000	996	490,000	20,000	—
328	450,000	080	390,000	60,000	—
505	470,000	552	430,000	40,000	—
871	490,000	109	390,000	100,000	—
075	410,000	938	470,000	—	60,000
453	470,000	866	450,000	20,000	—
839	470,000	358	410,000	60,000	—
930	490,000	077	390,000	100,000	—
480	470,000	977	490,000	—	20,000
190	430,000	947	470,000	—	40,000
514	470,000	992	490,000	—	20,000
180	430,000	386	410,000	20,000	—
Total	—	—	—	580,000	160,000
Average	—	—	—	29,000	8,000

becomes a slow and clumsy aid. The efficient way is to turn to high-speed computers, which, when properly instructed, can generate 5000 three-digit random numbers in less than two minutes. Recently there has been a growing interest in simulation methods, due mainly to the development and popularization of computer technology.

Concluding Remarks

In the last section we presented some of the essential aspects of the queuing theory, which seeks optimal solutions to waiting-line problems. The analysis of queues is focused on the utilization parameter. The proportion of time that service facilities remain completely idle decreases as the utilization factor increases. The average waiting time of arrivals increases sharply as the arrival rate approaches the total capacity of the service facility. The mean length of the waiting line increases sharply, too, as the utilization factor approaches unity.

As to the comparative merits of analytical and Monte Carlo solutions to waiting-line problems, we note that the use of the former requires theoretical models whose strict assumptions may or may not be approximately met by real-life situations. The Monte Carlo method for simulation has the advantage of flexibility, since it can be applied to theoretical distributions as well as to empirical distributions that may not satisfy the properties of any theoretical model. In addition, the Monte Carlo technique is capable of solving many problems for which analytical methods have not yet been developed, or for which available mathematical solutions are far too involved. Above all, it is a powerful tool available to research workers who are not sophisticated mathematicians.

We must realize, however, that with Monte Carlo simulation we have the same problems of statistical variation and experimental design as with real-life sampling or direct experimentation. Frequently, very large samples are required and simulation must continue for a long time to reach stable conditions. Therefore, when a mathematical model is realistic and can be manipulated with simple algebra to obtain the optimal solution, it possesses a definite advantage over Monte Carlo simulation.

Note that while both simulation and direct experimentation are subject to sampling error, the former is more desirable than the latter on many counts. With direct experimentation, we must sample repeatedly under the same conditions—a thing quite difficult to achieve; the method may also disrupt actual operations and be very expensive. Monte Carlo simulation, however, can be conducted continuously "on paper" with little cost. It permits us, therefore, to evaluate many alternatives, which we could not attempt in real-life experimentation.

It may also be pointed out that Monte Carlo simulation is a very versatile technique, useful for a large variety of models and problems other than queuing models.

EXERCISES

7–1 A purchasing agent is considering an acceptance plan for incoming lots of a manufactured product that can be classified as either good or bad. The plan calls for taking a random sample of 20 items with replacement from each lot. If there are no defectives in the sample, the lot is accepted; otherwise the lot is rejected.

(a) Show that, according to this plan, the probability of accepting the lot with 1 percent defective is 0.8179, and the probability of accepting the lot with 10 percent defective is 0.1216.

(b) Show that the Poisson approximations of the probabilities above are 0.819 and 0.135, respectively.

7–2 The purchasing agent of an electronics firm must establish a rule for deciding on the acceptance or rejection of an incoming lot of 50,000 identical parts. He finally decides on this plan: sample 40 items without replacement and accept the lot if there are no defective items in the sample; otherwise, reject the lot. What are the operating characteristics of this plan? [*Hint*: The sampling plan here obeys the hypergeometric law. However, n is very small relative to N; the hypergeometric probabilities can be approximated by the binomial probabilities, which, in turn, owing to small p and large n, can be approximated by Poisson probabilities.]

7–3 A sequential decision rule calls for a sample of 30 with replacement from an incoming lot of a manufactured item. If no defectives are found, the lot is accepted; if two or more defectives are found, it is rejected; if one defective is found, a further sample of 30 is taken with replacement. If the cumulative number of defectives out of 60 is one, the lot is accepted; if it is three or more, the lot is rejected. Otherwise, a third sample of 30 is drawn with replacement. If the cumulative number of defectives in the combined sample is three or more, the lot is rejected, and if less than three, the lot is accepted. What is the approximate probability of rejection if the lot is 1 percent defective? If the lot is 8 percent defective?

7–4 Give a comparative evaluation of the sampling plan in the preceding problem with the following single-sampling plan: Take an unconditional random sample of 50; accept the lot if there are two or less defectives in the sample; otherwise, reject the lot.

7–5 A new product requires a total fixed cost of \$3 million to produce and to promote. Furthermore, the unit variable cost and selling price are estimated at 20 cents and 70 cents, respectively.

(a) State the total revenue, total cost, and net revenue functions for this problem.

(b) Determine the break-even volume of sales.

7–6 A distributor has an opportunity to market his product in a new territory. The total fixed cost for this venture is \$500,000 for facilities, advertising, and so forth. If the variable contributing rate is \$1.25, what is the break-even volume of sales? If the selling price is \$1.75, what is the average variable cost?

7–7 Suppose the sales volume in Exercise 7–5 is normally distributed with $\mu = 8$ million units and $\sigma = 3$ million units.

(a) What are the expected profits for abandoning and introducing the new products, respectively?

(b) What is the expected value of perfect information in this case?

(c) Is it wise to postpone the decision until a perfect forecaster can be made available if such information costs \$0.5 million to obtain?

7–8 Suppose the sales volume in Exercise 7–6 is normally distributed with $\mu = 500{,}000$ units and $\sigma = 250{,}000$ units.

(a) Give the expected profits for abandoning and initiating the venture considered.

(b) Compute the expected value of perfect information.

(c) Should the decision be postponed until perfect information is obtained?

7–9 A newly formed manufacturing firm finds that it can install either one of two types of production facilities. Type I requires a total fixed capital of only $0.5 million with a variable operating cost of $20 per hour. Type II requires $1 million to install with a variable operating cost of only $13 per hour. Owing to uncertainty in the demand for its output, the required number of hours of operating time on the facility is a random variable, which is thought to be normal with $\mu = 80{,}000$ hours and $\sigma = 30{,}000$ hours.

(a) Which production facility should be installed in terms of break-even analysis? In terms of expected total costs?

(b) How should the opportunity-cost functions for the two types of facilities be expressed?

(c) What is the value of perfect information?

[*Hint*: Here, the break-even point is where the two cost functions are equal. Why? The slope of nonzero loss functions is the difference between the two variable operating costs per hour. Why?]

7–10 The mean arrival rate of customers at a repair shop is 20 per hour, and the shop has a mean service rate of 40 per hour.

(a) What is the utilization factor of this system?

(b) What is the probability of forming a waiting line of length n in this system, for $n = 0, 1, 2, 3$, and 4?

(c) What is the average number of customers in the system at any instance?

(d) What is the mean time between arrivals?

(e) What is the mean service time per servicing?

7–11 The mean arrival rate of machine breakdowns in a factory is 80 per week, and the factory's service station can repair 120 machines per week.

(a) What is the proportion of time that the service station remains idle?

(b) What is the probability that there are n down machines in the queue, for $n = 0, 1, 2, 3$, and 4?

(c) What is the expected number of broken-down machines in the system?

(d) What is the mean waiting time of an arrival?

(e) What is the average time a broken-down machine spends in the system?

7–12 Arrivals at an information desk that has one information clerk are considered to be Poisson, with an average time of ten minutes between one arrival and the next. The time that customers spend in seeking information of one kind or another is assumed to be exponential with a mean of three minutes.

(a) What is the probability that a person arriving at the desk will have to wait?
(b) What is the average number of elements in the system?
(c) The manager of the organization will hire a second clerk if convinced that an arrival would expect to have to wait at least three minutes for receiving information. By how much must the flow of arrival be increased in order to justify a second clerk?

7–13 The owner of a gas station expects a customer every five minutes. Service takes, on the average, four minutes.
(a) What are the mean arrival rate and mean service time, respectively?
(b) What is the probability that a customer will not have to wait?
(c) What is the probability that there will be a waiting line?
(d) The owner would hire an additional attendant if he found that a customer had to wait two minutes or longer before being serviced. Would the current arrival rate justify such an addition?

7–14 In a manufacturing firm, machines break down at a rate of 0.8 per hour, and the nonproductive cost of each machine waiting to be repaired and being repaired is estimated at $25 per hour. Two repairing teams are available. Team A asks for $25 per hour and guarantees that it can easily repair 1.5 machines per hour. Team B asks for $20 per hour and claims to fix 1 machine per hour. Given that the service (time) rates of the two teams are reliable, which team should the firm employ?

7–15 In the preceding problem, if team A asks for $30 per hour and team B asks for $15 per hour, which team should be employed?

7–16 A clothing factory has a large number of identical machines; the mean breakdown rate is 1 per hour and the cost of idle time is $10 per hour. The manager is considering two organizations, X and Y, for the job of providing maintenance. Organization X is known to have a repairing rate of 2 machines per hour and a charge of $20 per hour; organization Y has a repairing rate of 1.5 machines per hour and a charge of $12 per hour. Which organization should be given the contract?

7–17 Arrivals at a watch repair shop occur at a rate of ten per eight-hour day, and the watch repairman has a mean service time of 30 minutes per repair. If he repairs watches in the order of arrivals, what is his expected idle time per day? How many watches are ahead of the average number of watches just arrived?

7–18 In the children's clothing department of a department store there are one salesclerk, one wrapper, and one cashier. The salesclerk services a customer in an average of 1.5 minutes. The wrapper takes 1 minute to wrap, on the average. The mean time of the cashier to receive payment is 0.5 minute. Customers arrive at a rate of 30 per hour.
(a) What is the mean number of customers being served by the three clerks?

(b) What is the mean number of customers in the waiting line of the whole system?
(c) What is the total number of customers in the whole system, including those waiting and those being served?
(d) What is the probability of having 3 customers waiting to be served by the salesclerk, 2 customers waiting for the wrapper, and 1 customer waiting for the cashier, at any instant?

7–19 Calculate relative frequencies and cumulative relative frequencies for the following data:

Class of monthly orders in 100-month period	*No. of months in which corresponding quantity of orders has been received*
000–199	10
200–399	21
400–599	32
600–799	15
800–999	10
1000–1199	8
1200–1399	4
	100

7–20 Using random numbers, simulate a possible distribution of orders for the next 20 months for data in the preceding exercise.

7–21 Use the simulated data in Table 7–1 to determine the mean waiting time and mean time spent in the system for three checkout counters. Furthermore, if the waiting time of a customer is estimated to cost 10 cents per minute and the cost of an additional checkout counter is estimated at $5 per hour, is it justifiable to have three checkout counters instead of two?

7–22 Simulate production data and quantity moved for 15 days with data in Tables 7–2 and 7–3, assuming 11 trucks are used. Also compute daily averages of quantities unmoved and idle truck capacities.

8

BAYES' THEOREM AND DECISION MAKING

It may be recalled that the application of the Bayesian strategy requires the establishment of a probability distribution for the state random variable. When a decision must be made immediately, the decision maker does not have the option of conducting an experiment or taking a sample to learn more about the variability of the states of nature. Instead, he can base the decision only on the prior probability distribution, which may consist either of weights assigned in accordance with the decision maker's degrees of belief or else of relative frequencies based on past data. Such probabilities, however, may be based upon very scanty prior information. This drawback becomes more serious in decision situations where the optimal Bayesian measure, or the expected value of the optimal act, is sensitive to the probability distribution of the state variable—where, that is, a different optimal act may result if a slightly different prior distribution is proposed.

The foregoing remarks indicate that the quality of a decision process can be improved if the decision maker's subjective judgment, as revealed by the prior probabilities, can be incorporated with additional information on the state. That is, if time permits and if it is judged worthwhile, research should be conducted, as suggested in Section 7–2, and the resulting outcome should be employed to revise the prior probabilities before a final decision is made. The mechanics of revising prior probabilities with additional information is called the *Bayes' theorem*, a simple conditional probability rule introduced in the eighteenth century by an English clergyman, Thomas Bayes. If the Reverend Mr. Bayes could witness the development and use of his relatively simple theorem today, he would be amazed indeed to find that in spite of controversy and heated debate, Bayes' theorem has become the foundation of a whole new school of thought in statistical analysis called

Bayesian statistics. Much of the material covered in the previous two chapters falls within this school of thought. We shall, in the present chapter, first introduce Bayes' theorem via an example and then proceed to see how it can be employed for decision making. We shall return, in Chapter 16, to discuss Bayesian statistics further after a detailed examination of classical, or traditional, decision procedures.

8-1 BAYES' THEOREM

A coin is tossed. If it turns up heads, two balls will be drawn from Urn I; otherwise, two balls will be drawn from Urn II. Urn I contains two black and six white balls. Urn II contains seven black balls and one white ball. In both cases, the selections are to be made with replacement. What is the probability that (a) Urn I is used, and (b) Urn II is used, given that both balls drawn are black?

To solve this problem, let us first denote the event that Urn I is used by θ_1 and the event that Urn II is used by θ_2. Here, θ_1 and θ_2 are states of nature that are mutually exclusive and exhaustive. According to the conditions of this experiment, we must have

$$P(\theta_1) = 1/2 \quad \text{and} \quad P(\theta_2) = 1/2,$$

because the choice of an urn depends upon the toss of a coin. These values are called *prior probabilities*, or the *prior probability distribution*, of the state variable. Prior probabilities are also called, sometimes, *marginal probabilities of the states*. Now, let E be the event that "Both balls are black." Here, E evidently stands for the experimental or sample result; it is the additional information obtained.

What we seek to find are the conditional probabilities

$$P(\theta_1 \mid E) = \frac{P(\theta_1 \cap E)}{P(E)}$$

and

$$P(\theta_2 \mid E) = \frac{P(\theta_2 \cap E)}{P(E)}.$$

To obtain these probabilities we need, first of all, the probabilities of the intersection sets $\theta_1 \cap E$ and $\theta_2 \cap E$. From information given in the problem, we must have

$$\begin{aligned} P(\theta_1 \cap E) &= P(\theta_1 \cap BB) = P(\theta_1)P(BB \mid \theta_1) \\ &= (1/2)(2/8)(2/8) = 4/128, \\ P(\theta_2 \cap E) &= P(\theta_2 \cap BB) = P(\theta_2)P(BB \mid \theta_2) \\ &= (1/2)(7/8)(7/8) = 49/128. \end{aligned}$$

In these joint probabilities the terms $P(BB \mid \theta_1)$ and $P(BB \mid \theta_2)$, or, in general, $P(E \mid \theta_j)$, are called likelihoods, because they indicate how likely the sample result E under consideration is to occur *given* each and every one of the possible states of nature, θ_j (urns, in our example).

Next, we need the probability of the event E, the denominator in the ratios for our desired conditional probabilities. By the theorem of total probability, we see that

$$\begin{aligned} P(E) &= P(\theta_1 \cap E) + P(\theta_2 \cap E) \\ &= 4/128 + 49/128 = 53/128. \end{aligned}$$

We note that $P(E)$ is not only a joint probability but is also called the *marginal probability* of a sample (or experiment) result.

Finally, we have

$$P(\theta_1 \mid E) = \frac{P(\theta_1 \cap E)}{P(E)} = \frac{4/128}{53/128} = 4/53;$$

$$P(\theta_2 \mid E) = \frac{P(\theta_2 \cap E)}{P(E)} = \frac{49/128}{53/128} = 49/53.$$

These two values, called *posterior probabilities*, constitute the *posterior probability distribution* of the state variable. They are so called because they are determined after the results of the experiment are known—that is, at least after the balls have been drawn and we know their colors. It is interesting to note how the evidence provided by the outcome of two black balls is revealed in the high posterior probability of using Urn II, where black balls predominate in contrast to the uniform prior distribution.

The procedure used for finding posterior probabilities from prior probabilities (which often represent a decision maker's belief in the states of nature) and likelihoods (which are probabilities of sample outcome conditional to the states) is called *Bayes' theorem*. From our understanding of the foregoing example, we may now state this theorem as follows: If $\theta_1, \theta_2, \ldots,$ and θ_n are mutually exclusive events whose union is the sample space of an experiment, and if E is an arbitrary sample outcome defined on this sample space such that $P(E) > 0$, *Bayes' theorem* says that probability of θ_j, given E, is

$$\begin{aligned} P(\theta_j \mid E) &= \frac{P(\theta_j \cap E)}{P(\theta_1 \cap E) + P(\theta_2 \cap E) + \cdots + P(\theta_n \cap E)} \\ &= \frac{P(\theta_j)P(E \mid \theta_j)}{\sum_j P(\theta_j)P(E \mid \theta_j)} \\ &= \frac{P(\theta_j)P(E \mid \theta_j)}{P(E)}. \end{aligned} \tag{8-1}$$

To gain more insight into the use of Bayes' theorem for revising probabilities, let us consider an example. We are told that there are in a bag two physically identical chips and that both chips may be red, or both may be white, or one may be of either color. Suppose that one chip is drawn and it turns out to be red; what are the respective probabilities now for the three possible states of nature? Suppose the chip is returned to the bag, another random selection is made, and again it turns out to be red; what are the respective probabilities for the three states of nature in view of this additional information?

We note that the three states of nature under consideration are

θ_1 = both chips are red,
θ_2 = both chips are white,
θ_3 = one chip is red and one is white.

Now, since we have no information whatsoever concerning the distribution of color of the chips, we may assume all the three states of nature are equally likely. With this assumption, the posterior probabilities given that a red chip results (E) can be easily computed by Bayes' theorem, as shown by Table 8–1.

Table 8–1

State of nature θ_j	*Prior* $P(\theta_j)$	*Likelihood* $P(E \mid \theta_j)$	*Joint* $P(\theta_j \cap E)$	*Posterior* $P(\theta_j \mid E)$
θ_1	$1/3$	1	$1/3$	$2/3$
θ_2	$1/3$	0	0	0
θ_3	$1/3$	$1/2$	$1/6$	$1/3$
	$3/3$		$P(E) = 3/6$	$3/3$

Thus, with the sample result of a red ball, the second state of nature becomes an impossible event and is eliminated from further consideration. Furthermore, this sample result indicates that θ_1 is more probable than θ_3.

Next, to revise the probabilities associated with the states using the additional information that the second draw also produces a red ball, we apply Bayes' theorem in the same fashion, except that we should now employ the posterior probabilities obtained before as the prior probabilities, as shown in Table 8–2.

Clearly, the probability associated with θ_1 becomes even larger than before.

If, instead of computing posterior probabilities with each draw, we had treated the two selections of a red ball as an unconditional random sample

Table 8-2

State of nature θ_j	Prior $P(\theta_j)$	Likelihood $P(E \mid \theta_j)$	Joint $P(E \cap \theta_j)$	Posterior $P(\theta_j \mid E)$
θ_1	$2/3$	1	$2/3$	$4/5$
θ_3	$1/3$	$1/2$	$1/6$	$1/5$
	$3/3$		$P(E) = 5/6$	$5/5$

of size 2, application of Bayes' theorem with the same uniform prior probabilities would have yielded the same revised probabilities.

The preceding example illustrates the main application of Bayes' theorem in revising probabilities. That is, a decision maker is often confronted with a set of mutually exclusive and exhaustive states of nature in connection with a phenomenon that is subject to test by experiment. Before an experiment is performed, the decision maker may be completely ignorant about the distribution of the state variable, and he often assumes that all states of nature are equally likely. Or, if he has grounds for an educated guess, he may assign prior probabilities in such a way that the weights assigned are proportional to the "degree of belief" he has in various states. Thus, different decision makers may assign different prior probabilities to the same set of states of nature.

For instance, to judge whether a coin is fair, one may initially assign equal probabilities to the occurrences of heads and tails. Another person, somehow believing that the coin is biased in favor of heads, may be inclined to give, say, $2/3$ as the prior probability of heads and $1/3$ as that of tails. Such initial differences, however, do not hinder the use of Bayes' theorem, since its very aim is to modify the prior probabilities, $P(\theta_j)$, by experiment. After the experiment is performed, we replace $P(\theta_j)$ by $P(\theta_j \mid E)$. Furthermore, we may conduct a new experiment by using posterior probabilities of the preceding experiment as prior probabilities. As we proceed with repeated experiments, evidence accumulates and modifies the initial prior probabilities, thereby modifying the intensity of a decision maker's belief in various states of nature. Repeated revisions will soon produce such low posterior probabilities for some states that they can be eliminated from further consideration. In other words, the more evidence we accumulate, the less important are the prior probabilities. The only restriction on the application of Bayes' theorem is that all states of nature must be tenable in a given situation and that none is assigned a prior probability of 0 or 1.

8-2 FURTHER COMMENTS ON BAYES' THEOREM

Owing to the importance of Bayes' theorem in modern decision theory, we shall offer several additional observations about its application and about the effects of prior probabilities and sample results on the eventual revised probabilities.

First of all, let us note that the prior probabilities of the state variable are called the prior distribution and may be denoted by $f_0(\theta_j)$; that the likelihood is the probability of the particular sample result given each and every state and may be denoted by $f(x \mid \theta_j)$; that the sum of the joint probabilities is the probability of the sample result under the given prior probability distribution and may be denoted by $f(x)$; and that the revised (posterior) probability distribution may be denoted as $f_1(\theta_j)$. In terms of these new and more general notations, equation (8–1) can be written as

$$f_1(\theta) = \frac{f(x \mid \theta) f_0(\theta)}{f(x)}. \tag{8–2}$$

This is the general expression of Bayes' theorem, which we shall use, for example, in Chapter 16.

Now, consider a quality control problem where the decision variable is the fraction defective in each production run of 5000 units. The prior distribution of this state variable is given in the first two columns of Table 8–3. To check whether the process is in control, a random sample of 100 is taken with replacement from each production run and the number of defectives is observed.

Suppose that in one sample four units are found defective; how does this information influence the prior beliefs concerning the probabilities of the states?

To answer this question, we first try to revise the prior probability distribution with the sample information. This is done with Bayes' theorem in Table 8–3. Here, the likelihoods obey the binomial law, and, owing to

Table 8–3

States θ_j	*Prior* $f_0(\theta_j)$	*Likelihood* $f(4 \mid \theta_j)$	*Joint* $f(\theta_j)f(4 \mid \theta_j)$	*Posterior* $f_1(\theta_j)$
0.01	0.05	0.015	0.00075	0.00508
0.02	0.10	0.090	0.00900	0.06102
0.03	0.15	0.168	0.02520	0.17085
0.04	0.20	0.195	0.03920	0.26576
0.05	0.30	0.176	0.05250	0.35593
0.06	0.10	0.134	0.01340	0.09085
0.07	0.05	0.091	0.00455	0.03085
0.08	0.05	0.059	0.00290	0.01966
	1.00		$f(x) = 0.14750$	1.00000

large n and small p, these binomial probabilities can be approximated by the Poisson probabilities. For example,

$$\begin{aligned}P(4 \mid p = 0.01) &= b(4; 100, 0.01)\\ &= p[4; \mu = 100(0.01)]\\ &= p(4; 1)\\ &= 0.015;\end{aligned}$$

$$\begin{aligned}P(4 \mid p = 0.02) &= b(4; 100, 0.02)\\ &= p(4; 2)\\ &= 0.090;\end{aligned}$$

and so on. The rest of the calculations in the table are self-evident.

Now, let us point out that the value $f(x) = 0.1475$ is the probability under the prior distribution of the sample obtained. Once the sample result, x, is known, the posterior probability of each state can be determined as that proportion of the probability of the sample result contributed by each state. An interesting way to interpret the value of $f(x)$ is this: If, say, 10,000 samples of size 100 were drawn from a situation described by the state variable and its corresponding prior distribution, there would be, on the average, 1475 samples with four defective units. Of these 1475 samples with $x = 4$, about eight would result for the state $\theta_1 = 0.01$, 90 would result for the state $\theta_2 = 0.02$, and so on. Hence, since 392 out of the 1475 occurrences of the sample result are due to the state $\theta_4 = 0.04$, the probability to be assigned to state θ_4, given $x = 4$, should be $392/1475 = 0.26576$, the proportion of the total occurrences attributable to the state 0.04.

A quick comparison between $f_0(\theta)$ and $f_1(\theta)$ shows that sampling has indeed modified the initial probabilities assigned to the states. Since the sample result indicates 4 percent defective, it is not surprising to find that prior probability for $\theta_4 = 0.04$ has been revised upward by about 30 percent. Not anticipated on the basis of intuition, however, is the upward revision of the prior probabilities for the joining states, θ_3 and θ_5, though the change is relatively less than for θ_4. Meanwhile, as expected, probabilities assigned to all other states, except the three just mentioned, have been modified downward. Furthermore, the further the state is from θ_4, the greater is the reduction. Thus, sampling has the effect of condensing the probabilities around the observed sample result.

Note that the largest posterior probability may still be attached to some other state, despite the sample result, if the prior probability of that state is much larger than that for the state that is identical to the sample result. Our example verifies this. Furthermore, it should be clear that a different posterior distribution would emerge, given the same sample result, if a different prior distribution were assumed. However, as we shall see, a different prior might not alter optimal act based on posterior probabilities.

Now, we have seen how, by way of Bayes' theorem, sampling information can be employed to revise a decision maker's initial, prior, subjective probabilities over the states of nature. But how does a decision maker go about making a decision by using both prior judgment and additional information?

Let us first note that in Bayesian statistics, the decision procedure of using prior state probabilities alone is called a *prior analysis*, and the procedure of using both prior and posterior probabilities is called a *posterior analysis*.

Returning to the problem of quality control for our current example, suppose the management, on the basis of experience, has established the criterion that if the lot just produced contains 5 percent or less defectives, the process should be considered as in control, a_1, and if it it contains more than 5 percent defectives, the process should be considered as out of control, a_2.

In this connection, we note that both the prior mean, $[E(\theta) = \sum\theta f_0(\theta) = 0.0435]$, and the posterior mean, $[E(\theta) = \sum\theta f_1(\theta) = 0.0440]$, are less than 5 percent. However, from these values alone, we cannot yet decide whether a_1 or a_2 is optimal, since we have not yet included the payoffs in our analysis.

Suppose the management, after carefully studying past cost data, has determined the payoffs in terms of opportunity losses as shown in the first three columns of Table 8–4. If the management is willing to make a decision on the prior information alone, then the optimal act is a_2 with an expected opportunity loss of $160. This value, as may be recalled, is also the expected

Table 8–4. Computations of Expected Opportunity Losses

			Weighted *OL*			
	OL		Prior		Posterior	
θ_j	a_1	a_2	a_1	a_2	a_1	a_2
0.01	$ 0	$200	$ 0	$ 10	$ 0	$ 1.02
0.02	0	200	0	20	0	12.20
0.03	0	200	0	30	0	34.17
0.04	0	200	0	40	0	53.15
0.05	0	200	0	60	0	71.19
0.06	750	0	75.00	0	68.14	0
0.07	850	0	42.50	0	26.22	0
0.08	950	0	47.50	0	18.68	0
		$EOL(a_i)$	$165.00	$160	$113.04	$171.73

value of perfect information. If the cost of sampling 100 units is less than this amount, a posterior analysis should be worthwhile. In terms of a posterior analysis, a_1 now becomes the optimal act with an expected opportunity loss of $113.04. All these computations are given in Table 8–4. Note that the prior expectation is computed with prior probabilities as weights and that the posterior expectation is computed with posterior probabilities as weights.

To summarize, in posterior analysis, the decision maker is required to

(1) establish a set of prior probabilities over the states of nature;
(2) take a sample or conduct an experiment and observe the outcome;
(3) revise the prior probabilities into posterior probabilities on the basis of additional information obtained by using Bayes' theorem;
(4) use the Bayesian strategy with posterior probabilities as the basis for selecting the optimal act, which is the Bayesian measure that maximizes expected payoffs or that minimizes expected opportunity losses.

The merit of decision making with additional information is that uncertainty associated with the state variable may be reduced. However, gathering information may be quite expensive, and research seldom, if ever, provides perfect information. In view of these considerations, the complete Bayesian decision procedure also involves what is called a *preposterior analysis*, an analysis conducted before information is to be gathered. Precisely, a preposterior analysis attempts to answer, on paper, the question: "Should a sample be taken, and if so, how large should it be?" In other words, we must determine whether the economic value of the information contained in the sample is worth the cost. We shall now take up the various aspects of both posterior and preposterior analyses.

8–3 BAYESIAN POSTERIOR ANALYSIS

Before each new product is marketed nationally, a certain cosmetics manufacturer always has a panel of executives forecast the market response. The manufacturer's product research staff has just perfected a new perfume with oriental ingredients, "Black Jade," which is ready for commercialization. The executives' composite ratings indicate a prior distribution for various levels of sales of Black Jade as follows:

θ_1: high level of sales with $P(\theta) = 0.4$,
θ_2: average level of sales with $P(\theta_2) = 0.4$,
θ_3: low level of sales with $P(\theta_3) = 0.2$.

For this decision situation, the two possible actions are

a_1: market the product nationally,
a_2: abandon the product.

The optimal act, as usual, depends upon which state of nature is to prevail and the payoff function for each act. Suppose that, according to estimates, the manufacturer's monetary payoffs are given as shown in Table 8–5.

Table 8–5. Payoff Matrix in Millions of Dollars for Black Jade Problem

Act	θ_1	θ_2	θ_3
a_1	4.0	1.0	−1.6
a_2	0.0	0.0	0.0

With the information available, a prior analysis can already be made. Furthermore, if the decision is to be made now, a_1 (market the product) is the optimal act, since

$$\begin{aligned} EP(a_1, \theta_j) &= 4.0(0.4) + 1.0(0.4) - 1.6(0.2) \\ &= \$1.68 \text{ million.} \end{aligned}$$

which is greater than

$$EP(a_2, \theta_j) = 0.$$

Now, if the manufacturer would like to make the final decision with additional information, he must consider first the expected value of perfect information and the estimated cost of obtaining information. For this problem we note that, assuming certainty, a_1 is the optimal act if either θ_1 or θ_2 will be the eventual state of nature, and a_2 is the optimal act if θ_3 actually emerges. Thus, the expected payoff with perfect information (EPPI) is

$$\begin{aligned} \text{EPPI} &= 4.0(0.4) + 1.0(0.4) + 0.0(0.2) \\ &= \$2 \text{ million.} \end{aligned}$$

With EPPI determined, the expected value of opportunity loss, or the expected value of perfect information, becomes

$$\begin{aligned} \text{EVPI} &= \text{EPPI} - \begin{pmatrix} \text{the expected value for the optimal} \\ \text{act of the prior analysis} \end{pmatrix} \\ &= 2 - 1.68 \\ &= \$0.32 \text{ million.} \end{aligned}$$

At this point, suppose the manufacturer estimates that additional information can be obtained by introducing Black Jade on a trial basis in a certain "typical" local market, and the cost involved is less than \$0.32 million; then he should proceed to conduct a posterior analysis.

Having decided on a posterior analysis, the first thing the decision maker should do is to establish the conditional probabilities—probabilities of possible sample results given various states of nature. Here, possible experimental results have the following relevance:

E_1: experimental result indicates θ_1—high level of sales;
E_2: experimental result indicates θ_2—average level of sales;
E_3: experimental result indicates θ_3—low level of sales.

According to the executives' past performance in forecasting, conditional probabilities are as revealed in Table 8–6.

Table 8–6

Possible research result E_k	*State of nature* θ_1	θ_2	θ_3
E_1	0.7	0.3	0.1
E_2	0.2	0.5	0.1
E_3	0.1	0.2	0.8

Note that each column in Table 8–6 refers to the conditional probabilities of each possible result given a specific forecasted state of nature. For instance, referring to the column headed by θ_1, we have: If θ_1 is the true state of nature, 70 percent of the products will have high levels of sales, 20 percent of the products will have average levels of sales, and 10 percent of the products will have low levels of sales. Also, each row gives the probabilities of a particular experimental result given each possible state of nature. These are the likelihoods for the particular result, and they are employed to compute posterior probabilities if that result is obtained.

Suppose Black Jade has just been test-marketed in the typical locality mentioned before, and sales in the test market were of a low level; what would be the optimal act now?

We must first revise the prior probability distribution in view of the new information. This is done in Table 8–7.

Then, applying the Bayes strategy to the posterior probabilities for the state variable, we have

$$EP(a_1) = 4(4/28) + 1(8/28) - 1.6(16/28).$$
$$= -\$0.057143 \text{ million;}$$

$$EP(a_2) = 0$$

Table 8–7

Prior $P(\theta_j)$	*Likelihood* $P(E_3 \mid \theta_j)$	*Joint* $P(\theta_j \cap E_3)$	*Posterior* $P(\theta_j \mid E_3)$
0.4	0.1	0.04	$4/28$
0.4	0.2	0.08	$8/28$
0.2	0.8	0.16	$16/28$
		0.28	$28/28$

Thus, a_2 now has become the optimal act in the posterior analysis, owing to an experimental outcome that indicated a low level of sales, in contrast to the conclusion reached on prior forecast alone. This clearly illustrates the importance of additional information in many decision situations.

It is interesting to recall that the *prior* EVPI of a decision sets the upper limit for the worth of obtaining additional information. After additional information is obtained through experiment or sampling, we can also compute the *posterior* EVPI in the same manner by deducting the posterior expected payoff for the optimal act under uncertainty from the posterior expected payoff under certainty. Thus, for our current example,

$$\begin{aligned}\text{Posterior EVPI} &= [4(4/28) + 1(8/28) + 0(16/28)] \\ &= 24/28 = \$0.857143 \text{ million,}\end{aligned}$$

which is greater than the prior EVPI of \$0.32 million. This increase in EVPI from the prior to the posterior analysis means that the cost of uncertainty has been increased; namely, in view of the additional information already gained from sampling, the availability of a perfect forecaster is more valuable than it was prior to the experiment. This happens, as in our current example, when the difference between prior and posterior probability distributions is so great that a reverse in the optimal act occurs after the incorporation of additional information. Clearly, a decrease in the EVPI from the prior to the posterior analysis might very well also occur. This could happen when both prior and posterior analyses lead to the same optimal act. Such a decrease in EVPI after the inclusion of additional information gained can be interpreted to mean that the doubt concerning the decision has been reduced because of the additional information.

8–4 BAYESIAN PREPOSTERIOR ANALYSIS

In Bayesian posterior analysis, as just discussed, the decision maker addresses the question: "Given that a sample of a certain size is drawn or an experimental test is made, what action should be taken on the basis of both prior and additional information?" That information is valuable, and that more information is always more valuable, cannot be denied. But information gathering is often very expensive, and information obtained by sampling can hardly eliminate all the uncertainty attached to the state variable. Therefore, under certain conditions, posterior analysis should thoroughly evaluate the possible worth of information and the cost of obtaining such information, thereby determining not only whether a posterior analysis should be conducted at all but also what is the best amount of information to obtain. Let us introduce the salient features of Bayesian preposterior analysis via the following decision problem.

Mr. Smith, the owner of a citywide express and van company in Chicago, is offered the five trucks of another line that is going out of business for 10 percent of interest in his company. Available to him are the following two acts:

a_1: accept the offer;
a_2: reject the offer outright.

The key determinant in this case is the number of trucks that may need major reconditioning. Mr. Smith does not possess full information on the conditions of the five trucks, except their ages, mileages, and makes. On the basis of what he knows, however, he is able to establish a prior probability distribution for the state variable, the number of trucks requiring major reconditioning, and to estimate the possible payoffs as shown in Table 8–8.

Table 8–8

Number of trucks requiring major reconditioning θ_j	*Prior probability* $P(\theta_j)$	*Payoff* a_1	a_2
0	0.35	\$10,000	0
1	0.15	6,000	0
2	0.10	2,000	0
3	0.05	0,000	0
4	0.10	−2,000	0
5	0.25	−5,000	0

If Mr. Smith is to make a terminal decision on the basis of prior probabilities alone, his optimal act is a_1, since $EP(a_2) = 0$ while

$$\begin{aligned} EP(a_1) &= (10{,}000)(0.35) + (6000)(0.15) + (2000)(0.10) \\ &\quad + (0)(0.05) + (-2000)(0.10) + (-5000)(0.25) \\ &= \$3150.00. \end{aligned}$$

For convenience, we shall call the prior analysis the no-sample case—that is, $n = 0$.

We note that the expected payoff with perfect information is $4600, yielding an expected value of perfect information of 4600 − 3150 = $1450, which is as large as 46 percent of the expected payoff for the optimal act in the no-sample case. Clearly, it may be desirable to obtain additional information before a final decision is made.

Suppose that sampling of a truck, which includes thorough inspection of the parts and a trial run of 50 miles, costs $60; then we could actually have 100 percent inspection with a total cost of 5(60) = $300, which is much less than $1450—the expected value of perfect information. However, this procedure is uneconomical and unnecessary in many decision situations, including the present problem, since the terminal expected payoff with 100 percent inspection may actually be smaller than that for partial inspection (sampling). This is indeed the main merit of a preposterior analysis, which, among other things, also includes the determination of the optimal sample size.

The main technical aspect of the preposterior analysis is the determination of what is called the terminal expected payoff, on paper, for each action with each possible sample result from each and every possible sample size. In the decision problem on hand, clearly, there are six possible sample sizes, ranging from 0 consecutively to 5. Let us take $n = 3$ for detailed analysis. Given this sample size, there would be the following four possible sample results:

E_0: no truck in the sample needs major reconditioning;
E_1: one truck in the sample needs major reconditioning;
E_2: two trucks in the sample need major reconditioning;
E_3: three trucks in the sample need major reconditioning.

Given the possible results of observation, as in the case of posterior analysis, the following three steps are to be carried out:

(1) Revise the prior probabilities, or compute the posterior probabilities, on the basis of each of the possible sample outcomes.
(2) Determine the terminal expected payoffs in terms of posterior probabilities for each act, given each and every specific result.
(3) Determine the overall terminal expected payoffs with the marginal probabilities of sample outcome for the particular sample, $P(E_k)$, as weights.

For example, if one truck in the conditional random sample (that is, selected without replacement) of three is found in need of major reconditioning, then the probability (likelihood) for E_1 to occur given that, say, θ_1 is the true state of nature, follows the hypergeometric law, and

$$P(E_1 \mid \theta_1) = \frac{\binom{k}{x}\binom{N-k}{n-x}}{\binom{N}{n}} = \frac{\binom{1}{1}\binom{5-1}{3-1}}{\binom{5}{3}} = \frac{(1)\dfrac{4!}{2!2!}}{\dfrac{5!}{3!2!}} = 0.6.$$

Other likelihoods of E_1, given other states of nature, are computed in the same fashion and are entered in the third column of Table 8–8. Note that $P(E_1 \mid \theta_0) = 0$, since we can never find a truck that needs major reconditioning if no truck in the population needs such service. Similarly, $P(E_1 \mid \theta_5) = 0$, because if all the trucks in the population need major reconditioning, then we can never have only one truck in the sample that needs no such service. With $P(\theta_j)$, $P(E_1 \mid \theta_j)$, $P(E_1)$, the marginal probability for a particular sample result E_1 and a particular sample with $n = 3$ given, $P(\theta_j \mid E_1)$ can readily be obtained. All these values are given in Table 8–9.

Table 8–9

θ_j	$P(\theta_j)$	$P(E_1 \mid \theta_j)$	$P(\theta_j \cap E_1)$	$P(\theta_j \mid E_1)$
0	0.35	$\binom{0}{1}\binom{5}{2}\Big/\binom{5}{3} = 0.0$	0.000	0.0000
1	0.15	$\binom{1}{1}\binom{4}{2}\Big/\binom{5}{3} = 0.6$	0.090	0.5455
2	0.10	$\binom{2}{1}\binom{3}{2}\Big/\binom{5}{3} = 0.6$	0.060	0.3636
3	0.05	$\binom{3}{1}\binom{2}{2}\Big/\binom{5}{3} = 0.3$	0.015	0.0909
4	0.10	$\binom{4}{1}\binom{1}{2}\Big/\binom{5}{3} = 0.0$	0.000	0.0000
5	0.25	$\binom{5}{1}\binom{0}{2}\Big/\binom{5}{3} = 0.0$	0.000	0.0000
	1.00		$P(E_1) = 0.165$	1.0000

In terms of the preceding posterior probabilities, the terminal expected payoff for a_1, given the particular sample outcome E_1, is

$$\begin{aligned} TEP(a_1 \mid E_1) &= (10{,}000)(0) + (6000)(0.5455) + (2000)(0.3636) \\ &\quad + (0)(0.0909) + (-2000)(0) + (-5000)(0) \\ &= \$4000.20. \end{aligned}$$

The terminal expected payoff for a_2, given E_1, as before, is 0.

The terminal expected payoffs for other sample outcomes E_0, E_2, and E_3 are obtained in exactly the same way and are given in Table 8–10.

Table 8–10

E_k	*Marginal probabilities of sample outcomes for* $n = 3$ $P(E_k)$	*Terminal expected payoffs* a_1	a_2
E_0	0.420	+\$9236.00	0
E_1	0.165	+\$4000.20	0
E_2	0.120	−\$500.00	0
E_3	0.295	−\$4570.00	0

A glance at Table 8–10 indicates that a_1 is the optimal act if $X \leq 1$, and a_2 is the optimal act if $X = 2$ or 3, where X stands for the number of trucks in the sample requiring major reconditioning. For this decision rule, the overall terminal expected payoff for our example becomes

$$\begin{aligned} \text{OTEP} &= (9236.00)(0.42) + (4000.20)(0.165) + (0)(0.120) + (0)(0.295) \\ &= \$4539.15. \end{aligned}$$

Note that the difference between the overall terminal expected payoff for a particular sample and the prior expected payoff for the optimal act is called the *expected value of sample information*, denoted by EVSI, for that sample. In our example, with $n = 3$,

$$\begin{aligned} \text{EVSI} &= 4539.15 - 3150.00 \\ &= \$1389.15, \end{aligned}$$

which is clearly much greater than the sampling cost of \$180. Also, that difference between OTEP and sample costs is called *net overall terminal expected payoff* for the particular sample, denoted by NOTEP. For our current example, with $n = 3$,

$$\begin{aligned} \text{NOTEP} &= 4539.15 - 180.00 \\ &= \$4359.15. \end{aligned}$$

The merit of the Bayesian decision rule is that it maximizes the terminal expected payoffs for a given set of prior probabilities and a particular sample size. Thus, while there exist other decision rules under the same conditions—such as "Take a_1 when there is no truck or one truck needing major reconditioning in a sample of three; otherwise take a_2"—none of these can yield a higher expected payoff than the Bayesian rule. Given the prior probabilities, however, the Bayesian decision rule would yield different expected terminal payoffs for different sample sizes. Thus, the complete Bayesian decision framework in preposterior analysis also includes the selection of the optimal sample size.

To find the optimal sample size, we ascertain the Bayesian rule for each possible sample size and compute the net overall terminal expected payoff (net of sampling costs). For the current example, we have already worked out the overall terminal expected payoffs for $n = 0$ and $n = 3$. The expected payoffs, sampling costs, and net overall terminal expected payoffs, NOTEP, for both these cases are summarized in Table 8–10, in which we have also given the corresponding results for $n = 1$ and $n = 2$. Calculations for these are not shown, since they are exactly the same as for $n = 3$.

Results in Table 8–11 indicate that the optimal sample size is $n = 2$, since it yields the highest NOTEP. Thus the Bayesian decision rule for preposterior analysis for this problem is: take a sample of $n = 2$, select a_1 if $X \leq 1$, and select a_2 if $X = 2$. Having completed the preposterior analysis, Mr. Smith may now go ahead and observe a sample of two trucks and decide according to the rule just stated.

An interesting question worth answering in this case is: "Why have we stopped with $n = 3$? In other words, why do we not need to show the results for $n = 4$ and $n = N = 5$?" The answer here is clearly related to the expected value of perfect information. We know that in a given decision problem, decision rules based on incomplete information cannot have expected payoffs higher than the expected payoff with perfect information, which in our case is \$4600. Thus, when we selected $n = 4$, the sampling costs would have been \$240, and then the net overall terminal expected payoff could not be expected to exceed \$4600 − \$240 = \$4360, which is

Table 8–11

Sample size	*Decision rule*		OTEP	*Sample cost*	NOTEP
	a_1	a_2			
0	Always	Never	\$3150.00	0	\$3150.00
1	$X = 0$	$X = 1$	4297.80	\$60.00	4237.80
2	$X \leq 1$	$X = 2$	4499.77	120.00	4379.77
3	$X \leq 1$	$X \geq 2$	4539.15	180.00	4359.12

less than the NOTEP for $n = 2$. Similarly, with $n = 5$, we can never expect to have an expected payoff greater than \$4600 − \$300 = \$4300, which is less than the NOTEP for $n = 2$. From this line of reasoning, it should be clear that we should actually have stopped at $n = 2$ instead of $n = 3$.

In conclusion, the Bayesian preposterior analysis is a very useful approach, but it is also very time-consuming and tedious. In determining the optimal size of sample, which is an indispensable part of selecting the optimal act, we need to calculate the terminal expected payoff for each and every relevant possible sample size. When a large number of samples must be considered, this could become a formidable and prohibitively costly venture. Fortunately, techniques are available for the special case of normal sampling (or the normal approximation to the conditional probabilities required) that reduce all these computations to a single formula. This we indicated in Section 7–2 in discussing the special form of the Bayesian strategy for nonequivalent state-action problems with linear payoffs and normal states. We shall take up this special case again in Chapter 16, where Bayesian inferential procedures will be considered. We have given the detailed procedure of preposterior analysis here merely to introduce the reader to the vital concept of the expected value of sample information and how it can be obtained. It may also be noted here that the use of a computer would simplify everything, making the computational burden a minor consideration.

EXERCISES

8–1 A manufacturing firm produces steel pipes in three plants with daily production volumes of 500, 1000, and 2000 units, respectively. From past experience it is known that the fraction of defective outputs produced by the three plants are, respectively, 0.005, 0.008, and 0.010. If a pipe is selected at random from a day's total production and found to be defective, verify that the probabilities that the pipe has been drawn from each of the three plants are $5/61$, $16/61$, and $40/61$, respectively.

8–2 Six bridge decks and four pinochle decks, all of identical designs and construction, are resting on a table. One deck is chosen at random and a card is then selected from it blindly. If the card drawn is a king of hearts, show that the probability that it came (a) from a bridge deck is 0.4091, and (b) from a pinochle deck is 0.5909. (Note: A pinochle deck contains 48 cards; two each of 9, 10, jack, queen, king, and ace in each of four suits: spades, hearts, diamonds, and clubs.)

8–3 "Mr. Smith's gardener is not dependable; the probability that he will forget to water the rosebush is $2/3$. The rosebush is in questionable condition anyhow; if watered, the probability for its withering is $\frac{1}{2}$; if not watered, the probability for its withering is $\frac{3}{4}$. Upon returning Smith finds that the rosebush has withered." Show that the probability that the gardener did not water the rosebush is $P(\text{not watered} \mid \text{withered}) = \frac{8}{11}$.

8–4 An automobile insurance company has insured 35,000 class A drivers (good risks), 50,000 class B drivers (median risks), and 15,000 class C drivers (bad risks). The probability that a class A, B, or C driver will have one or more accidents during any year is 0.01, 0.04, or 0.15. The company sells Mrs. Jones an insurance policy and within a year she has an accident. Verify that the probability of her being (a) a class A driver is 0.076, (b) a class B driver is 0.435, (c) a class C driver is 0.489.

8–5 A box containing eight TV transmitters is received from a supplier. In the past, 70 percent of all such boxes have had zero defectives, 20 percent have had one defective, and 10 percent have had two defectives. Three units are selected without replacement at random from this box, and one is found to be defective. Show that the probability of receiving a box with two defectives before the sample was taken was 0.10. Show also that, given the sample result, the probability that the box originally contained two defective parts is 0.417.

8–6 Consider the problem with payoff matrix Table 8–5 in the text. What is the optimal act if the market test indicates a high level of sales? An average level of sales?

8–7 Consider a problem with two states of nature, θ_1 as "rain" and θ_2 as "no rain," and three actions, a_1 as "stay at home," a_2 as "go out without an umbrella," and a_3 as "go out with an umbrella." Table 8P–1 gives opportunity losses (in utils) that represent the dilemma constituting the decision problem.

Table 8P–1

	State	
Act	θ_1	θ_2
a_1	0.4	0.4
a_2	0.5	0.0
a_3	0.2	0.5

(a) If the prior probabilities are $P(\theta_1) = 0.4$ and $P(\theta_2) = 0.6$, show that the expected losses, on prior information alone, are

$$EL(a_1) = 0.40, \quad EL(a_2) = 0.20, \quad \text{and} \quad EL(a_3) = 0.38,$$

where a_2 is the optimal act. Why?

(b) Now, suppose a single observation of a rain indicator (weather report) yields the likelihoods as in Table 8P-2, where E_1 indicates an observation of "rain" and E_2 of "no rain." If the rain indicator shows (or predicts) E_1, verify that, in terms of posterior analysis,

$$EL(a_1) = 0.400, \qquad EL(a_2) = 0.421, \quad \text{and} \quad EL(a_3) = 0.247,$$

where a_3 becomes the optimal act. Finally, if the rain indicator shows E_2, verify that a_2 is the optimal act with $EL(a_2) = 0.0645$.

Table 8P–2

	θ_1	θ_2
E_1	0.8	0.1
E_2	0.2	0.9

8–8 Professor Howard Raiffa, a leading American Bayesian, has offered an oil-drilling decision problem in his *Decision Analysis:* The oil wildcatter must decide either to drill, act a_1, or not to drill, act a_2, on a potential oil field. He is uncertain whether the hole is dry, θ_1, wet, θ_2, or soaking, θ_3. His payoffs are given in Table 8P–3. It is assumed that the cost of drilling is \$70,000. Here, payoffs are given as net returns associated with each act corresponding to the three possible states.

Table 8P–3

	Act	
State	a_1	a_2
θ_1	−\$70,000	0
θ_2	\$50,000	0
θ_3	\$200,000	0

At a cost of \$10,000, the wildcatter can take seismic soundings that will help determine the underlying geological structure at the site. The soundings will disclose whether the terrain below has (a) no structure (outcome *NS*)—that's bad, or (b) open structure (outcome *OS*)—that's so-so, or (c) closed structure (outcome *CS*)—that's really hopeful. Experts have provided joint probabilities as shown in Table 8P–4.

(a) What is the optimal act without experimentation?

(b) What is the expected value of perfect information?

(c) What is the optimal act with experimentation, given each and every one of the seismic outcomes?

Table 8P-4

State	Seismic outcome NS	OS	CS	Marginal probability of state
θ_1	0.30	0.15	0.05	0.50
θ_2	0.09	0.12	0.09	0.30
θ_3	0.02	0.08	0.10	0.20
Marginal probability of seismic outcome	0.41	0.35	0.24	1.00

[Note (1) that the marginal probability of state is the prior distribution of the state variable, and (2) that marginal probability of seismic outcome refers to $P(E_k)$ in the notations used in this text.]

8–9 For the decision problem discussed in Section 8–4, what is the optimal sample size if it cost $100 to sample a truck?

8–10 A manufacturer's purchasing agent must decide to accept or reject an incoming shipment of machine parts. He wishes to employ either of the following two acts for this situation:

a_1: conduct 100 percent inspection;
a_2: accept the lot as it is.

Inspection cost per unit is 50 cents, and the cost of a defective part that has been installed into a machine is $7. Furthermore, the prior distribution for the state variable, fraction defective of the lot, is assumed to be as shown in the second column of Table 8P–5. Explain why the costs for each action and the opportunity costs for each action in Table 8P–5 are correct. What would be the optimal act in this problem for a prior analysis? What is the expected value of perfect information?

Table 8P–5

		Cost		OC	
θ_j	$P(\theta_j)$	a_1	a_2	a_1	a_2
0.02	0.10	500	140	360	0
0.04	0.25	500	280	220	0
0.06	0.30	500	420	80	0
0.08	0.15	500	560	0	60
0.10	0.10	500	700	0	200
0.12	0.05	500	840	0	340
0.14	0.05	500	980	0	480

8–11 For the decision problem in Exercise 8–10, assuming a random sample of $n = 50$ has been drawn from the lot with replacement and two are found to be defective, show that, in terms of both prior and sample information, expected cost for the optimal act is \$375.75 and the corresponding expected opportunity loss is \$20.98.

8–12 An agent has agreed to receive, periodically, shipments of four large machines from one of the manufacturing firms he represents. The machine must be moved from the railroad yard upon arrival and installed in good working order in the customer's plants. If the agent accepts the responsibility for the shipment, a_1, he receives a flat fee of \$600 each time from the manufacturer, but he must reset, at his own expense, any machine that is damaged during the shipment. The cost of resetting is \$250 per machine on the average. If the agent rejects responsibility for the shipment, a_2, he receives no fee, and, of course, the manufacturer pays for the cost of resetting. According to the agent's experience, the prior probability distribution for the number of damaged machines per shipment may be assigned as follows:

θ_j:	0	1	2	3	4
$f(\theta_j)$:	0.45	0.25	0.10	0.10	0.10

(a) Determine the payoff functions for a_1 and a_2.
(b) Show that the prior analysis leads to the selection of a_1 with $EP(a_1) =$ \$312.50.
(c) What is the value of perfect information in this problem?

8–13 Suppose, for the preceding problem, the agent would like to decide with prior and sample information together, and suppose it costs \$20 to inspect one machine. Show that the optimal decision rule is: take a sample of $n = 1$, select a_1 if there is no defective in the sample and select a_2 if the machine sampled is defective. Furthermore, for this rule, the net overall terminal expected payoff is \$316.90. Explain also why, in this case, optimal sample size is $n = 1$.

9

ORGANIZING AND PRESENTING DATA

9-1 INTRODUCTION

By this time, the reader should be thoroughly aware that decision making under uncertainty requires appropriate probability models, since stochastic phenomena cannot be analyzed without specific models, such as those introduced in Chapters 4 and 5. Probability models often may be constructed from a priori reasoning; but it is also quite common that the precise features of the model in question are unknown and, possibly, can never be known. For the setting up of a useful model, these features must be known at least approximately. The way of discovering them is via experimentation.

The whole process of collecting, organizing, and describing data is motivated by a desire to know something about a random variable and its probability distribution. Data, the set of specific values observed in an experiment, comprise what is called a sample, from which we wish to deduce something about the set of all possible values that could have resulted—that is, about the population.

The information we wish to gather depends upon the problem on hand. It may range from some simple parameter such as the expectation or the variance to a complete description of the population as given by the distribution function.

Selection and application of particular descriptive methods often depend upon the nature of variation underlying the data on hand. It is therefore important for us to distinguish clearly the several types of data we may encounter in our work before introducing various statistical measures.

Numerical data may be classified as cross-section or time series. *Cross-section data* are those which are gathered at a given point of time, such as the number of defective items in a lot of radios, or tensile strengths of steel wires of certain make. A *time series* consists of a collection of observations recorded over a period of time, such as annual production figures of General Motors since 1945, or the changes in population in the United States every ten years since 1800.

Cross-section data, in turn, may be either quantitative or qualitative. This distinction has already been briefly noted in Chapter 1. A *quantitative variable* is one whose variates have numerical values and possess an inherent order, such as the diameter of a lot of iron pipes, or ages of a group of people. A *qualitative, or categorical, variable* is one whose variations fall into one or another of a set of mutually exclusive and exhaustive non-ordered classes. For example, defectives and nondefectives are the two possible categories of the variable, say, television picture tubes of a certain make. Also, the four possible categories of the variable undergraduate students are freshman, sophomore, junior, and senior.

There is another important way of classifying cross-section quantitative data. That is, they may be univariate, bivariate, or multivariate in nature. *Univariate data* refer to a single variable, such as the I.Q.s of a group of children of the same age, or the grades of students in an economics class. Alternatively, we see that univariate data are results of observing a specific characteristic of each element in a population or sample. Sometimes, we may wish to observe two distinct traits of each element under consideration, such as, say, weights and heights of adult men. Then our observations would produce two distinct variables, the weight and the height, which are appropriately called *bivariate data*. Similarly, three or more distinct properties may be observed simultaneously about each element, such as the family income, family consumption expenditure, and the size of family, so that *multivariate data* are obtained.

Data obtained from observation or measurement are usually recorded in the order they are gathered. In order to facilitate interpretation and analysis, they must first be orgainized into some orderly fashion and presented in a table or a graph. As to how data should be organized and presented, we first must know the nature of the data and the purpose of the investigation. In this chapter, we shall introduce methods of organizing and presenting qualitative and univariate quantitative data. Treatment of other types of data will be discussed in future chapters.

9–2 ORGANIZING QUALITATIVE DATA

Qualitative data are those that do not posses magnitudes but that can be classified into a set of mutually exclusive and completely exhaustive categories. The original observations, even in a single instance, often take the form of some listing of pairs, each consisting of an element and its attribute under investigation. For example, if we know balls contained in a bag are identical in every respect except in color, we may then proceed to determine the colors of the balls by drawing, say, 10 balls in some systematic way from the bag for examination. Suppose the results are those listed below:

1st ball:	red
2nd ball:	green
3rd ball:	green
4th ball:	white
5th ball:	red
6th ball:	black
7th ball:	green
8th ball:	red
9th ball:	red
10th ball:	black

From this listing, we see that there are only four different categories of observed colors. Furthermore, the data in this listing can be presented more vividly if arranged into a table such as Table 9–1. This type of

Table 9–1. The Distribution of Ten Balls by Colors

	Frequency	
Color	*Absolute*	*Relative*
red	4	0.4
green	3	0.3
white	1	0.1
black	2	0.2
Total	10	1.0

organization of data is called a *frequency distribution*, since it attempts to show the relationship between a set of mutually exclusive and exhaustive categories or classes and the number of observations in each.

Observe that frequencies may be given in terms of absolute numbers, denoted by f_i, or in terms of proportions, which are obtained by dividing each class frequency, f_i, by the total number of frequencies, $\sum f_i = n$; i.e., f_i/n. When proportions are used, we have what is called a *relative frequency distribution*. From this simple illustration we can already appreciate the value of a frequency distribution. It shows us, at a glance, the salient features of data on hand. In our example we immediately see that "red" is the dominating color of the balls, and the runner-up is the color green. Furthermore, red and green constitute 70 percent of the colors of the balls, and so on. Other important advantages of this type of data organization will be revealed as we proceed with our discussion.

It is relatively simple, as shown by our current example, to construct a categorical distribution (as we shall see later, a number of technical details complicate the case of numerical distributions). However, it may be quite troublesome, on occasions, to determine the number of categories to be used. For instance, in studying the various kinds of food sold by a grocery store we might adopt classes such as "meat and meat products," "frozen foods," and so forth; we would then have difficulty in deciding to which class an item such as "frozen meat" belongs. Thus, we must be very careful in designing a set of classes that are mutually exclusive and exhaustive.

9–3 ORGANIZING UNIVARIATE DATA WITH ONLY A FEW DISTINCT VALUES

A set of n observations or measurements collected from some physical experiment on a quantitative variable, say X, are assumed to be real numbers and are usually denoted as

$$x_1, x_2, x_3, \ldots, x_n.$$

If there are $k < n$ distinct values of X among the observations and k is small (such as the inventory of hats in a store, which can take on only a few finite values in size), then the frequency distribution of X can be simply and conveniently formed as

$$x_1, x_2, \ldots, x_k,$$

$$f_1, f_2, \ldots, f_k, \qquad f_1 + f_2 + \cdots + f_k = n,$$

where f_i is the number of times x_i occurred in the sample. When this is done, the observations, $x_1, \ldots, x_k$ are usually arranged in numerical order.

Thus, we see that a *frequency distribution* is a function whose domain is the set of classes x_i, $i = 1, \ldots, k$, and whose range is the set of class frequencies. In other words, in a frequency function, $f(x_i)$, each of a set of classes is paired with a number—its frequency.

A frequency distribution, like a function, may be expressed by an equation, a table, or a graph. In practice, it is usually presented in the last two forms.

As an illustration, suppose we have tossed a die 100 times and found that the number 1 spot has occurred 12 times, the number 2 spot 14 times, and so on; then the total of 100 outcomes can be summarized into a frequency distribution as in Table 9–2. Note, in this distribution, that

x_i = the classes, which are the six possible values the die can take on when it is thrown;
f_i = the absolute class frequencies, each of which is the number of times a given number spot on the die occurred in the 100 tosses;
f_i/n = the relative class frequencies;
F_i = the sample cumulative frequency distribution function;
F_i/n = the relative cumulative frequency distribution function.

Relative frequencies, as the reader knows, are estimates of probabilities. Furthermore, from a practical point of view, relative frequencies are much more efficient than absolute frequencies for the purpose of comparing two or more frequency distributions, especially in a situation where the variables differ greatly in their patterns of variability, or in the sizes of samples, or in units of measurements, or in a combination of these phenomena.

Table 9–2. Distribution of 100 Tosses of a Die

x_i	f_i	f_i/n	F_i	F_i/n
1	12	0.12	12	0.12
2	14	0.14	26	0.26
3	20	0.20	46	0.46
4	18	0.18	64	0.64
5	21	0.21	85	0.85
6	15	0.15	100	1.00
Total	100	1.00	—	—

The sample *cumulative frequency distribution function*, or simply the sample *distribution function*, $F_n(x_i)$, is defined as

$F_n(x_i)$ = the number (or proportion) of the sample values that satisfy $X \leq x_i$.

The importance of $F_n(x_i)$ is that it enables us to obtain from a distribution the number (or proportion) of items in the sample that are equal to or less than a specific value of the variable. A moment's reflection will show that from $F_n(x_i)$ we can obtain such as $X \geq x_i$ or, say, $x_2 \leq X \leq x_5$.

9–4 THE ORGANIZATION OF UNIVARIATE DATA WITH MANY DISTINCT VALUES

When the number of observations is large and when the number of possible classes of description is almost or just as large as the number of actual observations, as is often true for quantitative variables, the domain of a frequency function can no longer be described by a few distinct measurement categories. Instead, it must be now described by a set of interval categories.

Consider, for example, the data in Table 9–3, consisting of 200 measurements of the resistance in ohms of a certain kind of electrical product. A glance through this set of data reveals that while many of the x_i's have the same value, the number of distinct values is over 60. Clearly, if we organized this set of data into a distribution with so many distinct values, the result would be no more compact than the list of the original values and might reveal nothing about the distribution pattern. Thus, the only efficient and useful way of organizing data such as these is to construct a frequency distribution with a set of interval classes. This we do by dividing the whole range—the difference between the maximum and minimum values of the series—into a small number of subintervals (classes), and then distributing the original observations in accordance with these subclassifications. The whole range for data in Table 9–3 is divided into nine classes, as shown in Tables 9–4 and 9–5, for the purpose of constructing a frequency distribution for our illustrative data.

To distribute a mass of raw data over the classes that have already been established we employ either a tally sheet or an entry form. *Tallying*, as shown by Table 9–4, involves establishing classes and representing each item that falls in a class by one diagonal stroke; then the number of items in each class is counted and stated as the class frequency.

The use of an entry form is illustrated by Table 9–5. In this procedure, the classes are set up horizontally at the top in ascending order from left to right. Then the actual observations are entered into the appropriate classes. Finally the number of items falling into each class is determined and recorded at the bottom of the class.

The entry form is more laborious than tallying; however, it has certain advantages. First, we can easily discover wrong entries by scanning the columns. Second, if the original classification is not satisfactory, reclassification can be made readily with little extra effort. Third, we can find out how closely the mid-value of a class agrees with the average of the values of the observations in that class.

The results of the tally sheet or the entry form may now be presented in a frequency table, such as Table 9–6 for our illustrative data. The relative and cumulative frequencies of the same data are also given in the same

Table 9-3. Two Hundred Measurements of Resistance in Ohms of an Electrical Product

4.70	4.62	4.48	4.45	4.20
4.14	4.10	4.08	4.00	4.39
4.44	4.35	4.29	4.28	4.13
4.33	4.37	4.43	4.59	4.10
4.49	4.34	4.33	4.24	4.20
4.17	4.28	3.76	4.16	4.50
4.00	4.07	4.22	4.53	4.55
4.44	4.02	4.20	4.13	4.16
4.20	4.31	4.12	4.19	4.18
3.98	3.96	4.17	4.20	3.99
4.16	4.44	4.33	4.59	4.54
4.16	4.21	4.42	4.38	4.34
4.13	4.31	4.22	4.33	4.12
4.46	4.27	4.21	4.20	4.05
4.26	4.15	4.08	4.12	4.10
4.15	4.07	4.25	4.30	4.29
4.45	4.37	4.20	4.13	4.09
4.23	4.20	4.19	4.00	4.39
4.38	4.48	4.42	4.38	4.38
4.11	4.18	3.68	4.23	4.28
4.17	4.35	4.45	4.36	4.25
4.25	4.27	4.24	4.36	4.41
4.18	4.05	4.08	4.06	4.07
4.19	4.14	4.09	4.20	4.08
3.99	4.16	3.95	4.17	4.04
4.07	4.13	3.87	3.88	4.10
3.76	4.19	4.13	3.99	4.21
4.12	4.05	4.09	3.98	3.87
4.00	3.91	3.97	4.10	3.90
4.09	3.89	3.95	3.88	3.93
3.84	4.00	3.93	4.02	4.15
4.13	4.03	4.25	4.18	4.08
3.87	4.07	4.23	4.35	4.29
4.21	3.98	3.99	4.08	4.07
4.19	4.20	3.95	4.06	4.01
4.28	4.10	4.00	3.92	4.25
4.02	4.13	3.95	3.90	4.03
4.02	4.07	4.08	4.16	4.27
4.24	4.28	4.13	4.14	4.07
4.07	4.06	4.13	4.06	4.17

The total value of this series is 834.28.

Table 9–4. Tally Sheet for 200 Measurements of Resistance of an Electrical Product

Class	*Tally*	*Number of observations*
3.67–3.79	///	3
3.79–3.91	𝍸 ////	9
3.91–4.03	𝍸 𝍸 𝍸 𝍸 𝍸 ///	28
4.03–4.15	𝍸 𝍸 𝍸 𝍸 𝍸 𝍸 𝍸 𝍸 𝍸 𝍸 ////	54
4.15–4.27	𝍸 𝍸 𝍸 𝍸 𝍸 𝍸 𝍸 𝍸 𝍸 𝍸 /	51
4.27–4.39	𝍸 𝍸 𝍸 𝍸 𝍸 𝍸 /	31
4.39–4.51	𝍸 𝍸 𝍸 //	17
4.51–4.63	𝍸 /	6
4.63–4.75	/	1

table. It is interesting as well as important to note that the F_i in this table are located between the classes. The reason for this will be given shortly.

As the reader may have already noticed, the construction of a numerical frequency distribution for discrete data with only a few distinct values, such as the toss of a die mentioned before, is a relatively simple matter. However, with continuous data, as in the present case, or discrete data with many distinct values, as in the case represented by data in Table 9–7, construction of a frequency distribution is more difficult. The difficulty arises because the domain of the frequency function is organized as a set of intervals, although the data are continuous. Let us comment on this type of frequency distribution by using Table 9–6 as a frame of reference.

First, a number of technical terms need clarification. The groupings 3.67–3.79, 3.79–3.91, and so on are called *classes*. The numbers on the left of the classes are called *lower class limits* or *boundaries* and the numbers on the right, *upper class limits* or *boundaries*. The point halfway between the class boundaries of each class is called the *midpoint*, or *class mark*, of that class. Thus, the class marks in Table 9–6 are 3.73, 3.85, and so on. A class mark of a given class is obviously obtained by dividing the sum of the class boundaries of that class by 2. The class mark is of considerable theoretical and practical importance. It is used as the "typical value" of class frequency. In other words, we assume that all observations in a given class have the same value as the class mark of that class. This is called the *midpoint assumption*. This assumption is necessitated by the fact that in forming the frequency distribution we lose the values of individual observations. The

Table 9–5. Entry Form for 200 Measurements of Resistance of an Electrical Product

3.67–3.79	3.79–3.91	3.91–4.03	4.03–4.15	4.15–4.27	4.27–4.39	4.39–4.51	4.51–4.63	4.63–4.75
3.76	3.84	4.02	4.07	4.17	4.27	4.39	4.62	4.70
3.68	3.87	4.00	4.10	4.25	4.28	4.50	4.50	(1)
3.76	3.89	3.98	4.09	4.18	4.28	4.44	4.53	
(3)	3.87	4.00	4.07	4.21	4.28	4.49	4.59	
	3.88	3.99	4.14	4.16	4.29	4.44	4.55	
	3.87	3.91	4.07	4.25	4.28	4.46	4.54	
	3.88	3.96	4.13	4.15	4.29	4.45	(6)	
	3.90	4.02	4.05	4.16	4.28	4.44		
	3.90	3.98	4.10	4.18	4.29	4.48		
	(9)	4.02	4.05	4.17	4.27	4.48		
		3.95	4.12	4.20	4.38	4.43		
		3.97	4.10	4.17	4.27	4.42		
		3.95	4.10	4.26	4.34	4.42		
		3.93	4.03	4.16	4.35	4.45		
		3.99	4.14	4.23	4.36	4.45		
		3.95	4.07	4.16	4.36	4.41		
		3.95	4.13	4.25	4.38	4.39		
		3.99	4.07	4.21	4.30	(17)		
		3.98	4.03	4.24	4.33			
		4.00	4.07	4.17	4.38			
		3.92	4.11	4.18	4.33			
		4.00	4.06	4.21	4.33			
		3.99	4.09	4.19	4.35			
		4.00	4.08	4.15	4.37			
		3.93	4.08	4.20	4.31			
		4.02	4.12	4.15	4.31			
		4.01	4.08	4.19	4.34			
		4.00	4.13	4.20	4.37			
		(28)	4.09	4.18	4.35			
			4.10	4.22	4.38			
			4.08	4.20	4.33			
			4.14	4.16	(31)			
			4.06	4.22				
			4.13	4.21				
			4.08	4.19				
			4.06	4.25				
			4.10	4.20				
			4.06	4.24				
			4.13	4.25				
			4.05	4.17				
			4.12	4.23				
			4.09	4.19				
			4.07	4.24				
			4.08	4.20				
			4.13	4.20				
			4.04	4.16				
			4.13	4.23				
			4.08	4.20				
			4.13	4.19				
			4.07	4.20				
			4.13	4.20				
			4.07	(51)				
			4.12					
			4.13					
			(54)					

Table 9–6. Frequency Distribution of 200 Measurements of Resistance of an Electrical Product

Interval number i	*Class boundaries* (*ohms*)	*Midpoint* x_i	*Frequency* f_i	*Relative frequency* f_i/n	*Cumulative frequency* F_i	*Cumulative relative frequency* F_i/n
1	3.67–3.79	3.73	3	0.015	3	0.015
2	3.79–3.91	3.85	9	0.045	12	0.060
3	3.91–4.03	3.97	28	0.140	40	0.200
4	4.03–4.15	4.09	54	0.270	94	0.470
5	4.15–4.27	4.21	51	0.255	145	0.725
6	4.27–4.39	4.33	31	0.155	176	0.880
7	4.39–4.51	4.45	17	0.085	193	0.965
8	4.51–4.63	4.57	6	0.030	199	0.995
9	4.63–4.75	4.69	1	0.005	200	1.000
Total			200	1.000		

Table 9–7. Distribution of Employees of a Corporation by Annual Wages

Class	*Frequency*	*Frequency density*
Under $5,000	100	20.00
$5,000– 5,999	150	150.00
6,000– 6,999	223	223.00
7,000– 7,999	451	451.00
8,000– 9,499	764	509.33
9,500–10,999	912	608.00
11,000–12,499	807	538.00
12,500–13,999	764	509.33
14,000–15,999	810	405.00
16,000–17,999	653	326.50
18,000–19,999	421	210.50
20,000–22,999	82	27.33
23,000–25,999	53	17.67
26,000–29,999	21	5.25
30,000 and over	19	0.00
Total	6230	—

justification for this assumption is the belief that the class frequency is likely to distribute itself uniformly between the class boundaries. The number of units between two adjoining class limits is called the *class interval* and it may be denoted by c. When the same size of class interval is used for all classes, we have what is called a frequency distribution with *uniform class intervals*. Also, as has already been pointed out, the number of occurrences in each class is called the *class frequency*, f_i, and $\sum f_i = n$.

Next, a few "rough" rules need to be followed for efficiency in forming a frequency distribution of the class-interval type.

(1) The number of classes used should be neither too many nor too few. A large number of classes may not condense data sufficiently to be of any practical value, and a small number of classes tends to oversummarize data so that much valuable information is lost. The rule of the thumb is to have groupings varying from 5 to 15. Obviously, the actual number of classes to be employed depends upon the number of items in the sample and the range of actual observed values. Most important is the investigator's judgment that a certain number is about right for the data on hand.

In our illustrative data there are 200 items and the range is $4.70 - 3.68 = 1.02$. Since we have decided to use nine classes, the class interval should be $1.02/9 = 0.1133\ldots$. However, instead of using this exact but highly impractical class interval, we have selected a more convenient value of 0.12. To apply this practice, it is permissible to start with a smaller value than the minimum value in the series as the lower class boundary of the first class and to end with a larger value than the maximum value in the series as the upper class boundary of the last class. This is exactly what we have done in constructing a frequency distribution for the resistance data in Table 9–6.

(2) Sometimes a distribution may be formed without the lower class limit for the first class and/or without the upper class limit for the last class. Such classes are called *open-ended*. Also, class intervals need not be uniform. Open-ended classes and varying class intervals may be used to advantage in situations where there are a few extremely small and/or extremely large values, where data possess large gaps, and where the number of observations runs into thousands or even millions. For example, Table 9–7, which contains the distribution of all employees of a certain large corporation classified according to annual wages, is constructed by employing both practices.

When varying class intervals are used, it is necessary to compute what is called the *frequency density*. This is done by estimating what the class frequencies would be if uniform class intervals were utilized. Thus, in column 3 of Table 9–7 frequency densities are computed under the assumption of a uniform class interval of \$1000.00. Note that the frequency density for the first class is 20, since we have assumed a "zero" as the lower class boundary for that class, and $100/5 = 20$. Also note that the frequency density for the last class is zero, because we have assumed the upper class boundary for this class is ∞.

Introduction of the concept of frequency density enables us to define more precisely the quantity measured by the size-classes. "Frequency" is a loose descriptive term for this quantity. More precisely, this quantity represents the density of the observations—the number of cases—per unit of class-sizes. This idea will become more apparent when we consider graphic presentations of frequency distributions later.

(3) The class limits should be so designed as to possess mutual exclusiveness. Namely, given the set of class boundaries, we should have no doubt as to which group a given observation belongs. With discrete data, as in Table 9–7, there is no problem, since the values of the variable have definite breaks. With continuous data, however, artificial breaks must be created, and yet the reader of the table must be reminded of the continuity of data. To provide the appearance of continuity, we may employ seemingly overlapping class limits. For example, in Table 9–6 the class limits were designated as 3.67–3.79, 3.79–3.91, and so forth. These class limits, in reality, should be considered as simplifications for 3.67–3.78999 . . . , and 3.79–3.90999 . . . , indicating that the 9's continue indefinitely. For this reason we have adopted the convention that the classification 3.67–3.79 actually means $3.67 \leq x_i < 3.79$ and 3.79–3.91 means $3.79 \leq x_i < 3.91$. Thus, if an observation has a value of 3.79, it belongs to the second, not to the first, grouping. For the same reason we have adopted the convention (in the case of continuous distributions) of entering the cumulative frequencies between classes rather than centering them in the classes.

The determination of class intervals also differs from continuous to discrete data. In the case of continuous data, the class interval is simply the difference between the upper and lower class boundaries. For example, in the resistance data the class interval is 0.12, since $3.79 - 3.67 = 3.91 - 3.79 = \cdots = 0.12$. This is clearly consistent with our previous interpretation of class limits for continuous data. For discrete data with classifications such as, say,

$100–$199
200– 299
. . .

the class interval is $100 and not $99, since 100 is included as an observation in the first class and 200 is included as an observation in the second class.

9–5 GRAPHIC PRESENTATIONS OF FREQUENCY DISTRIBUTIONS

Properties of a distribution can often be brought out more vividly and effectively by a graph instead of a table. In addition, a graph is more suggestive as to the pattern of a distribution and the appropriate descriptive measures required.

Graphs for frequency distributions differ from each other with respect to types of data, whether they are quantitative or qualitative, and to forms of data organization, whether or not they are organized as cumulative distributions.

Graphs for Qualitative Distributions

A distribution classified according to attributes is usually presented in the form of a *horizontal bar chart,* such as Figure 9–1, which represents data from Table 9–1.

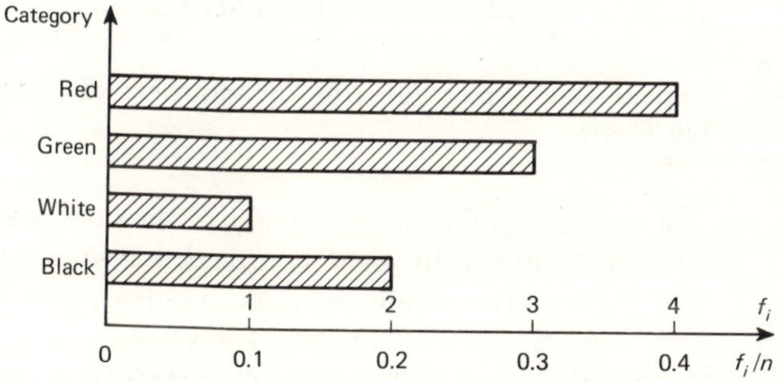

Figure 9–1 Horizontal Bar Chart of Ten Balls Distributed by Color

In a horizontal bar chart, as can be seen in Figure 9–1, the vertical scale indicates the categories and the horizontal scale measures the frequency, absolute or relative, in each category. The length of each bar is the frequency in that particular category. Widths of bars are entirely arbitrary and have no practical significance.

Graphs for Quantitative Distributions with a Few Distinct Values

A quantitative distribution with only a few discrete and distinct values is usually represented by a *vertical bar chart.* Thus, a graph for data in Table 9–2 would take the form of Figure 9–2.

The construction and properties of a vertical bar chart are the same as those of a horizontal bar chart, except that the scales are interchanged.

Graphs of Quantitative Distributions with Interval Groupings

There are three types of graphs for distributions in the grouped form: the histogram, the frequency polygon, and the smooth curve.

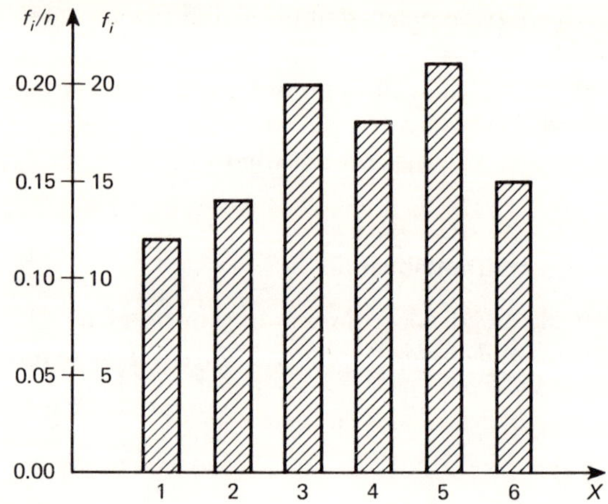

Figure 9–2 Vertical Bar Chart for the Distribution of Outcomes of Tossing a Die 100 Times

A *histogram* is constructed by using a segment of the horizontal scale to represent each class interval and by erecting a rectangle equal in height to the frequency in that class. Histograms representing data from Tables 9–6 and 9–7 are shown respectively by Figures 9–3 and 9–4.

Note that the vertical scale of a histogram must start from zero, though the horizontal scale may begin with any number. Notice also that there are

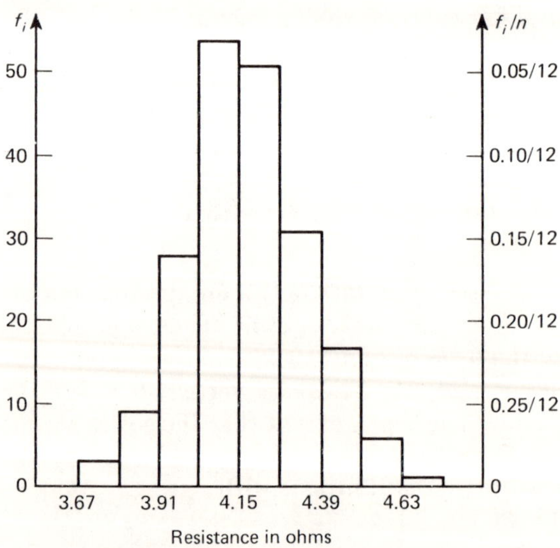

Figure 9–3 Histogram for Resistance Data

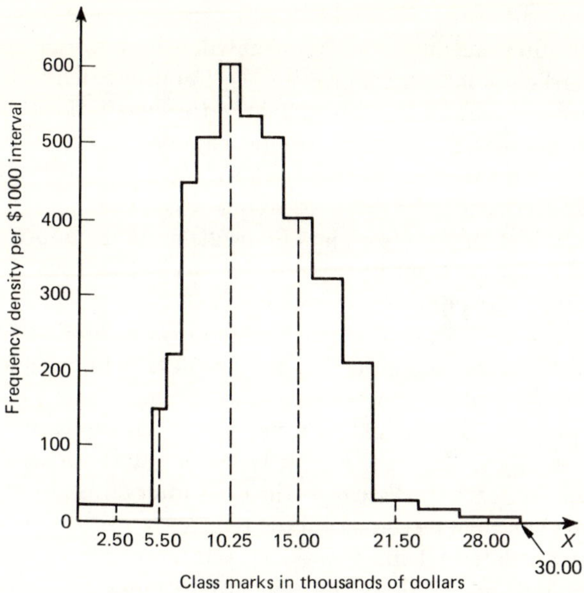

Figure 9–4 Histogram for Annual-Wages Data in Table 9–7

two vertical scales for Figure 9–3: the left one measures the absolute frequencies and the right one the relative frequencies.

In a histogram for absolute frequencies, the width of a bar measures the size of a class interval, while the height shows the frequency in that class on a per-interval basis. This concept of frequency per interval is identical with that of frequency density introduced before. Thus, in Figure 9–3, the size of class interval is 0.12 ohms, and the height for the first bar, for example, is 3, the absolute frequency in that class. This relationship should be interpreted as 3/0.12, with f_i/c being read as "frequency per interval." Now, when the frequency density, f_i/c, or the height of the bar is multiplied by the class interval, c, we have

$$\left(\frac{f_i}{c}\right)c = f_i,$$

which is the area of the bar—representing the frequency in a given class. From this we see that the total area of all the bars in a histogram equals the total number of items in the distribution. That is,

$$\sum\left(\frac{f_i}{c}\right)c = \sum f_i = n.$$

In the case of uniform class intervals, frequency density and class frequency are the same when the class-interval units are ignored. With varying class intervals, however, frequency density is obtained by employing a standard class interval. For example, frequency densities for data in Table 9–7 are determined by using $1000 as the standard interval. Clearly, frequency densities are in effect frequencies on a per-standard-interval basis. Since the height of the bar represents frequency density, we obviously must employ frequency densities, or frequencies per standard interval, in constructing a histogram when class intervals are not uniform. See Figure 9–4. Note also the alternative style of Figure 9–4 in contrast to that of Figure 9–3. Both are equivalent presentations.

Turning our attention to the construction of a histogram for a relative frequency distribution, we note that the problem is to construct one such that the total area of all the bars is equal to unity—the sum of all class relative frequencies. The height of a bar will equal the relative frequency for each class only if the class intervals are one unit wide, such as 1 oz., 1 inch, $1, and so forth. If, however, the class interval is not one unit wide, the height of a bar must be converted to "relative frequency per unit width." For instance, for data in Table 9–6 the class interval is 0.12 ohms, implying the measurements are made to the nearest 0.01 ohm. Thus, the interval is considered to consist of 12 such units. To reduce the relative frequencies to per unit width in this case, we simply divide each relative frequency by 12 and mark the vertical scale accordingly. (Study the vertical scale on the right of Figure 9–3 with data in Table 9–6.) When this is done, the same area of each bar becomes

$$\left[\frac{(f_i/n)}{12}\right]12,$$

and

$$\sum_i \left[\frac{(f_i/n)}{12}\right]12 = 1.$$

Note that the requirement of constructing a histogram for a relative frequency distribution on a per-unit-width basis is similar to that of plotting frequency densities with an absolute frequency distribution.

A *frequency polygon* is the line diagram of a frequency distribution. We construct it like a histogram by marking off the horizontal axis into class intervals and the vertical axis into numbers representing class frequencies. Instead of using a rectangle to show the frequency in each class, however, we plot a point at a height corresponding to the frequency density in each class at the class mark of each interval. Then these points are linked by straight lines. It is the practice to extend this line diagram to the class mark immediately before the first class on the left and to the class mark of the class immediately after the last class on the right. The two classes at both

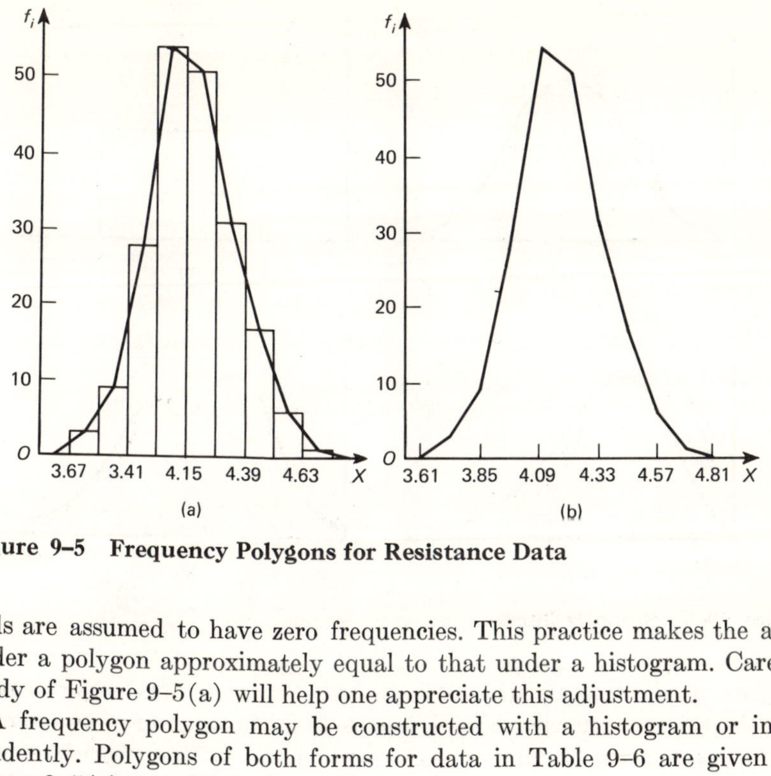

Figure 9–5 Frequency Polygons for Resistance Data

ends are assumed to have zero frequencies. This practice makes the area under a polygon approximately equal to that under a histogram. Careful study of Figure 9–5(a) will help one appreciate this adjustment.

A frequency polygon may be constructed with a histogram or independently. Polygons of both forms for data in Table 9–6 are given by Figure 9–5(a) and (b).

A *smooth curve* is a curve fitted to a histogram or polygon either freehand or in accordance with an appropriate mathematical expression under the assumption that the distribution is based on a very large number of observations on a potentially continuous scale.

A smooth curve performs two important functions. First, a smooth curve, based on a histogram that usually represents sample data of a limited number of observations, is considered as an approximation of the underlying population distribution from which the sample has been drawn. In this sense we sometimes call a smooth curve a *population model*. Second, the function rule for a histogram, because of the irregularities produced by limited observations, is often difficult or even impossible to formulate. However, the function rule for a smooth curve, approximately the same as the distribution under consideration, is relatively easy to discover. For these reasons, whenever we wish to discuss the "shapes" or "patterns" of distributions, we usually have in mind smooth curves of one kind or another.

It is easy to imagine that, theoretically, there must be an infinitely large number of population models. Fortunately, however, only a few are frequently encountered in practice. We shall mention some of them here.

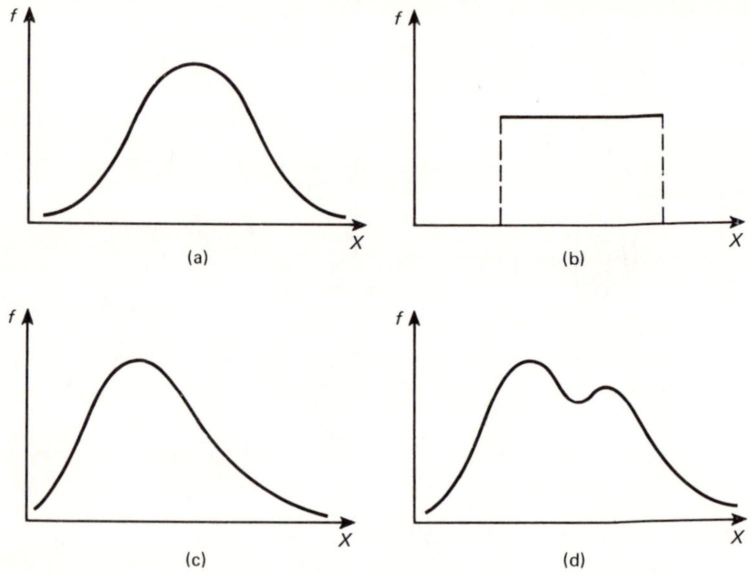

Figure 9–6 Smooth Curves

If we fitted a smooth curve to Figure 9–3, we would obtain a symmetric, bell-shaped curve as shown by Figure 9–6(a). Such a curve represents a normal population distribution, or a normal curve. Many variables, such as the weights of certain kinds of animals, the heights of men, the IQ's of school children, the characteristics of manufactured outputs, and so on are known to be normally distributed.

Data in Table 9–2 suggest what is called a *uniform*, or *rectangular, distribution*. As can be seen in Figure 9–6(b), a uniform distribution has the same frequency for each possible value or each group of values.

A distribution may be "skewed." A *skewed distribution*, the reader may recall, is one that lacks symmetry from the central value of a variable. For example, a smooth curve fitted to wage data as represented by Figure 9–4 would look like the picture in Figure 9–6(c), with frequencies concentrated at a relatively low value and a long tail on the right. This model is said to be *positively skewed*, or *skewed to the right*. Sometimes a variable, such as the ratio of sales cost to net sales of an industry, may have a *negatively skewed distribution*—a distribution with a long tail on the left and with the greatest concentration of frequencies at a relatively high value.

In closing this section, let us note that when a distribution has only one peak, it is said to be a *unimodal distribution*. A distribution with two peaks, as shown by Figure 9–6(*d*), is called *bimodal*. The distribution of deposit balances at commercial banks would be bimodal, with one "mode" for household accounts at a lower value and another "mode" for business accounts at a higher value. A distribution with three peaks is called *tri-modal*, and so on.

Graphs for Cumulative Distributions

There are also three types of graphic presentations of cumulative frequency distributions, corresponding to those for distributions in the non-cumulative form. They are the step diagram, the ogive, and the cumulative curve.

The *step diagram* graphs a cumulative distribution by a series of horizontal lines drawn at levels of the cumulated frequencies in the successive classes, with the vertical and horizontal axis scaled in the same fashion as for a histogram. The endpoints of the horizontal lines may or may not be connected by vertical lines. Cumulative data in Tables 9–2 and 9–6 are so graphed in Figures 9–7 and 9–8, respectively.

The ogive is a polygon that represents a cumulative distribution in the form of a line diagram. Figure 9–9 is the ogive for the resistance data given in Table 9–6. Note that the cumulative frequencies are plotted on the class boundaries.

An ogive is used mainly for interpolations, which can be made in two ways. First, if a point on the horizontal scale is selected, the corresponding number, or proportion, of observations in the distribution whose values are equal to or less than the value indicated by the selected point can be found from the vertical scale. For example, if we select the point of, say, 4.45 on the horizontal scale in Figure 9–9, draw a vertical line up to intersect the

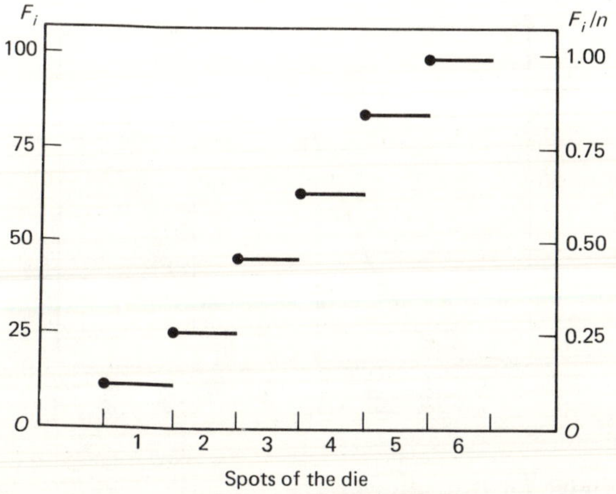

Figure 9–7 Step Diagrams for 100 Tosses of a Die

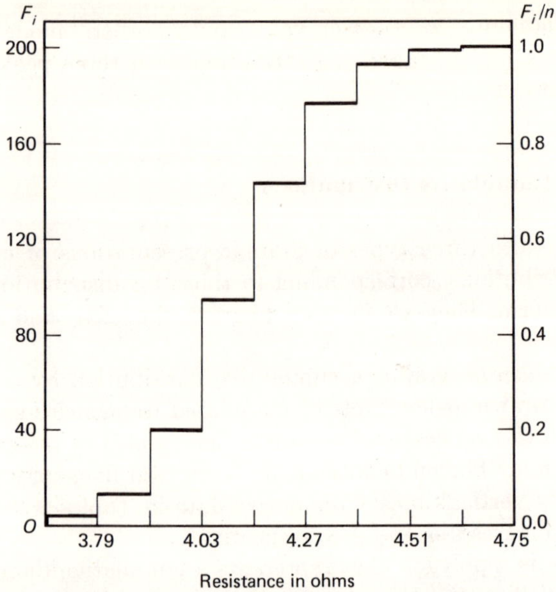

Figure 9–8 Step Diagram for Resistance Data Measured from 200 Units of an Electrical Product

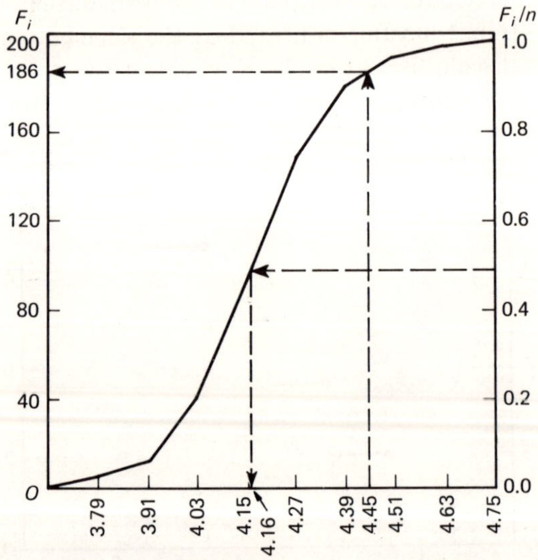

Figure 9–9 Ogive for Resistance Data

ogive, and from this intersection draw a horizontal line to the horizontal scale on the left, we get a value of 186. This means that 186 observations in the sample have values equal to or less than 4.45 ohms.

The second way of interpolation is just the first way reversed. This time we move from the vertical axis to the horizontal axis in order to find the value below which a given number, or proportion, of observations may fall. To illustrate, let us use the vertical scale on the right in Figure 9–9. Suppose we draw a horizontal line from the point 0.5 to intersect the ogive and then drop a perpendicular from the intersection to the horizontal scale; we then find a value of 4.16. This means that about 50 percent of the observations in the sample have values equal to or less than 4.16 ohms. This result also means that about 50 percent of the observations have values greater than 4.16 ohms. A value such as this, which divides the whole series into two equal parts, is called the *median*—a type of average to be discussed in the next chapter.

Just as a histogram or a polygon can be smoothed out to obtain a population model, so can a step diagram or ogive be smoothed out to represent the underlying population distribution function. Figure 9–10 shows four cumulative curves, or smoothed ogives, corresponding to (a) a uniform, (b) a symmetrical, (c) a positively skewed, and (d) a negatively skewed distribution.

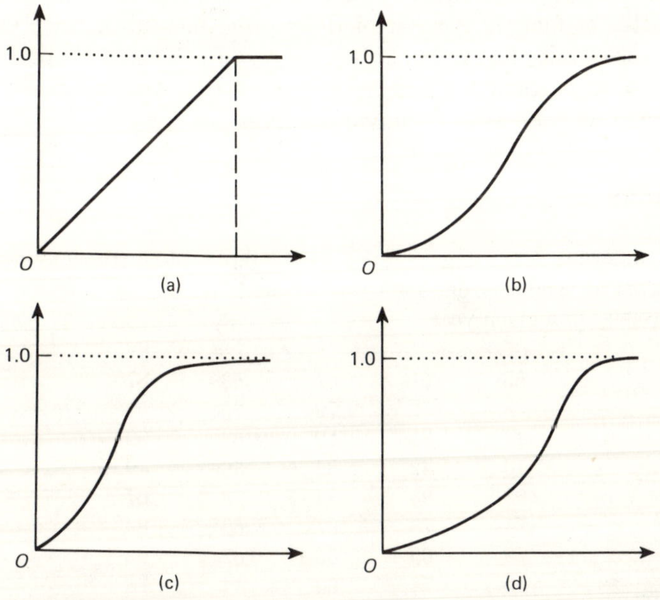

Figure 9–10 Cumulative Curves

9–6 CONCLUDING REMARKS

This chapter has been concerned with the organization and presentation of univariate data via the concept of "frequency distribution." The construction of frequency distributions is a relatively simple matter in principle, although the attainment of efficiency and of elegance in the end product requires considerable skill and personal judgment—which the reader will undoubtedly gain through experience.

There are a number of advantages to organizing a set of univariate data into a frequency distribution. First, a frequency distribution gives data an orderly form, facilitating further investigation. Second, the data are condensed to workable size without loss of their salient features. Third, a portrait is provided of the distribution pattern of data under study. Finally, as we shall see in the following chapter, it may be much more efficient to calculate descriptive measures from a frequency distribution than to do so from the ungrouped data originally recorded. These advantages become more sharply evident when we are concerned with large samples.

In concluding this chapter, we may also observe that any function, $x(x_1, x_2, \ldots, x_n)$, of a set of n observations is called a *sample statistic*, or *descriptive measure*. A frequency function, then, is a descriptive measure. As a matter of fact, it is a set of descriptive measures, since each class frequency is a trait of a group of observations. In the next chapter, we shall consider those particular summary measures that are commonly used to describe certain properties of univariate data.

EXERCISES

9–1 The following set of data represents final grades of 50 students in economics selected at random from among all students taking the course at a certain university in a given year.

66	61	67	53	87
58	72	79	36	51
47	81	42	62	92
40	42	82	56	49
93	95	34	98	67
74	78	67	68	89
70	65	65	67	62
56	74	60	44	78
34	86	75	76	55
66	44	85	88	76

Total for this sample is: 3312.

(a) Construct a frequency distribution (call it Table 9P–1) for the data above with $c = 10$, using an entry form.

(b) Determine f_i/n, F_i, and F_i/n for the results in (a).

9–2 The following set of data represents weekly per-capita food-consumption expenditures of 100 randomly selected families in San Francisco in 1971.

$24.05	7.65	8.18	11.25	13.20
15.65	7.42	13.03	13.20	8.63
19.60	22.38	13.67	5.50	11.12
15.10	6.41	14.95	23.62	15.10
13.70	11.03	17.00	12.87	21.00
21.15	14.32	12.14	17.21	13.58
19.10	21.00	18.78	19.05	5.08
14.24	10.10	14.25	13.84	12.72
14.73	14.85	7.44	5.45	5.15
18.24	16.81	6.68	27.87	17.07
10.81	12.00	19.13	33.11	4.85
14.93	29.82	10.62	20.71	19.89
23.96	21.09	23.50	13.55	21.78
29.33	8.67	26.81	12.24	13.50
20.27	7.51	17.02	14.34	13.85
20.90	9.12	9.26	6.61	17.44
15.41	7.60	5.91	11.23	5.55
13.10	8.59	10.62	9.61	8.63
17.55	21.84	22.13	21.90	14.59
9.50	10.15	8.78	17.20	16.85

Total value for this sample is: $1,469.52.

(a) Construct a frequency distribution (call it Table 9P–2) for the data above, using a tally sheet, with the first class designated as 4.50–7.49.

(b) Determine f_i/n, F_i, and F_i/n for the results in (a).

9–3 For the data in Table 9P–1, construct the following:

(a) Histogram, with both absolute and relative scales.

(b) Polygon, with both absolute and relative scales.

(c) Step diagram.

(d) Ogive.

9–4 Rework Exercise 9–3 for the data in Table 9P–2.

9–5 What underlying population model is suggested by the polygon of data in Table 9P–1?

9–6 How would a smoothed ogive for data in Table 9P–2 look? What kind of population model does it suggest?

10

DESCRIPTIVE MEASURES FOR UNIVARIATE DATA

As we noted in concluding Chapter 9, any function $x(x_1, x_2, \ldots, x_n)$ of n observations is called a sample statistic. Parallel measures constructed for populations are called parameters. We note that corresponding parameters and statistics are computed from population and sample data according to much the same equations; hence we adhere to the following conventions: (1) Population and sample sizes are denoted as N and n, respectively. (2) Parameters are represented by Greek letters or roman capitals, statistics by lower-case roman letters—for example, μ versus $\bar{x}$, σ^2 versus s^2, $X_{.5}$ versus $x_{.5}$, and so on. In reality, for reasons stated in Chapter 1, we seldom observe entire populations, and so in this chapter we shall define all summary measures of univariate data in terms of samples.

From the definition of a statistic above, there could be many functions of a sample we would like to have. In terms of theoretical analysis and of application, however, we are interested only in four broad geometric properties that characterize frequency distributions of univariate data:

(1) The location of the center of a distribution, or the degree of central tendency.
(2) The degree of variation of outcomes around the central point, or the interval of central tendency.
(3) The degree of skewness of a distribution—i.e., the lack of symmetry on both sides of the peak of a distribution.
(4) The degree of peakedness of a distribution, which describes how fast the curve falls from its peak.

Most of these properties have already been introduced in our discussion of random variables. Although they are more meaningful for unimodal distributions, they apply equally well for comparisons between different patterns of distributions. To facilitate interpretations, we shall relate our discussions mainly to unimodal distributions.

10–1 AVERAGES: MEASURES OF CENTRAL TENDENCY

As the reader already knows, a measure of central tendency is called an *average,* and there are many types of average, each possessing particular properties. Some, such as the median, the mode, and the mean, the reader has already encountered. Besides these, we shall introduce in this section two new averages—the geometric mean and the harmonic mean.

The Median

The *median,* denoted as $x_{.5}$, is, as its name indicates, the value of the middle item in a series when the items are arranged according to magnitude. For a series, say, \$2, \$4, \$5, \$7, and \$8, the median is the value of the third item, \$5. If there are six items in a series, say, 3, 4, 6, 7, 8, and 10 pounds, then any value between 6 and 7 pounds would divide the series into two equal parts; therefore, any such value could become the median. In practice, we usually assume the median to be halfway between the two central items. In our example, therefore, the median would be 6.5 pounds. The median may have values identical with its own on both sides of it. For instance, in a series of 1, 2, 3, 4, 5, 5, 5, 6, 7, 8, 9, the median is 5. Because of this characteristic, the *median* may now be formally defined as that value which divides a series in such a fashion that at least 50 percent of the items are equal to or less than it and at least 50 percent of the items are equal to or greater than it.

Clearly, the value of the median cannot be located from ungrouped data; we require an ordered array. This is considered the main defect of the median, since the construction of an array is often tedious and time-consuming.

As it is defined, the median is influenced only by the number of items and not by the value of items in the series. Thus it is a highly desirable measure of central location for skewed distributions, such as the distribution of income. Owing to this property, too, the median is especially meaningful in describing observations that are scored, such as rates, grades, and ranks, rather than counted or measured.

The median also has the interesting property that the sum of the absolute deviations of the observations from the median is less than the sum of the absolute deviations from any other point in the distribution. In symbols,

$\sum | x_i - x_{.5} | =$ a minimum. The median is often selected because of this property. A simple application is the decision on the optimal location of a plant. For example, suppose a restaurant chain organization has seven restaurants on a certain turnpike:

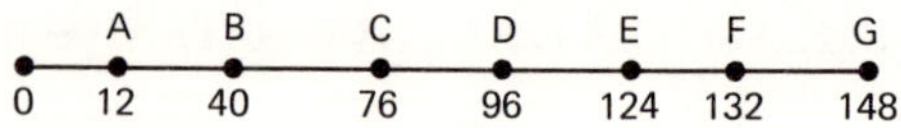

Each restaurant must send each day for fresh food from a central storehouse. To minimize the total distance from the seven restaurants to the central storehouse, where should it be located? Clearly, remembering that the sum of absolute deviations from the median is a minimum, we can easily decide that the central storehouse should be located at restaurant D or in its vicinity. The median distance is 96 miles and from it the total amount of travel would be minimized, at 276 miles each way. If the central storehouse is located at any other point, the sum of distances will be greater. At restaurant C, for instance, this sum is 296 miles. The student may check these computations in Table 10–1.

Table 10–1. Computation of Median Distance

		Distance	
Restaurant	*Location*	*From D*	*From C*
A	12	84	64
B	40	56	36
C	76	20	0
D	96	0	20
E	124	28	48
F	132	36	56
G	148	52	72
Total		276	296

The Mode

The *mode,* denoted by x_m, is that value of a series which appears more frequently than any other. This value can be discovered immediately when the data are arrayed. For example, in the series 1, 2, 3, 4, 4, 5, 6, and 7, the mode is 4. We can therefore consider the mode as typical in the sense that it is the most probable value in a series. The mode for a unimodal distribution (for ungrouped data) always coincides with an actual value in the series.

Many people have the mode in mind when they speak of "averages." When students say the average grade in their classes is C, for example, they usually mean that this, more than any other, is the grade students receive. The "average consumer" often means that type of consumer who appears most frequently with regard to a pattern of consumption. The term "typical wage" usually refers to the modal wage. The modal size of men's shoes is the typical size in the sense that more people buy this size than any other. Thus, we use the mode in preference to other averages if we like to point to the most typical value of the series.

An interesting application of the mode is the checking of calculations. If a calculation is repeated a number of times, the result accepted is the one that appears the greatest number of times.

Although the mode is a very simple and useful concept, its application presents many troublesome aspects. First, a distribution may reveal that two or more values repeat themselves an equal number of times, and in such a situation there is no logical way of determining which value should be chosen as the mode. Strictly speaking, with a discrete series, any value is called a mode if its number of appearances is not exceeded by either of the adjacent values. As long as the frequencies of the modal values are not equal, however, we might decide to select the value with the highest frequency as the mode for the series.

Second, in dealing with a continuous series whose values follow each other in a continuous sequence, there will be no mode unless the data are grouped. Also, even for discrete data, we may not find any value that appears more than once, as in the case of the populations of the large cities in the United States.

Third, the mode is a very unstable value. It changes radically if the method of rounding the data changes. For illustration, let us assume a series of weights with different methods of rounding as in Table 10–2.

Table 10–2

Actual weight	*Rounded to nearest pound*	*Rounded to last pound*
3.000	3	3
3.500	4	3
3.571	4	3
3.784	4	3
4.500	4	4
4.831	5	4
6.115	6	6
6.115	6	6

The mode for this series in actual weights is 6.115 pounds; when the data are rounded to the nearest pound, the mode becomes 4 pounds; when the data are rounded to the last pound, still another mode value appears—3 pounds. This erratic behavior may lead one to doubt the significance of the mode.

Finally, the mode could be an extreme value, as in the case of a triangle distribution, and then it could hardly be considered a measure of central tendency.

The first three difficulties can be overcome totally or in part when the data are arranged into a frequency distribution. In such a form, any tendency of the data to cluster around a certain value is more clearly shown, and the problem created by the first situation may be solved. Moreover, the erratic nature of the mode obtained from an array owing to different methods of rounding can be stabilized to a great extent by determining the mode from the frequency distribution of the same data. When they are grouped into classes, those items that differ in values only on the right of the decimal points may still fall in the same class irrespective of the method of rounding. Finally, a modal value may appear in grouped data even though there is no repetition of any value in the array, or because the data are continuous. A good illustration would be the series of the population of each large city in the United States, cited earlier; no repetition of any single value can be found from its array, but a modal population appears when the same data are grouped.

The Arithmetic Mean

The arithmetic mean, because of ease of computation and long usage, is the best known and most commonly used average. Sometimes it is simply referred to as "the mean" or "the average," but appropriate adjectives must be used when references are made to other types of means.

The *arithmetic mean*, denoted as $\bar{x}$, is the sum of the individual values in a sample divided by the number of observations in the sample:

$$\bar{x} = \frac{x_1 + x_2 + \cdots + x_n}{n} = \frac{1}{n}\sum_{i=1}^{n} x_i. \tag{10–1}$$

Applying the formula to the sample data in Table 9–3, we have

$$\bar{x} = \frac{1}{n}\sum x_i = \frac{1}{200}(834.28) = 4.17 \text{ ohms.}$$

The mean, as a computed average, has a number of interesting mathematical properties.

First, the mean is a typical value in the sense that it is the center of gravity—a point of balance. That is, the values of the observations on one side of the mean equal the values of the observations on the other side. The mean is typical also in that its value may be substituted for the value of each item in the series without changing the total. This property can easily be seen from the formula for the mean. Since $\bar{x} = \sum x_i/n$, we may derive, by multiplying both sides of the equation by n, the result $n\bar{x} = \sum x_i$. This feature often becomes the important consideration for us in deciding whether to use the mean instead of other types of averages that do not possess an analogous algebraic property.

Another property of the mean is that the algebraic sum of deviations from the mean (that is, taking due account of algebraic signs) is zero. This property can be written symbolically as $\sum (x_i - \bar{x}) = 0$. Algebraically, this property is derived from the fact that $n\bar{x} = \sum x_i$. Since the sum of variables, each reduced by a constant, is equal to the sum of the variables minus the constant multiplied by the number of terms in the sum, we can deduce from the equation that

$$\begin{aligned}(x_1 - \bar{x}) + (x_2 - \bar{x}) + \cdots + (x_n - \bar{x}) \\ &= x_1 + x_2 + \cdots + x_n - \underbrace{\bar{x} - \bar{x} - \cdots - \bar{x}}_{n \text{ times}} \\ &= \sum x - n\bar{x} = \sum x - \sum x = 0.\end{aligned}$$

The third mathematical property of the mean is that the sum of the squared deviations of the items from the mean is less than the squared deviations from any other point in the series. Symbolically, $\sum (x_i - \bar{x})^2 =$ minimum. (Readers may wish to verify this by way of some numerical illustrations of their own.) Because of this property the mean usually serves as a basis for measures of dispersion, and it is often used as a criterion for fitting curves to observed data. These points will be taken up in subsequent chapters.

The Geometric Mean

The *geometric mean*, denoted as g, may be defined as the nth root of the product of n items or values.

For ungrouped data, the formula is

$$g = \sqrt[n]{(x_1)(x_2)(x_3)\cdots(x_n)}. \tag{10–2}$$

For example, the geometric mean of 2 pounds, 4 pounds, and 8 pounds is

$$g = \sqrt[3]{(2)(4)(8)} = \sqrt[3]{64} = 4 \text{ lb}.$$

For the same series, $\bar{x} = 4.7$ pounds. It is always true that the arithmetic mean is greater than the geometric mean for any series of positive values, unless the items being averaged are of the same value, in which case the two averages are the same.

The computation of the geometric mean is made much easier if the formula is reduced to its logarithmic form, as

$$\log g = \frac{\log x_1 + \log x_2 + \cdots + \log x_n}{n}$$

$$= \frac{1}{n} \sum_i \log x_i. \qquad (10\text{–}3)$$

Thus, the logarithm of the geometric mean is equal to the arithmetic mean of the logarithms of the values. It may be noted that the geometric mean is meaningful only for sets of observations that are all positive. In addition, the geometric mean possesses two mathematical properties. The first is that the product of the values of a series will remain unchanged when the value of the geometric mean is substituted for each individual value. For example, the geometric mean for series 2, 4, 8, 16, and 32 is 8; therefore, we have

$$(2)(4)(8)(16)(32) = 32{,}768 = (8)(8)(8)(8)(8).$$

The other property of the geometric mean is that the sums of the deviations of the logarithms of the original observations above or below the logarithm of the geometric mean are equal; that is, $\sum (\log x - \log g) = 0$. Alternatively, we may say that the value of the geometric mean is such as to balance the ratio deviations of the observations from it. Thus, using the numbers just cited, we can see that

$$(8/2)(8/4) = 8 = (16/8)(32/8).$$

Because of this property, we may say that the geometric mean is typical in the sense that its value balances the ratio deviations of the items from it. Because of this property, too, the geometric mean is especially adapted to average ratios, rates of change, and logarithmically distributed series. In certain special cases of averaging ratios or percentages, such as the computation of the price index, the geometric mean can yield meaningful and logical results that the arithmetic mean cannot. For example, consider the data in Table 10–3, which shows the price changes of two commodities, A and B, from 1960 to 1970. During this period the price of A increased 100 percent and the price of B decreased by 50 percent. What was the relative average change in price?

The arithmetic mean is the wrong answer to that question. As indicated by the computations in Table 10–3, the application of $\bar{x}$ in this case has led to a nonsensical conclusion that the prices in 1970 were, on the average,

Table 10–3. Prices of Commodities A and B in 1960 and 1970

			Price relatives			
	Price		1960 = 100		1970 = 100	
Commodity	1960	1970	1960	1970	1960	1970
A	$5	$10	100	200	50	100
B	20	10	100	50	200	100
		Averages:	100	125	125	100

25 percent higher than those in 1960 when 1960 was chosen as the base, and that the prices in 1960 were 25 percent higher than those in 1970 when 1970 was chosen as the base. The arithmetic mean of ratios is inconsistent.

A consistent result is reached, however, when the geometric mean is applied in this situation, since

(a) if 1960 is chosen as the base, the prices of 1970 were 100 percent of the prices of 1960:

$$g = \sqrt[2]{200 \times 50} = \sqrt[2]{10{,}000} = 100;$$

(b) if 1970 is chosen as the base, the prices of 1960, as they should be, were also 100 percent of the prices in 1970:

$$g = \sqrt[2]{50 \times 200} = 100.$$

This method reduces the large upward-moving relative changes to ratios equal to the smaller decreasing relative changes by giving each equal weight when the nth root of the product of the relative is extracted. The arithmetic mean has an upward bias, because its method overemphasizes the large increasing relatives by treating them as absolute numbers instead of as the relative numbers they are.

The Harmonic Mean

If we take the reciprocal of the value of each item, compute the arithmetic mean of the reciprocals, and take the reciprocal of this mean, the result is known as the harmonic mean. Or, more compactly, the *harmonic mean* is the reciprocal of the arithmetic mean of the reciprocals of the observations. Symbolically,

$$h = \frac{1}{\dfrac{1/x_1 + 1/x_2 + \cdots + 1/x_n}{n}}$$

$$= \frac{1}{\dfrac{\sum_i (1/x_i)}{n}} = \frac{n}{\sum_i (1/x_i)}. \qquad (10\text{–}4)$$

To make computation easier, formula (10–4) may be written as

$$\frac{1}{h} = \frac{1/x_1 + 1/x_2 + \cdots + 1/x_n}{n}$$

$$= \frac{1}{n} \sum_i \frac{1}{x_i}. \qquad (10\text{–}5)$$

The harmonic mean of the three values 2, 4, and 8 is

$$\frac{1}{h} = \frac{1/2 + 1/4 + 1/8}{3}$$

$$= 7/24;$$

$$h = \frac{1}{7/24} = 24/7 = 3.43.$$

For the same data, the arithmetic mean is 4.7 and the geometric mean is 4. For any series, the values of which are not all the same and are all positive, the harmonic mean is always smaller than both the arithmetic mean and the geometric mean. Thus the harmonic mean is considered to have a downward bias, whereas the arithmetic mean has an upward bias.

The harmonic mean is useful mainly in processing ratio data that have physical dimensions, such as miles per gallon, cost per mile, miles per hour, and so on. For example, suppose you have spent one dollar for 3 dozen oranges in one shop, another dollar for 4 dozen oranges, and still another dollar for 5 dozen oranges in two other different stores; what is the "average price" per dozen of oranges? You may easily make the mistake of computing the arithmetic mean for your data. Since you may reason that the average prices in the three different stores are 33⅓ cents, 25 cents, and 20 cents, respectively, the average per dozen oranges in the three stores must be $(33\frac{1}{3} + 25 + 20)/3 = 26$ cents. You can see that this is incorrect by the following argument: you have spent only three dollars for 12 dozen oranges; if the average price were 26 cents, you would have spent (26) (12) = \$3.12!

Or you may note that your data are expressed in the form "so many dozens per dollar," but you wish to know the average price of "so many cents per dozen oranges." These two are inverse expressions. Consequently, the appropriate average can be obtained by the harmonic mean, as is done below:

$$\frac{1}{h} = \frac{1/33\frac{1}{3} + 1/25 + 1/20}{3}$$

$$= 0.04;$$

$$h = \frac{1}{0.04} = 25 \text{ cents.}$$

In concluding our discussion on averages, let us note the following relations among them:

(1) In samples whose values are all positive,

$$\bar{x} \geq g \geq h.$$

(2) For a symmetric distribution,

$$\bar{x} = x_{.5} = x_m.$$

(3) For a moderately skewed and unimodal distribution,

$$(\bar{x} - x_m) \doteq 3(\bar{x} - x_{.5}).$$

Relations (2) and (3) are illustrated by Figure 3–9.

10–2 MEASURES OF DISPERSION

The importance of averages is often exaggerated. An unqualified "average" may be virtually meaningless. One factor adding to the confusion is that with some distributions all the important averages fall closely together, while with others they fall far apart. If one says that average height of American soldiers is five feet eight inches, we get a very good idea of the stature of American soldiers. We don't need to ask whether the average is the median, the mode, or the mean; they would be the same or nearly so. However, if one says the average family income of a certain city is \$10,000 per year, we still don't know very much about the income distribution of that city unless we can find out which of the common kinds of average it is. In other words, an average can be very misleading unless it is identified and accompanied by other information telling us the range of things or their deviations from the average.

Perhaps the significance of variation is most vividly expressed by

Darrell Huff in his delightful little book *How to Lie with Statistics*: "Place little faith in an average or a graph or a trend when those important figures are missing. Otherwise you are as blind as a man choosing a camp site from a report of mean temperature alone. You might take 61 degrees as a comfortable annual mean, giving you a choice in California between such areas as the inland desert and San Nicolas Island off the south coast. But you can freeze or roast if you ignore the range. For San Nicolas it is 47 to 87 degrees but for the desert it is 15 to 104."

Let us now see how the degree of variation is measured.

The Range

The simplest measure of dispersion is the *range*—the difference between the highest and the lowest values in the sample. For a frequency distribution, the range may be considered either as the difference between the largest and smallest class boundaries, or as the difference between the last and the first class marks in a frequency table.

Judging from the method of obtaining the range, we see it as not only the simplest but also the crudest measure of dispersion. The range, therefore, has certain defects. Most important, the range, being a position measure and depending on the highest and lowest items alone, may be unduly influenced by one unusual value in the sample. For the same reason, the range is in no way a measure of scatter of the intervening items relative to the typical value. Finally, the range is highly sensitive to the size of the sample. The range tends to change, though not proportionately, in the same direction as the size of the sample varies. When the number is increased, it is possible that some item may have a greater value than the maximum and some item a smaller value than the minimum values in the previous sample. For this reason we cannot interpret the range properly without knowing the number of observations. This disadvantage is especially serious in comparing the variability of different distributions.

The range, despite its numerous shortcomings, may be used quite fruitfully as a measure of dispersion for many purposes. It is perhaps most useful in a situation where one desires to know only the extent of the extreme dispersion under "normal" conditions. If either the largest or the smallest item is unusual, the range reveals nothing about the "normal" distribution of the items. Thus, stock market reports are frequently stated in terms of their range, by quoting high and low prices of stocks over a period of time. When no exceptional movements of stock prices occur, the quoted range may measure the "normal" variation. When, however, exceptional movements are present, the range reveals the effects of temporary disturbing conditions in the market. Consequently, a comparison of price ranges of stocks through time may show the effects of exceptional factors upon different stocks as well as the fluctuations of stock prices caused by ordinary trading conditions. Sometimes the range may have certain distinct advantages in measuring the dispersion of a symmetric and continuous series; in such a case, the mean can be approximated by taking the average of the two extreme values, often called the *mid-range*. For example, it is the practice of meteorologists to derive the daily mean temperature by averaging only the maximum and minimum temperatures instead of using all the twenty-four readings of the day.

The range can also be used to advantage when the sample is small, especially when the same sampling operation is repeated often and an average of each successive result is utilized. For example, in quality control, since the same sample size is used repeatedly, comparisons between ranges are not affected by differences in sample size. Furthermore, the range reveals almost as much about the variability as do the values of all the individual observations when the sample size is small.

Finally, the range, being easy to compute and a common way of describing dispersion, is often used in engineering and medical reports.

The Average Deviation

The search for a measure of variability that would take all observed values into account and that would characterize the dispersion of the individual values from its central tendency naturally leads us to the idea of computing a measure such as $\sum (x_i - \bar{x})/n$; but this measure would always equal zero, since $\sum (x_i - \bar{x}) = 0$, and hence it can hardly be considered a measure of anything.

An obvious way of overcoming the difficulty is to find a mean of the deviations by ignoring the direction and attendant algebraic sign. In so doing, we would obtain what is called the *mean absolute deviation*, or simply the *average deviation*, of the sample. Let us denote this measure by *ad*, and we have

$$ad = \frac{1}{n} \sum_i | x_i - \bar{x} |. \tag{10–6}$$

The value of the average deviation is always interpreted in terms of the mean. For an approximately normal distribution, $\bar{x} \pm ad$ is the range that will include approximately 58 percent of the values of a distribution. Thus, if the average deviation has a relatively small value, the distribution is expected to be compact or uniform, since then more than half of the values of the variable are concentrated within a small range around the mean.

The average deviation is useful in dealing with small samples and in situations where no elaborate analysis is required. It is rarely used for large samples or grouped data, mainly because its calculation is rather time-consuming. We have introduced it here as the logical stepping-stone to a discussion of the variance, a superior measure of dispersion.

The Variance and the Standard Deviation

To by-pass the property $\sum (x_i - \bar{x}) = 0$, we can use squares of the deviations from the mean instead of taking their absolute values. Squared quantities not only prevent cancellations, they are also much more adaptable to mathematical analysis than absolute values.

The mean of the sum of squared deviations from the mean is called the *sample variance* and it is denoted by s^2. Symbolically,

$$s^2 = \frac{1}{n} \sum_{i=1}^{n} (x_i - \bar{x})^2. \tag{10–7}$$

The variance, from a practical point of view, is a rather awkward value, since units attached to it are squares, such as dollars², pounds², and so on. To convert this measure of variability into original units, we may take the positive square root of s^2, obtaining s—the *sample standard deviation.*

Needless to say, sample variance and standard deviation are interpreted in relation to the sample mean. Given $\bar{x}$, the greater is s^2 (or s), the more variable are the sample values. Furthermore, if the sample is quite large and the sample values are nearly normally distributed, the relationship between $\bar{x}$ and s is approximately the same as between population mean and standard deviation for a normal distribution.

Let us consider an example illustrating the computational procedure for these two statistics. Suppose there are two methods of assembling a certain mechanism, X and Y, and we wish to evaluate their relative efficiencies. For this we may select, say, ten workers to conduct an experiment. Each worker is to assemble the mechanism using both methods, and the time required for completion is observed for each method. Suppose the following data are obtained:

	Time (in minutes)	
Worker	x_i	y_i
1	35	27
2	36	28
3	49	53
4	44	52
5	43	48
6	37	29
7	38	34
8	42	47
9	39	40
10	40	45
Total	403	403

For the two sets of data above, we have

$$\bar{x} = \bar{y} = \frac{403}{10} = 40.3 \text{ minutes.}$$

Thus, both methods yield a mean time of 40.3 minutes. But can we conclude that both methods are equally efficient? To answer this question we need additional information about the dispersion in time for each method. Let us compute their variances and standard deviations. For X, we have

x_i	$(x_i - \bar{x})$	$(x_i - \bar{x})^2$	
35	-5.3	28.09	
36	-4.3	18.49	
49	$+8.7$	75.69	$s_X^2 = \frac{1}{n}\sum_i (x - \bar{x})^2$
44	$+3.7$	13.69	
43	$+2.7$	7.29	$= \frac{1}{10}(164.10)$
37	-3.3	10.89	$= 16.410;$
38	-2.3	5.29	
42	$+1.7$	2.89	$s_X = \sqrt{16.410}$
39	-1.3	1.69	$= 4.05$ minutes.
40	-0.3	0.09	
Total	0.0	164.10	

For Y, we have

y_i	$(y_i - \bar{y})$	$(y_i - \bar{y})^2$	
27	-13.3	176.89	
28	-12.3	151.29	
53	$+12.7$	161.29	$s_Y^2 = \frac{1}{n}\sum_i (y_i - \bar{y})^2$
52	$+11.7$	136.89	
48	$+7.7$	59.29	$= \frac{1}{10}(920.10)$
29	-11.3	127.69	$= 92.01;$
34	-6.3	39.69	
47	$+6.7$	44.89	$s_Y = \sqrt{92.01}$
40	-0.3	0.09	$= 9.59$ minutes.
45	$+4.7$	22.09	
Total	0.0	920.10	

While both methods yield the same mean time for completing the job, method X is much more uniform, or less variable, than method Y. Thus, we say $\bar{x}$ is more representative of X than $\bar{y}$ is of Y, since in the former case, the individual times are closer to the mean; in the latter, the individual times are scattered widely around the mean time. These observations also lead to the conclusion that method X is superior because it is more "dependable" by virtue of its smaller variance or standard deviation.

The sample variance, for reasons to be stated in Chapter 12, is sometimes defined as

$$\hat{s}^2 = \frac{\sum (x_i - \bar{x})^2}{n - 1}, \tag{10-8}$$

where $n - 1$ is called "$n - 1$ degrees of freedom," a concept explained in Chapter 14.

Also note that equations for s^2 and $\hat{s}^2$ are definitional equations. To facilitate computation, we may first expand the sum of squared deviations as below:

$$\begin{aligned}\sum (x - \bar{x})^2 &= \sum (x_i^2 - 2x_i\bar{x} + \bar{x}^2) \\ &= \sum x_i^2 - 2\bar{x} \sum x_i + n\bar{x}^2 \\ &= \sum x_i^2 - \frac{(\sum x_i)^2}{n}.\end{aligned}$$

Substituting this result into (10–7), we have

$$\begin{aligned}s^2 &= \frac{1}{n} \sum_i (x_i - \bar{x})^2 \\ &= \frac{1}{n}\left[\sum_i x_i^2 - \frac{(\sum_i x_i)^2}{n}\right] \\ &= \frac{\sum_i x_i^2}{n} - \left(\frac{\sum_i x_i}{n}\right)^2,\end{aligned} \tag{10-9}$$

and substituting it into (10–8), we obtain

$$\begin{aligned}\hat{s}^2 &= \frac{1}{n - 1}\left[\sum_i x_i^2 - \frac{(\sum_i x_i)^2}{n}\right] \\ &= \frac{n \sum_i x_i^2 - (\sum_i x_i)^2}{n(n - 1)}.\end{aligned} \tag{10-10}$$

With these last two equations, we can compute sample variance by merely cumulating the sum of the squares of the original values. We no longer need to compute the mean first, then take the deviations from the mean, and finally square those deviations as before.

The Coefficient of Variation

Often we are interested in comparing the variabilities between two or more sets of data. This can be done easily with their respective variances or standard deviations when the variables are given in the same units and when their means are approximately equal. When these conditions are missing, meaningful comparisons can be made only in terms of some relative measure of dispersion. One frequently used relative measure of variability is called the *coefficient of variation*, denoted as cv, which is simply the ratio of the standard deviation to the mean:

$$cv = \frac{\hat{s}}{\bar{x}}. \tag{10–11}$$

Suppose a scientist in India has obtained the following data concerning the weights of elephants and rats:

Elephants: $\bar{x}_1 = 24{,}000$ pounds,
$\hat{s}_1 = 1285$ pounds.

Rats: $\bar{x}_2 = 1.05$ pounds,
$\hat{s}_2 = 0.16$ pounds.

Clearly, just by a glance at the foregoing data, we can hardly tell which set has greater variability, since the two sets of measures in absolute numbers differ so greatly from each other. However, if we compute their respective coefficients of variation, we have

$$cv(X_1) = \frac{\hat{s}_1}{\bar{x}_1} = \frac{1285}{24{,}000} = 0.054 \text{ or } 5.4 \text{ percent},$$

$$cv(X_2) = \frac{\hat{s}_2}{\bar{x}_2} = \frac{0.16}{1.05} = 0.152 \text{ or } 15.2 \text{ percent}.$$

Thus, we see the variability of the weights of rats is almost three times greater than the variability of the weights of elephants. The former is as large as 15.2 percent of its mean, while the latter is only 5.4 percent of its mean.

10–3 DESCRIPTIVE MEASURES FOR GROUPED DATA

Locating the Median from Frequency Distributions

To save time and labor in locating the central values or computing other descriptive measures, we always work with grouped data when the sample is large. We shall begin with procedures for locating the median from a frequency distribution.

There are two popular procedures for locating the median from a distribution: the graphic method and the method of algebraic interpolation. In the *graphic method*, the median value is found by interpolation from the ogive of the distribution. This is done simply by drawing a horizontal line at $n/2$ from the absolute vertical scale or at the point of 0.5 from the relative scale to the ogive, and then dropping a perpendicular to the horizontal scale to locate the median value. Thus, from Figure 9–9, we find $x_{.5} = 4.16$ ohms.

With the method of *algebraic interpolation*, we first find the *median class*, the class whose cumulative frequency first exceeds the value of $n/2$. Thus, for the resistance data in Table 9–6, the median class is the fifth class, defined by the boundaries 4.15–4.27. Also, for the wage data in Table 9–7, as can be verified easily, the median class is the seventh class with boundaries \$11,000–\$12,499. When the median class is determined, the median can be located by the following formula for interpolation:

$$x_{.5} = L_{.5} + \left(\frac{n/2 - \sum f_{.5-1}}{f_{.5}}\right) c, \tag{10–12}$$

where

c = class interval of the median class,
$L_{.5}$ = the lower class boundary of the median class,
$\sum f_{.5-1}$ = the cumulative frequency in the class immediately before the median class, and
$f_{.5}$ = class frequency in the median class.

Applying this formula to the resistance data in Table 9–6, we find that

$$x_{.5} = 4.15 + \left(\frac{200/2 - 94}{51}\right) 0.12$$

$$= 4.16 \text{ ohms},$$

which is equal to that obtained by the graphic method but slightly smaller than 4.21, the class mark of the median class.

Similarly, for wage data in Table 9–7 we have

$$x_{.5} = 11{,}000 + \left(\frac{6230/2 - 2600}{807}\right) 1500$$

$$= \$11{,}957.$$

Locating the Mode from Frequency Distributions

The mode of a frequency distribution very often can be approximated by the class mark of the *modal class*—the class whose frequency density is a maximum. Thus, again, for the resistance data, $x_m = 4.09$, and for the wage data, $x_m = \$10{,}250$. (See Tables 9–6 and 9–7, respectively.) This method of locating the mode is quite satisfactory when the frequency densities in the classes immediately before (premodal class) and immediately after (postmodal class) the modal class are approximately equal. When this condition is not met, more satisfactory results can be obtained by algebraic interpolation with the following expression:

$$x_m = L_m + \left(\frac{\Delta_1}{\Delta_1 + \Delta_2}\right) c, \tag{10–13}$$

where

L_m = the lower class boundary of the modal class,
Δ_1 = the absolute difference between the frequency densities in the premodal and modal classes,
Δ_2 = the absolute difference between the frequency densities in the postmodal and modal classes, and
c = the class interval of the modal class.

Now, for the resistance data, we have

$$L_m = 4.03, \qquad \Delta_1 = |\,28 - 54\,| = 26,$$

$$c = 0.12, \qquad \Delta_2 = |\,54 - 51\,| = 3.$$

Therefore,

$$x_m = 4.03 + \left(\frac{26}{26 + 3}\right) 0.12$$

$$= 4.14 \text{ ohms.}$$

Also, for the wage data in Table 9–7 we have

$$x_m = 9500 + \left(\frac{98.67}{98.67 + 70.00}\right) 1500$$

$$= \$10{,}377.$$

It is interesting to observe that the median and the mode, as position averages, can be located for the frequency distributions constructed with open-ended classes, since for these measures the individual values of the series are not needed. Furthermore, these measures can be obtained with equal ease for distributions with uniform or varying class intervals, since the only class interval required is that which contains the median or the mode. But it must be noted that with unequal class intervals, modal value is located with reference to frequency densities.

Calculations of the Arithmetic Mean and Variance from Frequency Distributions

The arithmetic mean and variance are usually computed at the same time for grouped data, since, for both measures, the same work sheet can be used. Notice that both measures require all the individual values in the sample. However, we know that the individual values of a sample are lost in the process of organizing a frequency distribution. This difficulty is overcome by the midpoint assumption mentioned in the previous chapter. That is, if we use the class mark, x_i, of the ith class to represent each and every individual value in that class, then the total value of the observations falling in that class can be estimated as $f_i x_i$. Repeating this procedure for all classes yields the total value of the whole distribution. Consequently, the arithmetic mean for grouped data, with k groupings, can be defined as

$$\bar{x} = \frac{f_1 x_1 + f_2 x_2 + \cdots + f_k x_k}{\sum f_i}$$

$$= \frac{1}{n} \sum_{i=1}^{k} f_i x_i. \tag{10–14}$$

The definition for the sample variance is

$$s^2 = \frac{1}{n} \sum_{i=1}^{k} f_i (x_i - \bar{x})^2 \tag{10–15a}$$

or

$$\hat{s}^2 = \frac{1}{n-1} \sum_{i=1}^{k} f_i (x_i - \bar{x})^2. \tag{10–15b}$$

The foregoing expressions are definitional equations. Computations of these measures can be simplified by a process of coding. *Coding* consists of transforming data by

$$v_i = \frac{x_i - x_0}{a} \tag{10–16a}$$

or

$$x_i = a v_i + x_0, \tag{10–16b}$$

where x_0 and a are chosen such as to make computations on the values of the new variable v's easy. This can be thought of as *change of origin*, or *change of scale*.

To use coding for calculating $\bar{x}$ and s^2 from grouped data, it is most convenient to set x_0 to equal the class mark of one of the central classes and a to equal the size of the class interval. With these designations, by (10–16a), the class mark for the class whose midpoint is selected to be x_0 would be reduced to 0 and other class marks would become consecutive negative (for classes before x_0) and positive (for classes after x_0) integers, as shown by the third column of Table 10–4. With this transformation, too, the definitional equations given before for $\bar{x}$, s^2, and $\hat{s}^2$ would be reduced to

$$\bar{x} = av + x_0 = a\left(\frac{\sum_i f_i v_i}{n}\right) + x_0, \tag{10–17}$$

$$s^2 = a^2 s_V^2 = a^2\left[\frac{\sum_i f_i v_i^2}{n} - \left(\frac{\sum_i f_i v_i}{n}\right)^2\right], \tag{10–18a}$$

and

$$\hat{s}^2 = a^2 \hat{s}_V^2 = a^2\left[\frac{n\sum_i f_i v_i^2 - (\sum_i f_i v_i)^2}{n(n-1)}\right]. \tag{10–18b}$$

If we compute the mean and variance for resistance data by coding, as we should in order to save time and labor, we need a work sheet as shown by Table 10–4.

Table 10–4. Computing the Variance by Coding for Resistance Data

x_i	f_i	$v_i = \dfrac{x_i - 4.21}{0.12}$	$f_i v_i$	$f_i v_i^2$
3.73	3	−4	−12	48
3.85	9	−3	−27	81
3.97	28	−2	−56	112
4.09	54	−1	−54	54
4.21	51	0	0	0
4.33	31	+1	+31	31
4.45	17	+2	+34	68
4.57	6	+3	+18	54
4.69	1	+4	+4	16
—	200	0	−62	464

From results in Table 10–4, we have

$$\bar{x} = a\left(\frac{\sum_i f_i v_i}{n}\right) + x_0$$

$$= 0.12\left(\frac{-62}{200}\right) + 4.21$$

$$= 4.17 \text{ ohms.}$$

Also,

$$s^2 = a^2\left[\frac{\sum_i f_i v_i^2}{n} - \left(\frac{\sum_i f_i v_i}{n}\right)^2\right]$$

$$= (0.12)^2\left[\frac{464}{200} - \left(\frac{-62}{200}\right)^2\right]$$

$$= 0.032024,$$

$$s = \sqrt{0.032024},$$

$$= 0.1790 \text{ ohms,}$$

$$\hat{s}^2 = a^2\left[\frac{n \sum f_i v_i^2 - (\sum f_i v_i)^2}{n(n-1)}\right]$$

$$= (0.12)^2\left[\frac{200(464) - (-62)^2}{200(199)}\right]$$

$$= 0.032185,$$

$$\hat{s} = \sqrt{0.032185}$$

$$= 0.1794 \text{ ohms.}$$

Two observations should be made about the results above. First, in general, the mean computed from ungrouped data may differ slightly from that computed from grouped data. That they are identical in our example, after rounding, is an exception rather than the rule. However, our results do demonstrate the validity of using the midpoint assumption for making calculations from frequency distributions. Second, the difference between s^2 and $\hat{s}^2$ is very slight, as in our illustration, when n is large. This difference becomes more pronounced as n becomes smaller. From now on, however, unless noted otherwise, we shall always have $\hat{s}^2$ in mind when we mention sample variance.

For the sake of completeness note that, for grouped data,

$$\log g = \frac{f_1 \log x_1 + \cdots + f_k \log x_k}{n}$$

$$= \frac{1}{n}\sum_{i=1}^{k} f_i(\log x_i) \tag{10–19}$$

and

$$\frac{1}{h} = \frac{1/f_1x_1 + \cdots + 1/f_kx_k}{n}$$

$$= \frac{1}{n}\sum_{i=1}^{k}\left(\frac{1}{f_ix_i}\right); \tag{10-20}$$

these are seldom used in practice, since in situations where geometric and harmonic means are appropriate, the samples are usually quite small.

10-4 MOMENTS, SKEWNESS, AND PEAKEDNESS

Descriptive measures can be given in terms of *moment functions*. The *first moment* about the origin is the mean. The *kth moment* about the mean is given as $(1/N) \sum (x_i - \mu)^k$ for a population and as $(1/n) \sum (x_i - \bar{x})^k$ for the sample. The *second moment* about the mean is the variance. The mean and the variance (or the standard deviation) already give us a considerable amount of information. Additional insight into the relationship of observed values can be gained, however, if we take the *third* and *fourth moments* of, say, the sample, defined by

$$m_3 = \frac{1}{n}\sum_i (x_i - \bar{x})^3 \tag{10-21}$$

and

$$m_4 = \frac{1}{n}\sum_i (x_i - \bar{x})^4. \tag{10-22}$$

Measures, independent of the variance and also without dimensions, can be obtained by dividing by the appropriate power of the standard deviation. Thus, a measure of skewness will result if we divide m_3 by $\hat{s}^3$. That is,

$$a_3 = \frac{m_3}{\hat{s}^3}. \tag{10-23}$$

Note that a_3 measures the departure from symmetry. For a symmetrical distribution, $a_3 = 0$. Positive values of a_3 correspond to positively skewed distributions, long tails on the right; and, obviously, negative values of a_3 indicate negatively skewed distributions, long tails on the left.

Now, the quantity

$$a_4 = \frac{m_4}{\hat{s}^4} \tag{10-24}$$

is a measure of peakedness of a distribution. The degree of peakedness, or of *kurtosis*, is usually compared with a normal distribution for which

$a_4 = 3$. Distributions with $a_4 < 3$ are flatter at the center than the normal distribution; that is, individual observations will scatter widely about the mean. If, however, $a_4 > 3$, the distribution will be more peaked than the normal distribution; many values will be close to the mean. In this case the histogram has long tails and is narrow and tall near the mean. It may also be interesting to point out that, when $a_4 = 3$, the distribution is said to be *mesokurtic*; $a_4 < 3$, *platykurtic*; $a_4 > 3$, *leptokurtic* (see Figure 10–1).

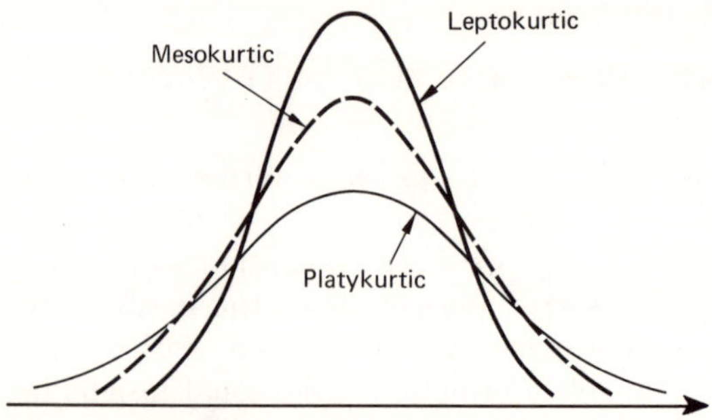

Figure 10–1 Illustration of Kurtosis

While a_3 and a_4 are mathematically precise measures, they are seldom employed in practice—for two reasons. First, they are quite laborious to compute, even with expansion or coding. Second, in contrast to $\hat{s}^2$, they yield relatively very little information about the underlying distribution. As a result, we are often content to measure skewness and peakedness by other devices that are much easier to compute.

The most frequently used measures of skewness in practice are the *Pearsonian coefficients of skewness*, which are defined as follows:

$$Sk_{p1} = \frac{3(\bar{x} - x_{.5})}{\hat{s}}, \tag{10–25}$$

and

$$Sk_{p2} = \frac{\bar{x} - x_m}{\hat{s}}. \tag{10–26}$$

Since we know that $(\bar{x} - x_m) \doteq 3(\bar{x} - x_{.5})$, the foregoing two expressions are roughly equivalent measures. However, x_m is a highly unstable value and, for grouped data, both $x_{.5}$ and x_m are subject to rounding errors; so these two measures may be expected to yield somewhat different results. Furthermore, it can be shown that the value of Sk_p ranges from -3 (indicating extremely negative skewness) through 0 (indicating symmetry) to $+3$ (indicating extreme positive skewness). Now, for the resistance data, we have

$$Sk_{p1} = \frac{3(4.17 - 4.16)}{0.1794} = +0.1672$$

and

$$Sk_{p2} = \frac{4.17 - 4.14}{0.1794} = +0.1672,$$

which reveal that the distribution of resistance data is very mildly skewed to the right, or nearly symmetrical.

One commonly used measure of peakedness is the *coefficient of kurtosis*, K, defined as

$$K = \frac{\frac{1}{2}(Q_3 - Q_1)}{x_{.90} - x_{.10}}$$

$$= \frac{\frac{1}{2}(x_{.75} - x_{.25})}{x_{.90} - x_{.10}}. \tag{10–27}$$

The numerator is called the *semi-interquartile range* and the denominator, the 90–10 *percentile range*. Both are sometimes employed as measures of variability.

It may be noted that both quartiles and percentiles, like the median, are measures that split a frequency distribution into a number of equal parts. Quartiles split a distribution into quarters, and percentiles divide it into hundredths. Quartiles and percentiles can be easily computed, as for the median, from the cumulative distribution. In computing quartiles, we are concerned with the classes that contain observations whose ranks in the ordered distribution are multiples of $n/4$. With percentiles, our attention should be on observations whose ranks are multiples of $n/100$.

To interpret the result from (10–27), we may note that, for a leptokurtic curve, most of the observations are concentrated near its center. As a result, the distances between $x_{.10}$ and $x_{.90}$ and between $x_{.25}$ and $x_{.75}$ tend to differ but little. That is, given $\hat{s}$, the sharper the peak, the closer these distances. Since the numerator of (10–27) is always less than the denominator, for a very peaked curve K approaches the limit of $\frac{1}{2}$. However, the more platykurtic the curve, the more the 90–10 percentile range tends to exceed the interquartile range. Therefore, as the range of the variable approaches infinity (where the curve is perfectly flat), K approaches zero. As the center of the region of mesokurtosis we are inclinded to take 0.25, the mid-range of K. This inclination is reinforced by the fact that, for the standardized normal distribution, $K = 0.263$. It can be easily verified (see Exercise 10–24) that for the resistance data, $K = 0.2500$, which is very mildly platykurtic, or nearly mesokurtic.

EXERCISES

10–1 A traveling salesman made five trips in two months, closing sales as tabulated below:

Trip	*Number of days*	*Value of sales*	*Sales per day*
1	3	\$100.00	\$33.33
2	5	250.00	50.00
3	7	400.00	57.14
4	4	90.00	22.50
5	6	210.00	35.00
Total	25	\$1050.00	\$197.97

The sales manager criticized the salesman's performance as not very good, since his mean daily sales were only \$39.59 (= \$197.97/5). The salesman called this an unfair statement, for his daily mean sales were as high as \$42.00 (= \$1050/25). What does each average here mean? Which mean seems to be more appropriate in this case?

10–2 The following are the starting monthly salaries in dollars of ten college graduates:

425, 500, 450, 436, 484

520, 500, 525, 419, 499

Determine the median, the mode, and the arithmetic mean for this set of data. Which measure seems best so far as this set of data is concerned?

10–3 Find $x_{.5}$, x_m, and $\bar{x}$ for the following sample, which consists of twelve measurements of a certain dimension in inches:

1.0015, 1.0016, 1.0017, 1.0018, 1.0018, 1.0018

1.0019, 1.0020, 1.0021, 1.0021, 1.0022, 1.0023

Which measure seems most appropriate for this set of data?

10–4 Compute $\bar{x}$, g, and h for the set of integers from 1 to 9 inclusive. Comment on your results.

10–5 The following data represent the December 1970 price indices of department store sales in the twelve largest cities in the United States:

111, 105, 113, 116, 133, 136

115, 118, 113, 113, 134, 118

Which measure would be the best for this set of data as a measure of central tendency? What is its value?

10–6 On a certain trip a motorist bought 10 gallons of gasoline at 29 cents per gallon, 15 gallons at 31 cents per gallon, 20 gallons at 33 cents per gallon, and 15 gallons at 28 cents per gallon. What is the average price per gallon of gasoline paid by the motorist?

10–7 Smith went from one city to another by plane at 450 miles per hour and returned by train at 40 miles per hour. The distance between the two cities is 1000 miles. What is the average speed at which Smith traveled?

10–8 If a certain operation can be completed by A, B, C, and D in 30, 20, 20, and 40 minutes, respectively, what is the average time required to complete the operation?

10–9 Select at random ten of your classmates and record their ages in months. With data so obtained, do the following:

(a) Determine the range.
(b) Compute the sample mean.
(c) Compute the sample variance and standard deviation—s^2, $\hat{s}^2$, s, and $\hat{s}$.

10–10 Select twenty of your classmates at random, record their final grades in points in their course in freshman mathematics, and determine

(a) The range.
(b) The mean.
(c) s^2, $\hat{s}^2$, s, and $\hat{s}$.

10–11 Which set of data in the preceding two problems has greater variability? Can you give reasons for such a difference?

10–12 What are the ranges for data in Tables 9P–1 and 9P–2, respectively? From these values, can you conclude whether one series is more or less variable than the other? Why or why not?

10–13 Compute the average deviations for data in Exercises 10–9 and 10–10. Are the results here in agreement with your comments on Exercise 10–11?

10–14 Two workers on the same job show the following results over a long period:

	A	B
Mean time of completing the job (minutes)	30	25
Standard deviation (minutes)	6	6

(a) Which worker appears to be more consistent in performance?
(b) Which worker is more efficient from an overall point of view?

10–15 Locate the median for data in Table 9P–1 by the graphic method and the method of algebraic interpolation.

10–16 Rework the preceding problem for data in Table 9P–2.

10–17 Locate the mode for data in Table 9P–1 by the method of algebraic interpolation.

10–18 Rework the preceding problem for data in Table 9P–2.

10–19 Compute the arithmetic mean, variance, and standard deviation for data in Table 9P–1.

10–20 Rework the preceding problem for data in Table 9P–2.

10–21 Determine the median, the mode, and the mean for data in Table 9P–2.

10–22 For data in Table 9P–1: Compute the coefficient of skewness, Sk_{p1} and Sk_{p2}.

10–23 Rework the preceding problem for data in Table 9P–2.

10–24 For resistance data in Table 10–4, show that

$$K = 0.2500$$

with

$$x_{.10} = 3.944,$$

$$x_{.25} = 4.052,$$

$$x_{.75} = 4.289,$$

and

$$x_{.90} = 4.418.$$

11

THEORY OF SAMPLING

11–1 INTRODUCTION

We have been concerned in the preceding two chapters with methods of computing descriptive measures for univariate sample data. Such sample statistics are never obtained for their own sake; instead, they serve as a basis for generalizing about population parameters that are unknown or can only be hypothesized.

Statistical methods that enable us to infer from limited data (samples) to the expected long-run behaviors (populations) are called by many names, such as *inductive statistics, analytical statistics, statistical inference,* or *classical decision procedures.* The process of inference always involves errors concerning the true state of nature. Thus, in order to have a measure of confidence in our inductive conclusions, we must devise methods to appraise such errors. Since these errors are the result of random or chance variations due to random sampling, they can be evaluated only in terms of probabilities.

A sample statistic, computed from a random sample, is a random variable; as such, it must have a probability distribution of its own. The distribution of a statistic, commonly known as the *sampling distribution* of a statistic, has well-defined properties as does any probability model. From these properties of the distribution of a statistic we can calculate the risks (errors due to chance) involved in statistical inference.

This chapter will introduce procedures for gathering evidence and for translating data into probabilistic terms amenable to inference about true states of nature. We shall be concerned with the construction of probability distributions for sample statistics and the evaluation of their properties. To this end, our discussion begins with a very fundamental concept called the sum of independent random variables, which is concerned with the law of large numbers and the central limit theorem—which, in turn, are essential for many arguments in the theory of sampling.

11–2 SUMS OF RANDOM VARIABLES

If X and Y are two *independent* random variables, then

$$E(XY) = E(X)E(Y). \tag{11–1}$$

Thus, the multiplicative property of probabilities extends to expectations. Furthermore, this property can be extended to any number of independent random variables. An important result of independence is additivity. Namely, if $X_1, \ldots, X_n$ are n independent random variables, then the expectation of the sum, $E(S)$, is the sum of expectations,

$$\begin{aligned} E(S) &= E(X_1 + \cdots + X_n) \\ &= E(X_1) + \cdots + E(X_n), \end{aligned} \tag{11–2}$$

and the variance of the sum, $V(S)$, is the sum of variances,

$$\begin{aligned} V(S) &= V(X_1 + \cdots + X_n) \\ &= V(X_1) + \cdots + V(X_n). \end{aligned} \tag{11–3}$$

Also, two random variables are said to be *identical* if they possess the same expectation, μ, and the same variance, σ^2. It follows that if $X_1, \ldots, X_n$ are identical, then

$$E(S) = \mu n; \tag{11–4}$$

and if X_i are also independent, then

$$V(S) = \sigma^2 n. \tag{11–5}$$

But why are we interested in the sum of independent random variables? First, a random variable under study may be inherently an additive combination of ingredients; for example, the binomial variable with parameters n and p is merely the sum of n independent and identical Bernoulli variables with a common parameter p. Second, the property of additivity holds for many special probability models. For example, the sum of two or more binomial variables is again a binomial variable; the sum of n independent Poisson variables or normal variables is again a Poisson or normal variable. Finally, many sample statistics are computed from sums of independent and identical random variables. Thus, the sample mean is the sum of n sample values multiplied by a constant $1/n$; the sample variance is computed with the sum of squared deviations from the sample mean; and so forth.

Computations involving a sum can be made directly from an n-fold integral or an n-fold sum. But such computations are usually extremely difficult, if not impossible, and the resulting distributions very complicated, except for a few exceptional cases such as the binomial, Poission, normal, and exponential models. Fortunately, thanks to two important theorems, the evaluation of variables that are sums becomes a relatively simple matter. These two theorems—the law of large numbers and the central limit theorem—we shall now introduce.

11-3 THE LAW OF LARGE NUMBERS

In essence, the *law of large numbers* says that the average of a sum of identically distributed independent random variables converges to the expected value of the individual distribution. This law can be formulated in a number of ways. The first general formulation, known as the *Bernoulli law of large numbers,* was stated by Jacob Bernoulli, and published posthumously in 1713. The Bernoulli law states: if S represents the number of successes in n identical Bernoulli trials, with probability p of success in each trial, and if $\delta > 0$, then

$$P\left(\left|\frac{S}{n} - p\right| \leq \delta\right) \longrightarrow 1 \qquad \text{as } n \longrightarrow \infty. \tag{11-6}$$

Here, S/n may be interpreted as the average number of successes in n trials and p is the expectation of success in a given trial. Intuitively, the Bernoulli law implies that if n is a sufficiently large and fixed integer, then the average number of successes would have a high probability of being close to p. This law can easily be proven by the Chebyshev inequality or, more appealingly, in numerical fashion as indicated by the discussion that follows.

Let a coin be tossed 100 times, and S be the sum of the number of heads; we wish to determine the probability that S is within 2 percent of 50 heads—that is, the probability that S will fall within the range from $50(1 - 0.02)$ to $50(1 + 0.02)$. This clearly is the probability that S will assume the value of 49, 50, or 51 heads. Such a probability can easily be computed by the binomial law; it is 0.24. Similarly, if a coin is tossed 1000, 10,000, and 100,000 times, what are the probabilities that the results are within 2 percent of 500, 5000, or 50,000 heads, respectively? By the binomial law again, we must have 0.47, 0.95, and 0.998. The deduction here is obvious: as $n \longrightarrow \infty$, the probability goes to one. As a matter of fact, this numerical proof can be stated as the law of large numbers for fair-coin tossing:

$$P\left(\left|\frac{S}{n} - \frac{1}{2}\right| \leq \delta\right) \longrightarrow 1 \qquad \text{as } n \longrightarrow \infty.$$

In this formulation, $|(S/n) - \frac{1}{2}| \leq \delta$ is the condition that S/n fall in the range $(\frac{1}{2} - \delta, \frac{1}{2} + \delta)$. If we set $\delta = 0.01$, then we get the 2 percent range that has been used here.

Clearly, the law of large numbers is the philosophical justification for all attempts to estimate probability empirically. It justifies, for instance, the relative-frequency theory of probability. Furthermore, it assures a high probability that a sample average is close to the unknown but constant expectation if n is large. This implies that the accuracy of estimates will always increase with increases in the sample size. We see, therefore, that much of analytical statistics could not exist without a theorem of this kind.

11-4 THE CENTRAL LIMIT THEOREM

We have just shown the validity of the law of large numbers by mathematical expression of the following fact: that by averaging an increasingly large number of observations of the value of a quantity we can obtain increasingly more accurate measurements of the expectation of that quantity. It must be noted now, however, that the center of the distribution of a sum S moves off to infinity, and the variance of a sum becomes larger and larger as n increases—since, as we know, $E(S) = \mu n$ and $V(S) = \sigma^2 n$. It is therefore quite necessary for us to learn something about the pattern of the distribution of a sum.

Under certain mild restrictions, the distribution of the sum of a large number of identically independent random variables, whatever distributions the summands possess, has approximately a normal distribution. This is the essence of what is called the *central limit theorem*, CLT, a name given it by G. Polya in 1920. The value of this theorem is that it requires virtually no conditions on distribution patterns of the individual random variables being summed. As a result it furnishes a practical method of computing approximate probability values associated with sums of arbitrarily distributed independent random variables. No wonder that among all probability laws this theorem is considered the most remarkable theoretical formulation from both the practical and the theoretical viewpoints.

The central limit theorem, first introduced by De Moivre early in the eighteenth century, has been expressed in many forms. One important version treats it as a consequence of sums of identically distributed independent random variables. In this version, if S is the sum of a large number of identical and independent random variables, each with mean μ and variance σ^2, then the variable $(S - \mu n)/(\sigma\sqrt{n})$ approaches the standard normal distribution when n becomes infinite. Here, $\sigma\sqrt{n}$ is the standard deviation of S.

The computational significance of the CLT is that, for large n, we can express the distribution function of S in terms of $N(0, 1)$ as follows:

$$P(S \leq s_0) \doteq N\left(\frac{s_0 - \mu n}{\sigma\sqrt{n}}\right), \tag{11–7}$$

where s_0 is a particular value of S.

As to how large n should be for (11–7) to hold, there is no standard answer. This depends upon the closeness of approximation required and the actual distribution forms of X_i. If the summands are normally distributed, then (11-7) provides exact probabilities no matter how small n is. If nothing is known about the distribution patterns of X_i, the rule of thumb is that n must be 25 or greater for satisfactory approximations. Of course, approximations improve when n becomes larger. The tendency of a sum toward normality is illustrated by Figure 11–1, in which distributions of the sums

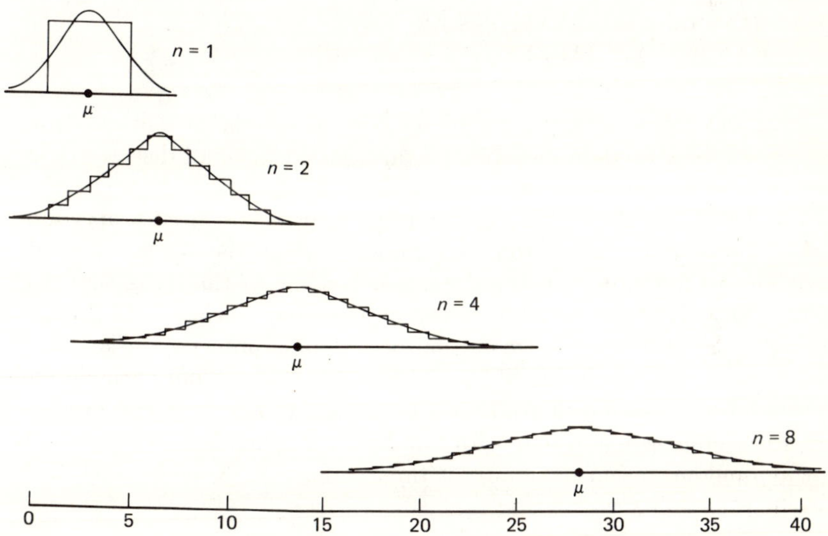

Figure 11–1 Distributions of Sums of Points on 1, 2, 4, and 8 Dice

of 1, 2, 4, and 8 dice are drawn. This figure shows, besides the evident tendency toward normality as n increases, that both the mean and the variance of the sum of identically distributed random variables increase with the increase in n.

Let us apply the central limit theorem in an example. A food-processing factory produces canned ground beef with a mean of 5 ounces and a variance of 0.3 ounces² per can. A box of 60 cans has a mean weight of (60) (5) = 300 ounces and a variance of (60) (0.3) = 18 ounces². By the CLT the weight of a box, denoted by W, would be approximately normally distributed.

This information would allow us to determine probabilities for various events concerning W, such as

$$P(W \leq 290 \text{ ounces}) = N\left(\frac{290 - 300}{\sqrt{18}}\right) = N(-2.36) = 0.0091,$$

$$P(W > 310 \text{ ounces}) = 1 - N\left(\frac{310 - 300}{\sqrt{18}}\right) = 1 - 0.9909 = 0.0091,$$

$$P(290 \leq W \leq 310) = N\left(\frac{310 - 300}{\sqrt{18}}\right) - N\left(\frac{290 - 300}{\sqrt{18}}\right) = 0.9818,$$

and so forth.

11-5 SAMPLE OBSERVATIONS AS RANDOM VARIABLES

Sampling refers to a process of selecting a sample of n items from a population or a random variable. In this section we shall discuss the consequences of random observations, thus providing a basis for later arguments on the derivation of sampling distributions and their related properties.

An observation made from a random variable is, again, a random variable. An individual observed value is random in the sense that it can assume any of all the possible values of a random variable. As a random variable, the probability distribution of a sample observation is identical with that of the population of measurements—the random variable under consideration. Imagine an urn that contains eight balls, two with number 1, two with number 3, and four with number 5. If a ball is selected from this urn at random, it can have any of three values—1, 3, or 5—each with a specific probability as follows:

Number on the ball	*Probability*
1	0.25
3	0.25
5	0.50
Sum	1.00

This evidently is identical with the population probability distribution.

Let us demonstrate this conclusion in another way. Suppose an unconditional simple random sample of two balls is drawn from the urn; then the possible samples of two balls and their associated probabilities become:

First ball	*Second ball*	*Probability*	
1	1	(0.25)(0.25) =	0.0625
1	3	(0.25)(0.25) =	0.0625
1	5	(0.25)(0.50) =	0.1250
3	1	(0.25)(0.25) =	0.0625
3	3	(0.25)(0.25) =	0.0625
3	5	(0.25)(0.50) =	0.1250
5	1	(0.50)(0.25) =	0.1250
5	3	(0.50)(0.25) =	0.1250
5	5	(0.50)(0.50) =	0.2500
		Sum	1.0000

where the first number drawn is a random variable X_1 and the second is a random variable X_2. The first assumes the value of 1 in the first three samples, the value of 3 in the second three samples, and the value of 5 in the last three samples. The probability that X_1 will assume the value of 1 is, theorefore, the sum of probabilities of the first three samples, and so on. The distribution of X_1 is thus

Value	*Probability*
1	0.2500
3	0.2500
5	0.5000
Sum	1.0000

Likewise, the probability distribution of X_2, the second number drawn, is identical with that of X_1 and is identical with the distribution of the parent population.

To generalize: When a simple random sample of size n is drawn from a finite population with replacement, or from an infinite population conditionally or unconditionally, we have n identically distributed random variables, $X_1, X_2, \ldots, X_n$, all possessing the same distribution as the parent population. Also, if the population is large, though finite, the sample observations, X_i, can in practice be treated as if they were identical and independent. From these findings we may finally define a *random sample* as one whose n observations can be described by the mathematical model of n independent random variables, each a replica of the population from which the sample is selected.

While sample observations are identically distributed when considered as random variables and are denoted by $X_1, \ldots, X_n$, the actual observations of a specific sample that has been drawn are all different and are denoted by $x_1, x_2, \ldots, x_n$. Thus, the term "random sample" refers to the observations thought of as random variables, not to a particular set of actual realizations. The latter may or may not be random, depending on whether the method employed to obtain a sample ensures the independence of observations. However, if X_i are identically distributed, then a specific collection of sample observations can be thought of as the particular values of a random variable X—that is, X: $x_1, x_2, \ldots, x_n$. In this connection, for a random sample of n items, we may treat the x_i as realizations of X_i:

$$X_1 = x_1, \quad X_2 = x_2, \ldots, \quad X_n = x_n.$$

Consequently, if the sample is random, then sample observations thought of as independent random variables or as a particular set of sample values are equivalent and can be used interchangeably.

The thought that sample observations are random variables leads us to the conclusion that any quantity computed from sample observations must also be a random variable. As a random variable, the computed quantity, or statistic, has a probability distribution of its own, called a *sampling distribution*, or the *probability distribution of a statistic*, which can be established directly from the parent population distribution. For example, in the previous illustration, the sum of numbers on the two balls drawn, $X_1 + X_2$, is a statistic; and, according to all the possible samples listed before, the possible sums and their associated probabilities are as follows:

$X_1 + X_2$	*Probability*		
2	0.0625	=	0.0625
4	2(0.0625)	=	0.1250
6	0.0625 + 2(0.1250)	=	0.3125
8	2(0.1250)	=	0.2500
10	0.2500	=	0.2500
	Sum		1.0000

This is called the *sampling distribution of the sum with* $n = 2$ *drawn from a population, with replacement, of eight balls, two with number* 1, *two with number* 3, *and four with number* 5. We may, and we usually do, call it simply the *distribution of a sum*. The full name, however, does reveal some important features of the probability distribution of a statistic. First, a sampling distribution is generated from a population distribution, known or assumed. Second, the same population may generate an infinite number of sampling distributions for the same statistic, each for a special sample size n. Finally, a population may generate sampling distributions for two or more different statistics. For example, a random sample of n may be taken from a shipment of steel wires of a certain make, and then a distribution for sample mean, another for sample variance, and still another for sample standard deviation can be derived. We shall now introduce sampling distributions for a number of the most frequently encountered statistics; others will be taken up in later chapters.

11–6 SAMPLING DISTRIBUTION OF THE NUMBER OF SUCCESSES

If a random sample of size n is taken from a population whose elements belong to two mutually exclusive categories—one containing elements that possess a certain trait and the other containing elements that do not—then the sampling distribution of the number of successes is a binomial distribution if sampling is made with replacement; and it is the hypergeometric distribution if sampling is made without replacement. Here, our discussion will assume that an unconditional random sample is taken, so that the binomial probability is appropriate.

The sampling distribution of the number of successes, being a binomial probability model, would have as its mean $\mu = np$ and as its variance $\sigma^2 = npq$. The standard deviation of a sampling distribution is usually called the *standard error* in the sense that it measures the sampling variability due to chance or to random forces. In this case, the standard error, denoted by σ, is $\sigma = \sqrt{npq}$.

Since we have thoroughly treated the binomial model before, we shall not include any illustrations of the distribution of the number of successes here. Instead, we shall move directly to the application of the central limit theorem to this sampling distribution. If X is the number of successes in a random sample of size n, the variable $(X - np)/\sqrt{npq}$ approaches the standard normal distribution as n approaches infinity. Consequently, the distribution function of the number of successes can be expressed as

$$P(X \leq x_0) \doteq N\left(\frac{x_0 - np}{\sqrt{npq}}\right), \tag{11–8}$$

where x_0 is a particular value of X. The left side of this equation, as revealed by Figure 11–2, is a step function that has jumps at nonnegative integers only. The right side, however, is continuous and increases as X increases from $-\infty$ to ∞. Yet the values of these functions are approximately equal at each argument for large n, with the continuous function passing through about the middle of each step of the discrete function. These approximations are good with $np > 5$ when $p \leq \frac{1}{2}$, and with $nq > 5$ when $p \geq \frac{1}{2}$.

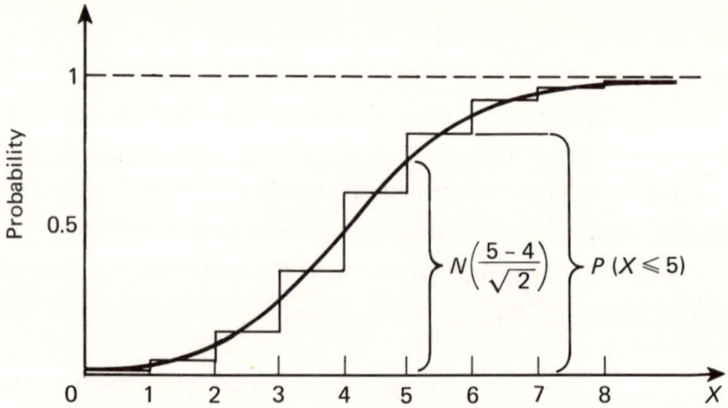

Figure 11–2 Normal Distribution Function Fitted to the Binomial Distribution Function with $n = 8$ and $p = 0.5$

Normal approximations become exact binomial probabilities when n becomes infinite.

Satisfactory normal approximations of the distribution of the number of successes can be made even when n is relatively small if we introduce what is called the *continuity correction factor*, *CCF*, whose value is a constant of ½. From Figure 11–2 we see that a better approximation to the value of, say, 5 can be made by taking the ordinate of the continuous curve ½ unit to the right of 5. Consequently, we write, in general,

$$P(X \leq x_0) \doteq N\left(\frac{x_0 + \frac{1}{2} - np}{\sqrt{npq}}\right). \tag{11–9}$$

Consider, for example, a distribution of the number of heads with $n = 8$ and $p = 0.5$, as shown in Figure 11–2. We have

$$\mu = np = 8(0.5) = 4$$

and

$$\sigma = \sqrt{npq} = \sqrt{(8)(0.5)(0.5)} = 1.41421.$$

Thus,

$$\begin{aligned} P(X \leq 5) &\doteq N\left(\frac{5 + \frac{1}{2} - 4}{1.41421}\right) \\ &= N(1.06) \\ &= 0.8554, \end{aligned}$$

which approximates the exact binomial probability of 0.8555;

$$\begin{aligned} P(X = 5) &= P(X \leq 5) - P(X \leq 4) \\ &\doteq N\left(\frac{5 + \frac{1}{2} - 4}{1.41421}\right) - N\left(\frac{4 + \frac{1}{2} - 4}{1.41421}\right) \\ &= N(1.06) - N(0.35) \\ &= 0.8554 - 0.6368 \\ &= 0.2186, \end{aligned}$$

which approximates the exact binomial probability of 0.2188;

$$\begin{aligned} P(X \geq 5) &= 1 - P(X \leq 4) \\ &\doteq 1 - N(0.35) \\ &= 0.3632, \end{aligned}$$

which approximates the exact binomial probability of 0.3633. Clearly, with the use of the CCF, the normal approximations are very close to the binomial probabilities with n as small as 8. These close approximations are obtained also because $p = 0.5$ in this case. In general, given n, the greater is the error of approximation the more p deviates from 0.5. Given p, the larger n is, the better the approximation becomes.

11–7 SAMPLING DISTRIBUTION OF A PROPORTION

A population proportion is defined as $\pi = x/N$, where x is the number of elements that possess a certain trait and N is the total number of items in the population. A sample proportion is defined as $p = x/n$, where x is the number of items in the sample that possess a certain trait and n is the sample size. Thus, a proportion may be considered as a proportion of successes, and it is obtained by dividing the number of successes by the sample size n. Because of this, we see that if a random sample of n is obtained with replacement, then the sampling distribution of p obeys the binomial probability law, and we can write its probability function, in terms of our new symbols, as

$$P(p \leq p_0) \doteq \sum_{x=0}^{np_0} \binom{n}{x} \pi^x (1 - \pi)^{n-x}. \tag{11–10}$$

The expected value and variance of the distribution of p, obtained by dividing n into the corresponding measures for the number of successes, are respectively

$$E(p) = \frac{n\pi}{n} = \pi \tag{11–11}$$

and

$$V(p) = \frac{n\pi(1 - \pi)}{n^2} = \frac{\pi(1 - \pi)}{n}. \tag{11–12}$$

The standard error of p, denoted by σ_p, which measures chance variations of sample proportions from sample to sample is

$$\sigma_p = \sqrt{V(p)} = \sqrt{\frac{\pi(1 - \pi)}{n}}. \tag{11–13}$$

If sampling is made without replacement, then the sampling distribution of p obeys the hypergeometric probability law. In this case, $E(p)$ is still identical with π but the standard error must be adjusted by a *finite population correction factor*, FPCF, which is equal to $\sqrt{(N - n)/(N - 1)}$:

$$\sigma_p = \sqrt{\frac{\pi(1 - \pi)}{n}} \sqrt{\frac{N - n}{N - 1}}. \tag{11–14}$$

If the sample is small relative to the population—say, the former is less than 5 percent of the latter—the FPCF approaches unity. Consequently, the introduction of FPCF would produce a practically insignificant difference. In other words, if a small sample is taken from a large population, we can treat sample observations as if they were independent variables, and σ_p can be computed without FPCF.

Now, it may be observed that the distribution of p for an unconditional random sample approaches the normal distribution when n becomes infinite. For a conditional random sample, under conditions specified in the last section, the distribution of p approaches the binomial model, which in turn approaches the normal model as a limit. In short, the distribution of p obeys the central limit theorem. Precisely, the variable $(p - \pi)/\sigma_p$ approaches $N(0, 1)$ when n becomes infinite. Thus, we state that the distribution function of p with large n is

$$P(p \leq p_0) \doteq N\left(\frac{p_0 - \pi}{\sigma_p}\right). \tag{11–15}$$

When n is rather small, as in the case of the distribution of the number of successes, satisfactory normal approximations can be obtained by introducing a continuity correction factor of $1/(2n)$. [Note: $1/(2n)$ is employed here instead of $\frac{1}{2}$ because a proportion of successes is the number of successes divided by n.] With the CCF, equation (11–15) becomes

$$P(p \leq p_0) \doteq N\left(\frac{p_0 + \dfrac{1}{2n} - \pi}{\sigma_p}\right). \tag{11–16}$$

We shall now use a simple example to show how a distribution of a proportion is derived, and to verify that the expectation of p is identical with π and that the standard error of p is defined by equation (11–14).

A class in advanced statistics consists of four female and six male students. A random sample of five students is drawn from this class without replacement. We wish to evaluate the sampling distribution of the proportion of female students in a sample of five.

In this case, $\pi = 4/10 = 0.4$. The number of all possible samples of five from a population of ten is

$$\binom{10}{5} = \frac{10!}{5!5!} = 252.$$

Furthermore, since there are only four female students in the class, there can be only five possible sample proportions:

$$p_0 = \frac{0 \text{ females}}{5} = 0.0,$$

$$p_1 = \frac{1 \text{ female}}{5} = 0.2,$$

$$p_2 = \frac{2 \text{ females}}{5} = 0.4,$$

$$p_3 = \frac{3 \text{ females}}{5} = 0.6,$$

and

$$p_4 = \frac{4 \text{ females}}{5} = 0.8.$$

To determine the probability for each of these possible sample results, we note that this distribution of p obeys the hypergeometric law, and its probability values can be obtained directly from Table B–4. To use this table, we have only to use x/n in the place of x—that is, $h(N, n, k, x) = h(N, n, k, x/n)$. For this problem, we have

$$\begin{aligned}
P(p = 0.0) &= h(10, 5, 4, 0/5) = 0.023810\\
P(p = 0.2) &= h(10, 5, 4, 1/5) = 0.238095\\
P(p = 0.4) &= h(10, 5, 4, 2/5) = 0.476190\\
P(p = 0.6) &= h(10, 5, 4, 3/5) = 0.238095\\
P(p = 0.8) &= h(10, 5, 4, 4/5) = \underline{0.023810}\\
\text{Sum} &\quad 1.000000
\end{aligned}$$

which is the sampling distribution of p with $n = 5$ drawn from a population of ten students with $\pi = 0.4$. The graph of this distribution of p is given by Figure 11–3.

To calculate the expected value and standard error for the distribution above we need the number of samples that will yield each of the possible sample proportions. This again is an easy matter, since the number of samples that yields a given sample proportion is the product of the total number of all possible samples and the probability of obtaining that sample proportion.

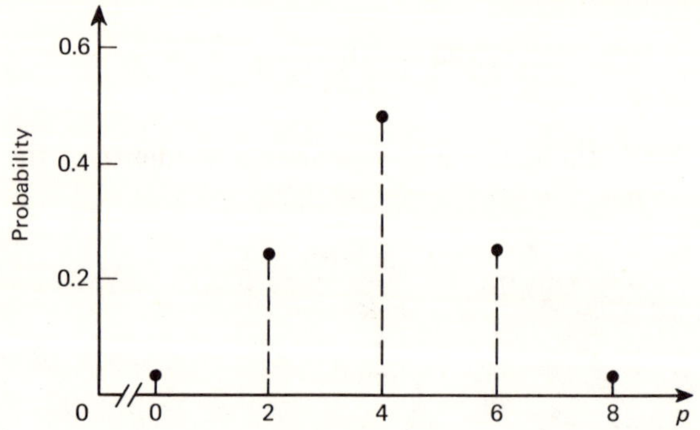

Figure 11–3 Sampling Distribution of p with a Conditional Random Sample of 4 from a Population with $N = 10$ and $\pi = 0.4$

Thus,

For $p = 0.0$,	there are	252(0.023810)	=	6 samples
For $p = 0.2$,	" "	252(0.238095)	=	60 "
For $p = 0.4$,	" "	252(0.476190)	=	120 "
For $p = 0.6$,	" "	252(0.238095)	=	60 "
For $p = 0.8$,	" "	252(0.023810)	=	6 "
		Sum		252 samples

Now, with the aid of Table 11–1, we find that

$$E(p) = \sum_i p_i f(p_i)$$

$$= (0.0)(6/252) + (0.2)(60/252) + \cdots + (0.8)(6/252)$$

$$= 0.4,$$

which, as promised, is identical with the population proportion π.

Table 11–1. Computations of the Expectation and the Standard Error of the Distribution of p with $n = 5$ and $\pi = 0.4$

p	f	$p - \pi$	$f(p - \pi)$	$f(p - \pi)^2$
0.0	6	−0.4	−2.4	0.96
0.2	60	−0.2	−12.0	2.40
0.4	120	0.0	0.0	0.00
0.6	60	+0.2	+12.0	2.40
0.8	6	+0.4	+2.4	0.96
	252			6.72

To compute the standard error directly from the standard formula, we have

$$\sigma_p = \sqrt{\frac{\sum f(p - \pi)^2}{\sum f}} = \sqrt{\frac{6.72}{252}} = 0.1634.$$

Let us now calculate the standard error by the theoretical formula to see if it yields the same result. If it does, we know that the equation for the standard error holds.

$$\begin{aligned} \sigma_p &= \sqrt{\frac{\pi(1 - \pi)}{n}}\sqrt{\frac{N - n}{N - 1}} \\ &= \sqrt{\frac{0.4(0.6)}{5}}\sqrt{\frac{10 - 5}{10 - 1}} \\ &= 0.1634, \end{aligned}$$

which is identical with the previous value.

As a second example, let us assume that in a shipment of electronic tubes, 30 percent are defective. If a random sample of 500 is drawn with replacement from this population, how should we go about evaluating the distribution of the defective proportion?

In this case, it is clearly impractical for us to list all the possible sample proportions and their associated probabilities. Fortunately, we need not do so. We need to know only the expectation and the standard error of the distribution of any statistic in order to have knowledge of the whole sampling distribution that is normal. For the distribution of p with $n = 500$ on hand, we have

$$E(p) = \pi = 0.3$$

and

$$\sigma_p = \sqrt{\frac{0.3(0.7)}{500}} = 0.0205.$$

Furthermore, the sample is large enough for this distribution of p to be satisfactorily approximated by the normal distribution. With this information, we can easily estimate the probability for any possible sample proportion for the distribution of p under consideration. For example, what is the probability that the sample proportion will be less than $150/500$? Equal to $145/500$? Between these two values? Greater than $165/500$?

To compute the first probability, we note that $150/500 = 0.30$ and that the continuity correction factor here is $1/[2(500)] = 0.001$. Thus,

$$\begin{aligned} P(p \leq 0.3) &\doteq N\left(\frac{0.3 + 0.001 - 0.3}{0.0205}\right) \\ &= N(0.05) = 0.5199. \end{aligned}$$

Next, the probability for p to be equal to $145/500$ is the same as between $145/500$ and $144/500$, or between 0.290 and 0.288. Therefore,

$$P(0.288 \leq p \leq 0.290) \doteq N\left(\frac{0.290 + 0.001 - 0.3}{0.0205}\right) - N\left(\frac{0.288 + 0.001 - 0.3}{0.0205}\right)$$
$$= N(-0.44) - N(-0.54)$$
$$= 0.3300 - 0.2946$$
$$= 0.0354.$$

Now, note that the probability for p to fall between $145/500$ and $150/500$ should be considered as between $144/500$ and $150/500$, or between 0.288 and 0.300. (Why?) We have, then,

$$P(0.288 \leq p \leq 0.300) \doteq N\left(\frac{0.3 + 0.001 - 0.3}{0.0205}\right) - N\left(\frac{0.288 + 0.001 - 0.3}{0.0205}\right)$$
$$= N(0.05) - N(-0.54)$$
$$= 0.2253.$$

Finally, the probability that the sample proportion is equal to or greater than $165/500$ is the probability of one minus that of $164/500 = 0.328$. Hence,

$$P(p \geq 165/500) \doteq 1 - N\left(\frac{0.328 + 0.001 - 0.3}{0.0205}\right)$$
$$= 1 - N(1.41)$$
$$= 1 - 0.9207 = 0.0793.$$

As another example, consider a conditional random sample of 20 drawn from a lot of 100 radio tubes, 80 of which are nondefective. What is the probability that the sample proportion of nondefective tubes will be less than 0.7?

Here, we have

$$E(p) = \pi = 0.8, \qquad \text{CCF} = \frac{1}{2(20)} = 0.025,$$

and

$$\sigma_p = \sqrt{\frac{0.8(0.2)}{20}}\sqrt{\frac{100 - 20}{100 - 1}} = 0.0804.$$

Consequently,

$$P(p \leq 0.7) \doteq N\left(\frac{0.7 + 0.025 - 0.8}{0.0804}\right)$$
$$= N(-0.93) = 0.1762.$$

11–8 SAMPLING DISTRIBUTION OF THE MEAN

If a random sample of size n is drawn from a population with mean μ and variance σ^2, then the sample observations are independent and identically distributed random variables. Furthermore, the sample mean, computed as

$$\bar{x} = \frac{1}{n}(x_1 + x_2 + \cdots + x_n),$$

is also a random variable, and the expected value and the variance of the sampling distribution of $\bar{x}$ can be derived simply. First, we note that

$$E(\bar{x}) = E\left[\frac{1}{n}(x_1 + x_2 + \cdots + x_n)\right]$$

$$= \frac{1}{n}[E(x_1) + E(x_2) + \cdots + E(x_n)]$$

$$= \frac{1}{n}(n\mu) = \mu. \tag{11–17}$$

That is, the expectation of the sample mean is the population mean.

Next, since sample observations are considered independent random variables, the additivity property holds for the variance. That is, the variance of the sum is the sum of the variances. Furthermore, since $V(x_i) = \sigma^2$, the population variance, we have

$$V(\bar{x}) = V\left(\frac{1}{n}\sum x_i\right)$$

$$= \frac{1}{n^2}[V(x_1) + V(x_2) + \cdots + V(x_n)]$$

$$= \frac{1}{n^2}(n\sigma^2)$$

$$= \frac{\sigma^2}{n}. \tag{11–18}$$

In this derivation we have employed the theorem that the variance of a constant times a variable is equal to the square of the constant times the variable.

The standard error of the mean, which measures the chance variability in sample means, is

$$\sigma_{\bar{x}} = \sqrt{V(\bar{x})} = \frac{\sigma}{\sqrt{n}}, \tag{11–19}$$

which reveals that $\sigma_{\bar{x}}$ is smaller than σ. Moreover, it indicates that when $n \longrightarrow \infty$, $\sigma_{\bar{x}} \longrightarrow 0$. Thus, the larger the sample, the less the fluctuation of sample means from sample to sample.

If samples are taken from a finite population without replacement, as in the previous cases, an FPCF must be introduced to compute the standard error of the mean. Namely,

$$\sigma_{\bar{x}} = \frac{\sigma}{\sqrt{n}}\sqrt{\frac{N-n}{N-1}}. \tag{11–20}$$

When the parent population is normal, then the sampling distribution of $\bar{x}$ is also normal, no matter how small the sample is. However, if the population distribution is anything but normal, then the variable $(\bar{x} - \mu)/\sigma_{\bar{x}}$ approaches the standard normal distribution when n becomes infinite. That the distribution of $\bar{x}$ approaches the normal probability model when n increases is illustrated by Figure 11–4. In this figure, the population distribution is given by $n = 1$ and $\mu = 0$. The density functions of $\bar{x}$ for $n = 3$, 6, 10, 50, and 100 are plotted against this population model in the same figure. This figure shows vividly the tendency for the distribution of $\bar{x}$ to approach normality rapidly when n is increased, even though the population is far from being normal itself.

By the CLT stated above, for large n the distribution function of $\bar{x}$ can then be expressed as

$$P(\bar{x} \leq \bar{x}_0) \doteq N\left(\frac{\bar{x}_0 - \mu}{\sigma_{\bar{x}}}\right). \tag{11–21}$$

Let us now give an example of a complete evaluation of the distribution of the mean. A conditional random sample of size two is drawn from a population that consists of five even numbers, 2, 4, 6, 8, and 10, written on identical chips. Under these conditions, we note first that the mean and the variance of the population are, respectively, 6 and 8. Next we observe that there are $\binom{5}{2} = 10$ equally likely samples, which are

(2, 4), (2, 6), (2, 8), (2, 10), (4, 6), (4, 8), (4, 10), (6, 8), (6, 10), (8, 10).

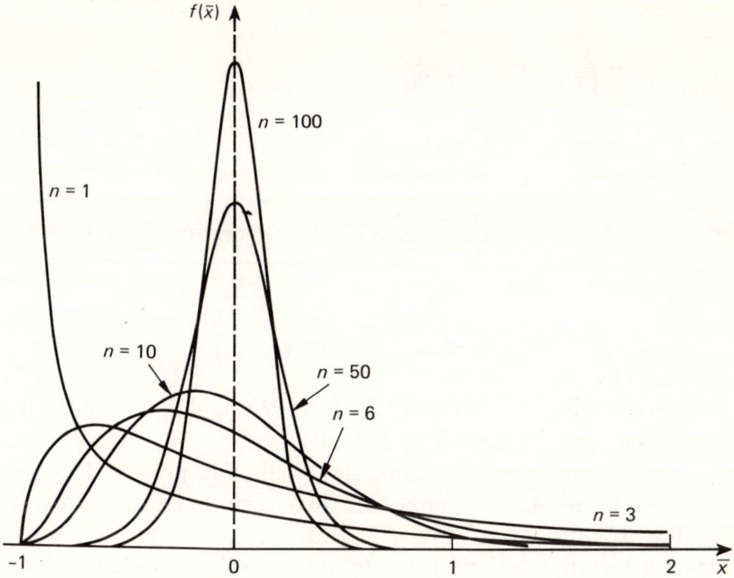

Figure 11–4 Illustration of the Effect of Sample Size on the Distribution of $\bar{x}$

Clearly, the sample means range consecutivity from 3 to 9, with the following probabilities:

Possible values of $\bar{x}$	*Probability*
3	$1/10 = 0.1$
4	$1/10 = 0.1$
5	$2/10 = 0.2$
6	$2/10 = 0.2$
7	$2/10 = 0.2$
8	$1/10 = 0.1$
9	$1/10 = 0.1$
Sum	$10/10 = 1.0$

For this distribution of $\bar{x}$, the expected value is

$$\begin{aligned} E(\bar{x}) &= \sum \bar{x}_i f(\bar{x}_i) \\ &= 3(0.1) + 4(0.1) + 5(0.2) + \cdots + 9(0.1) \\ &= 6, \end{aligned}$$

which is identical with the population mean stated before. Also, the standard error here is

$$\begin{aligned}\sigma_{\bar{x}} &= \sqrt{\sum_i [\bar{x} - E(\bar{x})]^2 f(\bar{x}_i)} \\ &= \sqrt{(3-6)^2(0.1) + (4-6)^2(0.1) + \cdots + (9-6)^2(0.1)} \\ &= \sqrt{3}.\end{aligned}$$

To see whether this result agrees with the theoretical formula for $\sigma_{\bar{x}}$, we compute

$$\sigma_{\bar{x}} = \sqrt{\frac{\sigma^2}{n}}\sqrt{\frac{N-n}{N-1}} = \sqrt{\frac{8}{2}}\sqrt{\frac{5-2}{5-1}} = \sqrt{3}.$$

Thus we see that the standard error of the mean is related to the population standard deviation as specified by (11–20).

Now, let us consider another example. Steel wires produced by a certain factory have a mean tensile strength of 500 pounds and a standard deviation of 20 pounds. If a random sample of 100 wires is drawn from the production line during a certain week, with a total output of over 100,000 units, what is the probability that the sample mean will differ from 500 pounds by 4 pounds or more? What is the probability that the sample mean will be less than 496 pounds? More than 504 pounds?

For this case, we have

$$E(\bar{x}) = \mu = 500 \text{ pounds}$$

and

$$\sigma_{\bar{x}} = \frac{\sigma}{\sqrt{n}} = \frac{20}{\sqrt{100}} = 2 \text{ pounds}.$$

(Why hasn't the FPCF been used to compute the standard error here?) With these values, the desired probabilities become

$$P(|\bar{x} - 500| \geq 4) \doteq 2N\left(\frac{-4}{2}\right) = 2(0.0228) = 0.0456,$$

$$P(\bar{x} \leq 496) \doteq N\left(\frac{496-500}{2}\right) = N(-2) = 0.0228,$$

and

$$P(\bar{x} > 504) \doteq 1 - N\left(\frac{504-500}{2}\right) = 0.0228.$$

It was reported by a certain state that on 125 farms the recorded average wheat yield was 16 bushels per acre, with a standard deviation of 5 bushels. If a random sample of 36 farms is selected without replacement, what is the probability that the sample mean will be less than, or equal to, 14.5 bushels?

Answer:

$$P(\bar{x} \leq 14.5) \doteq N\left(\frac{14.5 - 16}{\dfrac{5}{\sqrt{36}}\sqrt{\dfrac{125 - 36}{125 - 1}}}\right)$$

$$= N(-2.12) = 0.0170.$$

If the recorded average yield is correct, then the event of a sample with $n = 36$ having a mean less than 14.5 bushels per acre would be a very rare event.

11-9 DISTRIBUTION OF THE DIFFERENCE OF TWO PROPORTIONS

When two random samples drawn from two binomial variables are compared, it is possible to work only with the *proportion* of successes, not with the *number* of successes, unless both samples are of the same size. For example, during a presidential election, a sample of 100 voters is taken from one state and 40 are found to be in favor of candidate A; another sample of 150 voters is taken from a second state and 50 are found to be in favor of candidate A. Clearly, these two sets of figures cannot be evaluated unless they are reduced to proportions. Specifically, what we need here is a probability model of the difference of two proportions.

Two independent random samples are drawn. The first is of size n_1 from a binomial population with π_1 and the second is of size n_2 from a binomial population with π_2. Then the difference of the two sample proportions, $\Delta p = p_1 - p_2$, is again a sample statistic whose probability distribution has as its expected value

$$E(\Delta p) = \Delta\pi = \pi_1 - \pi_2, \tag{11-22}$$

and as its standard error

$$\sigma_{\Delta p} = \sqrt{\frac{\pi_1(1 - \pi_1)}{n_1} + \frac{\pi_2(1 - \pi_2)}{n_2}}. \tag{11-23}$$

Furthermore, if n_1 and n_2 are sufficiently large, then the sampling distribution of Δp is approximately normal. That is, by the CLT, we have

$$P(\Delta p \leq \Delta p_0) \doteq N\left(\frac{\Delta p_0 - \Delta\pi}{\sigma_{\Delta p}}\right). \tag{11-24}$$

A game of "heads or tails" is played by two friends, A and B. Each tosses a coin fifty times; one wins the game if he tosses five more heads than the other. What is the probability that A will win the game?

If we assume that the coins used are well balanced, the sampling distribution of Δp will have

$$E(p) = \pi_1 - \pi_2 = 0.5 - 0.5 = 0$$

and

$$\sigma_{\Delta p} = \sqrt{\frac{0.5(0.5)}{50} + \frac{0.5(0.5)}{50}} = 0.1.$$

Now, for A to win he must toss five more heads than B in fifty tosses, or A must have $5/50 = 0.1$ more heads than B. Thus, by the normal approximation,

$$P(\Delta p \geq 0.1) \doteq 1 - N\left(\frac{0.1 - 0}{0.1}\right)$$
$$= 1 - 0.8413 = 0.1587.$$

It is believed that 10 percent of the batteries produced by company A are defective, and that 5 percent of those produced by company B are defective. A random sample of 250 units is taken from the production line of company A and 20 are found to be defective; a random sample of 300 units is taken from the production line of company B and 18 are found to be defective. What is the probability of obtaining this or a smaller difference in sample proportions if the belief about the population parameters is correct?

If the belief were correct, then the distribution of Δp would be defined by

$$\Delta\pi = 0.1 - 0.05 = 0.05$$

and

$$\sigma_{\Delta p} = \sqrt{\frac{0.1(0.9)}{250} + \frac{0.05(0.95)}{300}} = 0.0228.$$

The desired probability, if we note that $\Delta p = 0.08 - 0.06 = 0.02$, becomes

$$P(\Delta p \leq 0.02) = N\left(\frac{0.02 - 0.05}{0.0228}\right)$$
$$= N(-1.32) = 0.0934.$$

In view of this rather small probability, it is highly doubtful that $\pi_1 = 0.1$ and $\pi_2 = 0.05$. Alternatively, this result indicates that it is highly likely that either $\pi_1 < 0.1$ or $\pi_2 > 0.05$, or both.

11-10 DISTRIBUTION OF THE DIFFERENCE OF TWO MEANS

In many fields of scientific investigation, we often desire to compare the means of two random variables, such as the effect of two conditions or treatments or production methods. Again, the comparison is made on the basis of two independent random samples: one with size n_1 drawn from one population with μ_1, another with size n_2 drawn from the other population with μ_2. Now, if $\bar{x}_1$ and $\bar{x}_2$ are the two sample means, then we can evaluate the possible difference between μ_1 and μ_2, $\Delta\mu = \mu_1 - \mu_2$, by the difference of the sample means, $\Delta\bar{x} = \bar{x}_1 - \bar{x}_2$. The problem here, therefore, is one of determining the properties of the sampling distribution of $\Delta\bar{x}$.

Let X_1, with μ_1 and σ_1^2, and X_2, with μ_2 and σ_2^2, be the two parent populations; then it is easy to see that the expected value of $\Delta\bar{x}$ must be

$$E(\Delta\bar{x}) = \Delta\mu = \mu_1 - \mu_2, \tag{11-25}$$

and the standard error of $\Delta\bar{x}$ is

$$\sigma_{\Delta\bar{x}} = \sqrt{\frac{\sigma_1^2}{n_1} + \frac{\sigma_2^2}{n_2}}. \tag{11-26}$$

We know that the sample mean is normally distributed when the sample size becomes infinite. There is reason to believe, then, that if $\bar{x}_1$ and $\bar{x}_2$ are independent and normally distributed, their difference, $\Delta\bar{x}$, must also be normally distributed when both n_1 and n_2 approach infinity. Thus, for sufficiently large integers of n_1 and n_2, the distribution function of $\Delta\bar{x}$ can be expressed as

$$P(\Delta\bar{x} \leq \Delta\bar{x}_0) \doteq N\left(\frac{\Delta\bar{x}_0 - \Delta\mu}{\sigma_{\Delta\bar{x}}}\right). \tag{11-27}$$

It may be noted here, as in the case of the distribution of the mean, that if X_1 and X_2 are normal variables, then $\Delta\bar{x}$ is also a normal variable, however small n_1 and n_2 are.

Spot welds of aluminum of high shearing strength produced by company A possess a mean of 4500 pounds and a variance of 40,000 pounds^2. Those produced by company B possess a mean of 4000 pounds and a variance of 90,000 pounds^2. If 50 welds of brand A and 100 welds of brand B are selected at random and tested, what is the probability that the sample mean strength of A will be at least 600 pounds more than B? At least 500 pounds more than B?

Since, for the distribution of $\Delta\bar{x}$ in this case, we have

$$\Delta\mu = 4500 - 4000 = 500 \text{ pounds}$$

and

$$\sigma_{\Delta\bar{x}} = \sqrt{\frac{40{,}000}{50} + \frac{90{,}000}{100}} = 41.23 \text{ pounds},$$

the desired probabilities become

$$P(\Delta\bar{x} \geq 600) \doteq 1 - N\left(\frac{600 - 500}{41.23}\right)$$

$$= 1 - N(2.43) = 0.0075,$$

$$P(\Delta\bar{x} \geq 500) \doteq 1 - N\left(\frac{500 - 500}{41.23}\right)$$

$$= 1 - N(0) = 1 - 0.5000 = 0.5000.$$

Steel pipes produced by a certain process have a mean diameter of 5 inches and a standard deviation of 0.1 inch; what is the probability that two lots, each with 25 pipes, will differ in mean diameter by (a) 0.01 inch or more? (b) 0.005 inch or more? (c) 0.005 inch or less?

Here, the distribution of $\Delta\bar{x}$ has

$$\Delta\mu = 5 - 5 = 0$$

and

$$\sigma_{\Delta\bar{x}} = \sqrt{\frac{(0.1)^2}{25} + \frac{(0.1)^2}{25}} = 0.0282 \text{ inch.}$$

Consequently,

(a) $$P(|\Delta\bar{x}| \geq 0.01) \doteq 2N\left(\frac{-0.01}{0.0282}\right)$$

$$= 2N(-0.35) = 0.7264,$$

(b) $$P(|\Delta\bar{x}| \geq 0.005) \doteq 2N\left(\frac{-0.005}{0.0282}\right)$$

$$= 2N(-0.18) = 0.8572,$$

(c) $$P(|\Delta\bar{x}| \leq 0.005) \doteq N\left(\frac{0.005}{0.0282}\right) - N\left(\frac{-0.005}{0.0282}\right)$$

$$= 0.5714 - 0.4286 = 0.1428.$$

11-11 DISTRIBUTIONS OF MEDIAN AND RELATED MEASURES

If the parent population distribution is normal, and if the sample size is greater than 30, the distribution of the median is approximately normal with expectation and standard error as follows:

$$E(x_{.5}) = \mu, \tag{11-28}$$

$$\sigma_{x_{.5}} = \frac{1.2533\sigma}{\sqrt{n}}. \tag{11-29}$$

We note that, while the expectation of the median is the same as the expectation of the mean, the standard error of the median is greater than the standard error of the mean by a multiplier of 1.2533.

Recall that the median is a position measure that divides a variable into two equal parts. Recall also that a variable may be divided into four equal parts by three *quartile* measures. The first quartile, Q_1, is a statistic whose value is equal to or less than 25 percent of the value of a variable. The second quartile, Q_2, is identical with the median. The third quartile, Q_3, is a measure that is equal to or greater than 75 percent of the values of a variable. Furthermore, for ungrouped data, we know that $Q_1 = n/4$, $Q_2 = 2n/4$, and $Q_3 = 3n/4$, where n is the sample size, as usual.

Distributions of Q_1 and Q_3 have the same standard error, which is defined as

$$\sigma_{Q_1} = \sigma_{Q_3} = \frac{1.3626\sigma}{\sqrt{n}}. \tag{11–30}$$

Furthermore, if the parent population is normal or approximately so and the sample size is greater than 30, distributions of Q_1 and Q_2 are nearly normal, with expectations very nearly equal to the first and second quartiles of the population, respectively.

Measures that divide a set of data into ten equal parts are called *deciles*. Clearly there are nine decile measures, with $D_1 = n/10$, $D_2 = 2n/10$, $D_3 = 3n/10, \ldots, D_9 = 9n/10$. Remarks made on the distribution of the median apply equally well to distributions of deciles. Expectations of deciles are very nearly equal to their corresponding population parameters. Standard errors of deciles are as follows:

$$\sigma_{D_1} = \sigma_{D_9} = \frac{1.7094\sigma}{\sqrt{n}}, \tag{11–31}$$

$$\sigma_{D_2} = \sigma_{D_8} = \frac{1.4288\sigma}{\sqrt{n}}, \tag{11–32}$$

$$\sigma_{D_3} = \sigma_{D_7} = \frac{1.3180\sigma}{\sqrt{n}}, \tag{11–33}$$

$$\sigma_{D_4} = \sigma_{D_6} = \frac{1.2680\sigma}{\sqrt{n}}. \tag{11–34}$$

Note that the standard error of D_5 is obviously the same as that of the median. Note also how the multiplier factor decreases as the decile value comes closer to the median.

11–12 SIGNIFICANCE OF SAMPLING DISTRIBUTIONS

The value of a sample statistic computed from a random sample of a given size drawn from a specific population depends upon which elements are included in the sample. A different sample will, usually, yield a different value of the statistic. This observation implies that a sample statistic is expected to be different from its corresponding population parameter, which is often an unknown constant; that is, measuring only a part instead of the entire population results in a margin of error, known as the *sampling error*—the difference between the value of a statistic and that of a population parameter. It is important to note, in addition, that a possible difference, although referred to as an error, might be due to chance, not to an error that has been made or will be made because of sampling. Furthermore, we must understand that sampling errors are not literally "errors" as the layman understands the word. It is perhaps more appropriate to call them "sampling variations" or "chance deviations," since their existence is due to the characteristic of chance inherent in random sampling. One basic value of sampling distributions is the assistance they give us in revealing the patterns of sampling errors and their magnitudes in terms of standard errors—that is, standard deviations of sampling distributions.

A sampling distribution shows that a statistic of a random sample may take on any of a set of values; but these values do not have the same probability of occurrence. We note that the closer a statistic is to its corresponding parameter, the greater the probability is for that statistic to happen. As we shall see, this property enables us to evaluate the validity of statistical inference.

In deducing a sampling distribution, we must first make an assumption about the appropriate population parameter. Inasmuch as any value can be assumed for a parameter, depending upon our knowledge or estimate of the population, there is no theoretical limit to the number of sampling distributions for the same sample size that can be taken from the same population. There is a sampling distribution for each assumed value of a parameter.

Furthermore, as we have already mentioned, given the assumed value of a parameter, there is a different sampling distribution of a statistic for each specific sample size.

Finally, under the same assumptions about a population and the same sample size, the sampling distribution of one statistic differs from that of another statistic. For example, the pattern of the distribution of $\bar{x}$ will differ from that of $\bar{s}^2$, even though both measures are computed from the same sample.

From previous observations, it is easy to appreciate why we should always bear in mind the assumption of the population, the name of the statistic, and the sample size when we think of a sampling distribution.

It will also be useful for us to remember that our discussion in this chapter might appropriately be called "theory of sampling for large samples." In short, when a sampling distribution with large n meets the requirements of the central limit theorem, we have applied the large-sampling method. How large n should be for the CLT to apply depends upon the particular statistic. For sample mean, for example, a large sample means that $n > 30$. In the case of sample proportions, the sample size required depends on the value of π. In general, the more π deviates from 0.5, the greater n should be for CLT to operate. In this connection, we may also note the distributions for $\hat{s}^2$ and $\hat{s}$ can also be approximated by the standard normal distribution if the parent population is normal and if $n \geq 100$. They are not introduced here, since inferences with σ^2 or σ are usually made with some special probability distributions introduced in Chapters 14 and 15.

It is important to note that when sample size is large enough for the CLT to operate, standard errors can be calculated with sample statistics in place of the corresponding population parameters when the latter are unknown; for instance, $\hat{s}$ can be used for σ, p can substitute for π, and so on. In other words, if the sample size is large enough, the sampling theory for large samples holds whether a standard error is computed with the known population parameter required or its sample estimate.

When samples are small and population parameters required to compute standard errors are unknown, we face what is called the "small-sampling method" or "exact-sampling theory." With exact-sampling theory, each statistic has a particular probability distribution other than the normal probability model, as shown in Chapters 14 and 15.

We turn next to a discussion of the traditional decision procedures of estimation and testing of hypotheses for the case of large samples.

EXERCISES

11–1 Suppose your class contains 100 students, numbered from 1 to 100. Select a simple random sample of 10 by using the table of random digits.

11–2 Show that the first number selected in the preceding problem comes from a uniform distribution with $f(x; 100) = 1/100$, $x = 1, 2, 3, \cdots, 100$.

11–3 A certain type of ground-to-ground missile has a probability of 0.25 of hitting the target. Ten such missiles are fired.
(a) Write the exact expression that exactly five will hit the target.
(b) Write the exact expression that at least five will hit the target.

11–4 A die is tossed sixty times.
(a) What is the mean of the sum of points?
(b) What is the variance of the sum of points?
(c) What is the probability that the total number of points will be less than 200?

11–5 The reliability of a rocket is the probability p that an attempted launching will be successful. It has been established that the reliability of a certain type of rocket is 0.9. A modification of this rocket is made and is being evaluated. Which set of evidence below throws more doubt on the hypothesis that the modified design is only 90 percent reliable?
(a) Of 121 modified rockets tested, 116 performed successfully.
(b) Of 81 modified rockets tested, 78 performed successfully.

11–6 Which of the following sets of evidence throws more doubt on the hypothesis that more voters are in favor of candidate A than in favor of candidate B in a presidential election?
(a) Of 10,000 voters sampled, 5100 are in favor of A.
(b) Of 1000 voters sampled, 510 are in favor of A.

11–7 An urn contains eight numbered balls, three with the number 1 and five with the number 2. Two balls are drawn at random without replacement from the urn.
(a) List all the possible samples and their associated probabilities.
(b) Derive the sampling distribution of the difference between the two numbers on the balls drawn.
(c) What comments are in order from your results in (a) and (b)?

11–8 A random sample of 4 is drawn from an infinite population with $p = 0.5$. Derive the sampling distribution of the number of successes (a) in terms of exact probabilities, and (b) in terms of normal approximations.

11–9 If a random sample of 100 is taken from the production line of a process in which 20 percent of the items are defective, what is the probability that 15 or less in the sample are defective?

11–10 Plot the distribution of the number of successes with $n = 20$ and $p = 0.4$ against its normal approximations. In your judgment, are the normal approximations good enough for practical situations?

11–11 Let $n = 3600$ and $p = 1/3$, and calculate the probability for the number of successes between $3600p - 20$ and $3600p + 20$.

11–12 Half-inch iron screws manufactured by a certain factory are occasionally not slotted. This occurs at random, and the compound probability of its happening and escaping inspection is 0.02. In a shipment of 2500 such screws, what is the probability that (a) 64 or more will lack slots? (b) 36 or less? (c) between 36 and 64?

11–13 Let the population be ten balls in an urn, five of which are black and five white. A random sample of six is drawn without replacement; evaluate the sampling distribution of the proportion of white balls. (Note: The evaluation of a sampling distribution involves the listing of all sample values with associated probabilities, and calculation of its expected value and standard error.)

11–14 Forty percent of the registered voters in a given state are Republicans. From this population a sample of five is drawn at random. Evaluate the sampling distribution of p, proportion of Republicans. [*Hint*: Since the sample is so very small relative to the population, independence may be assumed here.]

11–15 Let the population be the same as in the preceding problem, and let a random sample of 100 be drawn. Would you be surprised to find that there are 50 Republicans in the sample?

11–16 In a national election, 55 percent of the voters are in favor of a certain candidate; what is the probability that, in a sample of 100, the result will not show a majority in favor of the candidate?

11–17 If an unconditional random sample of 180 items is drawn from a population with $\pi = 1/3$, what is the probability that the sample proportion satisfies $50/180 < p < 70/180$?

11–18 On college faculties in New York City, 1/6 are women. If a random sample of 180 is drawn from this population, determine
(a) $P(p \geq 0.3)$.
(b) $P(0.1 \leq p \leq 0.25)$.

11–19 The variance of the distribution of p is found to be 0.0025 with $n = 100$; what is π?

11–20 Forty-six percent of union members are against trading with Red China. What is the probability that a poll would show a majority of this trait if the poll consisted of (a) 100 union members? (b) 1000 union members?

11–21 It is known that 5 percent of the radio tubes produced by a certain manufacturer are defective. If the manufacturer sends out 1000 lots, each containing 100 tubes, in how many of these lots can we expect to have (a) fewer than 90 good tubes? (b) 98 or more good tubes?

11–22 In a random sample of 1000 households selected in West Virginia in 1965, 96 of the heads of families were found to be unemployed. Was this sample result consistent with the assumption that 10 percent of the labor force there in that year were unemployed? Explain.

11–23 Evaluate the sampling distribution of $\bar{x}$ with $n = 2$ from a population of 10 digits, 0, 1, 2, 3, 4, 5, 6, 7, 8, and 9. Assume that sampling is made with replacement.

11–24 Rework the preceding problem under the assumption that a conditional sample of 2 is drawn.

11–25 The distribution of average weekly wages of 3000 secretaries in a Canadian city is found to have $\mu = \$68$ and $\sigma = \$3$. If 250 random samples, all of size 25, are drawn from this population without replacement, what would be the mean and standard error of the distribution of $\bar{x}$ formed by these 250 samples?

11–26 In the preceding problem, how many samples in the distribution can we expect to have means between \$66.80 and \$68.30?

11–27 A lot of 1000 fried chicken dinners has a mean weight of 12 ounces and a standard deviation of 0.6 ounce. What is the probability that in a random sample of 100 from this population the total weight will be (a) less than 1190 ounces? (b) more than 1195 ounces? (c) between 1190 and 1195 ounces? [*Hint*: That the combined weight will be between 1190 and 1195 ounces is the same as that the sample mean will be between 11.90 and 11.95 ounces.]

11–28 Packages to be delivered by a factory have a mean weight of 300 pounds and a standard deviation of $\sqrt{2500}$ pounds. What is the probability that 25 packages taken at random and loaded on a truck will exceed the specified capacity of the truck, known to be 8200 pounds?

11–29 A liberal arts college has 100 faculty members, 60 of whom have doctoral degrees. Two samples, with $n_1 = n_2 = 30$, are drawn from this faculty independently, with replacement, and whether or not they possess doctoral degrees is noted. What is the probability that the two samples will differ by 8 or more with doctoral degrees?

11–30 Rework the preceding problem under the assumption of nonreplacement.

11–31 A certain make of bearing balls has a mean weight of 0.5 ounces and a standard deviation of 0.02 ounce. Two unconditional random samples are taken independently from a certain day's production, with $n_1 = 500$ and $n_2 = 800$. What is the probability that the two sample means differ by more than 0.002 ounce? By less than 0.001 ounce?

11–32 Two different makes of TV picture tubes, A and B, possess the following parameters: $\mu_A = 1400$ hours, $\sigma_A{}^2 = 40{,}000$ hours^2, $\mu_B = 1200$ hours, and $\sigma_B{}^2 = 10{,}000$ hours^2. A random sample of 125 tubes is drawn from each make; determine the probability that (a) make A will have a mean life at least 160 hours longer than B; (b) make A will have a mean life at least 250 hours longer than B.

11–33 A normal population has $\mu = 100$ and $\sigma = 5$. If a random sample of 250 is drawn from this population, is it likely that this sample will have a median of 91 or less?

11–34 If the first quartile of a population with $\sigma = 10$ is 11, what is the probability for Q_1 with $n = 100$ to be less than 10?

11–35 If the third quartile of a random variable, with $\sigma^2 = 81$, is 75, what is the probability for Q_3 with $n = 81$ to be greater than 77?

12

CLASSICAL DECISION PROCEDURES: ESTIMATION

12–1 STATISTICAL INFERENCE

Having a knowledge of sampling distributions, we are prepared to discuss traditional decision procedures for scientific investigations. The scientist often knows, or is willing to assume, that the population or the random variable from which he samples is of a certain functional form whose parameter he attempts to evaluate. For example, from theoretical considerations or previous studies, it might be known that the IQ of school children possesses a normal function, $n(\mu, \sigma^2)$, where the mean IQ, μ, is the unknown parameter of interest. In general, when we say that a random variable X is distributed as $f(x; \theta)$, we mean that θ, (which stands for any parameter, such as μ, π, σ) is the unknown parameter in the frequency function, $f(x)$.

Statistical inference refers to methods by which one tries to select a random sample from a population and, on the basis of sample observations, attempts (1) to estimate the value of the unknown parameter θ, or (2) to decide whether θ is the same as some preconceived value of θ, say, θ_0. The first of these decision procedures is called the estimation of θ; the second is known as the testing of a hypothesis concerning θ.

In this chapter we shall be concerned with methods of statistical estimation, and the next chapter with methods of hypothesis testing. With respect to estimation of a parameter, there are two types to be considered: point estimation and interval estimation. We shall introduce them in that order.

12–2 POINT ESTIMATION

The procedure in point estimation is to select a random sample of n observations, $x_1, x_2, \ldots, x_n$, from a population $f(x; \theta)$, and then to use some preconceived method to arrive at a number, say $\hat{\theta}$ (read "theta hat"), that we accept as an estimator of θ. Note that the estimator, $\hat{\theta}$, is a single point on the real number scale—thus the name *point estimation*.

In the terminology common to estimation problems, a statistic is called an *estimator* and the value of the statistic is called an *estimate*. Thus, the decision function is the statistic, or estimator, and the value of the function is the value of the statistic, or the estimate. It is important to observe that an estimator is a function of n independent sample values. Considering this fact, we should select a function, say, $\hat{\theta} = g(x_1, \ldots, x_n)$, that will give us the best estimator of θ. Such a function, unfortunately, is often difficult and sometimes impossible to determine in practice. This difficulty arises from the fact that the classical procedure is based on sample information alone. The classical procedure, in other words, does not involve the use of subjective probability through the use of prior probability law, unlike the decision procedure we considered in Chapters 6, 7, and 8. Furthermore, the classical analysis does not include an operational loss-function concept that can be integrated into the analysis. These two omissions prevent the classical analysis from being a complete theory in the sense that its use establishes a unique decision. (Statisticians who incorporate both concepts of subjective prior probability law and loss functions are called Bayesian decision theorists.)

Failing in selecting a best estimator, classical statisticians are forced to fall back on heuristic rules of thumb in judging whether an estimator is a good one. A good estimator, as common sense dictates, is "close" to the parameter being estimated. More precisely, the quality of an estimator is evaluated in terms of several heuristic criteria, including (1) unbiasedness, (2) consistency, (3) efficiency, and (4) sufficiency. Let us comment on these criteria first before we move on to see how they can be incorporated into an estimator problem.

Unbiasedness

An estimator is said to be *unbiased* if its expected value is identical with the population parameter being estimated. That is, if $\hat{\theta}$ is an unbiased estimator of θ, then we must have $E(\hat{\theta}) = \theta$.

Thus, we see that the sample mean $\bar{x}$, the sample proportion p, the difference between sample means $\Delta\bar{x}$, and the difference between two sample proportions Δp are all unbiased estimators, since in each case the expectation of the estimator is equal to the corresponding parameter being estimated. However, the sample variance, computed with the multiplying constant $1/n$, is a biased estimator of σ^2 because

$$E(s^2) = \left(\frac{n-1}{n}\right)\sigma^2 = \sigma^2 - \frac{\sigma^2}{n}.$$

The bias here is the quantity $-\sigma^2/n$. Clearly this negative bias will vanish when n becomes infinite, and therefore s^2 becomes an unbiased estimator of σ^2. Furthermore, when the sample variance is computed with the multiplier $1/(n-1)$, it will be an unbiased estimator of σ^2 however small n is, since

$$\hat{s}^2 = \frac{1}{n-1} \sum_i (x_i - \bar{x})^2$$

or

$$(n-1)\hat{s}^2 = \sum_i (x_i - \bar{x})^2.$$

Subtracting a constant μ from x_i and $\bar{x}$ in the expression above, we have

$$(n-1)\hat{s}^2 = \sum_i (x_i - \mu)^2 - n(\bar{x} - \mu)^2.$$

Now, taking expectation on both sides of this, we obtain

$$\begin{aligned}(n-1)E(\hat{s}^2) &= \sum_i E[(x_i - \mu)^2] - nE[(\bar{x} - \mu)^2] \\ &= \sum_i \sigma^2 - n\left(\frac{\sigma^2}{n}\right) \\ &= (n-1)\sigma^2,\end{aligned}$$

or

$$E(\hat{s}^2) = \sigma^2.$$

The criterion of unbiasedness assumes significance when many investigations produce estimates of the same parameter. These estimates, if unbiased, can be combined to produce a still better estimate of the parameter.

Consistency

An estimator $\hat{\theta}$ is said to be consistent for θ if, for any $\delta > 0$,

$$P(|\hat{\theta} - \theta| < \delta) \to 1 \qquad \text{as } n \to \infty.$$

In other words, an estimator is said to be consistent if the probability that it equals the parameter being estimated approaches 1 as n approaches infinity. This is clearly true for $\bar{x}$ or p, or $\hat{s}^2$, as a consequence of the law of large numbers.

Recall that $E(\bar{x}) = E(x_{.5}) = \mu$; but are the sample mean and the sample median equally satisfactory to estimate μ? The answer is yes if $f(x\colon \mu)$ is symmetrical and if consistency is the only criterion under consideration. If, however, the population distribution is skewed so that the population mean and median are not identical, we would expect the sample mean to come closer to the population mean and the sample median to come closer to the population median when the sample size is increased. Thus, when n approaches infinity or approaches the size of the population, the sample mean will approach the population mean in value, and the sample median will approach the population median (which is different from the population mean). So we see that the sample mean is, but the median is not, a consistent estimator of μ when the population distribution is not symmetrical. Furthermore, the sample mean is an unbiased estimator of μ no matter what form the population distribution assumes, while the sample median is an unbiased estimator of μ only if the population distribution is symmetrical. We conclude, therefore, that the sample mean is better than the sample median as an estimate of μ in terms of both unbiasedness and consistency.

Efficiency

Given that estimators are unbiased or at least consistent, then an estimator $\hat{\theta}_1$ is said to be more efficient than another estimator $\hat{\theta}_2$ for θ if the variance of the first is less than the variance of the second. This criterion seems to be an intuitively clear concept. Obviously, the smaller the variance of an estimator is, the more concentrated is the distribution of the estimator around the parameter being estimated and, therefore, the better this estimator is.

The best illustration of relative efficiency in estimators is the estimation of μ by the sample mean and by the sample median. If the population is normally distributed, then both the sample mean and the sample median are consistent and unbiased estimators of μ. Yet, we can claim that the sample mean is better than the sample median as an estimator of μ. This claim is made in terms of efficiency. We know that $V(\bar{x}) = \sigma^2/n$ and that

$$V(x_{.5}) = (1.2533\sigma/\sqrt{n})^2 = 1.57076\sigma^2/n.$$

That is, $V(\bar{x}) < V(x_{.5})$. As a matter of fact, $V(\bar{x}) = 0.64V(x_{.5})$. Thus, we conclude that the sample mean is more efficient than the sample median as an estimator of μ.

When we say that one estimator is more efficient than another, it seems reasonable to think of an estimator that is the *most* efficient for the parameter being estimated. The most efficient estimator, in terms of our understanding of relative efficiency, must possess a variance as small in value as possible. That is, we think of the most efficient estimator as a *minimum-variance estimator*. An estimator $\hat{\theta}$ is said to be a minimum-variance estimator of θ if

$$E[\hat{\theta} - E(\hat{\theta})]^2 \leq E[\hat{\theta}^* - E(\hat{\theta}^*)]^2,$$

where $\hat{\theta}^*$ is any estimator of θ other than $\hat{\theta}$.

The lower bound of a minimum variance can be found by what is called the Cramer-Rao inequality, according to which, for example, if $\hat{\theta}$ is an estimator of μ, then the variance of $\hat{\theta}$ cannot be less than σ^2/n—that is, $V(\hat{\theta}) \geq \sigma^2/n$. Here, $\hat{\theta}$ could be the sample mean, the sample median, or some other statistic. Whatever it is, its variance has a lower bound of σ^2/n. But we know that $V(\bar{x}) = \sigma^2/n$. Thus, besides being unbiased and consistent, $\bar{x}$ is a minimum-variance estimator of μ.

Sufficiency

A rigid mathematical statement of this last desirable property, sufficiency in estimation, is rather involved. Fortunately, we find it entails an accurate intuitive meaning. An estimator is said to be sufficient if it conveys as much information as possible about the parameter contained in the sample so that no additional information will be supplied by any other estimator. This rather obvious yet vague statement may be clarified by an illustration.

Consider, for example, a normal population with a unit variance but an unknown mean, μ. If we wish to estimate this unknown mean, we may take a random sample of, say, two observations. These observations, X_1 and X_2, are assumed to be independent random variables, and each is distributed identically with the population. Now, when we investigate the collection (X_1, X_2), we investigate all the information this sampling experiment yields. Furthermore, the sample set (X_1, X_2), obtained from the parent random variable whose mean we seek to estimate, furnishes all the information that is available for us to compute various estimators.

Now, suppose we select the following two functions of (X_1, X_2) for evaluation:

$$\hat{\theta}_1 = \frac{X_1 + X_2}{2} \quad \text{and} \quad \hat{\theta}_2 = \frac{X_2 - X_1}{2}.$$

We see that the set $(\hat{\theta}_1, \hat{\theta}_2)$ contains the same information as (X_1, X_2), since one is functionally related to the other. Next we note that in $(\hat{\theta}_1, \hat{\theta}_2)$, $\hat{\theta}_2$ is normally distributed with an expectation of zero and a variance of 0.5. Clearly, the distribution of $\hat{\theta}_2$ does not involve μ, and therefore it does not yield any useful information on the parameter we seek to estimate. Again, in $(\hat{\theta}_1, \hat{\theta}_2)$, $\hat{\theta}_1$ is normally distributed with an expectation identical with μ and a variance of 0.5. Clearly, $\hat{\theta}_1$ is a sufficient estimator of μ, since it provides all the information on μ we can obtain from the sample data, and since any other estimator, such as $\hat{\theta}_2$, cannot yield any more information for the task of estimation on hand.

The significance of sufficiency lies in this: if a sufficient estimator exists, it is absolutely unnecessary to consider any other nonsufficient estimators; a sufficient estimator ensures that all information that a sample can furnish with respect to the estimation of the parameter of interest is being utilized. It may now be mentioned that $\bar{x}$, p, $\Delta\bar{x}$, and Δp are sufficient estimators of the corresponding parameters μ, π, $\Delta\mu$, and $\Delta\pi$, respectively.

Having decided upon the properties desirable for an estimator, we must now devise methods of estimating functions that will provide estimators satisfying these properties. Among the many important methods, we shall consider only two in this text: the *maximum likelihood function* and the *least square method.* The first is our immediate concern; the latter we shall take up when we treat time series and association analysis.

12-3 MAXIMUM LIKELIHOOD FUNCTION

We shall develop the concept of the maximum likelihood function by an example. Suppose, after taking a sample of ten students from colleges and universities, we find that three of them are against the admission of Red China to the United Nations. We would like to know, given this sample evidence, how likely is it that the population proportion will be 0.1? 0.2? . . . 0.9? 1.0?

To answer this question, we note that the sample proportion in this case is a binomial variable. If the true proportion π is, say, 0.1, then we have

$$P(p = 0.3) = \binom{10}{3}(0.1)^3(0.9)^7 = 0.0574.$$

If the true proportion π is, say, 0.2, then we have

$$P(p = 0.3) = \binom{10}{3}(0.2)^3(0.8)^7 = 0.2013.$$

Similarly, we can compute the probability for the sample proportion to be 0.3 given any other possible values of π specified in the query. Let us tabulate these probabilities as follows:

π	0.0	0.1	0.2	0.3	0.4	0.5	0.6	0.7	0.8	0.9	1.0
$P(p = 0.3)$	0.0	.0574	.2013	.2668	.2150	.1172	.0425	.0090	.0008	.0000	0.0

Now, if we think that the true proportion can assume any value between 0 and 1 instead of just those given above, we are thinking of the function $L(x) = P(p = 0.3 \mid \pi = x)$, which is defined for all x's satisfying the inequality $0 \leq x \leq 1$.

$$L(x) = \binom{10}{3}\pi^3(1 - \pi)^7$$

is called the *likelihood function* for π and the observed result $p = 0.3$ with $n = 10$. This function clearly gives the conditional probability of $p = 0.3$, given that $\pi = x$. We can graph this function by plotting the values shown in the preceding table and connecting them by a smooth curve. In doing

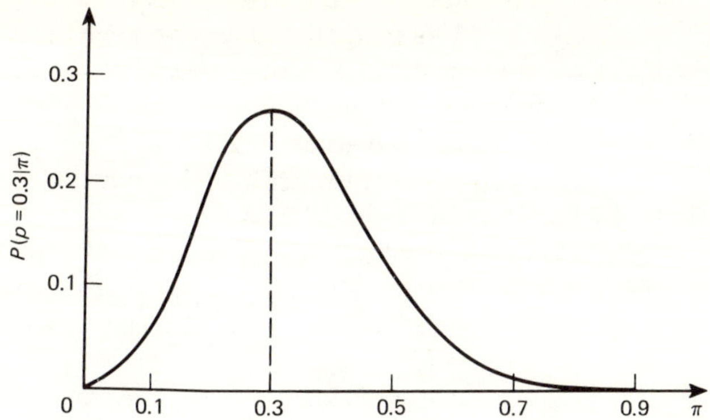

Figure 12–1 Likelihood Curve for π Given p = 0.3 and n = 10

so we obtain what is called the *likelihood curve* for π given $p = 0.3$ with $n = 10$, as shown by Figure 12–1.

We ask: what value of π seems to make the sample outcome observed most likely? The answer is: the value of π must make the probability $P(p = 0.3)$ as high as possible. Looking at Figure 12–1, we see that this value is $\pi = 0.3$. We may now draw a conclusion. In our sample of ten we found that three were against the admission of Red China to the United Nations. Among all values of π, the value of $\pi = 0.3$ seems to make this sample result most likely. We therefore estimate π to be 0.3. Having done this, we have made what is called a *maximum likelihood estimate* for π.

To generalize the foregoing discussion, suppose that the population is distributed as $f(x;\theta)$, where θ is the parameter we attempt to estimate. Although θ is unknown, we may well know a set S of numbers to which θ must belong. That is, S is the collection of possible values of the parameter θ. If x_0 is an observed estimate computed from a sample taken from the variable X (the population), we can consider the function

$$L(y) = P(\hat{\theta} = x_0 \mid \theta = y)$$

as the *likelihood function for θ, given $\hat{\theta} = x_0$*. Furthermore, if

$$L(\hat{\theta}) = \max\left[L(y) : y \in S\right],$$

then we say that $\hat{\theta}$ is the *maximum likelihood estimator of θ*.

Thus, with the maximum likelihood method, we treat a parameter θ as a variable and select a value of θ such that the likelihood (probability) that a set of sample observations obtained at random is a maximum. The process seems to be straightforward and simple—a consequence, unfortunately, of the simplicity of our illustration. As a matter of fact, mathematical methods devised by R. A. Fisher, which enable us to find estimators based on maximum likelihood reasoning, are actually beyond the reach of most readers of this text. But the reader may be assured that most statistics introduced in the preceding chapters are obtained by maximum likelihood methods. Let us now mention, without proof, the results of the maximum likelihood method for several situations.

First, if a random sample of size n is drawn from a binomial population, than the MLE of π is $p = x/n$. We know that $E(p) = \pi$; p is therefore an unbiased estimator of π. Furthermore, since $\sigma_p = \sqrt{(\pi)(1 - \pi)/n}$, we see that σ_p approaches 0 as n approaches infinity. Thus, p is also a consistent estimator of π. Generally, we add a "hat" to the parameter to indicate its estimator. Here, we have $\hat{\pi} = p$.

Second, the MLE of μ is the sample mean $\bar{x} = \sum x/n$, or $\hat{\mu} = \bar{x}$. We have already observed that $\bar{x}$ is an unbiased, consistent, efficient, and sufficient estimator of μ.

Third, the MLE of the difference between two population proportions is the difference between two independent random sample proportions; that is, $\Delta\hat{\pi} = p_1 - p_2 = (x_1/n_1) - (x_2/n_2)$. We note that Δp possesses all the four desirable properties of an estimator.

Fourth, the MLE of the difference between two population means is the difference between two independent sample means:

$$\Delta\hat{\mu} = \bar{x}_1 - \bar{x}_2 = \sum x_1/n_1 - \sum x_2/n_2.$$

Is $\Delta\bar{x}$ an unbiased, consistent, efficient, and sufficient estimator for $\Delta\mu$?

Fifth, for a normal population the MLE of σ^2 is the sample variance; that is,

$$s^2 = [\sum (x - \bar{x})^2]/n.$$

It has already been pointed out that s^2 is a biased estimator of σ^2 and that this bias can be corrected by introducing a division constant of $(n - 1)$, or this bias disappears when n becomes infinite. Quite often, σ^2 or σ may be known from previous studies. Also, owing to the fact that if a constant is added to or subtracted from every value of a variable, the distribution of the variable changes only its scale (not its variance), the value of σ known previously can be used even though μ has increased or decreased. And when σ is needed for computation purposes but is unknown, $\hat{s}$ can be used in its place, as mentioned in the previous chapter, without sacrificing much accuracy if n is large.

In concluding our discussion on point estimation, let us observe that the maximum likelihood method provides estimators that are consistent, efficient, and sufficient. It does not, however, provide unbiased estimators in many cases. It is also interesting to note that maximum likelihood estimators are functions of sufficient statistics but they need not be sufficient themselves.

12-4 INTERVAL ESTIMATION

As we have seen, the procedure in point estimation is to select a function of a sample random variable that will represent "best" the parameter being estimated. Its merit is that it provides a specific value for the parameter under investigation. This merit, however, is also the very deficiency of a point estimate. Being a single point on the real number scale, a point estimate does not tell us how close it is to the parameter being estimated. In order to have an explicit probabilistic expression of the possible difference between an estimate and the value of the parameter being estimated, the use of interval estimation is suggested.

Internal estimation refers to the estimation of a parameter by a random interval, called the *confidence interval*, whose endpoints, L and U with $L < U$, are functions of the observed random variables such that the probability that the inequality $L < \theta < U$ is satisfied is expressed in terms of a predetermined number, $1 - \alpha$. This formulation may be stated as follows:

$$P[|\hat{\theta} - \theta| < kSE(\hat{\theta})] = 1 - \alpha,$$

where $SE(\hat{\theta})$ stands for the standard error of the estimator $\hat{\theta}$, $1 - \alpha$ is called the confidence coefficient, k is the confidence multiplier corresponding to the value of $1 - \alpha$, and α is the probability that the interval does not include the true parameter value. We may now note that the inequality in the expression above can be written as

$$|\theta - \hat{\theta}| < kSE(\hat{\theta})$$

or

$$-kSE(\hat{\theta}) < \theta - \hat{\theta} < +kSE(\hat{\theta}).$$

Adding $\hat{\theta}$ to all members of this extended inequality, we obtain

$$\hat{\theta} - kSE(\hat{\theta}) < \theta < \hat{\theta} + kSE(\hat{\theta}).$$

We can now express the general statement of a confidence interval for estimating θ as

$$P[\hat{\theta} - kSE(\hat{\theta}) < \theta < \hat{\theta} + kSE(\hat{\theta})] = 1 - \alpha. \tag{12-1}$$

Here,

$\hat{\theta} - kSE(\hat{\theta}) = L$, the lower confidence limit,

and

$\hat{\theta} + kSE(\hat{\theta}) = U$, the upper confidence limit.

For obvious reasons, we call L and U confidence limits, which are random endpoints of a random interval. The adjective "random" employed here is indicated by P; the P refers to the thing that is random inside the brackets of (12–1), and that is the estimator $\hat{\theta}$.

When the parent population is normal, or when the central limit theorem applies, the sampling distribution of the estimator $\hat{\theta}$ is normal. As a result, the confidence multiplier k turns out to be the value of a standardized normal variable, $Z_{\alpha/2}$. The most frequently used confidence coefficients and their corresponding confidence multipliers for the normal case are given below:

Confidence coefficient $1 - \alpha$	*Normal confidence multiplier* $Z_{\alpha/2}$
0.90	1.65
0.95	1.96
0.9545	2.00
0.98	2.33
0.99	2.58

Thus, if it is decided that $1 - \alpha = 0.95$, then we can write

$$P[\hat{\theta} - 1.96SE(\hat{\theta}) < \theta < \hat{\theta} + 1.96SE(\hat{\theta})] = 0.95,$$

which is commonly called a 95 percent confidence interval.

We note from (12–1) that a confidence interval gives a probability of $1 - \alpha$ that the parameter being estimated will fall into the interval bounded by L and U. If the same estimating function is employed for various samples of the same size, then each sample statistic, $\hat{\theta}$, will give a different interval and, on the average, $1 - \alpha$ of all these intervals will cover the parameter θ. In practice, when we construct only one confidence interval to estimate the parameter, the parameter may or may not fall within the interval. Nevertheless, we have a *level of confidence* of $1 - \alpha$ that it will. More obviously, if we have constructed a 99 percent confidence interval, we say that the probability (confidence) is 0.99 that the interval will include the parameter being estimated. That is, if we construct such an interval again and again many times, in the long run 99 percent of these intervals will cover the unknown parameter and 1 percent of them will not. Now, since 99 percent of the intervals will cover the parameter, we can certainly afford to *behave* as if each of our intervals would cover the parameter.

The preceding arguments are demonstrated in Figure 12–2. This figure shows that the distribution of the estimator $\hat{\theta}$ is assumed to be normal. Each dot under the density curve is thought of as a possible value of the estimator, or a group of estimators of equal values. The horizontal scale measures possible values of $\hat{\theta}$, and the vertical scale measures probabilities

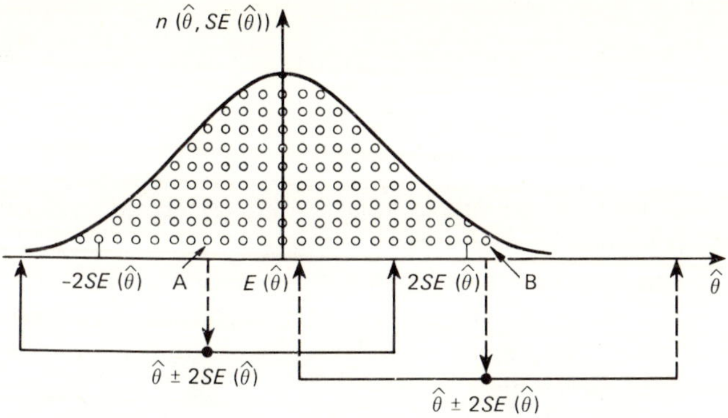

Figure 12–2 Illustration of 95.45 Percent Confidence Intervals

associated with these values. Now, if we wish to construct, say, a 95.45 percent confidence interval, then $Z_{\alpha/2} = 2$. Clearly, if $\hat{\theta}$ assumes a value represented by any dot within the limits $E(\hat{\theta}) \pm 2SE(\hat{\theta})$, then the interval $\hat{\theta} \pm 2SE(\hat{\theta})$ will cover θ, which is equal to $E(\hat{\theta})$. (See, for example, the random interval identified as A.) But we know that $\hat{\theta}$ would fall within the interval $E(\hat{\theta}) \pm 2SE(\hat{\theta})$ 95.45 percent of the time. Now, if $\hat{\theta}$ happens to be one of the extreme values that are outside the interval of two standard errors from the expectation, then $\hat{\theta} \pm 2SE(\hat{\theta})$ will not cover θ. (See, for example, the random interval identified as B.) But $\hat{\theta}$ deviates from its expectation by 2 standard errors or more only 4.55 percent of the time. Thus, we see that we have a confidence level of 0.9545 in estimating θ with a 95.45 percent confidence interval. We can make similar demonstrations and statements for confidence intervals with other confidence coefficients by using Figure 12–2 in the same fashion.

In interval estimation we naturally wish to have a shorter rather than a longer random interval, since the shorter the interval the more precise is our estimate. However, we must note that the actual width of a confidence interval is dictated by the confidence coefficient and the sample size. Given n, and therefore $SE(\hat{\theta})$, the shorter the interval, the smaller is $1 - \alpha$, the level of confidence. If we insist upon a given width of interval and a given high degree of confidence, the size of the sample must be increased so that $SE(\hat{\theta})$ is reduced.

12–5 CONFIDENCE LIMITS FOR μ

We know that the sampling distribution of $\bar{x}$ is normal if the parent population is normal or if the size of the sample is greater than 30. We know also that $E(\bar{x}) = \mu$, and $\sigma_{\bar{x}} = \sigma/\sqrt{n}$. An interval estimate of μ involves the construction of confidence limits for μ on the basis of $\bar{x}$ as follows:

$$P(\bar{x} - Z_{\alpha/2}\sigma_{\bar{x}} < \mu < \bar{x} + Z_{\alpha/2}\sigma_{\bar{x}}) = 1 - \alpha,$$

or, more compactly,

$$\bar{x} \pm Z_{\alpha/2}\sigma_{\bar{x}}. \tag{12–2}$$

Thus the procedure follows the general statement on interval estimation made in the previous section: First, obtain a random sample. Second, compute the sample mean. Third, compute the random endpoints with the sample mean as implied by (12–2). Let us now consider an illustration.

The Chamber of Commerce in Miami Beach wishes to estimate the mean expenditure per tourist per visit in that city. For this purpose a random sample of, say, 100 tourists has been selected for investigation, and it has been found that $\bar{x} = \$800$. Now, suppose the standard deviation of expenditures for all tourists is known to be \$120. Then we have $\sigma_{\bar{x}} = 120/\sqrt{100} = \$12$. Finally, suppose the Chamber wishes to construct a 95 percent confidence interval. Then the confidence multiplier is $Z_{\alpha/2} = 1.96$, and with data at hand, we have

$$\begin{aligned}\text{lower confidence limit} &= L = \bar{x} - 1.96\sigma_{\bar{x}}\\ &= 800 - 1.96(12) = \$776.48,\end{aligned}$$

and

$$\begin{aligned}\text{upper confidence limit} &= U = \bar{x} + 1.96\sigma_{\bar{x}}\\ &= 800 + 1.96(12) = \$823.52.\end{aligned}$$

So we state: the mean expenditure of tourists per person per visit at Miami Beach falls in the interval from \$776.48 to \$823.52, with a level of confidence at 0.95.

Now, we may observe that there is an error δ involved in the interval estimate of μ. This *error* may be defined as $\delta = \bar{x} - \mu$, which is clearly half the length of a given confidence interval. Furthermore, this error must be less than $Z_{\alpha/2}\sigma_{\bar{x}}$ at a confidence level of $1 - \alpha$. That is, if we construct, say, a 95 percent confidence interval for μ, we are 95 percent sure that the error $\delta < 1.96\sigma_{\bar{x}}$. Or, we say that the probability is 0.95 that $\delta < 1.96\sigma_{\bar{x}}$.

The last interpretation leads us to consider what is called the *risk* of an interval estimate. Such a risk is defined as the probability that the error of estimation will be equal to or greater than $Z_{\alpha/2}\sigma_{\bar{x}}$— that is, $P(\delta \geq Z_{\alpha/2}\sigma_{\bar{x}})$. For instance, if $1 - \alpha = 0.95$, then the risk is $1 - 0.95 = 0.05$ or α that the error will be equal to or greater than $1.96\sigma_{\bar{x}}$.

An interesting consequence of the relationship between error and risk is that it gives us a clue to the determination of a proper sample size, so that a specified width of interval at a given level of confidence can be satisfied. To see why, let us note, for confidence limits for μ, that

$$\delta \leq Z_{\alpha/2}\left(\frac{\sigma}{\sqrt{n}}\right),$$

and upon solving this equation for n, we have

$$n \geq \frac{Z_{\alpha/2}{}^2\sigma^2}{\delta^2}. \tag{12-3}$$

In our previous example, the error of estimation was

$$\delta = 1.96(12) = \$23.52$$

at a confidence level of 0.95. Now, if the Chamber of Commerce would like to increase the level of confidence to 0.99 without increasing the error of estimation, the sample size would have to be enlarged from 100 to

$$\begin{aligned} n &\geq \frac{Z_{\alpha/2}{}^2\sigma^2}{\delta^2} \\ &= \frac{(2.58)^2(120)^2}{(23.52)^2} \\ &\doteq 173. \end{aligned}$$

On the other hand, if the Chamber wishes to reduce the error of estimation to, say, \$20, with the level of confidence remaining at 0.95, then the required sample size becomes

$$n \geq \frac{(1.96)^2(120)^2}{(20)^2} \doteq 138.$$

As an exercise, the reader might try to find the required sample size in this illustration if it is specified that $1 - \alpha = 0.99$ and $\delta = \$20$.

When we are confronted with a completely new situation where the population variance is unknown, we may use the sample variance $\hat{s}^2$ from a pilot study as an estimate of σ^2 in determining the desired sample size. Suppose, for example, a new drug for the reduction of temperature has been perfected and a sample variance $\hat{s}^2 = \hat{\sigma}^2 = 0.1$ has been found. The manufacturer wishes to have an interval estimate of the mean effectiveness (average reduction in temperature) and he is willing to assume an error of 0.1 degrees at $1 - \alpha = 0.99$. How large a sample is required to satisfy these conditions?

Using $\hat{\sigma}^2$ to replace σ^2 in equation (12-3), we obtain

$$n \geq \frac{Z_{\alpha/2}{}^2\hat{\sigma}^2}{\delta^2} = \frac{(2.58)^2(0.1)}{(0.1)^2} = 67.$$

When population variance is not known, the standard error can be computed with sample variance if $n > 30$. Thus the approximate confidence limits for μ are

$$\bar{x} \pm Z_{\alpha/2}\hat{\sigma}_{\bar{x}} = \bar{x} \pm Z_{\alpha/2}\left(\frac{\hat{\sigma}}{\sqrt{n}}\right). \tag{12-4}$$

If a random sample is drawn from a finite population without replacement, and if $n \geq 0.05N$, then the finite population correction factor of $\sqrt{(N - n)/(N - 1)}$ should be introduced to calculate $\sigma_{\bar{x}}$ or $\hat{\sigma}_{\bar{x}}$. All formulas in this section have been designed under the assumption of either infinite population or sampling with replacement from a finite population.

12–6 CONFIDENCE LIMITS FOR π

The sampling distribution of p is asymptotically normal with $E(p) = \pi$ and $\sigma_p = \sqrt{(\pi)(1 - \pi)/n}$. In this case a confidence interval may be constructed for π in the same way as that described in the previous section. Corresponding to a confidence coefficient $1 - \alpha$, the approximate confidence limits for π, assuming n is large, are

$$p \pm Z_{\alpha/2}\hat{\sigma}_p, \tag{12–5}$$

where $\hat{\sigma}_p$ is the unbiased estimator of σ_p defined as

$$\hat{\sigma}_p = \sqrt{\frac{p(1 - p)}{n - 1}},$$

or the biased estimator of σ_p defined as

$$\hat{\sigma}_p = \sqrt{\frac{p(1 - p)}{n}}.$$

These two estimators differ but very slightly when n is large. We shall use the biased estimator because it is easier to work with.

We use the estimator of the standard error because the computation of σ_p requires π, the unknown parameter that we try to estimate. We may be assured, however, that when n is large, p is a very good point estimator of π.

The parameter π is undefined for an infinite population. (Why?) Thus, when we talk about a proportion we are thinking of a finite population. Therefore, when sampling is made without replacement, the estimator of σ_p must be calculated with a finite population correction factor as follows:

$$\hat{\sigma}_p = \sqrt{\frac{p(1 - p)}{n}}\sqrt{\frac{N - n}{N - 1}}.$$

Confidence limits defined by (12–5) require a large sample size. Intuitively, we can see that the more the sample proportion deviates from 0.5, the larger n should be. W. G. Cochran has given us working rules on this point as presented in Table 12–1. According to this table, for example, if $p = 0.1$, the sample size should be at least 600 to use normal approximation.

Table 12-1

If p equals:	*Use normal approximation only if n is at least equal to*
0.5	30
0.4 or 0.6	50
0.3 or 0.7	80
0.2 or 0.8	200
0.1 or 0.9	600
0.05 or 0.95	1400

SOURCE: Reprinted from W. G. Cochran, *Sampling Techniques* (New York: John Wiley & Sons, Inc., 1953), p. 41.

An unconditional random sample of $n = 500$ manufacturing firms has been selected; 80 of these firms report that they are planning to increase investment expenditures during the next two years. What is the 99 percent confidence interval for π?

We have, for this problem,

$$p = {}^{80}\!/_{500} = 0.16$$

and

$$\hat{\sigma}_p = \sqrt{\frac{(0.16)(0.84)}{500}} = 0.0164.$$

Consequently, the desired confidence limits are

$$p \pm Z_{\alpha/2}\hat{\sigma}_p = 0.16 \pm 2.58(0.0164) = 0.16 \pm 0.042.$$

Or, the 99 percent confidence interval is

$$0.118 < \pi < 0.202.$$

As an exercise, the reader might construct a 95 percent confidence interval for the previous illustration under the assumption that a conditional random sample was taken from a population of 2500 manufacturing firms.

The error of estimating π, as in the case of the mean, is the difference between the sample proportion and the population proportion—namely, $\delta = p - \pi$. Furthermore, the probability is $1 - \alpha$ that this error is less than $Z_{\alpha/2}\sigma_p$. With this concept of maximum error, we can again determine the sample size required in order to satisfy a given degree of precision in estimating π. Now, since

$$\delta = Z_{\alpha/2}\sigma_p = Z_{\alpha/2}\sqrt{\frac{\pi(1 - \pi)}{n}},$$

upon solving for n, we obtain

$$n \geq \pi(1 - \pi)\left(\frac{Z_{\alpha/2}}{\delta}\right)^2. \tag{12–6}$$

An unfortunate aspect of this result is that we need the value of π, a value we wish to estimate, in order to determine n. A practical solution to this difficulty is to use any information we have, such as, say, a sample proportion obtained in the past, as an estimate of π. When no information is available on π, we may assume $\pi = \frac{1}{2}$. The logic for this assumption is that

$$\begin{aligned}\pi(1 - \pi) &= \pi - \pi^2 \\ &= \tfrac{1}{4} - (\tfrac{1}{4} - \pi + \pi^2) \\ &= \tfrac{1}{4} - (\tfrac{1}{2} - \pi)^2,\end{aligned}$$

showing that the maximum value $\pi(1 - \pi)$ can have is $\frac{1}{4}$ at $\pi = \frac{1}{2}$.

Suppose that the producer of a new TV program wishes to ascertain the proportion of all viewers watching TV who are tuned in to his show. He is willing to assume an error of ± 5 percent at a confidence level of 0.95; how large a sample should be taken?

By (12–6) and assuming $\pi = \frac{1}{2}$, we find

$$n \geq (\tfrac{1}{2})(\tfrac{1}{2})\left(\frac{1.96}{0.05}\right)^2 \doteq 384.$$

This result indicates that if we take a random sample of 384 or larger, the probability is 0.95 that the sample proportion will differ from the true value by 5 percent or less.

Now, suppose the producer knows that in the past programs similar to his new one shared approximately 25 percent of the TV audience; he may then use 0.25 as the value of π, and the sample size required becomes

$$n \geq (0.25)(0.75)\left(\frac{1.96}{0.05}\right)^2 \doteq 288.$$

Thus, we see once again the value of information about past experience. In this case, possessing some information about π, we can achieve the same degree of precision with a much smaller sample. This advantage is especially important when sampling is expensive.

Confidence limits for π can be improved, especially when n is relatively small, in two ways. First, since the distribution is discrete, a continuity correction factor of $1/(2n)$ may be introduced for p to determine a confidence interval for π as follows:

$$\left(p - \frac{1}{2n}\right) - Z_{\alpha/2}\hat{\sigma}_p < \pi < \left(p + \frac{1}{2n}\right) + Z_{\alpha/2}\hat{\sigma}_p, \tag{12–7}$$

which, when n is quite large, reduces to (12–5).

The second procedure is to obtain what is called the *quadratic* confidence interval for π. This is done by solving the inequality

$$|\pi - p| < Z_{\alpha/2}\sqrt{\frac{\pi(1-\pi)}{n}}$$

for π, yielding the following confidence limits for π:

$$\frac{1}{1 + Z_{\alpha/2}^{2}/n}\left(p + \frac{Z_{\alpha/2}^{2}}{2n} \pm Z_{\alpha/2}\sqrt{\frac{p(1-p)}{n} + \frac{Z_{\alpha/2}^{2}}{4n^{2}}}\right). \qquad (12\text{–}8)$$

This rather involved formula, when the sample is very large, reduces to the simpler form of (12–5). Furthermore, these and the last limits are still approximate, since the relationship between $Z_{\alpha/2}$ and $1 - \alpha$ is determined under the assumption that p is normally distributed, which is only approximately so. However, when n is relatively small, (12–7) and (12–8) do yield a more satisfactory approximation of confidence limits for π. As an exercise, the reader might work out the example of the manufacturing firms, using both these formulas; he will see that they yield practically the same results as those obtained before.

12–7 CONFIDENCE LIMITS FOR $\Delta\mu$

When two independent random samples of $n_1 > 30$ and $n_2 > 30$ are taken, then the distribution of the difference between the two sample means, $\Delta\bar{x}$, is approximately normal with

$$E(\Delta\bar{x}) = \Delta\mu \quad \text{and} \quad \sigma_{\Delta\bar{x}} = \sqrt{\sigma_1^2/n_1 + \sigma_2^2/n_2}.$$

Given the confidence coefficient, the approximate confidence limits for $\Delta\mu$ may be expressed as follows:

$$\Delta\bar{x} \pm Z_{\alpha/2}\sigma_{\Delta\bar{x}}. \qquad (12\text{–}9)$$

An automobile manufacturing company has recorded observations of two different makes of batteries, both of which are being used in its cars. Forty observations of make A showed a mean life of 32 months. Forty-five observations of make B showed a mean life of 30 months. Past experience indicates that the standard deviations for both makes of batteries are the same—4 months. What are the confidence limits for the true difference in mean life between the two types of batteries with $1 - \alpha = 0.95$?

For this problem, we have $\Delta\bar{x} = \bar{x}_A - \bar{x}_B = 32 - 30 = 2$ and

$$\sigma_{\Delta\bar{x}} = \sqrt{\frac{4^2}{40} + \frac{4^2}{45}} = 0.87.$$

Substituting these data into (12–9), we find

$$2 \pm 1.96(0.87) = 2 \pm 1.7.$$

Or, the 95 percent confidence interval in this case is

$$0.3 < \Delta\mu < 3.7.$$

Two cities are separated by a river. The chamber of commerce in city A claims that the average family income of that city is \$500 more than that in city B. The chamber of commerce in city B disputes this, and a statistician is hired to settle the argument. He attacks the problem by estimating the true difference between the average family income in the two cities. One random sample is drawn from each city, and the following results are obtained:

Sample from city A	*Sample from city B*
$n_1 = 100$	$n_2 = 120$
$\bar{x}_1 = \$5900$	$\bar{x}_2 = \$5800$
$\hat{s}_1^2 = 9050$	$\hat{s}_2^2 = 8700$

Since population variances are unknown, the statistician uses sample variances to compute an estimate of the standard error of $\Delta\bar{x}$. This is valid, since sample sizes in this case are quite large.

$$\hat{\sigma}_{\Delta\bar{x}} = \sqrt{\frac{\hat{s}_1^2}{n_1} + \frac{\hat{s}_2^2}{n_2}} = \sqrt{\frac{9050}{100} + \frac{8700}{120}} = 12.7671.$$

He proceeds then to construct a 99 percent interval for $\Delta\mu$ as follows:

$$\Delta\bar{x} - 2.58\hat{\sigma}_{\Delta\bar{x}} < \Delta\mu < \Delta\bar{x} + 2.58\hat{\sigma}_{\Delta\bar{x}},$$

which, in terms of sample values, turns out to be

$$100 - 2.58(12.7671) < \Delta\mu < 100 + 2.58(12.7671)$$

or

$$\$67.06 < \Delta\mu < \$132.94.$$

This result indicates that while the average family income in city A may be higher than that in city B, the true difference is far below the \$500 claimed by the chamber of commerce in city A. Furthermore, we can state with a confidence level of 0.99 that the mean incomes in the two cities differ only from a little over \$67 to a little less than \$133.

12–8 CONFIDENCE LIMITS FOR $\Delta\pi$

The sampling distribution of Δp, given that p_1 and p_2 are determined from independent random samples and that n_1 and n_2 are sufficiently large as specified by values in Table 12–1, is approximately normal with $E(\Delta p)$

$= \Delta\pi$ and

$$\sigma_{\Delta p} = \sqrt{\frac{\pi_1(1-\pi_1)}{n_1} + \frac{\pi_2(1-\pi_2)}{n_2}}\,.$$

No new knowledge is required for the construction of confidence limits for $\Delta\pi$. However, we are faced with the same difficulty now as in the case of interval estimation for π: we need the values of π_1 and π_2 (which are unknown) for the calculation of $\sigma_{\Delta p}$. Inasmuch as $\Delta\pi$ is unknown, we are forced to use p_1 and p_2 as point estimators for π_1 and π_2, respectively, and find the estimator for $\sigma_{\Delta p}$ as follows:

$$\hat{\sigma}_{\Delta p} = \sqrt{\frac{p_1(1-p_1)}{n_1} + \frac{p_2(1-p_2)}{n_2}}\,.$$

With the estimate for $\sigma_{\Delta p}$ available, if we assume large n_1 and n_2, the approximate confidence limits for $\Delta\pi$ corresponding to a confidence coefficient $1 - \alpha$ are

$$\Delta p \pm Z_{\alpha/2}\hat{\sigma}_{\Delta p}. \tag{12-10}$$

Prior to a gubernatorial election in a certain state, two random samples are taken, one for male voters with $n_1 = 150$ and one for female voters with $n_2 = 120$. It is found that 70 men in the first sample are in favor of candidate X and 60 women in the second sample are in favor of the same candidate. What are the confidence limits for $\Delta\pi$, the true difference in voting preferences between men and women in that state, given that $1 - \alpha = 0.95$?

For the problem at hand, we note that

$$p_1 = 70/150 = 0.47,$$

$$p_2 = 60/120 = 0.50,$$

$$\Delta p = 0.47 - 0.50 = -0.03,$$

and

$$\hat{\sigma}_{\Delta p} = \sqrt{\frac{(0.47)(0.53)}{150} + \frac{(0.50)(0.50)}{120}} = 0.061.$$

With these data, we have as the desired confidence limits

$$-0.03 \pm 1.96(0.061) = -0.03 \pm 0.11956.$$

Or, the 95 percent confidence interval for $\Delta\pi$ in this case is

$$-0.15152 < \Delta\pi < 0.09152.$$

Had we computed Δp as $p_2 - p_1 = 0.50 - 0.47 = 0.03$, then the desired confidence interval would be

$$-0.09152 < \Delta\pi < 0.15152.$$

The negative value for the lower confidence limit here should not be a surprise to us, since the error of estimate for this problem is 1.96(0.061) = 0.11956, which is much larger than the observed difference between sample proportions. The interesting point is that the negative value for the lower confidence limit in conjunction with the positive upper limit implies that $\Delta\pi$ may very well be zero. In other words, the sample result does not indicate that men's voting preferences are different from women's in the state. It might be that there is a real difference—that is, $\Delta\pi \neq 0$—but the samples may be too small to bring out this difference. In any event, our proper inference from the confidence interval above should be this: given the stated conditions of the sampling experiment and the confidence coefficient of 0.95, a difference of 3 percent between two sample percentages does not support the contention that the voting preferences are different between men and women in the said state.

12–9 SUMMARY AND CONCLUSIONS

In this chapter we have been concerned with the traditional decision procedures of estimation. Estimation is the statistical method of making inferences about parameter values on the basis of sample statistics. An estimator of a parameter given by a single point derived from sample observations is called a point estimator. An estimator is said to be good if it possesses the properties of unbiasedness, consistency, efficiency, and sufficiency. The maximum likelihood method provides estimators that are consistent, efficient, and sufficient; but it does not always provide estimators that are unbiased. It can be shown that all estimators (except $\hat{s}^2$) introduced so far in this text are maximum likelihood estimators.

An estimator of a parameter given by a random interval whose endpoints are functions of sample observations is called an interval estimator. Interval estimators indicate the precision or accuracy of an estimate, and are therefore preferable to point estimators. Each interval estimator gives a probability—a confidence level—that the confidence limits constructed will cover the parameter being estimated.

In interval estimation, the error of estimation, the level of confidence, and the sample size are closely related. The error here is defined as the difference between the estimator and the parameter being estimated. With a probability of $1 - \alpha$, this error cannot be greater than the product of the confidence multiplier and the standard error of the estimator. The probability that the error of estimation is equal to or greater than this product is thought of as the *risk* in estimation—that is, the probability that the confidence interval will not cover the parameter being estimated. Given the sample size, the more precise or the shorter the confidence interval is, the lower must be the level of confidence; or the more the confidence coefficient is increased, the less precise (or the greater the error of) estimation becomes. In general, if the goal is to have a small error and a high level of confidence, the sample size must be increased. It is possible to determine the minimum size of sample required to yield predetermined precision and confidence. Usually, this is done by equating the error of estimation to the maximum error, $Z_{\alpha/2}SE(\hat{\theta}) = \delta$, and solving for n, which is always involved with the calculation of the standard error of an estimator under investigation.

Our discussion of estimation so far has assumed that the sampling distribution of an estimator is normally distributed in accordance with the central limit theorem. Insofar as many sampling distributions are only approximately normal, confidence limits constructed with normal confidence multipliers possess only approximate values. Such approximate confidence limits are quite satisfactory for estimating parameters in many types of investigations.

Finally, it may be noted that although our treatment covers estimations for only μ, π, $\Delta\mu$, and $\Delta\pi$, the same general procedure of interval estimation can be applied to other parameters—such as the variance, the median, the mean of binomial variables, and other position measures—provided that n is large enough to allow the central limit theorem to operate. These few parameters have been singled out for evaluation because they are by far the most important for decisions involving univariate data. We shall continue to be concerned with them in the next chapter as we take up the traditional decision procedure of hypothesis testing.

EXERCISES

12–1 A shipment contains a very large number of radio tubes of which a proportion π is defective. A random sample of 10 tubes contains 2 defectives. Plot the likelihood function for π and give the maximum likelihood estimate for π.

12–2 Let π be the proportion of people in a certain city who watch a certain TV program on Friday night. In a random sample of 20 people from this city, 6 are viewers of this program. Graph the maximum likelihood function for π and find a maximum likelihood estimate for π.

12–3 The standard deviation of average weekly wage rates for elevator operators is known to be \$2.79. An unconditional random sample of 100 elevator operators yields a mean average weekly wage rate of \$87.45. Construct (a) 95 percent and (b) 99 percent confidence intervals for the true mean average weekly rates of elevator operators.

12–4 Suppose there are 5000 elevator operators in a large city, and suppose a sample mean of \$87.45 is obtained from a conditional random sample of $n = 100$. What are the (a) 95 percent and (b) 99 percent confidence intervals for μ now?

12–5 Measurements of the diameters of a conditional random sample of 50 from a lot of 200 bearing balls show as a mean $\bar{x} = 1.05$ inches, and as a standard deviation $\hat{s} = 0.04$ inch.
(a) What are the 95 percent confidence limits for the mean diameter of the 200 bearing balls?
(b) At what level of confidence could we say that the mean diameter of all 200 bearing balls is 1.05 ± 0.005?

12–6 A conditional random sample of 45 grades taken from a class in statistics of 221 students shows a mean of 70 points and a standard deviation of 9 points. What are the 98 percent confidence limits for the mean of all 221 grades?

12–7 In measuring assembly time for a certain mechanism, the engineer estimates that the standard deviation is 0.05 second. How many times should the assembly be performed so that he can be (a) 95 percent and (b) 99 percent certain that the error will not exceed 0.01 second?

12–8 A department store wishes to estimate, with a confidence coefficient of 0.98 and a maximum error of \$5, the true mean dollar value of purchases per month by its customers with charge accounts. How large a sample should the store take from its records to satisfy the specifications if the standard deviation is known to be \$15?

12–9 A sample of 35 steel wires selected at random from a very large shipment shows a mean tensile strength of 1500 pounds and a standard deviation of 20 pounds.
(a) If this sample mean is used to estimate the mean tensile strength of the whole shipment, what can be said for the possible error with a probability of 0.95?
(b) If the error estimated in (a) is to be reduced by half without reducing the level of confidence, what size sample is required?
(c) If the estimate requires that the error estimated in (a) be reduced by half and the probability be 0.99 that the reduced error is a maximum, how large should the sample be?

12–10 In a random sample of 25 college instructors in the United States, 16 are found to hold Ph.D. degrees.
(a) Construct an approximate 95 confidence interval for the proportion of all U.S. college instructors holding Ph.D. degrees by using equation (12–5).

(b) Do the same by using equation (12–8).
(c) Comment on the difference in results for (a) and (b).

12–11 Ten out of 400 articles produced by a certain process are found to have defects. Construct a 99 percent confidence interval for the probability that a single article produced in the process is defective.

12–12 A sample poll of 100 members of the AFL-CIO chosen at random reveals that 55 are against trade with Red China.
(a) Find the 95 percent confidence limits for the proportion of all the union members against trade with Red China.
(b) What is the minimum size of sample required in order to be 95 percent sure that at least 50 percent of the union members are against trade with Red China?

12–13 During a certain week, it was observed and recorded by a small department store that 5750 of the 12,500 persons who entered the store made at least one purchase. Treating this as a random sample of all potential customers, find the 99 percent confidence limits for the actual proportion of persons entering the store who will make at least one purchase.

12–14 The auditor of a large bank wishes to estimate the proportion of monthly statements for the bank's demand depositors that have errors of various kinds, and he specifies a confidence coefficient of 0.99 and a maximum error of estimation of 0.25 percent (or 0.0025).
(a) What is the minimum sample size required if there is no information available on the true proportion of the monthly statements that are in error?
(b) What is the minimum sample size required if the auditor, according to his past experience, believes the true proportion is near 0.01?

12–15 The same graduate examination was given to 50 seniors at university A and to 50 seniors at university B. The mean and standard deviation of grades received by students at A are, respectively, 73.6 and 10; the sample mean and standard deviation of grades made by students at B are, respectively, 72.4 and 8. Construct a 90 percent confidence interval for the true difference in average grades between the seniors at universities A and B.

12–16 A random sample of 200 brand A radio batteries shows a mean life of 140 hours and a standard deviation of 10 hours. A random sample of 120 brand B radio batteries yields a mean life of 125 hours and a standard deviation of 9 hours. Give (a) 95 percent and (b) 99 percent confidence limits for the difference in the mean lifetimes of the populations of brands A and B.

12–17 Two randomly selected groups, each of 50 students, at a secretarial school are taught shorthand by two different systems and then tested for performance in taking dictation. It is found that the first group averages 120 words per minute with a standard deviation of 11 words, while the second group averages 110 words per minute with a standard deviation of 10 words. Construct a 99 percent confidence interval for $\Delta\mu$.

12–18 The mean and the standard deviation of the maximum loads supported by 100 cables produced by Alpha Company are found to be 20 tons and 1.1 tons, respectively. The mean and the standard deviation of the maximum loads supported by 60 cables produced by Beta Company are 16 tons and 0.8 ton, respectively. Find (a) 95 percent and (b) 98 percent confidence limits for the difference in mean maximum loads between the two brands of cables.

12–19 In a random sample of 600 women, 300 indicate that they favor American intervention in South Vietnam. In a random sample of 400 men, 100 indicate that they favor the same. Construct (a) 95 percent and (b) 99 percent confidence limits for the difference in proportions of all women and all men who favor American intervention in South Vietnam.

12–20 Twenty-five out of 250 TV picture tubes produced by process I are defective, and 15 out of 180 produced by process II are defective. Construct (a) 95 percent and (b) 99 percent confidence intervals for the difference in proportions of defective tubes produced by the two processes. Comment on your results.

12–21 Of two groups of patients suffering from chronic headaches, A and B, consisting of 200 and 100 individuals, respectively, the first are given a new kind of headache pill and the second are given conventional pills; 15 of group A find immediate relief, as do 12 of group B. Construct a 99 percent confidence interval for the difference in proportions of immediate relief afforded by the two kinds of medicine; comment on the result.

13

FURTHER CLASSICAL DECISION PROCEDURES: STATISTICAL TESTING

13–1 A GENERAL PROCEDURE OF TESTING

The classical theory of testing as a decision procedure refers to the process of reaching a conclusion with respect to various hypothesized states of nature or values of population parameters on the basis of sample observations.

In discussing testing, we shall consider problems that confront the decision maker with only two alternative courses of action. Such problems, sometimes called *two-action* testing problems, involve such choices as buying or not buying a given commodity, changing or not changing a process of production, accepting or rejecting a lot of merchandise, continuing or discontinuing production, using or not using a more expensive channel of advertising, and so forth. We shall try first to develop a general procedure of testing for two-action problems. In principle, all statistical tests are similar to those we are about to discuss. Thus, when the reader fully understands the basic ideas underlying this general procedure, he should have no further difficulties with the tests introduced in this and subsequent chapters.

Formulating Hypotheses

A *statistical hypothesis,* in formal terms, is an assumption about the distribution of a random variable. When being applied, a hypothesis is usually given in a form that specifies one or more of the parameters associated with the population under consideration.

In two-action problems, the decision maker needs to focus his attention on two subsets of the states of nature. The action a_0 is optimal for some subsets of states H_0. The action a_1 is optimal over the remaining subset of states H_1. H_0 and H_1 are sometimes called the *null* and *alternative hypotheses,* respectively. Furthermore, the union of H_0 and H_1 partitions the set of states over which the actions a_0 and a_1 are optimal.

The first step in testing is the formulation of hypotheses. In other words, hypotheses must be "predesignated." If we were to use sample information in formulating hypotheses, we would be using the same data to formulate as well as to test the hypotheses. Using such a procedure, we could hardly expect to demonstrate anything.

On the other hand, the requirement that hypotheses be predesignated does not mean that we should never use sample data to suggest hypotheses that had not been anticipated. Such a hypothesis, of course, cannot be tested rigorously by the data that suggest it. It can, however, be regarded as a new hypothesis to be tested in the future by sample data collected independently. We should realize that useful new hypotheses or theories may be suggested by data employed to test the predesignated hypothesis.

There are various ways to formulate hypotheses, depending upon the problem on hand. In general, a null hypothesis is often a precise statement, such as $\theta = \theta_0$. Sometimes, however, we may find a less precise form more appropriate, such as $\theta \leq \theta_0$ or $\theta \geq \theta_0$.

Usually, the alternative hypothesis is stated in a nonspecific form, such as $\theta < \theta_0$, $\theta > \theta_0$, or $\theta \neq \theta_0$. The first two types of designations for H_1 are called *single-tail tests*. More precisely, when H_1 is given as $\theta < \theta_0$, we have a *left-tail test*. When H_1 is stated as $\theta > \theta_0$, we have a *right-tail test*. The last form of an alternative hypothesis, $\theta \neq \theta_0$, indicates what is called a *two-sided*, or *two-tail, test*. Which of these designations we should use depends on the nature of the decision problem at hand. This point will become clear after we have presented examples of a few statistical tests.

The various ways of formulating hypotheses lead us to distinguish between what are called simple and composite hypotheses. A *simple hypothesis* is a complete specification of a probability distribution. For example, the hypotheses H_0: $\pi = 0.6$ and H_1: $\pi = 0.4$ are both simple. A simple hypothesis contains only one state of nature. A hypothesis that is not simple is a *composite hypothesis*; thus, it is convenient to think of a composite hypothesis as made up of a number of simple hypotheses. For example, the hypothesis $\mu \geq 1000$ hours is composite, since μ could be any value ranging from 1000 to ∞ hours. Thus, a composite hypothesis contains more than one state of nature. As we shall see, if we know a simple hypothesis to be true, then the probability distribution of the testing statistic is completely defined. If we know a composite hypothesis to be true, we still do not know the exact distribution of the "testing statistic." Also, when we say that a composite hypothesis is true we mean that one of the states that make it up is true.

We should note that even though common terminology calls H_0 the hypothesis being tested and H_1 the alternative hypothesis, in reality we are testing to decide between two hypotheses, H_0 and H_1. That is, in a statistical test concerning a population, we try to find a rule that will enable us to choose between two courses of action.

Specifying the Level of Significance

A moment's reflection will reveal that, in testing a null hypothesis against an alternative hypothesis on the basis of a particular sample result, we may make two kinds of errors owing to random fluctuations in sample data. On the one hand, H_0 may be in fact true but sample data appear to be inconsistent with it and, consequently, it is rejected. The error of rejecting a true H_0 is called a *type I error*, or an *error of the first kind*. The probability of making such an error is referred to as the *level of significance* and is universally denoted by α. When a type I error is committed, we wrongly decide to take a_1 or accept H_1 when H_0 is true. Thus,

$$\alpha = P(I) = P(a_1 \mid H_0) = P(H_1 \mid H_0). \tag{13–1}$$

On the other hand, H_0 may in fact be false; yet, on the basis of sample observations, we are led to accept it as true. The error of accepting a false H_0 is called a *type II error*, or an *error of the second kind*. The probability of committing a type II error is denoted as β, and

$$\beta = P(\text{II}) = P(a_0 \mid H_1) = P(H_0 \mid H_1). \tag{13–2}$$

Note that the classical decision procedure of testing is based on sample information only. It neither uses subjective judgment on the states nor explicitly evaluates the losses for the action-state pair, $L(a, \theta)$. For each state, the decision maker is concerned only with whether a loss from a given action is zero or positive. Since only two actions are available and the losses are opportunity losses, he knows which action is the best for each state. He does not know, however, the magnitude of the loss involved in selecting between actions. The two types of errors and the implied losses for them may be summarized as in Table 13–1.

The weakness just mentioned must be recognized along with the strengths of the classical methods. These methods are usually simpler and therefore

Table 13–1. Type I and Type II Errors and Their Losses

Action taken on basis of sample	*State of nature* $\theta \in H_0$	$\theta \in H_1$
a_0	$L(a_0, \theta) = 0$ correct	$L(a_0, \theta) > 0$ (α)
a_1	$L(a_1, \theta) > 0$ (β)	$L(a_1, \theta) = 0$ correct

faster and less expensive than their decision-theory counterparts. Also, under certain circumstances, the conclusions reached by the classical procedures are strong enough to satisfy the decision maker that the more complete but more costly methods possible under the decision-theory approach would lead to the same conclusions. These are some of the main reasons why the classical approach is still in wide use.

We note that in classical testing procedure we may often specify α alone without explicit evaluation for β. Reasons for this practice will be given in connection with our evaluation of β in the last section of this chapter.

Thus, the second step of our general procedure is to select a level of significance: the specification of the value of α with the sample size given. The α value is commonly set at 0.05 or 0.01, occasionally, α values larger than 0.05 or smaller than 0.01 are also employed. If $\alpha = 0.05$ and H_0 is rejected, for example, we say that the sample results are significantly different from H_0 at the $\alpha = 0.05$. That is, if we use this procedure to make decisions, in the long run the sample results could be expected to cause rejection of true null hypotheses 5 percent of the time.

When the null hypothesis is rejected at $\alpha = 0.05$, the test result is said to be "significant." When the null hypothesis is rejected at $\alpha = 0.01$, the test result is said to be "highly significant." These two phrases are used without explicit definitions, but in statistical terms, the word "significant" does have a precise meaning. The concept "level of significance" is employed to indicate merely that a statistical test evaluates the "significance" of the sample result as compared with the null hypothesis H_0.

The question now arises: How does a decision maker decide on the size of risk to be taken? The answer depends upon the following two considerations.

(1) *The decision maker's degree of belief in the null hypothesis*: the greater the prior degree of belief in the truth of H_0, the smaller should be the probability of rejecting it, e.g., α should be set at 0.01 or even 0.001 since, if one believes strongly that H_0 is true, it should be accepted unless there is very strong evidence to the contrary.

(2) *The decision maker's judgment on the relative costs and consequences of type I and type II errors*: when erroneous rejection of H_0 would be costly or risky, α should be small; when erroneous acceptance of H_0 would be costly or serious, β should be low (i.e., α may be set at a higher level, such as 0.05 or a much higher value). (We shall see that α and β vary inversely given the sample size.) This point, unfortunately, is only a vague, implicit notion in classical analysis, owing to its failure to determine explicitly the value of $L(a, \theta)$. However, the classical methods are often consistent with certain loss-function forms, such as the quadratic and piecewise loss functions, even though they are not taken explicitly into account.

Both of the foregoing considerations are rather complex; we shall discuss them in a more detailed fashion in Chapter 16, where we shall consider Bayesian inference procedures—procedures that recognize both costs of erroneous decisions and prior degrees of belief in the null hypothesis. With reference to the second consideration we shall explain briefly here the rational of specifying α alone.

Suppose it has been established with little doubt by medical scientists that if the average nicotine content of cigarettes is 25 milligrams or more, lung cancer is likely to develop in a smoker. If, however, the average nicotine content is less than 25 milligrams, the user is relatively safe. Suppose that you were a habitual smoker and were willing to risk your life; you would certainly try to find a brand of cigarettes whose average nicotine content is less than 25 milligrams. In this case, you should formulate your hypotheses as follows:

$$H_0\colon \mu \geq 25, \qquad H_1\colon \mu < 25.$$

For this test, a type I error would be made if you concluded that the mean was less than 25 when in fact it was equal to or greater than 25. Clearly, this is a serious mistake, since you would then select a brand of cigarettes that would endanger your life. You should try to avoid such a mistake by employing a small α risk of say, 0.01, 0.001, or even 0.0001. Here, if a type II error were committed, you would conclude wrongly that the mean was greater than or equal to 25 when actually it was less than 25. You would then reject this brand of cigarettes even though it could be considered safe. A type II error in this case would not be serious, since it would cost you only the trouble of testing another brand.

Thus, where the consequence of committing a type I error is serious, we should specify α at 0.01, or even lower if the occasion demands it. In such a case, clearly, we need not be deeply concerned with β risk.

Now, suppose a cannery specifies that 10 ounces of green asparagus tips be put in each can and adjusts the automatic measuring device accordingly. If the average weight is below this figure, the cannery may run into trouble with government inspectors and incur heavy fines. But if the average weight is higher than this figure, the cannery will make less profits. From time to time the cannery should test the following hypotheses to see whether the measuring device is appropriately adjusted:

$$H_0\colon \mu = 10; \qquad H_1\colon \mu \neq 10.$$

In this test, a type I error seems to be less important than a type II error. If the null hypothesis is rejected when it is true, all that is wasted is the time spent in checking the measuring device. If, however, H_0 is accepted when H_1 is actually true, the cannery is confronted with the risk of large fines and possible loss of goodwill, and, if μ is really greater than 10, there is also the risk of lower profits. Consequently, it is advisable to set α at 0.05 or at an even higher level. Note that, with n fixed, the larger α is specified the smaller β becomes—a desirable condition in the present case.

Situations may arise, of course, where errors of the first and the second kinds are equally serious. In the preceding example, for instance, we may find that the time lost in examining the measuring device is costly because it entails stoppage of production. If this is the case, then it would be desirable to have both α and β small. Probabilities of committing both types of error can be set at any desired level, provided the sample size is permitted to vary.

Selecting the Testing Statistic

Having formulated hypotheses and determined the level of significance, we now turn to the third step of our general testing procedure: the selection of a *testing statistic* whose sampling distribution is known under the assumption that the null hypothesis is true. The common practice is to choose as the testing statistic the measure that is used to estimate the parameter in the terms of which H_0 and H_1 are stated. For example, if H_0 and H_1 are stated in terms of population mean, then a sample mean would be selected as the testing statistic.

When we have selected, say $\hat{\theta}$ to test hypotheses concerning θ, we may use $\hat{\theta}$ to test H_0 against H_1 directly, or we may use its equivalent, $Z = (\hat{\theta} - \theta_0)/SE(\hat{\theta})$, where θ_0 is the assumed parameter in H_0, to do the job. The employment of Z as the testing statistic is especially convenient if the sampling distribution of $\hat{\theta}$ is normal, since Z then has the standard normal distribution.

Establishing Decision Criteria

At this point—as a fourth step—a decision criterion must be established for the purpose of rejecting or accepting H_0. This involves the division of the sampling distribution of the testing statistic into two parts: the *critical*, or *rejection*, *region*, and the *acceptance*, or *nonrejection*, *region*. This division depends upon the designation of the alternative hypothesis, the level of α, and the sampling distribution of the testing statistic—all of which have already been predetermined. Throughout our discussion in this chapter we shall assume that the parent population is normal or the sample is large enough for the central limit theorem to operate; that is, we shall assume the sampling distribution of the testing statistic is normal. Then the task becomes one of discovering a *critical value*, a standard normal deviate, that would separate the critical region from the region of acceptance. We shall now show how such a critical value can be found for each of the three most frequently used forms of alternative hypotheses.

First, if we test $H_0: \theta = \theta_0$ against $H_1: \theta > \theta_0$, then only large values of $\hat{\theta}$, the testing statistic, would support the validity of H_1. Consequently, given α, we try to determine the critical value, K, so that $P(\hat{\theta} > K) = \alpha$, where $K = \theta_0 + Z_{1-\alpha}SE(\hat{\theta})$. That is, if we are concerned with a right-tail test, the decision rule is: reject H_0 if and only if $\hat{\theta} > K$; accept H_0 if and only if $\hat{\theta} \leq K$. Since Z is a standard normal deviate, this decision rule can also be stated, more conveniently, as this: reject H_0 if and only if

$$Z = \frac{\hat{\theta} - \theta_0}{SE(\hat{\theta})} > Z_{1-\alpha}. \tag{13–3}$$

The critical region for a right-tail test is shown in Figure 13–1.

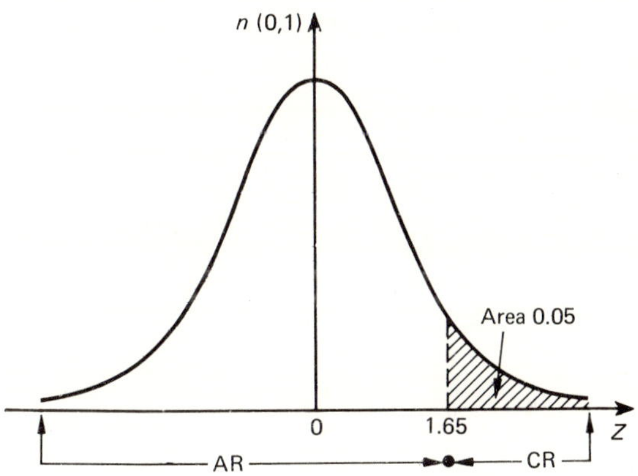

Figure 13–1 A One-Sided Critical Region for Testing $H_0: \theta = \theta_0$ against $H_1: \theta > \theta_0$ at $\alpha = 0.05$

Second, if we are concerned with a left-tail test—that is, if we are testing $H_0: \theta = \theta_0$ against $H_1: \theta < \theta_0$—then H_1 is supported only by small values of $\hat{\theta}$. Now, the critical value should be so chosen that $P(\hat{\theta} < K) = \alpha$, where $K = \theta_0 - Z_{\alpha}SE(\hat{\theta})$. In terms of the standard normal deviate, the decision rule becomes: reject H_0 if and only if

$$Z = \frac{\hat{\theta} - \theta_0}{SE(\hat{\theta})} < Z_{\alpha}. \tag{13–4}$$

The critical region for a left-tail test is shown by Figure 13–2.

Finally, in the case of a two-tail test, we are testing $H_0: \theta = \theta_0$ against $H_1: \theta \neq \theta_0$. Here, clearly, both small and large values of the testing statistic tend to support H_1. There is reason to expect that the critical region now, as shown by Figure 13–3, is divided equally between both ends of the density function of $\hat{\theta}$. Thus, we have two critical values in this case, and H_0

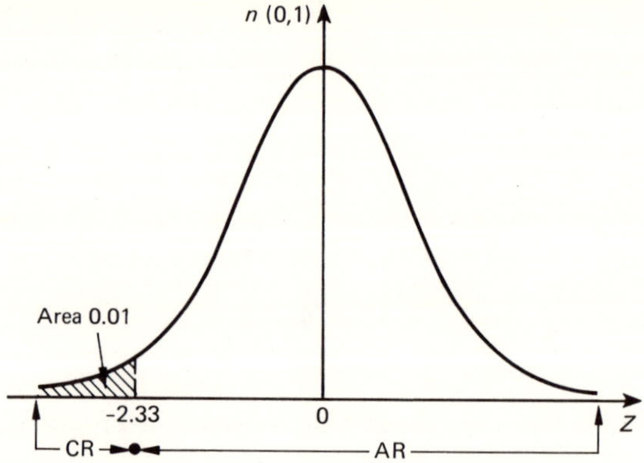

Figure 13–2 A One-Sided Critical Region for Testing H_0: $\theta = \theta_0$ against H_1: $\theta < \theta_0$ at $\alpha = 0.01$

will be rejected if

$$\hat{\theta} < K_1 = \theta_0 - Z_{\alpha/2} SE(\hat{\theta})$$

or if

$$\hat{\theta} > K_2 = \theta_0 + Z_{1-\alpha/2} SE(\hat{\theta}).$$

In terms of Z values, this decision rule can be stated as follows: Reject H_0 if and only if

$$Z = \frac{\hat{\theta} - \theta_0}{SE(\hat{\theta})} < Z_{\alpha/2} \qquad (13\text{–}5a)$$

or

$$Z = \frac{\hat{\theta} - \theta_0}{SE(\hat{\theta})} > Z_{1-(\alpha/2)}. \qquad (13\text{–}5b)$$

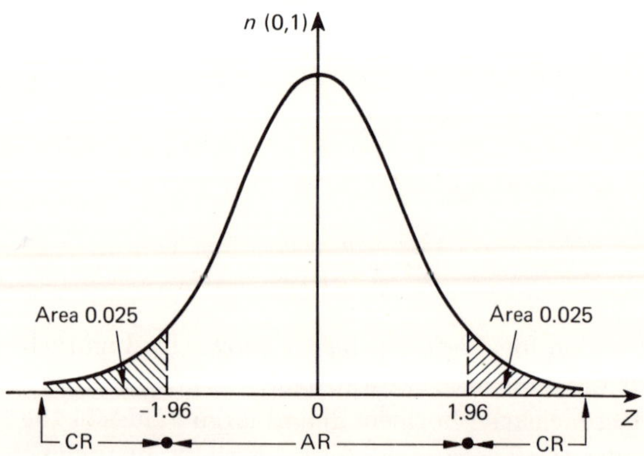

Figure 13–3 A Two-Sided Critical Region for Testing H_0: $\theta = \theta_0$ against H_1: $\theta \neq \theta_0$ at $\alpha = 0.05$

For convenience, the preceding decision criteria, under the assumption that the sampling distribution of the testing statistic is normal, are summarized in Table 13-2 for the two most frequently used α values of 0.05 and 0.01.

Table 13-2. Decision Rules for Testing $\theta = \theta_0$ against Various Alternatives in Terms of the Standard Normal Deviate

	Reject H_0 if and only if	
H_1	$\alpha = 0.05$	$\alpha = 0.01$
$\theta > \theta_0$:	$Z = \dfrac{\hat{\theta} - \theta_0}{SE(\hat{\theta})} > 1.65$	$Z = \dfrac{\hat{\theta} - \theta_0}{SE(\hat{\theta})} > 2.33$
$\theta < \theta_0$:	$Z = \dfrac{\hat{\theta} - \theta_0}{SE(\hat{\theta})} < -1.65$	$Z = \dfrac{\hat{\theta} - \theta_0}{SE(\hat{\theta})} < -2.33$
$\theta \neq \theta_0$:	$Z = \dfrac{\hat{\theta} - \theta_0}{SE(\hat{\theta})} < -1.96$	$Z = \dfrac{\hat{\theta} - \theta_0}{SE(\hat{\theta})} < -2.58$
	or	or
	$Z = \dfrac{\hat{\theta} - \theta_0}{SE(\hat{\theta})} > 1.96$	$Z = \dfrac{\hat{\theta} - \theta_0}{SE(\hat{\theta})} > 2.58$

The values of Z for other values of α can easily be found by the reader from Table B-7.

Performing Computations

Having taken the first four steps, we have designed a statistical test completely. We may now proceed to the fifth step—performing various computations from a random sample of size n necessary for the test. These computations include the testing statistic, the standard error of the testing statistic, and the critical value of the testing statistic.

Making Decisions

Finally, as a sixth step, the decision maker may draw statistical conclusions and select the optimal act. On the basis of previous computations, he would reject H_0 if the testing statistic fell in the critical region and accept H_0 if the testing statistic fell in the acceptance region. Then, in view of the statistical conclusion he would select an appropriate course of action.

If the testing statistic falls in the critical region, the difference between the sample statistic and the hypothetical parameter is considered to be significant. In other words, we think the sample result is so rare that it cannot be explained by chance variations alone. Then H_0 should be rejected and we may state: "The null hypothesis is false," or "The sample observations are not consistent with the null hypothesis." (Remember that the rejection of H_0 is in effect the acceptance of H_1.) On the other hand, if the testing statistic falls in the region of nonrejection, we reason that the difference between the sample result and the hypothetical parameter can be explained by chance variations and therefore is not significant statistically. Consequently, we decide not to reject H_0 and state: "The sample result is not inconsistent with the null hypothesis."

The final decision is associated with the statistical conclusion. Generally, statistical conclusion is the basis on which a decision maker selects an appropriate course of action. His rule is to associate his decision with testing before the testing is conducted; he decides in advance that a_0 is optimal and will be taken if H_0 is accepted and that a_1 is optimal if H_0 is rejected.

Summary

The general procedure of statistical testing consists of six steps: (1) formulating the null and the alternative hypotheses, (2) specifying the level of significance, (3) selecting the testing statistic, (4) establishing decision criteria, (5) performing computations, and (6) making decisions. We shall apply this procedure, in the next three sections, to a number of statistical tests concerning μ, π, $\Delta\mu$, and $\Delta\pi$ under the assumption that the sampling distribution of the testing statistic is normal in each case. Then in the last section we shall move on to discuss procedures of evaluating both α and β.

It is not necessary for one to go through all these six steps in each and every test. However, it is advisable for the reader to follow this outline until he has gained a feeling for the underlying logic of testing.

13–2 TESTING THE HYPOTHESIS $\mu = \mu_0$

The population mean is a very important decision parameter. Often we wish to know if a population mean has increased, or decreased, or remained unchanged. Or we may be interested in determining whether a population mean is significantly greater than, or less than, some assumed value. Has the real per-capita income in Burma increased since its independence? Can a new assembly procedure really reduce the mean time of assembling a mechanism? Has there been no change in the mean lifespan of American males during the past ten years? Each of these questions can be answered by testing $\mu = \mu_0$ against some appropriate alternative hypotheses.

To test whether a specified μ_0 is reasonable in view of sample results, the testing statistic used is the sample mean or, more conveniently for our general testing procedure, the critical ratio $Z = (\bar{x} - \mu_0)/\sigma_{\bar{x}}$.

An Upper-Tail Test

A manufacturing process used by a factory for a number of years yields an average output of 100 units per hour with a standard deviation of 8 units. A new machine for making the same product has just been put on the market. While it is very expensive compared to the one now in use, its adoption would be quite profitable, if its mean output is more than 150 units per hour. To decide whether the new machine should be purchased to replace the old, the management of the factory buys 35 of the new machines as an experiment. The final decision is to be made according to the following procedure:

1. Formulating Hypotheses: Let us assume that management believes the average output per hour of the new machines to be greater than 150 and also that it is anxious to avoid the error of buying when it should not; then the hypotheses should be stated as

$$H_0\colon \mu \leq 150 \text{ units}; \qquad H_1\colon \mu > 150 \text{ units}.$$

2. Specifying the Level of Significance: In this situation the rejection of H_0 if it is true would lead to serious consequences, since then the wrong action (buying the expensive machines) would be taken. In order to guard against the purchase of new machines that might not actually be as productive as expected, management decides to set α at 0.01, a low probability of rejecting H_0 when it is true.

3. Selecting the Testing Statistic: To test the hypotheses stated in item 1, the testing statistic is the critical ratio $Z = (\bar{x} - 150)/\sigma_{\bar{x}}$, which is a standard normal variable.

4. Establishing the Decision Criterion: At $\alpha = 0.01$, and for a right-tail test, the decision rule is: Reject H_0 if and only if $Z > 2.33$; otherwise, accept H_0.

5. Performing Computations:

(a) Assume that the 35 new machines constitute a random sample, and that it has been found that $\bar{x} = 160$ units per hour.

(b) Assuming that the standard deviation of outputs for the new machines is identical with that for the old (that is, $\sigma = 8$ units), then, with $n = 35$, we have

$$\sigma_{\bar{x}} = \frac{8}{\sqrt{35}} = 1.35.$$

(c) The critical ratio is

$$Z = \frac{\bar{x} - \mu_0}{\sigma_{\bar{x}}} = \frac{160 - 150}{1.35} = 7.41.$$

6. Making Decisions: Since Z falls in the critical region, the difference between the sample mean and the hypothetical mean is significant. H_0 is not true. The rejection of H_0 is the acceptance of H_1. According to management's predesignated criterion, the new machine should be purchased to replace the old. (Note: When $Z = 7.41$, we know that if H_0 were true, the probability of obtaining a difference as large as 10 between $\bar{x}$ and μ_0 would be practically zero; that is, the significance of this test is actually that $\alpha = 0$. What does this observation mean or imply?)

A Lower-Tail Test

The average distance for stopping a certain make of automobile while it is traveling 30 miles an hour is 65 feet. The engineering department of the company has designed a new brake system thought to be more effective. To test this invention, the new brake system is installed in 64 cars; tests show that the average distance for stopping a car at a speed of 30 miles an hour is 63.5 feet, with a standard deviation, $\hat{s}$, of 4 feet. Does the difference of 1.5 feet prove that the new brake system is more effective than the old?

Now, if the management wishes to avoid the mistake of not adopting the new brake system if it really should, the question can be answered by the following test:

1. Hypotheses: H_0: $\mu = 65$ feet; H_1: $\mu < 65$ feet.

2. Level of Significance: $\alpha = 0.05$.

3. Testing Statistic: $Z = (\bar{x} - \mu_0)/\hat{\sigma}_{\bar{x}}$.

4. Decision Criterion: Reject H_0 if and only if $Z < -1.65$.

5. Computations: The sample mean, with $n = 64$, was given before as 63.5 feet. The standard deviation of the population is unknown in this case, but the sample is large enough for us to estimate it by $\hat{s}$, which was given as 4 feet. Thus,

$$\hat{\sigma}_{\bar{x}} = \frac{\hat{s}}{\sqrt{n}} = \frac{4}{\sqrt{64}} = 0.5$$

and

$$Z = \frac{63.5 - 65}{0.5} = -3.00.$$

6. Decision: Since $Z = -3.00 < -1.65$, H_0 is rejected. The new

brake system is more effective than the old. The management should seriously consider its adoption.

A Two-Tail Test

A floor-wax factory has nationwide selling outlets. In the past, its salesmen earned a mean commission of \$600 per month, with a standard deviation of \$80. Recently, new brands of floor wax have entered the market. This factor tends to reduce the volume of sales per salesman and, thus, to reduce the salesmen's income. Meanwhile, the list price of the firm's output has increased along with general price inflation; this factor tends to increase the salesmen's commission per sale. The management of the firm is anxious to discover the net effect of these two factors on its salesmen's commissions. To achieve this, the following decision procedure is adopted:

1. HYPOTHESES: H_0: $\mu = \$600$; H_1: $\mu \neq \$600$.

2. LEVEL OF SIGNIFICANCE: $\alpha = 0.05$.

3. TESTING STATISTIC: $Z = (\bar{x} - \mu_0)/\sigma_{\bar{x}}$, which, as before, is a standard normal variable.

4. DECISION CRITERION: With $\alpha = 0.05$, for a two-tail test the decision rule becomes: Reject H_0 if and only if $|Z| > 1.96$. That is, reject H_0 if and only if $Z < -1.96$ or $Z > 1.96$.

5. COMPUTATIONS: Suppose that an unconditional random sample of $n = 100$ is drawn from all the salesmen's commission accounts, and it is found that $\bar{x} = \$585$. Also, since $\sigma = \$80$, we have

$$\sigma_{\bar{x}} = \frac{80}{\sqrt{100}} = 8$$

and

$$Z = \frac{585 - 600}{8} = -1.88.$$

6. DECISIONS: $Z = -1.88$, which falls in the acceptance region; therefore, H_0 is not to be rejected. This result indicates that the difference of \$15 may be attributed to random fluctuations in the sample statistic and that it is not significant at an α risk of 0.05. Since sample results lead us to conclude that the salesmen's monthly mean income has not changed, the management need no longer be concerned with this problem.

It is perhaps more appropriate for this problem to state the alternative hypothesis as $\mu < \$600$, since the management would have no trouble if the mean commision were equal to or greater than \$600. Had the alternative hypothesis been so designated, we would reject H_0, for the same level of significance of 0.05, if and only if $Z < -1.65$. Recall that the computed Z value for the present test is -1.88, which is less than -1.65. Thus, the same difference that is insignificant for a two-tail test may become significant for a single-tail test. Further, had we decided to conduct a left-tail test, the managerial decision would be entirely different for the same sample results. Instead of choosing to do nothing (because H_0 was accepted), the management might (because of the acceptance of H_1) act in one or more ways to keep its sales force intact: step up its advertizing campaign in order to increase its sales, increase the percentage of commission paid, or improve the quality of its output in order to improve its competitive strength.

These observations indicate the importance of formulating hypotheses and specifying level of significance, thereby establishing the decision rule, before the collection and observation of data. Unless the required procedure is followed, we may be led to conduct a statistical test within the framework of an improper decision procedure, which is both suggested and tested by the same sample results.

13–3 TESTING THE HYPOTHESIS $\pi = \pi_0$

We have seen that situations often arise in which sampling is concerned with proportions rather than the mean value of a population. In such a case, observations are qualitative in nature. The sampling of attributes, as we already know, yields information as to whether workers are males or females, whether products are defective or nondefective, and so forth. A population, of course, may contain more than two mutually exclusive categories. However, it is often possible to group the elementary units in a population into two mutually exclusive qualitative classes. For instance, voters may be Republican, Democratic, Liberal, or independent, but all of those who are eligible to vote can be classified as being for or against a given candidate.

In any event, when we analyze qualitative data, we are interested in verifying an assumption made about the population proportion of successes, π. Here, we would like to test the hypothesis $\pi = \pi_0$ with the sample proportion p as the testing statistic. Also, in such a test, we compute the standard error of p by using π_0 in the null hypothesis. This is a logical practice, because in our test the sampling distribution of the testing statistic is determined under the assumption that the null hypothesis is true.

The manufacturer of a patent medicine claims that its product is more than 70 percent effective in giving immediate relief from skin itches, from any cause. The American Medical Association, hoping to verify this claim, appoints a committee of seven skin specialists to evaluate the effectiveness of the medicine. Some members of the committee are confident that the manufacturer's claim could not possibly be true (owing to their past experience with similar drugs), but others suspect that the claim might be valid, so it is finally agreed that both alternatives should be considered. Also, after some thought, it is decided to set α at 0.01 and to try the medicine on 200 patients who have skin disorders. In other words, the testing procedure is designed as follows:

1. HYPOTHESES: $H_0: \pi = 0.70$; $H_1: \pi \neq 0.70$.

2. LEVEL OF SIGNIFICANCE: $\alpha = 0.01$.

3. TESTING STATISTIC: $Z = (p - \pi_0)/\sigma_p$, which is a standard normal variate if H_0 is true.

4. DECISION CRITERION: Reject H_0 if and only if $Z < -2.58$ or $Z > 2.58$.

5. COMPUTATIONS: Suppose that among the 200 patients treated by this medicine, 125 experienced immediate relief; then $p = 0.625$. Also, for this test,

$$\sigma_p = \sqrt{\frac{\pi_0(1 - \pi_0)}{n}} = \sqrt{\frac{0.7(0.3)}{200}} = 0.0324$$

and

$$Z = \frac{0.625 - 0.700}{0.0324} = -2.31.$$

6. DECISIONS: Since $Z = -2.31 > -2.58$, H_0 cannot be rejected. A difference of -0.075 between the sample proportion and the hypothetical value of π is not sufficient to disprove the manufacturer's claim. If the new drug does not have any undesirable side effects, it should be recommended for use.

As a second example, suppose that one of Philadelphia's diaper services has advertised in a newspaper as follows: "There are more than ten diaper services in Philadelphia, but 30 percent of the families who use diaper services are our customers.—The Smile Diaper Service." Suppose then that other diaper services complain to the Better Business Bureau, challenging the validity of this claim.

Before the BBB can act on the complaint, it must decide about the truth of the claim made by The Smile Diaper Service. As an aid in arriving at this decision, the Bureau selects a random sample of 150 families, without replacement, from the thousands currently using diaper services. It finds that 39, or 26 percent, of the 150 families are customers of the Smile Diaper

Service. Is the difference of 4 percent between the sample result and the claimed share of the market statistically significant?

To answer this question, the BBB adopts the following decision procedure:

1. Hypotheses: H_0: $\pi = 0.30$; H_1: $\pi < 0.30$. Note that the alternative hypothesis here suggests a left-tail test. It is appropriate because the BBB is interested primarily in verifying Smile's claim against the alternative that its market share is not as large as claimed.

2. Level of Significance: In deciding the value of α, the BBB is confronted with two conflicting forces. On the one hand, Smile would like the Bureau to adopt a decision criterion that would minimize the probability of its incorrectly rejecting the null hypothesis. This course demands a small α. On the other hand, the Bureau itself and other diaper services are more concerned with an unwarranted acceptance of the null hypothesis. Thus, they would like to have a small β or large α. After thoughtful consideration, the Bureau finally decides on the 5 percent level of significance as a balanced criterion, namely, $\alpha, = 0.05$.

3. Testing Statistic: $Z = (p - \pi_0)/\sigma_p$.

4. Decision Criterion: With $\alpha = 0.05$ for a left-tail test, reject H_0 if and only if $Z < -1.65$. (We are assuming here that the BBB wants to reach a decision one way or another. Otherwise, the decision rule may be stated as: Reject H_0 if and only if $Z < -1.65$, but reserve judgment or conduct another test if $Z \geq -1.65$.)

5. Computations:

$$p = 39/150 = 0.26,$$

$$\sigma_p = \sqrt{\frac{0.3(0.7)}{150}} = 0.0374,$$

$$Z = \frac{0.26 - 0.30}{0.0374} = -1.07.$$

6. Decision: The value of Z falls in the acceptance region. If $\pi = 0.3$, it is not unlikely for $p = 0.26$ to occur. The claim made by The Smile Diaper Service is not unfounded in terms of sample observations. There are, therefore, no grounds for censure.

13-4 TESTING THE HYPOTHESIS $\theta_1 = \theta_2$

One often meets such decision problems as those suggested by the following questions: Is the difference between the mean income of lawyers and that of physicians significant? Has a population of smokers a greater susceptibility to lung cancer than a population of nonsmokers? Is there any real difference between voting habits of men and women? Does one TV program have more appeal to the public than others? Is the new method of therapy for a certain class of patients more effective than the old? Has one make of aluminum channels a greater variability in stiffness than another make? Has the rate of economic growth in Red China been considerably reduced because of the Mao Tse-tung's "cutural revolution?" Has teachers' real income in the United States changed since 1945?

These questions are concerned with the comparison of two populations. One action is called for if they can be thought of as the same population; another is appropriate if there is a difference, or if some change has taken place from one population to another. Our interest here may focus on the difference between population means, between population proportions, or between population variances. We shall discuss statistical tests concerning the first two types of difference in this section, taking up the last kind in Chapter 15.

Comparison of Two Population Means

To test whether there is a significant difference between two population means, μ_1 and μ_2, we formulate the null hypothesis that there is no difference between them—that is, H_0: $\mu_1 = \mu_2$. The testing statistic used here is the difference between two sample means computed from two independently drawn random samples. Under the assumption that H_0 is true, the critical ratio

$$Z = \frac{\Delta\bar{x} - \Delta\mu}{\sigma_{\Delta\bar{x}}} = \frac{\bar{x}_1 - \bar{x}_2}{\sigma_{\Delta\bar{x}}}$$

has the standard normal distribution. Note that if the null hypothesis $\mu_1 = \mu_2$ is true, then $\Delta\mu = \mu_1 - \mu_2 = 0$.

To ascertain whether a new fertilizer is more effective than the old for wheat production, a tract of land was divided into 100 squares of equal areas, all of the same quality. The new fertilizer was applied to 50 squares and the old fertilizer to the other 50. The mean number of bushels of wheat harvested per square of land using the new fertilizer was 25.5 with a variance of 22. The corresponding mean and variance for the squares using the old fertilizer were 24.6 and 19, respectively. Is the new fertilizer more efficient than the old at $\alpha = 0.01$?

1. HYPOTHESES: H_0: $\mu_1 = \mu_2$; H_1: $\mu_1 > \mu_2$.

2. LEVEL OF SIGNIFICANCE: $\alpha = 0.01$.

3. TESTING STATISTIC: $Z = (x_1 - x_2)/\sigma_{\Delta\bar{x}}$.

4. Decision Criterion: Reject H_0 if and only if $Z < +2.33$.

5. Computations: Since both n_1 and n_2 are quite large (both are greater than 30), sample variances can be used as a pooled estimator of population variance to compute the standard error of $\Delta\bar{x}$. That is,

$$\hat{\sigma}_{\Delta\bar{x}} = \sqrt{\frac{\hat{s}_1^2}{n_1} + \frac{\hat{s}_2^2}{n_2}} = \sqrt{\frac{22}{50} + \frac{19}{50}} = 0.906.$$

With this value of $\hat{\sigma}_{\Delta\bar{x}}$, we have

$$Z = \frac{\bar{x}_1 - \bar{x}_2}{\hat{\sigma}_{\Delta\bar{x}}} = \frac{25.5 - 24.6}{0.906} = 0.99.$$

6. Decisions: Z falls in the acceptance region, and H_0 cannot be rejected. The new fertilizer is not more efficient than the old at the 1 percent level of significance.

An automobile manufacturing company has recorded observations of two different makes of batteries, both of which are being used on its cars: 40 observations of make A indicated a mean life of 32 months; 45 observations of make B showed a mean life of 30 months. Also, past experience indicates that the standard deviation for both makes of batteries is the same, 4 months. Is there a real difference in the mean life between the two makes of batteries at $\alpha = 0.05$?

The appropriate hypotheses in this case are $H_0: \mu_A = \mu_B$ and $H_1: \mu_A \neq \mu_B$. Also,

$$Z = \frac{\bar{x}_A - \bar{x}_B}{\sigma_{\Delta\bar{x}}} = \frac{32 - 30}{\sqrt{\frac{4^2}{40} + \frac{4^2}{45}}} = 2.30,$$

which is greater than 1.96; therefore, H_0 is rejected. The difference of 2 months in mean life cannot be explained by chance variations alone.

For this problem, it may be interesting to recall that the 95 percent confidence interval discussed in the previous chapter is $0.3 < \Delta\mu < 3.7$, which does not include $\Delta\mu = 0$. This indicates that a confidence interval with $1 - \alpha$ corresponds to a two-tail test at α. In general, for example, the inequality

$$\hat{\theta} - 1.96SE(\hat{\theta}) < \theta < \hat{\theta} + 1.96SE(\hat{\theta})$$

states that before a sample is selected and $\hat{\theta}$ determined, the probability is 0.95 that the interval computed from $\hat{\theta}$ will include the unknown but constant θ. When we test the null hypothesis $\theta = \theta_0$ against the alternative $\theta \neq \theta_0$ and assume that the variance σ^2 for H_0 is the same as that for H_1, we can actually adopt the decision rule of rejecting H_0 if θ_0 does not lie in the confidence interval. That is, H_0 is rejected if

$$|\theta_0 - \hat{\theta}| > 1.96SE(\theta). \tag{13-6}$$

This is the same as saying, however, that the testing statistic $\hat{\theta}$ falls farther away from θ_0 than a certain quantity; this is actually a two-tail test we have been conducting. In the confidence coefficient $1 - \alpha$, α is the level of significance of a test. To show this relationship for a test against a single-tail alternative, we have, of course, to construct a one-sided confidence interval as introduced in the next chapter.

Comparison of Two Population Proportions

To compare two population proportions, we conduct a test for H_0: $\pi_1 = \pi_2$ against some appropriate alternative. The testing statistic employed is usually $(p_1 - p_2)/\sigma_{\Delta p}$, where the sample proportions are derived from two independent random samples. Under the assumption that $\pi_1 = \pi_2 = \pi$,

$$\sigma_{\Delta p} = \sqrt{\frac{\pi(1-\pi)}{n_1} + \frac{\pi(1-\pi)}{n_2}} = \sqrt{\pi(1-\pi)\left(\frac{n_1+n_2}{n_1 n_2}\right)}.$$

But the population proportions are unknown to us; we take the weighted mean of the two sample proportions as the estimator. Call this estimator $\bar{p}$; we then have

$$\bar{p} = \frac{n_1 p_1 + n_2 p_2}{n_1 + n_2}. \tag{13-7}$$

Substituting $\bar{p}$ into the previous expression, we have as an estimator of $\sigma_{\Delta p}$ the following:

$$\hat{\sigma}_{\Delta p} = \sqrt{\bar{p}(1-\bar{p})\left(\frac{n_1+n_2}{n_1 n_2}\right)}. \tag{13-8}$$

In the spring of 1954 the Federal Reserve System and the Research Center of the University of Michigan jointly sampled 3000 spending units in a survey of consumer finances. A similar study was made in the spring of 1955, when 3120 spending units were interviewed. These two surveys, among other things, revealed that in 1954, 7.7 percent of those interviewed intended to purchase television sets, whereas in 1955, only 5.9 percent of those questioned planned to do so. Had there been a significant change in consumers' intentions to buy television sets between the spring of 1954 and that of 1955? To answer this question, we conduct the following test:

1. Hypotheses: H_0: $\pi_1 = \pi_2$; H_1: $\pi_1 \neq \pi_2$.

2. Level of Significance: $\alpha = 0.05$.

3. TESTING STATISTIC: $Z = (p_1 - p_2)/\sigma_{\Delta p}$, which has the standard normal distribution if H_0 is true.

4. DECISION CRITERION: Reject H_0 if and only if $Z < -1.96$ or $Z > 1.96$; otherwise accept H_0.

5. COMPUTATIONS: Sample data give

1954	1955
$n_1 = 3000$	$n_2 = 3120$
$p_1 = 0.077$	$p_2 = 0.059$

We have also

$$\bar{p} = \frac{(3000)(0.077) + (3120)(0.059)}{3000 + 3120} = 0.0678.$$

By (13–8),

$$\hat{\sigma}_{\Delta p} = \sqrt{0.0678(1 - 0.0678)\left[\frac{3000 + 3120}{(3000)(3120)}\right]} = 0.00643.$$

Therefore,
$$Z = \frac{0.077 - 0.059}{0.00643} = 2.80.$$

6. DECISIONS: Z falls in the rejection region; between the springs of 1954 and 1955 there had been a significant change in consumers' buying intentions toward TV sets.

A sample poll of 200 college students and 250 high school students showed that 59 percent and 55 percent, respectively, were against the draft system as it applied in 1967. Was this difference of 4 percent in sample proportions sufficient evidence to support the contention that a greater percentage of college students was dissatisfied with the draft system at a level of significance of 0.01?

For this problem,

$$\bar{p} = \frac{(200)(0.59) + (250)(0.55)}{200 + 250} = 0.568,$$

$$\hat{\sigma}_{\Delta p} = \sqrt{0.568(1 - 0.568)\left[\frac{200 + 250}{(200)(250)}\right]} = 0.0469,$$

and
$$Z = \frac{0.59 - 0.55}{0.0469} = 0.85.$$

Since the value of the critical ratio is less than 2.33, the null hypothesis, $\pi_1 = \pi_2$, cannot be rejected in favor of the alternative, $\pi_1 > \pi_2$. In other words, the difference of 4 percent is not sufficient evidence to support the contention that the college students were more dissatisfied with the draft system than the high school students in 1967, at $\alpha = 0.01$ with $n_1 = 200$ and $n_2 = 250$. However, there might be a real difference; if so, this difference might be revealed statistically by larger samples.

13-5 POWER AND OC FUNCTIONS

Power and OC Functions Defined

The general testing procedure discussed so far assumes a predetermined sample size and inexact H_1. As such, we can only specify α, leaving β uncontrolled. Before closing this chapter, we shall consider an explicit evaluation of both α and β.

First, let us note that in two-action testing problems, one and only one of the following four consequences can occur:

(1) We may reject a true H_0. The probability of this wrong decision is α.
(2) We may accept a true H_0. The probability of this correct decision is $1 - \alpha$.
(3) We may accept a false H_0. The probability of this wrong decision is β.
(4) We may reject a false H_0. The probability of this correct decision is $1 - \beta$.

Here, $1 - \beta$ is called the *power of test*, since it is the ability of the test to detect the alternative hypothesis when it is in fact true. The power of test, as the probability of rejecting a false H_0 (or of accepting a true H_1), depends on the value of β, which can be determined only if the sampling distribution of the testing statistic, $\hat{\theta}$, is correctly defined. This requirement implies that the parameter covered by H_1 must be specific, or that H_1 should be a simple hypothesis. Thus, if we are testing

$$H_0\colon \theta = \theta_0 \quad \text{against} \quad H_1\colon \theta = \theta_1,$$

then

$$\begin{aligned}\text{power of test} &= P(\hat{\theta} \text{ falls in CR} \mid \theta = \theta_1)\\ &= 1 - \beta,\end{aligned}$$

where CR stands for the critical region, which depends on the value of α, given n.

When we have a composite H_1, the parameter is not specified; therefore the sampling distribution is undefined. To find the power of a test in this case, we treat $1 - \beta$ as a function, termed as the *power function* (PF), that relates the probability of rejecting a null hypothesis to all possible values of the parameter covered by the null and alternative hypotheses at a given level of significance and a given sample size.

Now, suppose we are testing the following pair of hypotheses:

$$H_0\colon \theta = \theta_0, \qquad H_1\colon \theta \in W,$$

where W is the set of all possible values, θ_i, for θ in H_1. For this test, if H_0 is false—that is, $\theta \in W$—the power function is defined in terms of a likelihood function:

$$\begin{aligned} \text{PF}(\theta_i) &= P(\hat{\theta} \in \text{CR} \mid \theta = \theta_i) \\ &= 1 - \beta. \end{aligned} \tag{13–9}$$

However, if H_0 is true, then we have

$$\begin{aligned} \text{PF}(\theta_0) &= P(\hat{\theta} \in \text{CR} \mid \theta = \theta_0) \\ &= \alpha. \end{aligned} \tag{13–10}$$

The complement of the power function is called the *operating-characteristic*, or *performance, function*. It is also defined in terms of a likelihood function. If H_0 is true, the OC function is given as

$$\begin{aligned} \text{OC}(\theta_0) &= P(\hat{\theta} \in \text{AR} \mid \theta = \theta_0) \\ &= 1 - \alpha, \end{aligned} \tag{13–11}$$

where AR stands for the acceptance region. This is the probability of making the correct decision of accepting the true H_0.

When H_0 is false, or $\theta \in W$, we have

$$\begin{aligned} \text{OC}(\theta_i) &= P(\hat{\theta} \in \text{AR} \mid \theta = \theta_i) \\ &= \beta, \end{aligned} \tag{13–12}$$

the probability of committing a type II error—accepting a false H_0 given that the true parameter value is covered by H_1.

Note that OC and PF are complementary functions:

$$\text{OC}(\theta_0) = 1 - \text{PF}(\theta_0)$$

and

$$\text{OC}(\theta_i) = 1 - \text{PF}(\theta_i).$$

Note also that the power of the test is identical with α only when H_0 is true; otherwise, it is equal to $1 - \beta$. Finally, the operating-characteristic function always has β as its value except where H_0 is true, since then $\text{OC}(\theta_0) = 1 - \alpha$.

To clarify previous comments and to bring out the use of power and OC functions in testing, let us turn to consider some examples.

Power and OC Functions for Single-Tail Tests

A production process has been in control for some time with percent defectives $p = 0.10$. To check whether the process remains in control, simple random samples of 20 are inspected, and if more than four defectives are observed, the process is regarded as "out of control." For this decision rule, we are actually continuously deciding between the two hypotheses

$$H_0\colon p \leq 0.10, \qquad H_1\colon p > 0.10$$

at $\alpha = 0.04317$, since

$$\begin{aligned}\alpha &= P(a_1 \mid H_0)\\ &= P(X \geq 5 \mid p = 0.10)\\ &= 1 - B(4; 20, 0.10)\\ &= 1 - 0.95683\\ &= 0.04317.\end{aligned}$$

Now, for various possible alternative values of p_1, the power function of this test becomes

$$\mathrm{PF}(\theta_i) = P(X \geq 5 \mid p_1 = \theta_i).$$

For example, if $p_1 = 0.20$, the power of test, or the probability of rejecting $p_0 = 0.10$, becomes

$$\begin{aligned}\mathrm{PF}(0.20) &= P(X \geq 5 \mid p_1 = 0.20)\\ &= 1 - B(4; 20, 0.20)\\ &= 1 - 0.62965\\ &= 0.37035,\end{aligned}$$

which is $1 - \beta$.

The corresponding OC function—the probability of accepting $p_0 = 0.10$ when in fact $p_1 = 0.20$—is

$$\begin{aligned}\mathrm{OC}(0.20) &= 1 - \mathrm{PF}(0.20)\\ &= 1 - 0.37035\\ &= 0.62965,\end{aligned}$$

which is β.

Table 13-3 shows the power and OC functions for a few selected values of the parameter covered by H_0 and H_1.

Table 13-3. Power and OC Functions for Testing $p_0 \leq 0.10$ against $p_1 > 0.10$ with $n = 20$ and $\alpha = 0.04317$

p	$\mathrm{PF}(\theta_i) = 1 - \beta$	$\mathrm{OC}(\theta_i) = \beta$
0.01	0.00000	1.00000
0.05	0.00257	0.99743
0.10	0.04317 $(= \alpha)$	0.95683 $(= 1 - \alpha)$
0.20	0.37035	0.62965
0.25	0.58516	0.41484
0.30	0.76249	0.23751
0.40	0.94905	0.05095
0.50	0.99409	0.00591
0.60	0.99968	0.00032
0.70	1.00000	0.00000

The graphs of the power and OC functions are called the *power* and OC *curves,* respectively. The power and OC curves for our illustrative test are shown, as suggested by the values in Table 13–2, in Figure 13–4(a) and (b).

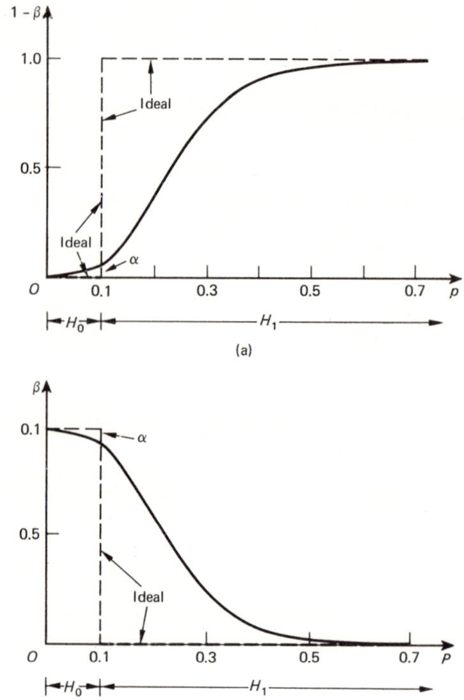

Figure 13–4 Power and OC Curves for Testing $p \leq 0.10$ against $p > 0.10$ at $\alpha = 0.04317$ with $n = 20$ Drawn against Ideal Power and OC Curves

It is interesting to note that the power curve is, as it should be, low over H_0 and high over H_1. This indicates, given α and n, that the power of the test increases, or the probability of detecting a true alternative parameter value increases, when the alternative parameter value deviates more and more from the parameter value covered by H_0. As a matter of fact, an "ideal" power function for a right-tail test should be

$$PF(\theta_i) = 1 - OC(\theta_i) = 0, \qquad \text{when } \theta \leq \theta_0;$$

$$PF(\theta_i) = 1 - OC(\theta_i) = 1, \qquad \text{when } \theta > \theta_0.$$

That is, if we had perfect information concerning θ, we would always reject H_0 when H_1 is true and never reject H_0 otherwise. In the absence of perfect information, such an ideal power function cannot be obtained.

Returning to our illustrative case, we find the power function indicates that when the process is operating at a level equal to or less than 10 percent defectives, the sample result rarely concludes that the process is "out of control." The probability is less than 0.05. However, when the fraction defective is as large as 0.20, we can only detect the process to be out of control less than 40 percent of the time. If the fraction defective is very large, say 0.30 or greater, the sample is almost certain to detect this at once. Now, if the manager wishes to steepen the power curve so as to distinguish more quickly and sharply between the production of 10 percent defectives and, say, 20 percent defectives, given the same α, he can increase n in order to change the shape of the power curve to conform more to the ideal.

It is interesting to note that given α and n, one decision rule is said to be better or more powerful than another if the power curve for the former is everywhere higher than that for the latter. In this connection, we say that a decision rule is the *uniformly most powerful test* if it yields the highest power curve possible.

It turns out, other things being equal, that the power function of a test depends upon the selection of the critical region (decision rule). If the critical region is at one tail end (if the test is one-sided) of the distribution of the testing statistic, then the test will be uniformly most powerful, since then β is the smallest possible, given α and n. If, however, the critical region consists of any other portion of the area under the density curve of the testing statistic, the test will be less powerful, since then we have a larger β than the minimum possible. These statements are demonstrated by Figure 13–5, which shows that while α values in (a) and (b) are identical, the critical region in (b) is unwisely chosen and the effect on β is disastrous.

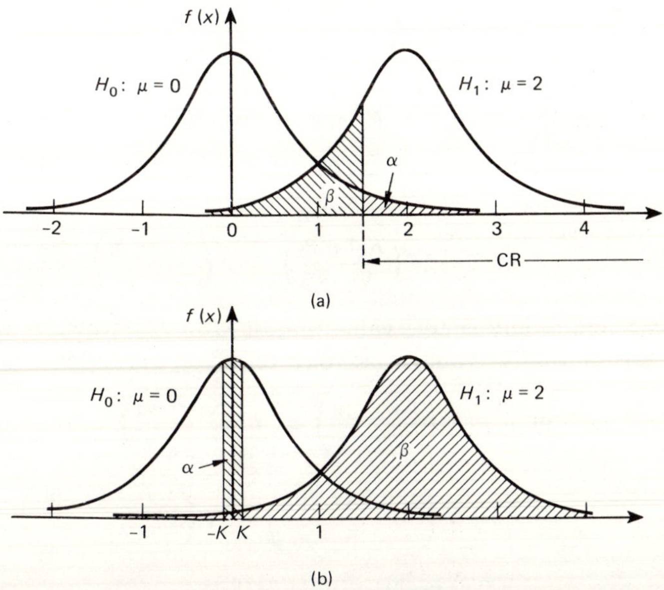

Figure 13–5 (a) Proper and (b) Improper Selection of Critical Region, CR

The OC function, as the probability of making the wrong decision of accepting H_0 given that H_1 is true, or as the complement of the power function, should be high over H_0 and low over H_1. Ideally, for a right-tail test, we should have

$$OC(\theta_i) = 1, \qquad \text{when } \theta_i \leq \theta_0;$$

$$OC(\theta_i) = 0, \qquad \text{when } \theta_i > \theta_0.$$

Observe that the actual power and OC curves for our illustrative test are drawn against the ideal power and OC functions as represented by the heavy lines.

Power and OC Functions for Two-Sided Tests

The mean monthly family income in a certain city was known to be \$400, with a standard deviation of \$50, during the late 1950's. The municipal government now wonders whether there has been any significant change in this average, and plans to find out by taking a random sample of 100 families and testing the following hypotheses at $\alpha = 0.05$:

$$H_0\text{: } \mu = 400, \qquad H_1\text{: } \mu \neq 400.$$

Under the conditions specified, we have the following decision rule: The null hypothesis will be rejected if and only if

$$\bar{x} < K_1 = \mu_0 - 1.96\sigma_{\bar{x}} = 400 - 1.96\left(\frac{50}{\sqrt{100}}\right) = \$390.20$$

or $$\bar{x} > K_2 = \mu_0 + 1.96\sigma_{\bar{x}} = 400 + 1.96(5) = \$409.80.$$

In general, for a two-sided alternative hypothesis, we would accept H_0: $\theta = \theta_0$ if and only if $K_1 < \hat{\theta} < K_2$ and, therefore,

$$\begin{aligned} OC(\theta_i) &= \beta = P(H_0 \mid H_1) \\ &= P(K_1 < \hat{\theta} < K_2) \\ &= N\left(\frac{K_2 - \theta_1}{SE(\hat{\theta})}\right) - N\left(\frac{K_1 - \theta_1}{SE(\hat{\theta})}\right), \end{aligned} \tag{13–13}$$

where θ_1 refers to any simple value contained in H_1. Now, in the present case, if the alternative value is assumed to be, say, $\mu_1 = 395$, then

$$\begin{aligned} OC(395) &= N\left(\frac{K_2 - \mu_1}{\sigma_{\bar{x}}}\right) - N\left(\frac{K_1 - \mu_1}{\sigma_{\bar{x}}}\right) \\ &= N\left(\frac{409.8 - 395}{5}\right) - N\left(\frac{390.2 - 395}{5}\right) \\ &= N(2.96) - N(-0.96) \\ &= 0.9985 - 0.1685 = 0.8300. \end{aligned}$$

If $\mu_1 = 415$, then

$$\text{OC}(415) = N\left(\frac{409.8 - 415}{5}\right) - N\left(\frac{390.2 - 415}{5}\right)$$

$$= N(-1.04) - N(-4.96) = 0.1492.$$

Other values of β in Table 13-4 have been computed in the same fashion. The last column of this table contains a few values of the power function

Table 13-4

H_1	β	$1 - \beta$
375.0	0.001	0.999
382.5	0.062	0.938
387.5	0.295	0.705
392.5	0.677	0.323
397.5	0.921	0.079
400.0	0.950 $(= 1 - \alpha)$	0.050 $(= \alpha)$
402.5	0.921	0.079
407.5	0.677	0.323
412.5	0.295	0.705
417.5	0.062	0.938
425.0	0.001	0.999

of this test. Figure 13-6 has been designed to give the reader a feeling of β in terms of areas under the density curves of the distributions of $\bar{x}$ given alternative values of μ. The power and OC curves for this test are shown by Figures 13-7 and 13-8, respectively. It is interesting to note that both β and $1 - \beta$ are symmetrical about μ_0 in the case of two-sided alternatives.

In general, as we have already noted, single-tail tests are uniformly most powerful tests. There is no uniformly most powerful test with the two-sided alternative, but a two-tail test is a reasonable one on many counts. First of all, from the OC curve (Figure 13-8) we see that this test is unbiased in the sense that the largest probability of accepting H_0 is at μ_0. This property is not possessed by single-tail tests. Furthermore, as Figure 13-9 shows, the two-tail test has some chance of detecting excesses of the true value of the parameter above θ_0, though not as good a chance as the right-tail test. However, unlike the right-tail test, the two-tail test has a fairly good chance of detecting shortages of the true parameter below θ_0. Similar comparisons can be made between a left-tail test and a two-tail test. Thus, in a given situation, the test selected should be the one whose OC curve comes closest to giving the proper decisions for various values of θ that might prevail.

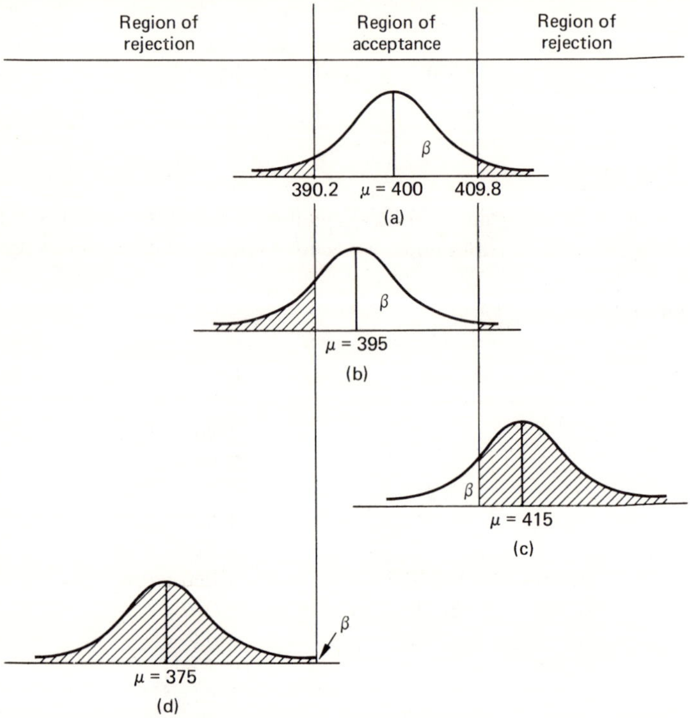

Figure 13–6 Probability of Committing Type II Error in Testing $\mu = 400$ against $\mu \neq 400$ at $\alpha = 0.05$ with $n = 100$, Where Shaded Areas Represent $1 - \beta$

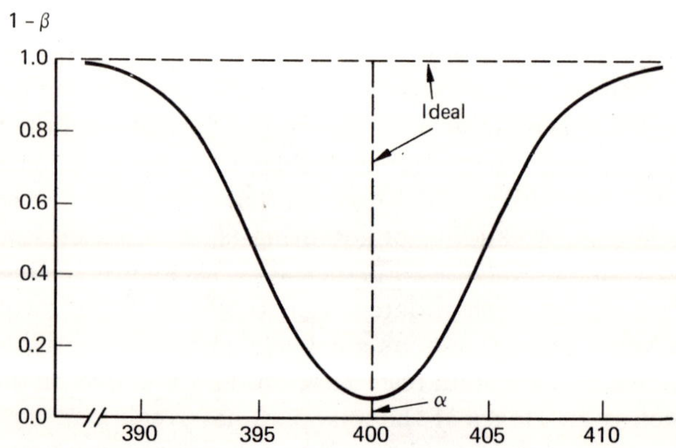

Figure 13–7 Power Curve for H_0: $\mu = \$400$ and H_1: $\mu \neq \$400$ Drawn against Ideal Power Curve

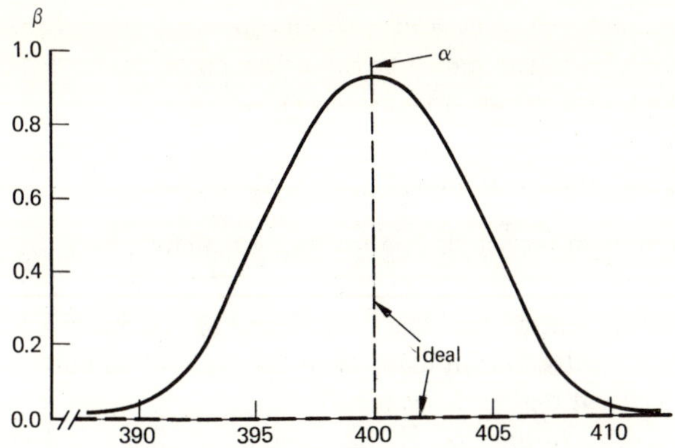

Figure 13–8 OC Curve for H_0: μ = \$400 and H_1: $\mu \neq$ \$400 Drawn against Ideal OC Curve

Now we may note that the ideal power function for a two-sided test is

$$1 - \text{OC}(\theta_i) = 0, \qquad \text{when } \theta_i = \theta_0;$$

$$1 - \text{OC}(\theta_i) = 1, \qquad \text{when } \theta_i \neq \theta_0.$$

Similarly, the ideal OC function should satisfy

$$\text{OC}(\theta_i) = 1, \qquad \text{when } \theta_i = \theta_0;$$

$$\text{OC}(\theta_i) = 0, \qquad \text{when } \theta_i \neq \theta_0.$$

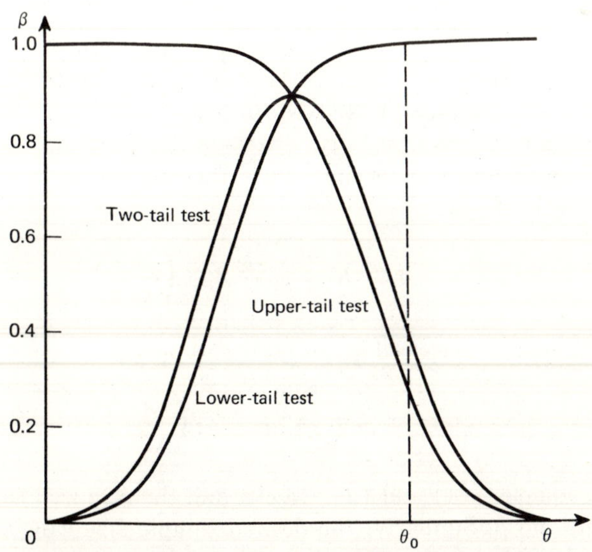

Figure 13–9 OC Curves of Three Tests with H_0: $\theta = \theta_0$ at $\alpha = 0.01$

Furthermore, just in the case of a single-tail test, given n, the larger is α the more powerful is the test. On the other hand, given α, the larger is n the closer the power curve is to the ideal.

Relationship between α, β, and n

Our discussion so far yields the following conclusions:

(1) For a given sample size, the probability of committing a type II error is automatically determined, once the level of significance is selected or the critical region determined.

(2) For a given sample size, if the error of the first kind is reduced, the error of the second kind is increased.

(3) A type I error is completely under control in a given test because it is specified. Furthermore, the common practice of choosing $\alpha = 0.05$ or 0.01 indicates that the error of rejecting a true null hypothesis is seldom made. A type II error is frequently made if the true value of a parameter is close to θ_0. In such a situation, however, the mistake of accepting a false null hypothesis is not serious. When the true parameter deviates greatly from θ_0, a type II error is a very serious matter. Fortunately, however, in such a situation the probability of accepting a false null hypothesis is very slight.

(4) Similarly, given n, the power of a test varies directly with α, since $\text{PF}(\theta_i) = 1 - \text{OC}(\theta_i)$, where $\text{OC}(\theta_i) = \beta$, which is inversely related with α.

(5) Also, given α, the greater is n, the more powerful is the test. In this connection we note that α and β can be fixed as desired if n is allowed to vary. For example, if we wish to test

$$H_0\colon \mu = C \quad \text{against} \quad H_1\colon \mu = C + A, \qquad \text{with } A > 0,$$

by setting α and β at specific values, then H_0 will be rejected given H_0 if and only if

$$\text{(1)} \qquad \bar{x} \geq K = \mu_0 + Z_{1-\alpha}\left(\frac{\sigma}{\sqrt{n}}\right),$$

and H_0 will be accepted given H_1 if and only if

$$\text{(2)} \qquad \bar{x} \leq K = \mu_1 - Z_\beta\left(\frac{\sigma}{\sqrt{n}}\right).$$

Now, equating (1) and (2), we can find the required sample size for fixed α and β. Similarly, the desired sample size can be found for a two-sided test or for hypotheses involving other parameters. The reader is referred for illustrations to Exercises 13–24 through 13–27.

We have been concerned in this and earlier chapters with classical inferential procedures for the large-sample case. In the next two chapters we shall discuss inferences from small samples.

EXERCISES

13–1 We suspect that the IQ's of the students at a certain college tend to run above the national average of 100. It is also known that, according to past knowledge, IQ is normally distributed with $\sigma = 10$.

(a) If we wish to verify this by statistical testing, how should the null and alternative hypotheses be formulated?

(b) If a random sample of size $n = 25$ is drawn from the student body of the college and a mean IQ of 104 is observed, what conclusion can we reach at $\alpha = 0.01$?

13–2 Medical reports have linked cigarette smoking with lung cancer and other ill effects. Suppose the cigarette producers know that before these reports the adult smokers in this country had smoked 10 cigarettes per day on the average with a standard deviation of 1.5 cigarettes. The producers are anxious to evaluate the effects of the medical reports upon the cigarette consumption of adult smokers. To do this, suppose that they have drawn a simple random sample of 144 adult smokers throughout the country and found the sample mean to be 8.5 cigarettes per day. How should the null and alternative hypotheses be stated in this case? What conclusion can you draw at $\alpha = 0.01$ in view of the sample results?

13–3 The Transit Authority of New York City uses tens of thousands of light bulbs every year. The brand that has been used up to now has a mean life of 1000 hours with a standard deviation of 90 hours. A new brand is offered to the Transit Authority at a price much lower than what it has been paying. The Authority decides that the new brand should be bought now, unless it has a mean life of less than 1000 hours at a level of significance of 0.05. Subsequently, 100 bulbs of the new brand are tested, yielding an average of 990 hours. Assuming that the standard deviation for the new brand is the same as that for the old, what should be the decision of the Transit Authority?

13–4 Forty-nine American soldiers observed at random yield a mean weight of 160 pounds, with a standard deviation of 11 pounds. Are these observations consistent with the assumption that the mean weight of all American soldiers is 170 pounds, at $\alpha = 0.01$?

13–5 In order for a manufactured part to fit other parts, a certain critical dimension is designed to be 5 inches. The variance of the manufacturing process is known to be 0.0064 inches^2. If a given day's sample of 49 yields $\bar{x} = 5.05$, what is the 95 percent confidence interval for μ? In view of this interval, should the null hypothesis, $\mu = 5$, be rejected in favor of the alternative hypothesis, $\mu \neq 5$, at $\alpha = 0.05$?

13–6 Certain steel pipes are supposed to have a mean outside diameter of 10 inches with a standard deviation of 0.018 inches. We wish to test H_0: $\mu = 10$ against $H_1: \mu \neq 10$. If $\bar{x} = 9.94$ for a given day's sample of 36 pipes, what is the corresponding 99 percent confidence interval for μ? Does this interval indicate the rejection or acceptance of the null hypothesis at $\alpha = 0.01$?

13–7 A manufacturer claimed that at least 90 percent of the machine parts that it supplied to a factory conformed to specifications. An examination of 200 such parts revealed that 160 parts were not faulty. Determine whether the manufacturer's claim is legitimate at the 1 percent level of significance.

13–8 A dairy is contemplating a change of its milk containers from glass to paper. The change will not be made, however, unless at least 70 percent of its customers so prefer. When a survey is made of 200 of its customers, 120 of them are in favor of the change. How should the dairy decide if it is willing to assume an α risk of 0.05?

13–9 A pair of dice is tossed 100 times, and "sevens" turn up 23 times. Test the hypothesis that the dice are fair at $\alpha = 0.01$.

13–10 A man who plans to open a restaurant in a certain city tells his bank, from which he wishes to borrow the necessary capital, that at least 50 percent of the residents in a nearby residential district will patronize his restaurant from time to time after it opens. Suppose you are the loan officer in the bank and wish to verify this claim at $\alpha = 0.05$. Furthermore, suppose that in a random sample of 50 residents in the district, 44 percent indicate their intention to patronize the proposed restaurant.

(a) What conclusions would you draw from the result above?

(b) Suppose the sample had been 200 instead of 50, and the sample proportion were still 0.44; would you conclude differently?

(c) Discuss the limitations of statistical testing in the light of your answers to (a) and (b).

13–11 A sample of 80 steel wires produced by factory A yields a mean breaking strength of 1230 pounds, with a standard deviation of 120 pounds. A sample of 100 steel wires produced by factory B yields a mean breaking strength of 1190 pounds, with a standard deviation of 90 pounds. Is there a real difference in mean strengths of the two makes of steel wires at $\alpha = 0.05$? $\alpha = 0.01$?

13–12 An examination was given to 50 seniors at college A and to 60 seniors at college B. At A, the mean grade was 75 with a standard deviation of 9. At B, the mean grade was 79 with a standard deviation of 7. Is there a significant difference between the performances of the students at A and those at B, given that $\alpha = 0.05$? $\alpha = 0.01$?

13–13 For data in Exercise 13–11, is the mean breaking strength of make A steel wire significantly greater than that of make B steel wire at the 5 percent level of significance?

13–14 For data in Exercise 13–12, is the performance of students at college A significantly worse than that of students at college B, with $\alpha = 0.05$? with $\alpha = 0.01$?

13–15 A firm's packaging machinery is known to pour dry cereal into economy-size boxes with a standard deviation of 0.6 ounce. Constant checks are made of the net weights in the boxes in order to maintain the adjustment of the machinery that controls the net contents. Two samples taken on two different dates yield the following information:

First sample	*Second sample*
$n_1 = 30$	$n_2 = 35$
$\bar{x}_1 = 18.7$ ounces	$\bar{x}_2 = 21.9$ ounces

Use $\alpha = 0.05$ throughout in answering the following questions:
(a) Test the hypothesis that, on the first date, the machine was adjusted to fill 20 ounces.
(b) Test the hypothesis that, on the second date, the machine was adjusted to fill 20 ounces.
(c) Test the hypothesis that there was no change in the adjustment of the machine between the two dates.
(d) What is the relationship between your answer to (c) and your answers to (a) and (b)?

13–16 A random sample of 100 men was selected from New York State in 1969 and 60 were found to be in favor of a "more modern" divorce law. A random sample of 100 women selected from the same state at the same time revealed that 40 were in favor of such a new law. Is the proportion of men favoring a new divorce law greater than that of women in New York State, given that $\alpha = 0.05$? given that $\alpha = 0.01$?

13–17 To determine whether a certain type of fertilizer is effective, 100 plants out of 500 are left unfertilized. Of the 100, 52 are found to have satisfactory growth, and out of 400 fertilized plants, 275 are found satisfactory. What conclusion can you draw at $\alpha = 0.05$? 00.1?

13–18 Two groups of certain types of patients, A and B, each consisting of 200 people, are used to test the effectiveness of a new serum. Both groups are treated identically except that group A is given the serum while group B is not. It is found that 140 and 120 of groups A and B, respectively, recover from the disease. Is this observed result sufficient evidence for the conclusion that the new serum helps to cure the disease if we are willing to assume an α risk of 0.01?

13–19 During the spring of 1968, students in one of this author's classes were instructed to conduct a survey in New York City of the voters' reactions toward various policies of the national Democratic administration, and the following results were obtained:

Satisfied with the Administration's:	*Men* ($n_1 = 500$)	*Women* ($n_2 = 400$)
Vietnam policy	210	145
Civil rights policy	397	232
Space policy	311	227
Economic policy	278	189
Defense policy	286	192

(a) Assuming that the samples are random, do you think there is a significant difference, at $\alpha = 0.05$, between the reactions of men and women toward each of the policies mentioned above?

(b) What conclusions can you make insofar as all New Yorkers' reactions toward these policies are concerned?

13–20 Suppose that a purchasing agent of a department store is interested in buying a certain make of small cardboard container that, according to the manufacturer's specifications, has a mean crushing strength of 30 pounds and a standard deviation of 3 pounds. These specifications are exactly what the purchasing agent desires. In deciding to buy or not to buy, suppose he does not want to forego the opportunity of buying a useful product. Explain why the hypotheses should be stated as

$$H_0: \mu \geq 30 \text{ pounds} \quad \text{and} \quad H_1: \mu < 30 \text{ pounds}.$$

Now, if the decision is to be made at $\alpha = 0.05$, with $n = 36$, what would the power and OC functions of this test be?

13–21 Suppose that the purchasing agent in Exercise 13–20 wishes to avoid the mistake of buying when he should not. How should the hypotheses be stated now? Again, assume $\alpha = 0.05$ and $n = 36$. Evaluate the power and OC functions.

13–22 In a manufactured part that is to fit other parts, a certain critical dimension is designed to be 4.6 inches. The variability in manufacturing is indicated by a variance of 0.25 inch^2. The process is considered to be in control and is contained if it produces this part with a mean of 4.6 inches for the critical dimension. Otherwise, the process is considered out of control and it is stopped. To check whether the process is in control, the plant statistician is satisfied with $\alpha = 0.05$ and $n = 25$, taken every other hour.

(a) Formulate the appropriate hypotheses.

(b) Show the α and β errors schematically.

(c) Establish the decision rule.

(d) Construct the power and OC curves.

13–23 The U. S. Navy decides to accept a shipment of 200,000 raincoats if the percentage of defectives in the lot is less than 10. Furthermore, the Navy is anxious to accept the lot and is willing, by observing a random sample of 1500 selected from the lot, to assume a risk of 0.05 of rejecting a shipment that meets its specifications. How should the hypotheses be stated in this case? What is the decision rule for this test? What are the values of the OC function for $\pi_1 = 0.12$? 0.11? 0.10? 0.09? 0.08? 0.07? 0.05?

13–24 In planning whether or not to have a branch in a certain city, the management of a department store has set up the following criteria: Build the branch if the mean monthly family income in that city is \$500; do not build it if the mean monthly income is \$450; $\alpha = 0.05$ and $\beta = 0.10$. Establish the decision rule (that is, find the critical value and sample size) for this case (assuming $\sigma = \$90$).

13–25 The variability in manufacturing a certain part for wall clocks is indicated by a standard deviation of 0.02 inch, which, according to past experiènce, remains stable. This part is specified to have an outside diameter of 2.06 ± 0.03 inch. When this specification is not met, the production process is considered out of control, and it must be stopped for adjustment. If $\alpha = 0.01$ and $\beta = 0.10$, how large a sample is required for the purpose of checking whether or not the process is in control?

13–26 Suppose, with reference to a binomial population, we wish to test H_0: $\pi = 0.4$ against H_1: $\pi = 0.6$. Furthermore, suppose we would like to have $\alpha = 0.01$ and $\beta = 0.10$; how large should the sample be? What is the critical value?

13–27 Suppose that the President of the United States would introduce legislation for socialized medicine if sample results indicated that 70 percent of the voters desire it, and would not introduce such legislation if sample results indicated that 50 percent of the voters desire it. Furthermore, suppose that he would like to assume an α risk of 0.01 and a β risk of 0.05. How large a sample is required? What is the critical value for this decision rule?

14

CHI-SQUARE DISTRIBUTIONS

14–1 EXACT-SAMPLING THEORY

Up to now, we have been concerned with making inferences from large samples. The concept "large sample" relates, of course, to the sampling distribution of the statistic under consideration. In general, large-sampling theory refers to a class of situations in which the probability distribution of a sample statistic is normal or approximately normal, either because the population under investigation is normal or because the sample size is sufficiently large for the central limit theorem to be operative.

Specifically, as pointed out earlier, in the case of $\bar{x}$ a sample is considered large if $n > 30$; in the case of position measures, such as the median and quartiles, a sample is considered large if $n > 30$ and if the population is approximately normal; in the case of sample proportions, a large sample ranges from 30, when the value of sample proportion p is equal to 0.50, to 1400, when the value of p shifts from 0.50 to 0.05 or to 0.95. It may also be mentioned, in case sample variance or sample standard deviation, the central limit theorem would apply if $n \geq 100$ and if the population is normal or approximately so.

Sometimes, however, we may be called upon to make inferences when the underlying assumptions for large-sampling theory cannot be met. Situations may arise in which either the population standard deviation is unknown because we are confronted with a new problem or a new theory, or the sample size is small owing to physical limitations such as those related to medical research or to such practical limitations as the high costs involved in sample observations.

Although with small samples we do not have as much information as we would like, we are not completely without resources. The proper course is to draw conclusions or to make decisions from the sample by taking into account the scanty nature of the evidence. As a matter of fact, procedures of statistical inference with small samples are the same as those presented in the preceding two chapters, except that now we cannot apply the central limit theorem and assume that sampling distributions are normal. The study of statistical inferences with small samples is called *small-sampling-theory*, or *exact-sampling theory*. As already stated, the main difference between large and small sampling is between sampling distributions: for large samples, sampling distributions are assumed to be normal; for small samples, sampling distributions may take a number of forms. In addition to the binomial distribution (which has been discussed in detail before) there are three probability distributions that a statistic with small n often assumes: the chi-square, Student's t, and F distributions. All three models are related to the normal probability model and are defined in terms of the number of degrees of freedom.

We shall introduce chi-square distributions and their applications in this chapter, t and F distributions in the next. Since all these distributions are defined in terms of "degrees of freedom," we should have, therefore, a clear idea of this concept before we proceed.

14–2 DEGREES OF FREEDOM

The notion of "degrees of freedom" was first mentioned in connection with our definition of $\hat{s}^2$ in Chapter 10, and it was brought up again when we discussed the properties of point estimation in Chapter 12. On neither occasion did we explain the meaning of this concept.

The concept of degrees of freedom is a mathematical one. It is the name given to the number of linearly independent observations occurring in a sum of squares. A rigorous discussion of this definition uses moderately advanced ideas from linear algebra. Rather than devote space to these topics, we shall merely approach the definition heuristically.

Suppose we try to estimate a population mean by a sample of size 1. The single value in the sample would constitute the estimate, and any inference drawn must necessarily be based on this single value. However, this value must be known. Now, if we attempt to compute the sample variance with the same sample, the result will be zero. This is true irrespective of the single value observed—that is, whether the value is known or not. This is the same as saying that one observation provides no information about the variance. In order to know anything about the variance at all, we need at least a sample with $n = 2$. Now, either of the two values alone tells us nothing about the variance; only a second value furnishes any information. Thus, the variance is based on only one item of the two in the sample. Thus in this case we say there is only $2 - 1 = 1$ degree of freedom. Here, in the computation of the variance, one value provides no information and hence one degree of freedom is lost. Generally, one degree of freedom is lost from a sample of size n, hence the number of degrees of freedom is $n - 1$.

Alternatively, we note that when we calculate the sample variance in terms of deviations measured from the sample mean $\bar{x}$ rather than from the unknown population mean μ, we get a sum of n squared deviations, $\sum (x_i - \bar{x})^2$, which must obey the side relation or constraint $\sum x_i = n\bar{x}$. This implies that, if we have the base $\bar{x}$ and any $n - 1$ deviations therefrom, the nth deviation is automatically determined. It is linearly dependent on others and, therefore, is not "free" to vary randomly. Had the deviations been taken from a known μ, the nth deviation would have remained linearly independent of the $n - 1$ other observations. Consequently, a sample of n squared deviations taken from $\bar{x}$ furnishes less useful information than the same number of deviations measured from μ. Also, when we say that a sample has $n - 1$ degrees of freedom, we speak of it as if it consisted of one item less.

When two or more sample variances computed from independent random samples (taken from the same population) are available, we often wish to utilize all information to estimate the population variance by computing a *weighted*, or *pooled*, variance. Consider, for example, the case of two independent samples. An unbiased estimator of the population variance can be obtained as

$$\hat{s}_w^2 = \frac{(n_1 - 1)\hat{s}_1^2 + (n_2 - 1)\hat{s}_2^2}{n_1 + n_2 - 2}$$

for which the number of degrees of freedom is $n_1 + n_2 - 2$. This is because $\hat{s}_1^2$ has n_1 linear comparisons but a given mean, and thus $n_1 - 1$ linearly independent comparisons; similarly for $\hat{s}_2^2$. Therefore, the number of independent linear comparisons for $\hat{s}_w^2$ defined above is $(n_1 - 1) + (n_2 - 1)$.

To generalize: the *number of degrees of freedom*, denoted by δ, may be thought of as the number of elements that can be chosen freely; or as the number of variables that can vary freely; or as the number of independent variables. Given the sample size, the number of degrees of freedom is $\delta = n - k$, where k is the number of constraints for the computations of a statistic $\hat{\theta}$ involving sums of squares, and the constraints may be, for example, the number of estimators required to calculate $\hat{\theta}$ in question.

Let us see how chi-square distributions are defined and how they can be employed for statistical inference.

14-3 PROPERTIES OF CHI-SQUARE DISTRIBUTIONS

If $X_1, X_2, \ldots, X_n$ are independent standard normal variables, then the sum of the squares of these variables is said to have a chi-square, χ^2, distribution with $\delta = n$. That is,

$$\chi_n^2 = X_1^2 + X_2^2 + \cdots + X_n^2. \tag{14–1}$$

There are infinitely many χ^2 distributions, one corresponding to each positive integer n. The χ^2 distribution corresponding to the number δ will be denoted as χ_δ^2.

We shall now state the important properties of chi-square distributions, without proof, for our later use. Some of these properties can be deduced by reflective study of Figure 14–1, which gives the density curve of chi-square distributions for a number of selected δ.

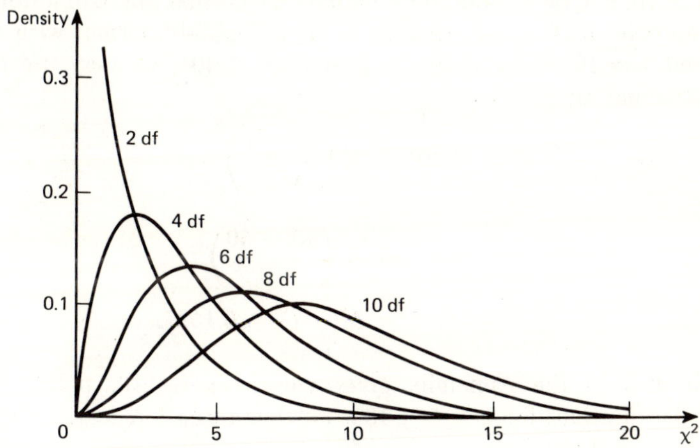

Figure 14–1 Chi-Square Distribution Curves for Selected Degrees of Freedom

(1) If X has a standard normal distribution, then X^2 has the distribution of χ_1^2—that is, chi-square distribution with 1 df.

(2) If X_1 is $\chi_{\delta_1}^2$ and X_2 is $\chi_{\delta_2}^2$ and if X_1 and X_2 are independent, then $X_1 + X_2$ is $\chi_{\delta_1+\delta_2}^2$. This is commonly called the *additive property* of chi-square distributions.

(3) If X is a standard normal variable and $x_1, x_2, \ldots, x_n$ are n observations of a random sample, drawn from X, then

$$\sum_1^n x_i^2 \qquad \text{has the distribution } \chi_n^2.$$

(4) If X is any normal variable and if $x_1, x_2, \ldots, x_n$ are n observations of a random sample, drawn from X, then

$$\sum_1^n \left(\frac{x_i - \mu}{\sigma}\right)^2 \qquad \text{has the distribution } \chi_n^2.$$

Notice that properties (3) and (4) are equivalent statements. In (3) a random sample is drawn from $n(0, 1)$; in (4) a random sample is drawn from $n(\mu, \sigma^2)$, and $(x_i - \mu)/\sigma$ transforms (4) into (3).

(5) The value of χ^2 ranges from 0 to infinity, since it is the sum of squared values.

(6) A χ^2 distribution is completely defined by the number of degrees of freedom. If X is χ_δ^2, its mean and variance are $\mu = \delta$ and $\sigma^2 = 2\delta$, respectively.

(7) Chi-square distributions are positively skewed. As δ gets large, however, χ_δ^2 approaches the normal distribution with $\mu = \delta$ and $\sigma = \sqrt{2\delta}$. In practice, probabilities for chi-square distributions, when $\delta > 30$, can be computed by employing normal approximations. For example, if X is χ_{50}^2, then X is approximately normal with $\mu = 50$ and $\sigma = 10$. If we desire to find $P(X \leq 40)$, we may use the approximation:

$$P(X \leq 40) \doteq N\left(\frac{40 - \delta}{\sqrt{2\delta}}\right)$$

$$= N\left(\frac{40 - 50}{10}\right)$$

$$= N(-1) = 0.1587.$$

Table B-9 in the Appendix gives various percentile points for chi-square distributions for $1 \leq \delta \leq 100$. For instance, if X is χ_{15}^2, then from Table B–9 we find that

$$P(X > 7.26) = P(7.26 < \chi_{15}^2 < \infty) = 0.95,$$

$$P(X > 5.23) = P(5.23 < \chi_{15}^2 < \infty) = 0.99,$$

$$P(X > 25.00) = P(25.00 < \chi_{15}^2 < \infty) = 0.05,$$

and so forth. Thus, this table is in the cumulative form and gives upper-tail probabilities for chi-square distributions.

14–4 STATISTICAL INFERENCE ABOUT POPULATION VARIANCE

The unbiased point estimator of variance with a sample of size n was defined as

$$\hat{s}^2 = \frac{1}{n - 1} \sum (x - \bar{x})^2.$$

If we divide both sides of this expression by the population variance σ^2, then we would have

$$\frac{\hat{s}^2}{\sigma^2} = \frac{1}{n-1} \Sigma \left(\frac{x-\bar{x}}{\sigma}\right)^2,$$

which can be written as

$$\frac{(n-1)\hat{s}^2}{\sigma^2} = \Sigma \left(\frac{x-\bar{x}}{\sigma}\right)^2.$$

The right side of this result, from property (4) of chi-square distributions, would have a chi-square distribution with $n - 1$ df, provided that the population is normally distributed.

Thus, we can formally state: If $\hat{s}^2$ is the variance of a random sample with size n taken from a normal population, the ratio $(n - 1)\hat{s}^2/\sigma^2$ is χ_{n-1}^2. Or, we can write

$$\chi_\delta^2 = \frac{(n-1)\hat{s}^2}{\sigma^2}, \tag{14-2}$$

with $\delta = n - 1$. This ratio is an appropriate testing statistic for hypotheses about population variances.

Suppose that the Transit Authority of New York City plans to purchase light bulbs for its subway system. The Authority wishes to have bulbs that possess not only long life but also a high degree of uniformity. It decides, on the basis of past experience, that the variance should not exceed 250 hours^2. Now, a test of 20 bulbs of a certain make yields a mean life of 1000 hours, which is considered to be satisfactory, but with a variance of 300 hours^2. Does this result indicate that the population variance exceeds 250?

Following the general procedure of testing developed in Chapter 13, we have now the following test:

(1) Hypotheses: H_0: $\sigma^2 \le 250$; H_1: $\sigma^2 > 250$.

(2) Level of Significance: Arbitrarily set at 0.05.

(3) Testing Statistic: $\chi_\delta^2 = (n - 1)\hat{s}^2/\sigma^2$, which has a chi-square distribution with $\delta = n - 1 = 19$.

(4) Decision Rule: Since $P(\chi_{19}^2 > 30.14) = 0.05$, H_0 will be rejected if and only if the observed chi-square value is greater or equal to 30.14, and H_0 will be accepted if and only if the observed chi-square value is less than 30.14.

(5) Computations:

$$\chi_{19}^2 = \frac{(20-1)300}{250} = 22.80.$$

(6) DECISIONS: Since the observed χ^2 falls in the acceptance region, H_0 cannot be rejected. This sample result, in other words, is not sufficient evidence for us to conclude that $\sigma^2 > 250$. The Authority may, therefore, consider this make of light bulb satisfactory and decide on its purchase.

Tests on variances are typically of the upper-tail variety, since we are usually concerned with the fact that the variance may be too large. When we wish to conduct a two-sided test with chi-square distributions, we divide the region of rejection into two equal parts. For instance, if $\delta = 19$ and $\alpha = 0.05$ as in the previous test but the alternative is stated as $\sigma^2 \neq 250$, we would reject H_0 if $\chi^2 < 8.91$ or if $\chi^2 > 32.85$. In other words, we find the χ^2 value that has 97.5 percent of the area to the right of it (8.91) and the χ^2 value that has 2.5 percent of the area to the right of it (32.85). Note that $P(8.91 < \chi_{19}{}^2 < 32.85) = 0.95$.

To have an interval estimate of σ^2, as before, we try to construct a confidence interval such that it will contain σ^2 at a specified confidence probability, $1 - \alpha$. Such an interval may be made by using the information that if equation (14–2) is solved for σ^2, we would have

$$\sigma^2 = \frac{(n-1)\hat{s}^2}{\chi_\delta{}^2}.$$

Now, inserting the values of $(n - 1)$, $\hat{s}^2$, and χ^2, we can estimate the lower and upper confidence limits for σ^2. Specifically, the confidence interval for σ^2 is

$$P\left(\frac{(n-1)\hat{s}^2}{\chi^2_{n-1,\alpha/2}} < \sigma^2 < \frac{(n-1)\hat{s}^2}{\chi^2_{n-1,1-\alpha/2}}\right) = 1 - \alpha. \qquad (14\text{–}3)$$

That is, for σ^2, the lower (L) and upper (U) confidence limits are

$$L = \frac{(n-1)\hat{s}^2}{\chi^2_{n-1,\alpha/2}} \quad \text{and} \quad U = \frac{(n-1)\hat{s}^2}{\chi^2_{n-1,1-\alpha/2}}.$$

These chi-square values depend upon the number of degrees of freedom and the confidence level. These are indicated by the subscripts to χ^2 in the foregoing expressions. Note once more that the χ^2 table gives the probability that a given χ^2 value is exceeded. Thus, if we desire to have a 95 percent confidence interval, we should choose the value of $\chi_{\delta,0.025}{}^2$ for the lower confidence limit and the value of $\chi^2_{\delta,0.975}$ for the upper confidence limit. Similarly, if $1 - \alpha = 0.99$, then values of $\chi_{\delta,0.005}{}^2$ and $\chi_{\delta,0.995}{}^2$ should be used for the lower and upper confidence limits, respectively. It is interesting to observe that the larger chi-square value is used for L and the smaller chi-square value is used for U in constructing a confidence interval for σ^2 because, as can be seen from equation (14–2), χ^2 and σ^2 are inversely related.

As an example, let us now construct a 95 percent confidence interval for σ^2, using the illustrative material from the preceding test. Recalling that, in this instance, $n = 20$ and $\hat{s}^2 = 300$, we have, therefore,

$$L = \frac{(20-1)(300)}{\chi^2_{19,0.025}} = \frac{(19)(300)}{32.85} = 173.5$$

and

$$U = \frac{(20-1)(300)}{\chi^2_{19,0.975}} = \frac{(19)(300)}{8.91} = 639.7,$$

or

$$173.5 < \sigma^2 < 639.7.$$

The confidence interval for the standard deviation can be derived directly from that for the variance. This is done by taking the square roots of the limits for σ^2. Thus, the corresponding 95 percent confidence interval for our illustrative problem is

$$\sqrt{173.5} < \sigma < \sqrt{639.7}; \qquad \text{that is, } 13.2 < \sigma < 25.3.$$

Sometimes, especially in the case of estimating σ^2, we are concerned only with an upper confidence limit, not with the lower. In such a situation, we state that the variance is equal to or less than some specific value. Namely, we are interested in constructing a confidence interval of the form $0 < \sigma^2 < U$ with a confidence probability of $1 - \alpha$. Here, U can be found by the expression

$$U = \frac{(n-1)\hat{s}^2}{\chi^2_{n-1,1-\alpha}}, \tag{14-4}$$

where the $\chi_\delta{}^2$ is the value that has $1 - \alpha$ of the area to the right of it. Here, $0 < \sigma^2 < U$ is called a *one-sided confidence interval*, with $1 - \alpha$, and corresponds to an upper-tail test at α.

The 95 percent one-sided confidence interval for the light-bulb illustration has

$$U = \frac{(20-1)(300)}{\chi^2_{19,0.95}} = \frac{(19)(300)}{10.12} = 563.$$

With this value of U, we see that $0 < \sigma^2 < 563$, which contains the hypothetical value of 250 in the null hypothesis for the previous test. This explains once again why H_0 was not rejected.

A corresponding confidence interval of the form $L < \sigma^2 < \infty$, which is simply called the other one-sided confidence interval, is of little practical significance.

In conclusion, we point out that chi-square is an amazingly versatile statistic. In addition to its use for inference with variances, it can also be used to test any hypothesis concerning categorical data, of which the binomial form is but a special case. With categorical data, observations consist mainly of frequencies occurring in the various classes or compartments of a table of simple or multiple classifications. Owing to the differences in classification and in reasons for testing, chi-square tests can be identified as those most useful to determine frequencies of one-way classification within two or more categories, the goodness-of-fit of a specific probability or density function to sample data grouped into several classes, the independence of cross classifications (or contingency tables), and homogeneity. These chi-square tests, constituting a part of what is called nonparametric statistics, are introduced in the remainder of this chapter. Further comments on the scope and nature of nonparametric statistics are given in Chapter 20.

14–5 THE CHI-SQUARE TEST ON FREQUENCIES

Generally, with a chi-square test, we first formulate the null hypothesis under which expected, or theoretical, frequencies (that is, of a theoretical probability distribution) are determined. Next, we investigate sample data for the purpose of establishing observed frequencies. Then we compare these two sets of frequencies by taking differences between them. Finally, on the basis of these differences, we may specify a decision criterion to judge whether the observed frequencies, on the average, diverge significantly from the expected frequencies. We wish to determine, from an appropriate criterion, whether these differences are due to chance variations in random sampling. If so, the null hypothesis is accepted; otherwise, it is rejected.

For this group of problems, it can be shown that the sum of the ratio of the squared differences between observed frequencies, denoted as o_i, and expected frequencies, denoted as e_i, to the expected frequencies is distributed approximately as a chi-square variable. That is,

$$\chi_\delta^2 = \sum \frac{(o_i - e_i)^2}{e_i} \tag{14–5}$$

is approximately distributed as chi-square, with degrees of freedom equal to the number of classes or categories compared, less the number of restrictions imposed on the comparison. More precisely, for (14–5) $\delta = k - 1$ if we can compute expected frequencies without having to estimate population parameters from sample statistics. Here, k is the number of classes; thus if we know $k - 1$ of the expected frequencies, the remaining frequency is uniquely determined because we must satisfy the condition that $\sum o_i = \sum e_i = n$, where n is the sample size or the total frequency. In this case, however, we have $\delta = k - 1 - m$ if we can compute the expected frequencies only by estimating m population parameters by sample statistics.

Two observations may be made here in connection with the use of (14–5) for various types of chi-square tests. First, whenever $\delta = 1$, especially when the total number of frequencies is rather small, say, less than 50, it is advisable to introduce a continuity correction factor of ½ in computing the chi-square value. That is, when $\delta = 1$, we modify equation (14–5) as follows:

$$\chi_\delta^2 = \sum \frac{(|o_i - e_i| - \frac{1}{2})^2}{e_i} \tag{14–6}$$

Second, there must be at least five items in each theoretical frequency class. The purpose here is to avoid inflating chi-square values by dividing the squared differences by a small size of expected frequency. In a situation where one or more classes have theoretical frequencies less than 5, we may combine them into a single category before calculating the difference, $o_i - e_i$. In doing so, it is important to remember that the number of degrees of freedom is determined with the number of classes after the regrouping. For example, if we start out with ten classes and three of these have small theoretical frequencies, we may pool these three classes into one and we then have $k = 8$ classes to compare.

We shall now consider the application of chi-square to test hypotheses about frequencies. Two examples will be given: one is concerned with one-way classification with two classes; the other, with one-way classification with more than two classes.

A Test on Frequencies of Two Classes

In Chapter 13 the Better Business Bureau had to make a decision about the claim by the The Smile Diaper Service that it shared 30 percent of the diaper market in Philadelghia. The sample data showed that among the 200 families surveyed, 52 were customers of Smile. The BBB set out to verify whether 3 out of every 10 users of diaper service in Philadelphia used Smile, and made the test under the assumption that the sample proportion was normally distributed. To verify the same claim by a chi-square test, we have

(1) Hypotheses: H_0: Customers are distributed in the ratio 3:7; H_1: The ratio is not 3:7.

(2) Level of Significance: $\alpha = 0.05$.

(3) Testing Statistic:

$$\chi_\delta^2 = \sum \frac{(|o_i - e_i| - \frac{1}{2})^2}{e_i},$$

which is distributed as χ_1^2. (*Note*: For this test, $k = 2$; therefore, $\delta = 2 - 1 = 1$. Because $\delta = 1$, the continuity correction factor is employed.)

(4) Decision Rule: $P(\chi_1^2 > 3.84) = 0.05$; reject H_0 if and only if the computed chi-square value is greater than 3.84. Otherwise, accept H_0. (*Note*: Any departure from the ratio 3:7 will result in an inflation of

chi-square. Thus the entire region of rejection must be at the upper end of the chi-square distribution.)

(5) COMPUTATIONS:

Smile's customers	o_i	e_i	$\lvert o_i - e_i \rvert - \frac{1}{2}$	$(\lvert o_i - e_i \rvert - \frac{1}{2})^2/e_i$
Yes	52	60	7.5	0.94
No	148	140	7.5	0.40
Total	200	200	—	1.34

(6) DECISION: The computed chi-square value is 1.34, which is less than 3.84. The null hypothesis cannot be rejected at $\alpha = 0.05$. The sample data are not sufficient evidence for doubting Smile's claim. It is interesting to note that this conclusion is identical with the one reached for the same problem when we used the normal probability law as the decision criterion. This should not be a surprising result, since we know that a chi-square statistic is the square of a normal statistic. The two critical regions correspond to one another, since for $\alpha = 0.05$ and $\delta = 1$,

$$P(Z < -1.96) + P(Z > 1.96) = P(\chi_1^2 > 3.84) = 0.05.$$

It may be observed also that for the present example, the hypotheses can be stated in the binomial notation or multinomial notation. In binomial notation, we have $H_0: p = 0.3$ against $H_0: p \neq 0.3$. In multinomial notation, we have H_0: $p_1 = 0.3$ and $p_2 = 0.7$ against H_1: $p_1 \neq 0.3$ and $p_2 \neq 0.7$.

A Test on Frequencies of More Than Two Classes

While the chi-square test can be considered as equivalent to the normal test when there are only two classes, as revealed by the preceding example, it is much more convenient to use the chi-square test when there are more than two categories, as the following example shows.

Five coins are tossed 100 times for the purpose of determining whether all five are fair. For convenience, we conduct this test by observing the number of samples that contain 0, 1, 2, 3, 4, or 5 heads. The appropriate chi-square test with six categories is as follows:

(1) HYPOTHESES: H_0: All colns are fair; that is, the probability of obtaining heads in any random toss of any of the five coins is $\frac{1}{2}$; H_1: Not all coins are fair.

(2) LEVEL OF SIGNIFICANCE: $\alpha = 0.01$—arbitrarily assigned.

(3) TESTING STATISTIC:

$$\chi_\delta^2 = \sum \frac{(o_i - e_i)^2}{e_i},$$

which is approximately distributed as χ_3^2. Here $\delta = 4 - 1 = 3$, as shown by the table in step (5), since the first class is combined with the second and the fifth class is combined with the last, owing to small theoretical frequencies in the first and the last classes. Actually we have, after the combinations, $k = 4$.

(4) DECISION RULE: The critical region is separated from the acceptance region by 11.34 because $P(\chi_3^2 > 11.34) = 0.01$.

(5) COMPUTATIONS: First, the five coins are tossed 100 times and this total number of tosses is distributed according to the number of heads, as recorded in the first two columns in the accompanying table; the second column, of course, contains the observed frequencies. Next, if all the five coins are fair, then in any toss of the five coins the number of heads should be distributed as a binomial distribution. Such a theoretical distribution is given in column three. With this theoretical distribution, the expected frequencies are automatically determined, as shown in the fourth column. In the fifth column we have the computation for the chi-square value under consideration, and it is 26.88.

No. of heads	o_i	$b(x; 5, 0.5)$	$e_i = 100[b(x)]$	$(o_i - e_i)^2/e_i$
0	1 } 11	0.0312	3.12 } 18.74	3.20
1	10	0.1562	15.62	
2	20	0.3125	31.25	4.05
3	36	0.3125	31.25	0.72
4	23 } 33	0.1562	15.62 } 18.74	10.85
5	10	0.0312	3.12	
Total	100	1.0000*	100.00*	18.82

* These totals are not exactly 1 and 100, respectively, because of rounding.

(6) DECISIONS: The observed value of χ_3^2 is greater than 11.34 and falls in the critical region. We therefore reject H_0 that all the five coins are fair. This implies that at least some of the coins are biased.

14-6 TESTING GOODNESS-OF-FIT

In many problem situations prevailing conditions make it seem reasonable that the population follows a specific probability model; nonetheless, it is beneficial to have a statistical testing procedure to substantiate the adequacy of such a model. Statistical inference methods discussed in the preceding two chapters were designed under the assumption that the population is normal or that samples are large enough for the central limit theorem to be operative. The assumption of normality is not a crucial one insofar as inferences about means are concerned, but it can seriously affect the drawing of conclusions about variances. While experience may tell us that a given random variable is normally or nearly normally distributed, we may still doubt the normality assumption and seek an appropriate procedure to prove or disprove it. The statistical procedure employed for this purpose is called a *test of goodness-of-fit.*

It begins, as does our general testing procedure, by formulating the null hypothesis that a given population has a specific probability or density function, such as the uniform, the binomial, the Poisson, or the normal model. Then a random sample is taken from the population and observed, which furnishes us with observed frequencies (empirical distribution) in terms of chi-square tests. Then the theoretical distribution, as specified in the null hypothesis, is fitted to the empirical distribution. The theoretical probability values, when multiplied by the total number of frequencies in the sample, become expected frequencies. After this, the chi-square test of goodness-of-fit follows exactly the same procedure as that on frequencies, except that theoretical frequencies are now usually computed with sample statistics as estimators for the corresponding population parameters.

Testing the Fit of a Uniform Distribution

The simplest, and a useful, test of goodness-of-fit is that with a discrete uniform distribution, such as the toss of a coin or the throw of a die. Let us consider an example.

It is known that the number of sales made by a salesman usually depends upon the number of calls he makes. Thus, salesmen of any firm are always encouraged to make as many calls as possible. If salesmen are equally ambitious, energetic, resourceful, and so on, all should be expected to make the same number of calls during the same period of time. The sales manager of a certain company is anxious to test the validity of this hypothesis. He selects at random six salesmen's records for a given week and finds that a total of 210 calls were distributed as follows:

Salesman	*Number of calls*
A	35
B	20
C	47
D	32
E	51
F	25
Total	210

The test:

(1) HYPOTHESES: H_0: The number of calls is uniformly distributed; that is, $f(x; n) = 1/n$, $x = 1, 2, \ldots, n$; H_1: The number of calls is not uniformly distributed.

(2) LEVEL OF SIGNIFICANCE: $\alpha = 0.01$.

(3) TESTING STATISTIC:

$$\chi_\delta^2 = \sum \frac{(o_i - e_i)^2}{e_i},$$

which is distributed approximately as χ_5^2. Here, $\delta = 6 - 1$, since we can compute expected frequencies without estimating parameters by sample statistics.

(4) DECISION RULE: H_0 shall be rejected if and only if the computed chi-square value is greater than 15.09, since $P(\chi_5^2 > 15.09) = 0.01$.

(5) COMPUTATIONS:

Salesman	o_i	$f(x; 6)$	$e_i = 210[f(x)]$	$(o_i - e_i)^2/e_i$
A	35	1/6	35	0.00
B	20	1/6	35	6.43
C	47	1/6	35	4.11
D	32	1/6	35	0.26
E	51	1/6	35	7.31
F	25	1/6	35	2.86
Total	210	6/6	210	20.97

(6) DECISION: Reject H_0, since the computed chi-square value falls in the rejection region. This conclusion means that the fit of a uniform distribution to the calls made by salesmen is bad. This inference also implies that salesmen may be different in ambition, energy, resourcefulness, and so on.

Testing the Fit of a Binomial Distribution

Let us try to fit a binomial distribution to the data of tossing five coins 100 times discussed in the previous section. To judge whether such a fit is good or bad, we have the following chi-square test:

(1) HYPOTHESES: H_0: The number of heads is distributed as a binomial distribution, $b(x; n, p)$; H_1: The number of heads is not distributed as a binomial.

(2) Level of Significance: $\alpha = 0.01$.

(3) Testing Statistic:

$$\chi_\delta^2 = \sum \frac{(o_i - e_i)^2}{e_i},$$

which is seen to be distributed as χ_3^2. Here, $\delta = k - 1 - m = 5 - 1 - 1 = 3$. The number of classes is five, since the first two of the six original classes are combined into a single category. Furthermore, $m = 1$, because we need the sample estimate of p, the probability of heads in any single toss of each of the five coins, to compute expected frequencies.

(4) Decision Rule: H_0 will be rejected if and only if the computed chi-square value is greater than 11.34, since $P(\chi_3^2 > 11.34) = 0.01$.

(5) Computations: For this problem the mean number of heads is $np = 5p$. For the actual or observed frequency distribution, the mean number of heads is

$$5\hat{p} = \frac{\sum fx}{\sum f} = \frac{0(1) + 1(10) + 2(20) + 3(36) + 4(23) + 5(10)}{100} = 3.$$

Or, $\hat{p} = 0.6$, which may be considered as the best estimate of the binomial parameter under investigation. With this estimate, the fitted binomial distribution is given as

$$b(x; 5, 0.6) = \binom{5}{x}(0.6)^x(0.4)^{5-x}.$$

With this probability distribution, the expected frequencies are then determined and the chi-square value is computed.

Number of heads	o_i	$b(x; 5, 0.6)$	$e_i = 100[b(x)]$	$(o_i - e_i)^2/e_i$
0	1 } 11	0.0102	1.02 } 8.70	0.61
1	10	0.0768	7.68	
2	20	0.2304	23.04	0.40
3	36	0.3456	34.56	0.06
4	23	0.2592	25.92	0.33
5	10	0.0778	7.78	0.63
Total	100	—	—	2.03

(6) Decision: Since the computed chi-square value is less than 11.34, H_0 is accepted. This conclusion means that the fit of the data is good. (*Note*: The fit here is made with p estimated as 0.6. This conclusion is, therefore, consistent with our previous decision that not all of the five coins are fair in view of the same sample data.)

Testing the Fit of a Poisson Distribution

The number of clients who visited the office of a young lawyer each day during his first 100 days of practice is distributed as follows:

Number of clients	0	1	2	3	4	5	6
Number of days	40	34	16	7	2	1	0

Determine the goodness-of-fit of these data by a Poisson distribution.

First, to compute the expected frequencies we must have the Poisson probabilities, which in turn require the value of the mean number of visitors per day. For the data above, we have

$$\hat{\mu} = \frac{T}{N} = \frac{0(40) + 1(34) + 2(16) + 3(7) + 4(2) + 5(1) + 6(0)}{100}$$

$$= 1 \text{ client per day.}$$

With this estimate, the Poisson distribution to be fitted to our sample data becomes

$$p(x; 1) = e^{-1}\frac{1^x}{x!}.$$

Next, we have the necessary computations:

Number of clients	o_i	$p(x; 1)$	$e_i = 100[p(x)]$	$(o_i - e_i)^2/e_i$
0	40	0.36788	36.788	0.28
1	34	0.36788	36.788	0.21
2	16	0.18394	18.394	0.31
3 or more	10	0.08030	8.030	0.48
Total	100	—	—	1.28

For this problem, $k = 4$ and $m = 1$; therefore, $\delta = 2$. We know that $P(\chi_2^2 > 5.99) = 0.05$ and $P(\chi_2^2 > 9.21) = 0.01$, so the fit here is good at $\alpha = 0.05$ as well as at $\alpha = 0.01$.

Testing the Fit of a Normal Distribution

Measurements of resistance for 360 units of a certain type of electrical product result in the frequency distribution shown in the first two columns of the accompanying table. For this sample, we find that $\bar{x} = 3.65$ ohms and $\hat{s} = 0.1334$ ohms. Suppose we desire to test the hypothesis that the population from which the sample is drawn is normal; then we may proceed by fitting a normal distribution to the sample data by using $\bar{x}$ and $\hat{s}$ as estimators of the corresponding population parameters. Using the normal probability table, we can easily compute expected frequencies under this null hypothesis. For example, let the variable under consideration be X; then

$$\begin{aligned} P(3.4 < X < 3.4) &= N\left(\frac{3.4 - 3.65}{0.1334}\right) - N\left(\frac{3.3 - 3.65}{0.1334}\right) \\ &= N(-1.87) - N(-2.62) \\ &= 0.0307 - 0.0044 \\ &= 0.0263, \end{aligned}$$

and $e_i = 360(0.0263) = 9.468$, and so forth.

Resistance (*ohms*)	o_i	$n(3.65; 0.1334)$	e_i	$(o_i - e_i)^2/e_i$
3.3–3.4	10	0.0263	9.468	0.03
3.4–3.5	37	0.1028	37.008	0.00
3.5–3.6	73	0.2222	79.992	0.61
3.6–3.7	114	0.2886	103.896	0.98
3.7–3.8	81	0.2222	79.992	0.01
3.8–3.9	31	0.1028	37.008	0.98
3.9–4.0	14	0.0263	9.468	2.17
Total	360	—	—	4.78

Thus the computed chi-square value is 4.78, with $\delta = k - 1 - m = 7 - 1 - 2 = 4$. This fit is good at both the 5 percent and 1 percent levels of significance, since $P(\chi_4^2 > 9.49) = 0.05$ and $P(\chi_4^2 > 13.28) = 0.01$. Note that the result of applying the chi-square test to frequencies with class intervals is dependent on the choice of intervals.

14–7 TESTS OF INDEPENDENCE: CONTINGENCY-TABLE TESTS

Tests on frequencies and on goodness-of-fit are both concerned with multinomial populations. In both cases, populations as well as samples are classified in accordance with a single attribute. We shall now see that the same technique of chi-square tests can be applied to data other than multinomial distributions.

When the population and the sample are classified according to two or more attributes, we may use *tests of independence* to determine whether the principles or criteria employed for cross classification are meaningful or effective. For instance, a random sample of n retail stores may be classified by size of capitalization and by type of ownership. The proportion of each of the two classes in the population is unknown. Our interest would be to establish whether there is any dependency relationship between a store's capitalization and its type of ownership. Clearly, in a case such as this, we would want to test the hypothesis that the size of capitalization is independent of the type of ownership against the hypothesis that they are related. In this test of independence, we try to verify the significance of a capitalization difference in type of ownership. If the difference is not significant, we conclude that the type of ownership is independent of the size of capitalization. Otherwise, we say that these two criteria of classification are related or dependent.

Tests of independence are also called *contingency-table tests*. So far, we have been concerned with only *one-way classification tables* since, in each case, the observed frequencies have occupied a single row or a single column. Also because the observed frequencies are distributed in k classes (columns or rows), one-way classification tables are called $1 \times k$ (read 1 *by* k) or $k \times 1$ *tables*. Extending these ideas, we can arrive at 2×2, or *two-way classification tables* and, generally, $R \times C$ tables in which the observed frequencies occupy R rows and C columns. Such tables are often called *contingency tables*. Corresponding to each observed frequency in an $R \times C$ table, there is an expected frequency computed under the specified null hypothesis. Frequencies, observed or expected, that occupy the cells of a contingency table are called *cell frequencies*. The total frequency in each row or each column is called the *marginal frequency*.

To evaluate differences between observed and expected frequencies contained in contingency tables, we employ the same statistic as that for tests discussed in the previous section,

$$\chi_\delta^2 = \sum \frac{(o_i - e_i)^2}{e_i}.$$

The sum is taken over all cells in a contingency table; and, in general,

$$\delta = (R - 1)(C - 1).$$

In testing independence with a 2×2 contingency table, we would have $\delta = (2 - 1)(2 - 1) = 1$. As recommended before, the continuity correction factor of ½ should be used to compute the chi-square value. For large samples, corrected and uncorrected chi-square values may be practically the same, and so the continuity correction factor may be ignored. Let us consider an example.

There is reason to believe that high-income families usually send their children to private colleges and low-income families often send their children to city or state colleges. To verify this, 1600 families are selected at random from California and the following results are obtained:

Observed Data

	College		
Income	*Private*	*Public*	*Total*
Low	506	494	1000
High	438	162	600
Total	944	656	1600

Here we have two classifications: income, and type of college. At a glance, we see that, relatively, a greater number of high-income families send their children to private colleges than do low-income families. But is such a difference in proportions significant or is it merely the result of chance variations in random sampling? The chi-square test of independence can be used to answer questions such as this by comparing the set of observed frequencies with that of expected frequencies.

Under the assumption that family income and type of college are independent, we would expect the proportion of all families that send their children to private colleges to be $944/1600$. The expected number of low-income families that send their children to private colleges is then

$$(944/1600)(1000) = 590.$$

Now, since the expected frequencies must agree with the marginal frequencies of the observed frequencies, and since there are four cells in a 2×2 contingency table, we can automatically fill in the remaining three expected numbers once any one of the four expected numbers is determined. For this reason only 1 degree of freedom is associated with a 2×2 contingency table. The expected frequencies for our example are as follows:

Expected Data

	College		
Income	*Private*	*Public*	*Total*
Low	590	410	1000
High	354	246	600
Total	944	656	1600

Following our general testing procedure, the chi-square test of independence for data on hand becomes

(1) HYPOTHESES: H_0: Income and type of college are independent; H_1: They are related or dependent.

(2) LEVEL OF SIGNIFICANCE: $\alpha = 0.01$.

(3) TESTING STATISTIC:

$$\chi_\delta^2 = \sum \frac{(o_i - e_i)^2}{e_i},$$

which is approximately distributed as χ_1^2.

(4) DECISION RULE: The critical region of this test is $6.63 < \chi_1^2 < \infty$. Thus, accept H_0 if and only if the computed chi-square value is less than 6.63.

(5) COMPUTATIONS: From observed and expected data given before, we have

$$\begin{aligned}\chi_1^2 &= \sum \frac{(o_i - e_i)^2}{e_i} \\ &= \frac{(506 - 590)^2}{590} + \frac{(494 - 410)^2}{410} + \frac{(438 - 354)^2}{354} + \frac{(162 - 246)^2}{246} \\ &= 77.78.\end{aligned}$$

(*Note*: The computation is made here without the continuity correction, because the sample is large.)

(6) DECISIONS: Reject H_0. Family income and type of college are not independent. A greater proportion of high-income families send their children to private colleges.

As a second example, let us consider this: Are men different from women with respect to their concept of a successful life defined in terms of material well-being, M, spiritual well-being, S, and fame, F? Suppose that a random sample of 430 people has been selected across the United States and the following results have been obtained:

Observed Data

	Successful life			
Sex	*M*	*S*	*F*	*Total*
Men	80	95	60	235
Women	60	70	65	195
Total	140	165	125	430

In this case, we are interested in testing the hypothesis that a person's concept of a successful life is independent of his sex against the hypothesis that the two are related. The null hypothesis here says that men and women are of the same opinion about the meaning of a successful life. Under this assumption, we would expect, for example, $140/430$ of the 430 people interviewed to define a successful life as material well-being. The expected numbers of men and of women who consider this a criterion for a successful life, respectively, are

$$(140/430)(235) = 76.51,$$
$$(140/430)(195) = 63.49.$$

Expected frequencies for other cells can be obtained in the same manner; they are recorded below:

Expected Data

	Successful life			
Sex	*M*	*S*	*F*	*Total*
Men	76.51	90.17	68.32	235
Women	63.49	74.83	56.68	195
Total	140.00	165.00	125.00	430

There are six cells for a 2×3 contingency table. The requirement that cell frequencies must agree with marginal frequencies, however, gives us but 2 degrees of freedom. We are free to place one number in one of the three columns and to place another number in any one of the four cells of the remaining two columns. After these two numbers are given, all numbers in the empty cells are uniquely determined from the indicated marginal frequencies. This conclusion is in agreement with the fact that $\delta = (R-1)(C-1) = (2-1)(3-1) = 2$.

For data given before, we find that

$$\chi_2^2 = \frac{(80-76.51)^2}{76.51} + \frac{(95-90.17)^2}{90.17} + \frac{(60-68.32)^2}{68.32} + \frac{(60-63.49)^2}{63.49} + \frac{(70-74.83)^2}{74.83} + \frac{(65-56.68)^2}{56.68}$$

$$= 3.16.$$

Since $P(\chi_2^2 > 5.99) = 0.05$ and $P(\chi_2^2 > 9.21) = 0.01$, we accept the null hypothesis that the concept of a successful life is independent of a person's sex at both $\alpha = 0.05$ and 0.01. Men and women seem to interpret a successful life in essentially the same way with respect to the three principles.

14-8 EFFICIENT FORMULAS FOR COMPUTING χ^2

Efficient formulas are available to compute chi-square values from 2 × 2 and 2 × 3 continuency tables that involve only observed frequencies.

Denote cell frequencies as A, B, C, and D, marginal frequencies as m_1, m_2, m_3, and m_4, and total number of observations as n, for a 2 × 2 contingency table as given below:

Observed Data

A	B	m_3
C	D	m_4
m_1	m_2	n

then it can be shown that

$$\chi_1^2 = \frac{n(\text{AD} - \text{BC})^2}{m_1 m_2 m_3 m_4}. \tag{14-7}$$

For instance, applying this formula to our first example for test of independence, we have

$$\chi_1^2 = \frac{1600((506)(162) - (494)(438))^2}{(944)(656)(1000)(600)} = 77.78,$$

as obtained before.

When the sample is small, say less than 50, and when some or all cell frequencies are less than 5, it is desirable to compute χ_1^2 with a continuity correction factor. In this case, the efficient formula becomes

$$\chi_1^2 = \frac{n(|\,\text{AD} - \text{BC}\,| - \tfrac{1}{2}n)^2}{m_1 m_2 m_3 m_4}. \tag{14-8}$$

For a 2 × 3 contingency table, given the following notations:

Observed Data

A	B	C	m_4
D	E	F	m_5
m_1	m_2	m_3	n

it can be shown that

$$\chi_2^2 = \frac{n}{m_4}\left(\frac{A^2}{m_1} + \frac{B^2}{m_2} + \frac{C^2}{m_3}\right) + \frac{n}{m_5}\left(\frac{D^2}{m_1} + \frac{E^2}{m_2} + \frac{F^2}{m_3}\right) - n. \tag{14-9}$$

Applying this expression to data in our second example of test on independence, we have

$$\chi_2^2 = \frac{430}{235}\left(\frac{80^2}{140} + \frac{95^2}{165} + \frac{60^2}{125}\right) + \frac{430}{195}\left(\frac{60^2}{140} + \frac{70^2}{165} + \frac{65^2}{125}\right) - 430$$
$$= 3.16,$$

as before.

In general, we note that

$$\chi_\delta^2 = \sum \frac{(o_i - e_i)^2}{e_i}$$
$$= \sum \frac{(o_i^2 - 2o_ie_i + e_i^2)}{e_i}$$
$$= \sum \frac{o_i^2}{e_i} - 2\sum \frac{o_ie_i}{e_i} + \sum \frac{e_i^2}{e_i}$$
$$= \sum \frac{o_i^2}{e_i} - n. \tag{14-10}$$

This formula is valid for contingency tables of any dimension. Its use, however, requires expected frequencies. Hencc it is recommended for only 3 × 3 or greater dimension tables. To see the generality of this expression, we may apply it to the data in our 2 × 3 contingency table test and find that

$$\chi_2^2 = \sum \frac{o_i^2}{e_i} - n$$
$$= \frac{80^2}{76.51} + \frac{95^2}{90.17} + \frac{60^2}{68.32} + \frac{60^2}{63.49} + \frac{70^2}{74.83} + \frac{65^2}{56.68} - 430$$
$$= 3.16,$$

as before.

14–9 TESTS OF HOMOGENEITY

Tests of homogeneity are designed to determine whether two or more independent random samples are drawn from the same population or from different populations.

The chi-square test of homogeneity is an extension of the chi-square test of independence. In both cases we are concerned with cross-classified data. As we shall immediately see, too, the same testing statistic used for tests of independence is used for tests of homogeneity. These two types of test are, however, different in a number of ways. First, they are associated with different kinds of problems. Tests of independence are concerned with the problem of whether one attribute is independent of another, while tests of homogeneity are concerned with whether different samples come from the same population. Second, the former involve a single sample taken from one population; but the latter involve two or more independent samples, one from each of the possible populations in question. This second point also implies that, in the case of independence, all marginal frequencies are chance quantities and that, in the case of homogeneity, row totals are sample sizes that are chosen numbers.

To illustrate this type of test, let us suppose that three samples are taken: one consists of 115 professional people, one of 110 businessmen, and one of 125 farmers. Each individual chosen is asked to select, say, one of the three categories that best represents his feeling toward a certain national policy. Suppose these three categories are (1) in favor of the policy, F; (2) against the policy, A; (3) indifferent toward the policy, I. Assume that the results of the interviews are distributed as follows:

Observed Data

	Reaction			
Occupation	*F*	*A*	*I*	*Total*
Professionals	80	21	14	115
Businessmen	72	15	23	110
Farmers	69	31	25	125
Total	221	67	62	350

From the way the problem is posed, an appropriate null hypothesis to be tested seems to be: The three samples come from the same population; that is, the three classifications are homogeneous insofar as the opinion of three different groups of people about the national policy under consideration is concerned. This also means there exists no difference in opinion among the three classes of people on the issue. From the alternative expression of the null hypothesis for this problem, we can see why it is called a test of homogeneity. (When we say things are homogeneous we mean they have something in common or they are the same or they are equal.)

We note that if the null hypothesis stated before is true, then the best estimates for proportions specifying "in favor of the policy," "against the policy," and "indifferent toward the policy," respectively, should be $^{221}/_{350}$, $^{67}/_{350}$, and $^{62}/_{350}$. Thus, of the 115 professional people, expected frequencies for the three categories become

$$(^{221}/_{350})(115) = 72.61 \text{ in favor,}$$

$$(^{67}/_{350})(115) = 22.01 \text{ against,}$$

and

$$(^{62}/_{350})(115) = 20.37 \text{ indifferent.}$$

Expected frequencies for the other two groups of people can be computed in the same way. Expected data for the whole problem are as follows:

Expected Data

	Reaction			
Occupation	*F*	*A*	*I*	*Total*
Professionals	72.61	22.01	20.37	115
Businessmen	69.46	21.06	19.49	110
Farmers	78.93	23.93	22.14	125
Total	221.00	67.00	62.00	350

Again following our general procedure of testing, we have this test of homogeneity for the problem:

(1) HYPOTHESES: H_0: The three samples are drawn from the same population; H_1: They are drawn from different populations.

(2) LEVEL OF SIGNIFICANCE: $\alpha = 0.05$, arbitrarily selected.

(3) TESTING STATISTIC:

$$\chi_\delta^2 = \sum \frac{(o_i - e_i)^2}{e_i},$$

which is distributed approximately as χ_4^2. Here, $\delta = (3 - 1)(3 - 1) = 4$.

(4) DECISION RULE: The rejection region is defined for this test as $9.49 \leq \chi_4^2 < \infty$, since $P(\chi_4^2 > 9.49) = 0.05$.

(5) COMPUTATIONS: From observed and expected data given before,

we have

$$\chi_4^2 = \frac{(80 - 72.61)^2}{72.61} + \frac{(21 - 22.01)^2}{22.01} + \frac{(14 - 20.37)^2}{20.37}$$

$$+ \frac{(72 - 69.46)^2}{69.46} + \frac{(15 - 21.06)^2}{21.06}$$

$$+ \frac{(23 - 19.49)^2}{19.49} + \frac{(69 - 78.93)^2}{78.93}$$

$$+ \frac{(31 - 23.93)^2}{23.93} + \frac{(25 - 22.14)^2}{22.14}$$

$$= 0.75 + 0.05 + 1.99 + 0.09 + 1.74 + 0.63 + 1.25 + 2.09 + 0.37$$

$$= 8.96.$$

This computation can also be made with (14–10), the more efficient formula, where we need not obtain the differences between observed and expected frequencies. Using this formula here, we consider n as the total number of observations in all the samples. Thus,

$$\chi_4^2 = \sum \frac{o_i^2}{e_i} - n$$

$$= \frac{80^2}{72.61} + \frac{21^2}{22.01} + \frac{14^2}{20.37} + \frac{72^2}{69.46} + \frac{15^2}{21.06}$$

$$+ \frac{23^2}{19.49} + \frac{69^2}{78.93} + \frac{31^2}{23.93} + \frac{25^2}{22.14} - 350$$

$$= 8.96.$$

(6) Decisions: The computed chi-square value falls in the region of acceptance. The three samples are drawn from the same population. The views of professional people, businessmen, and farmers are homogeneous insofar as the national policy under discussion is concerned.

14–10 SUMMARY AND CONCLUSIONS

This chapter begins by distinguishing between the large- and small-sampling theories. The notion of "large sample" is a relative one: relative to the sample statistic under consideration.

So far as the general procedures of estimation and testing are concerned, the same logic applies whether we are dealing with large or small samples. The only difference is that, with large samples, inferences are made with the normal probability law, while with small samples, sampling distributions vary from one statistic to another. The most frequently encountered continuous probability models for small samples are the chi-square, Student's t, and F distributions. All of these belong to a family of distributions that are generated from independent standard normal variables and are defined in terms of degrees of freedom. The number of degrees of freedom may be thought of as the number of elements in a sample that can be chosen freely; or as the number of variables that can vary freely; or as the number of independent variables.

The remainder of this chapter is devoted to the properties and applications of chi-square distributions. A chi-square distribution has been defined as the sum of the squares of n independent standard normal variables.

While the main topic of this chapter has been the making of inferences about the population variance with chi-square distributions, most of the material has been concerned with chi-square tests regarding frequencies, goodness of fit, independence, and homogeneity. These chi-square tests constitute a part (not necessarily the most important part) of nonparametric methods in contrast to the classical parameteric statistics. We have introduced these tests here merely for the purpose of the broad application of the chi-square model. As we shall see in Chapter 20, nonparametric statistics involves many other testing procedures.

We shall now proceed to discuss the nature and applications of t and F distributions.

EXERCISES

14–1 In the past, scores on tests of elementary statistics have been found to be approximately normal with a variance of 100. A class of 31 students now taught by an outstanding statistician whose method of teaching seems to favor the bright students and to penalize the poor students is expected to have a wider dispersion of test scores. Suppose that this class can be considered a random sample and suppose that the variance of student scores in the final examination is found to be 150; can we conclude that this instructor's method of teaching results in a wider dispersion at the 5 percent level of significance? At $\alpha = 0.01$?

14–2 Using data from the preceding problem, construct a 95 percent two-sided and a 95 percent one-sided confidence interval for σ^2.

14–3 Sixteen steel wires tested for tensile strength yielded a variance of 140 pounds^2. Is this result consistent with the belief that the population variance is greater than 125 pounds^2 at $\alpha = 0.10$? $\alpha = 0.05$?

14–4 Using data from the preceding problem, construct a 95 percent two-sided and a 95 percent one-sided confidence interval for σ^2.

14–5 Diameters of a certain make of tube are assumed to be normally distributed. Test the hypothesis $\sigma^2 = 0.2$ against $\sigma^2 > 0.2$, at $\alpha = 0.10$, given the ten observations 5.5, 5.4, 5.4, 5.6, 5.8, 5.7, 5.4, 5.5, 5.4, and 5.6 inches. Would you reach the same conclusion at $\alpha = 0.05$? $\alpha = 0.01$?

14–6 A random sample of 250 units of canned tomatoes is drawn and the variance for weights of the cans is found to be 5 ounces^2. What is the 99 percent confidence interval estimate for σ^2?

14–7 The variance of the lifetimes of a sample of 200 electric light bulbs is 90,000 hours^2. Find (a) 95 percent and (b) 99 percent confidence limits for σ^2.

14–8 Using data in the preceding two problems, find confidence intervals for σ for $1 - \alpha = 0.99$.

14–9 Out of 5000 portable radios examined, 200 had defective parts. Is this result in accord with the advertised standard that defectives represent 3 percent or less of total output in terms of a normal test? In terms of a chi-square test?

14–10 A department store employs 300 men and 400 women. In a given year 3100 days of absence were recorded for men and 4600 days of absence for women. Is the difference in rate of absenteeism between men and women significant at $\alpha = 0.05$ in terms of a chi-square test?

14–11 Suppose that a die is tossed 120 times with the following result:

Spots showing	1	2	3	4	5	6
Observed number	10	19	30	29	21	11

Is this result consistent with the hypothesis that the die is fair at $\alpha = 0.05$?

14–12 Suppose that 100 tosses of ten coins in each toss yielded the following result:

No. of heads	0	1	2	3	4	5	6	7	8	9	10
Observed data	0	1	2	8	25	28	22	4	6	3	1

Would you conclude that the coins are fair?

14–13 Suppose it is hypothesized that there are three times as many automobile accidents on Saturday and Sunday as on any other day of the week; what is the probability distribution of auto accidents throughout a week? From the record, 100 accidents are chosen independently of one another, and the distribution of accidents according to the days of the week is found to be

Mon.	*Tue.*	*Wed.*	*Thu.*	*Fri.*	*Sat.*	*Sun.*
5	9	10	8	11	30	27

Do the observed data contradict or confirm the hypothesis if $\alpha = 0.01$?

14–14 In Mendel's experiments with peas he observed 315 round and yellow, 108 round and green, 101 wrinkled and yellow, and 32 wrinkled and green. According to his theory of heredity, the numbers should be in the proportion 9:3:3:1. In view of his observed data, can we accept his theory if $\alpha = 0.01$?

14–15 A survey of 200 families with three children gave the following results:

Male births	0	1	2	3
Families	40	58	62	40

Are these data consistent with the hypothesis that male and female births are equally likely? [*Hint*: If this hypothesis is true, then the probabilities of 0, 1, 2, 3 male births are respectively $\frac{1}{8}$, $\frac{3}{8}$, $\frac{3}{8}$, $\frac{1}{8}$.]

14–16 Five coins are tossed 1000 times and, at each toss, the tails are counted. Results of this experiment are given below:

No. of tails	0	1	2	3	4	5
No. of tosses	38	144	342	287	164	25

Fit a binomial distribution to these data and test the goodness-of-fit at $\alpha = 0.05$.

14–17 Fit a binomial distribution to data in Exercise 14–12 and test whether the fit is good.

14–18 The daily number of automobile accidents in a certain city is observed during a 50-day period and the following information is obtained:

No. of accidents	0	1	2	3	4	5
No. of days	25	15	8	0	1	1

Fit a Poisson distribution to these data and test whether the fit is good.

14–19 In the accounting department of a bank 100 accounts are selected at random and examined for errors. Suppose the following result has been obtained:

No. of errors	0	1	2	3	4	5	6
No. of accounts	36	40	19	2	0	2	1

Does this information verify that the errors are distributed according to the Poisson probability law?

14–20 It is often stated that instructors tend to grade their students according to the "normal curve." Grades of 162 students in freshman mathematics classes during a certain year are found to be distributed as in Table 14P–1. Assuming that these grades constitute a random sample, can we conclude that the sample has been drawn from a population of grades that are normally distributed at $\alpha = 0.01$?

Table 14P–1

Class	*Frequency*
30 and under 40	5
40 and under 50	10
50 and under 60	28
60 and under 70	52
70 and under 80	31
80 and under 90	26
90 and under 100	10
Total	162

14–21 Four hundred independent random samples, each with 200 units, are drawn from a population, and sample variances are computed. This experiment gives the following information. From these data can we conclude that the population is normally distributed?

Class	100–105	105–110	110–115	115–120	120–125	125–130	130–135	Total
f	21	56	61	125	70	47	20	400

[*Hint*: If the population is normally distributed, we expect the distribution of sample variance to be approximately normal with n as large as 200.]

14–22 A random sample consisting of 100 men and 100 women gives the following results in connection with their attitudes toward the New York City income tax introduced in 1966:

Sex	*For*	*Against*	*Total*
Men	25	75	100
Women	35	65	100
Total	60	140	200

In view of this information, can you conclude that New Yorkers' attitude toward the tax is independent of their sex at $\alpha = 0.05$?

14–23 A market research firm desires to determine whether the inclusion of a coin in a questionnaire would increase the number of responses. Three hundred questionnaires, half with coins and half without, are sent to 300 persons selected at random; the following results are obtained:

	Responded	*Not responded*	*Total*
Coin included	97	53	150
Coin not included	80	70	150
Total	177	123	300

Is there significant evidence to indicate that the inclusion of a coin is related to the number of responses to the questionnaire?

14–24 A sample of 100 small machine parts was taken from outputs of three different operators to determine whether they tended to produce the same distribution of output quality as defined by two quality classifications; the results follow:

	Operator		
Quality	1	2	3
Good	20	27	32
Average	8	6	7

How should you state the hypothesis? What is your conclusion?

14–25 A firm selling four products wishes to determine how sales of these products are distributed among four general classes of customers. A random sample of sales provides the following information:

Customer group	Product 1	2	3	4
Professionals	85	23	56	36
Businessmen	153	44	128	75
Factory workers	128	26	101	45
Farmers	34	7	15	44

What conclusions can you draw?

14–26 For the purpose of comparing two manufacturing processes, 100 units are selected from outputs produced by each process; the following results are obtained:

Process	*Nondefectives*	*Defectives*
A	94	6
B	90	10

What are your conclusions? [*Hint*: This should be a test of homogeneity. Also, it is advisable to determine the chi-square value with the continuity correction factor in this case.]

14–27 Three random samples of students at a certain university are taken. The first contains 100 graduate students, the second 100 upperclassmen, and the third 100 lowerclassmen. Each is asked whether he considers the instruction he is receiving excellent, good, or average. The following results are gathered:

Classification	*Instruction* *Excellent*	*Good*	*Average*
Graduate	77	12	11
Upperclassmen	73	7	20
Lowerclassmen	85	10	5

What conclusions can you draw?

14–28 On a particular proposal of national importance, a random sample of 180 Democrats and a random sample of 150 Republicans yield the following information:

	In favor	*Opposed*	*Undecided*
Democrats	89	65	26
Republicans	62	54	34

In view of these data, can we conclude that there is no difference between the two parties insofar as this proposal is concerned at $\alpha = 0.05$? $\alpha = 0.01$?

15

STUDENT'S t AND F DISTRIBUTIONS

The Student's, or the t, distribution results from the study of sampling distributions of $\bar{x}$ under the conditions that the population standard deviation is unknown and that the sample size is 30 or less. To evaluate the distributions of $\bar{x}$ under these conditions, attention should be paid to the fluctuations in $\hat{s}$, which is used to compute the estimator for $\sigma_{\bar{x}}$, as well as those in $\bar{x}$. This technical refinement was introduced in 1908 by William S. Gosset, a statistician employed by Guinness, an Irish brewery. Guinness did not then permit members of its staff to publish their research findings. Gosset evaded this restriction by publishing his work under the pen name "Student." This explains the name "Student's distribution." Today, few have heard of Gosset, but "Student" is a celebrated name in the history of statistics, since the Student's t has become one of the most important probability models in analytical statistics.

While the t distribution is a very useful inferential tool, it cannot be employed to make inferences concerning more than two means. In his attempt to develop a theoretical model for the evaluation of three or more means in the early 1920's, R. A. Fisher succeeded in providing us with the F distribution. But, as we shall see, not only are both t and F distributions related to standard normal variables and degrees of freedom, but the square of the t variable can also be considered as a special case of the F variable.

In this chapter we shall first give the properties of both t and F distributions. Then we shall consider applications of t distributions to inferences about the mean and the equality between two means. Finally, we shall take up the employment of F distributions to study equality between two variances and equality among three or more means. Other uses of these probability models will be given in future chapters.

15-1 PROPERTIES OF STUDENT'S DISTRIBUTIONS

If $X_0, X_1, X_2, \ldots, X_n$ are $n + 1$ independent standard normal variables, then the statistic

$$t_\delta = \frac{X_0}{\sqrt{\frac{1}{n}(X_1^2 + X_2^2 + \cdots + X_n^2)}} \tag{15-1}$$

is said to have a *Student's*, or *t*, *distribution* with $\delta = n$ df. Notice that t is a ratio of a standard normal variable to the square root of a chi-square variable divided by its degrees of freedom. That is, (15-1) is equivalent to the expression $t_\delta = Z/\sqrt{\chi^2/\delta}$. There is a t distribution corresponding to each positive integer δ. Density curves for two t distributions, one with $\delta = 2$ and one with $\delta = 20$, are plotted against the standard normal distribution in Figure 15-1.

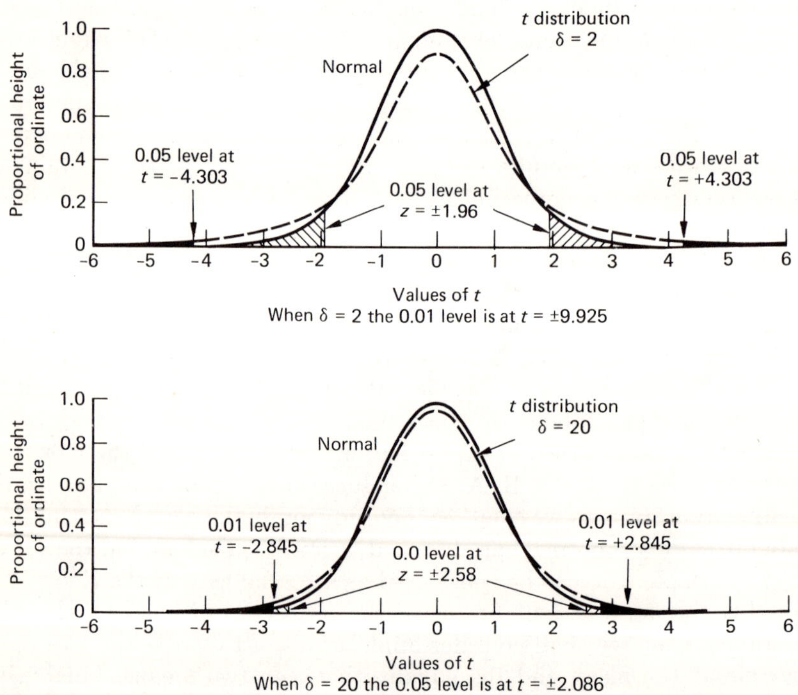

Figure 15-1 Comparison of t Distributions with the Standard Normal Distribution for $\delta = 2$ and $\delta = 20$

A number of important properties of t distributions are worth noting. First, like a normal variable, a t variable ranges in value from $-\infty$ to ∞. Second, like the standard normal distribution, a t distribution is symmetrical, having mean, mode, and median equal to zero, except that t_1 (t distribution with $\delta = 1$) has no mean. Third, the variance of t_δ, with $\delta > 2$, is $\delta/(\delta - 2)$; when $\delta \leq 2$, t_δ possesses no variance. Fourth, a t distribution has a greater dispersion than the standard normal distribution—namely, $[\delta/(\delta - 2)] > 1$. However, when δ gets large, t_δ approaches the standard normal distribution as a limit. This last property can be seen from the standard deviation of t_δ, which is $\sqrt{\delta/(\delta - 2)}$. This quantity is always greater than 1, but it comes closer to 1 as δ becomes larger. In practice, we may treat t_δ as $n(0, 1)$ when $\delta > 30$.

Table B–10 in the Appendix gives critical values at both tail ends of t distributions for a number of selected percentage points. For example, if X is t_2, then

$$P(-4.303 \leq X \leq 4.303) = P(-4.303 \leq t_2 \leq 4.303) = 0.95.$$

From this we see that

$$P(t_2 < -4.303) = 0.025$$

and

$$P(t_2 > 4.303) = 0.025.$$

See Figure 15–1 for $\delta = 2$.

Similarly, if X is t_{20}, we have

$$P(-2.845 \leq t_{20} \leq 2.845) = 0.99,$$

$$P(t_{20} < -2.845) = 0.005,$$

$$P(t_{20} > 2.845) = 0.005,$$

$$P(t_{20} > 2.528) = 0.01,$$

$$P(t_{20} < -1.725) = 0.05,$$

and so on. As a matter of fact, the column headings of Table B–10 are levels of significance for two-tail tests. Entries in the body are the critical values of t_δ for such tests. If we would like to have a single-tail test with. say, $\alpha = 0.05$, the critical value is found in the column with $2\alpha = 0.10$ for the appropriate number of df. For example, for $P(t_{20} < -1.725) = 0.05$, the number -1.725 is found in the column headed by 0.10 corresponding to $\delta = 20$.

15–2 PROPERTIES OF F DISTRIBUTIONS

If $X_1, X_2, \ldots, X_m$ and $Y_1, Y_2, \ldots, Y_n$ are all independent standard normal variables, then the statistic

$$F = \left(\frac{n}{m}\right)\frac{X_1^2 + X_2^2 + \cdots + X_m^2}{Y_1^2 + Y_2^2 + \cdots + Y_n^2} \tag{15–2}$$

is said to have the F distribution with (m, n) degrees of freedom. It turns out that an F distribution is the ratio of two chi-square distributions with δ_1 as the number of df for the numerator and with δ_2 as the number of df for the denominator. It may be denoted as $F_{\delta_1\delta_2}$. The most important thing for us to note is that an F distribution has two parameters, δ_1 and δ_2, the numbers of df. Density curves for $F_{2,10}$, $F_{2,2}$, $F_{10,2}$, and $F_{10,10}$ are shown in Figure 15–2.

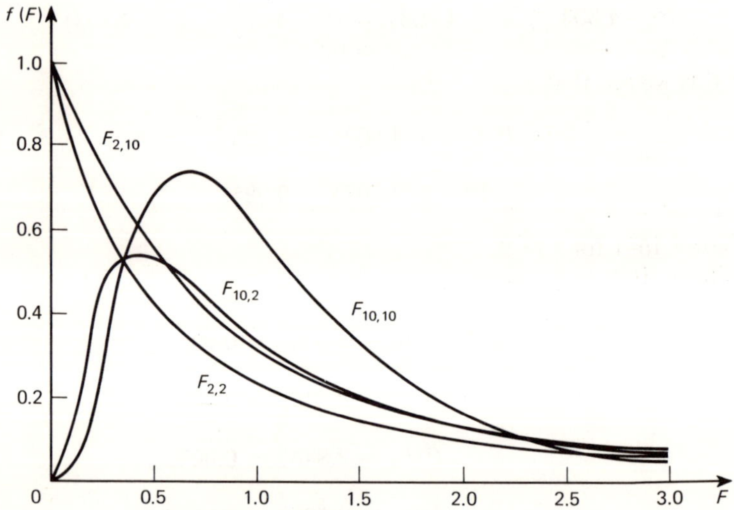

Figure 15–2 F Distribution Curves

Let us now observe some of the properties of F distributions. First of all, since an F distribution is a ratio of two squared quantities, the value of F ranges from 0 to infinity. Second, there is an F distribution for each pair of positive integers, δ_1 and δ_2. Third, an F distribution is positively skewed, but its skewness decreases with increases in the number of df. Fourth, the mean of $F_{\delta_1\delta_2}$ is

$$\mu = \frac{\delta_2}{\delta_2 - 2} \qquad \text{when } \delta_2 > 2,$$

and the variance of $F_{\delta_1\delta_2}$ is

$$\sigma^2 = \frac{2\delta_2^2(\delta_1 + \delta_2 - 2)}{\delta_1(\delta_2 - 2)^2(\delta_2 - 4)} \qquad \text{when } \delta_2 > 4.$$

An F distribution has no mean when $\delta_2 \leq 2$, and it has no variance when $\delta_2 \leq 4$. Fifth and last, if X is $F_{\delta_1\delta_2}$, then $Y = 1/X$ is $F_{\delta_2\delta_1}$. This is the famous *reciprocal property* of F distributions.

Percentage points for the right tail of F distributions at 10, 5, and 1 percent levels are given in Table B–11 in the Appendix. When X is, say, $F_{10,7}$, then

$$P(X > 3.64) = P(F_{10,7} > 3.64) = 0.05.$$

To say that $P(F_{10,7} > 3.64) = 0.05$ is the same as to say that

$$P(0 < X < 3.64) = P(0 < F_{10,7} < 3.64)$$
$$= 1 - 0.05 = 0.95.$$

See, for example, Figure 15–3.

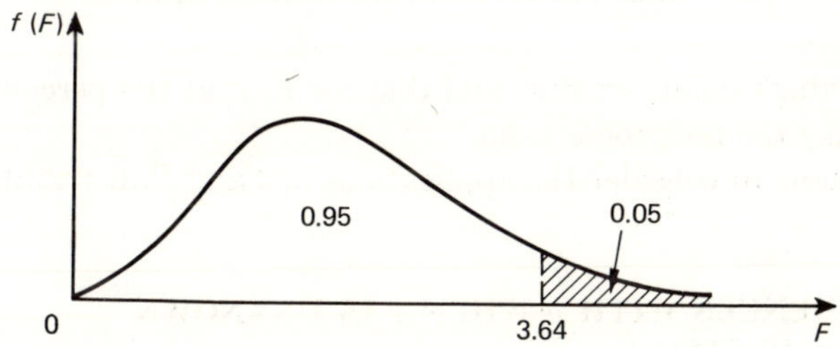

Figure 15–3 Upper-Tail Probability of $F_{10,7}$

Similarly,

$$P(F_{10,7} > 6.62) = 0.01$$

and

$$P(0 < F_{10,7} < 6.62) = 0.99.$$

Table B–11 can also be employed to find probabilities for the left tail points of F distributions by following the reciprocal property of F distributions. Suppose X is $F_{15,10}$; we may let Y be $F_{10,15}$. From Table B–11, we see that

$$P(Y > 2.54) = P(F_{10,15} > 2.54) = 0.05.$$

Then, by the reciprocal property,

$$P\left(\frac{1}{Y} < \frac{1}{2.54}\right) = 0.05,$$

and since $1/Y$ has the same distribution as X, we have

$$P\left(X < \frac{1}{2.54}\right) = P\left(F_{15,10} < \frac{1}{2.54}\right)$$

$$= P(F_{15,10} < 0.394) = 0.05.$$

By the same procedure, of course, the 1 percent left-tail point can be found (see Figure 15–4). In general, to find the left-tail critical value for $F_{\delta_1\delta_2}$ at a

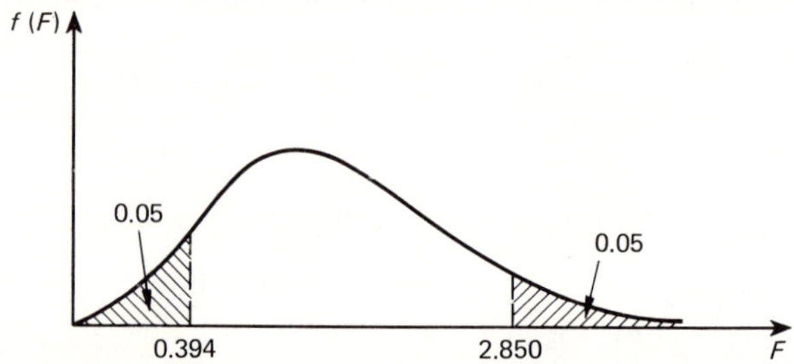

Figure 15–4 Upper- and Lower-Tail Probabilities of $F_{15,10}$

given percentage point, we first find that for $F_{\delta_2\delta_1}$ at the percentage point, and then take the reciprocal value.

We turn now to consider the applications of t and F distributions.

15–3 INFERENCES WITH μ WHEN σ IS UNKNOWN AND n IS SMALL

When we test hypotheses about μ we often assume, because of large sample size, that the unknown population standard deviation may be replaced by the sample standard deviation; that is, instead of $(\bar{x} - \mu)/(\sigma/\sqrt{n})$, we use $(\bar{x}-\mu)/(\hat{s}/\sqrt{n})$ as the testing statistic. The latter expression is assumed to be distributed approximately as a standard normal variable. The denominator in this ratio, $\hat{s}/\sqrt{n}$, however, is a random variable and it can hardly be considered as a constant when samples are small. But noting that $(\bar{x} - \mu)$ is a normal variable with zero mean, and that $\hat{s}/\sqrt{n} = \hat{\sigma}_{\bar{x}}$ is the square root of a chi-square variable with $\delta = n - 1$, by definition, then the ratio

$$t_\delta = \frac{\bar{x} - \mu}{\hat{s}/\sqrt{n}} \qquad (15\text{–}3)$$

is distributed as t_{n-1}.

To illustrate the t test concerning μ, suppose a medical research center announces it has developed a treatment for high blood pressure that can reduce a patient's blood pressure by more than 20 points. Learning about this, a doctor tries the treatment on ten of his patients who are suffering from high blood pressure. As a result, he finds that their blood pressure has been reduced by 21.5 points on the average, with $\hat{s} = 1.5$ points. If the doctor is interested in finding out whether the new treatment is as effective as claimed by the research organization, by using the following t test he could determine this on the basis of the information he has obtained from his trials.

(1) HYPOTHESES: H_0: $\mu \leq 20$; H_1: $\mu > 20$.

(2) LEVEL OF SIGNIFICANCE: $\alpha = 0.01$.

(3) TESTING STATISTIC: $t_\delta = (\bar{x} - \mu)/(\hat{s}/\sqrt{n})$, which is distributed as t_9.

(4) DECISION RULE: Since $P(t_9 < 2.821) = 0.01$, he would accept H_0 if $t < 2.821$. (*Note*: The critical value 2.821 is found from the column headed by 0.02 and the row headed by 9 df in the t table. The column under 0.02 is used for our upper-tail test at $\alpha = 0.01$ because, as stated before, the t table is constructed with two-sided critical values.)

(5) COMPUTATIONS: $n = 10$, $\bar{x} = 21.5$, $\hat{s} = 1.5$, $\hat{\sigma}_{\bar{x}} = 1.5/\sqrt{10} = 0.474$, and

$$t_9 = \frac{\bar{x} - \mu}{\hat{\sigma}_{\bar{x}}} = \frac{21.5 - 20.0}{0.474} = 3.165.$$

(6) DECISIONS: The difference, 1.5, between the sample mean and the hypothesized population mean is significant. H_0 is rejected in favor of H_1. If there are no undesirable side effects from the use of the new treatment, the doctor may decide to use it exclusively henceforth.

The confidence interval for μ with confidence coefficient $1 - \alpha$ in terms of t distributions is given as

$$P\left(-t_{n-1;\,\alpha/2} < \frac{\bar{x} - \mu}{\hat{\sigma}_{\bar{x}}} < t_{n-1;\,\alpha/2}\right) = 1 - \alpha,$$

which can be rearranged to yield

$$P(\bar{x} - t_{n-1;\,\alpha/2}\hat{\sigma}_{\bar{x}} < \mu < \bar{x} + t_{n-1;\,\alpha/2}\hat{\sigma}_{\bar{x}}) = 1 - \alpha.$$

From this, we see that the desired confidence limits for μ can be written as

$$\bar{x} \pm t_{n-1;\,\alpha/2}\hat{\sigma}_{\bar{x}}. \tag{15-4}$$

As an illustration, suppose we wish to have an interval estimate of the mean tensile strength of malleable iron castings produced by a certain factory. We may first draw a random sample of, say, 17 castings for observation. Suppose, furthermore, we find that, for this sample,

$$\bar{x} = 57{,}000 \text{ psi}, \qquad \hat{s} = 1600 \text{ psi}.$$

With these data, a 95 percent confidence interval may be constructed in accordance with (15–4):

$$\bar{x} \pm t_{16;0.025}\hat{\sigma}_{\bar{x}} = 57{,}000 \pm 2.12\left(\frac{1600}{\sqrt{17}}\right)$$

or

$$56{,}175 < \mu < 57{,}825.$$

Confidence intervals constructed with t distributions can be interpreted in the same way as those with the normal distribution. It is important, however, to note two points. First, the confidence interval constructed with a small sample is wider or less precise than that with a large sample. This is not only because a small sample has a larger standard error but also because the t multiple is larger than the z multiple (for the normal case). Second, if σ is known, confidence intervals for many different random samples of the same size would have the same width, even though they might have different midpoints, which are the different sample means. The reason here is that the standard error is a constant when the population standard deviation is known. When σ is unknown, the standard error is itself a random variable, which varies in value from sample to sample. As a result, confidence intervals for different sample means, computed from samples of the same size, will have not only different midpoints but also unequal widths. This last observation, however, should not lead us to infer that confidence intervals constructed with estimated standard errors from small samples have a different meaning. They are still probability statements and should be interpreted in the same way as those constructed under the assumption of normality.

15–4 INFERENCES CONCERNING TWO POPULATION MEANS WITH INDEPENDENT SMALL SAMPLES

In making inferences about two population means when samples are small and population variances are not known, it is again appropriate to employ the t distribution. To test the hypothesis that two population means are equal, we use the following ratio as the testing statistic:

$$t_{\delta} = \frac{\bar{x}_1 - \bar{x}_2}{\hat{\sigma}_{\Delta\bar{x}}}. \tag{15–5}$$

The denominator is the unbiased estimator of the standard error of a difference between two sample means made under the assumption that the two population variances are equal. It is given by

$$\hat{\sigma}_{\Delta\bar{x}} = \hat{s}_w \sqrt{\frac{n_1 + n_2}{n_1 n_2}}, \tag{15-6}$$

where $\hat{s}_w$ is the pooled, or weighted, estimator of the common population standard deviation and is defined as

$$\hat{s}_w = \sqrt{\frac{(n_1 - 1)\hat{s}_1^2 + (n_2 - 1)\hat{s}_2^2}{n_1 + n_2 - 2}}. \tag{15-7}$$

Note that in (15–5) the numerator is a normal variable and the denominator is the square root of a chi-square distribution with $n_1 + n_2 - 2$ degrees of freedom. Therefore, this ratio is distributed as $t_{n_1+n_2} - 2$.

Suppose we are interested in discovering whether or not there is any real difference between the average annual earnings of carpenters and house painters in a certain city. We draw two independent samples, which yield the following information:

Carpenters	*House painters*
$n_1 = 12$ men	$n_2 = 15$ men
$\bar{x} = \$6000$	$\bar{x} = \$5400$
$\hat{s}_1^2 = 565{,}000$	$\hat{s}_2^2 = 362{,}500$

Based on these data, we may conduct the following t test:

(1) HYPOTHESES: H_0: $\mu_1 = \mu_2$; H_1: $\mu_1 \neq \mu_2$.

(2) LEVEL OF SIGNIFICANCE: $\alpha = 0.05$.

(3) TESTING STATISTIC: $t_\delta = (\bar{x}_1 - \bar{x}_2)/\hat{\sigma}_{\Delta\bar{x}}$, which is distributed as $t_{n_1+n_2-2} = t_{25}$.

(4) DECISION RULE: With $\alpha = 0.05$ for a two-sided test, the acceptance region is defined as $-2.06 \leq t_{25} \leq 2.06$. Reject H_0 if $t < -2.06$ or if $t > 2.06$.

(5) COMPUTATIONS: Estimate of pooled standard deviation:

$$\hat{s}_w = \sqrt{\frac{(n_1 - 1)\hat{s}_1^2 + (n_2 - 1)\hat{s}_2^2}{n_1 + n_2 - 2}}$$

$$= \sqrt{\frac{11(565{,}000) + 14(362{,}500)}{12 + 15 - 2}}$$

$$= 697.$$

Estimate of standard error of the difference between two means:

$$\hat{\sigma}_{\Delta\bar{x}} = \hat{s}_w \sqrt{\frac{n_1 + n_2}{n_1 n_2}} = 697 \sqrt{\frac{12 + 15}{(12)(15)}} = 269.74.$$

Critical ratio:

$$t_{25} = \frac{\bar{x}_1 - \bar{x}_2}{\hat{\sigma}_{\Delta\bar{x}}} = \frac{6000 - 5400}{269.74} = 2.224.$$

(6) Decisions: The critical ratio is significant. The null hypothesis that carpenters and house painters have the same annual earnings is rejected at the 5 percent level of significance in favor of the alternative hypothesis that their annual average earnings are different.

As another example, suppose an agricultural experiment showed that seven test plots planted with one variety of corn yielded, on the average, 95.5 bushels per acre with a standard deviation of 6 bushels per acre; whereas six plots planted with another variety of corn yielded 86.7 bushels per acre with a standard deviation of 5.5 bushels per acre. In order to ascertain whether the mean yield of the first variety is significantly greater than that of the second, the research scientist conducted the following test:

(1) Hypotheses: H_0: $\mu_1 \leq \mu_2$; H_1: $\mu_1 > \mu_2$.

(2) Level of Significance: $\alpha = 0.05$.

(3) Testing Statistic: $(\bar{x}_1 - \bar{x}_2)/\hat{\sigma}_{\Delta\bar{x}}$, which is distributed as t_{7+6-2}.

(4) Decision Rule: Reject H_0 if $t > 1.796$, since $P(t_{11} > 1.796) = 0.05$.

(5) Computations: Sample information gives

$$n_1 = 7, \qquad n_2 = 6,$$

$$\bar{x}_1 = 95.5 \text{ bu.}, \qquad \bar{x}_2 = 86.7 \text{ bu.},$$

$$\hat{s}_1 = 6.0 \text{ bu.}, \qquad \hat{s}_2 = 5.5 \text{ bu.}$$

We have then

$$t_{11} = \frac{95.5 - 86.7}{\sqrt{\frac{6(6)^2 + 5(5.5)^2}{7 + 6 - 2}} \sqrt{\frac{7 + 6}{(7)(6)}}} = 2.521.$$

(6) Decision: Reject H_0. The first variety of corn seems to have a greater mean yield.

In our first illustration of t tests concerning two population means, we concluded that the average annual earnings between carpenters and house painters are different. If we desire to know the range within which such a difference would fall at a given level of probability, we need to construct a confidence interval for $\Delta\mu$. In general, with a confidence coefficient of $1 - \alpha$, the confidence interval for $\Delta\mu$ is given as

$$\Delta\bar{x} \pm t_{\delta;\alpha/2}\hat{\sigma}_{\Delta\bar{x}}, \tag{15-8}$$

where $\delta = n_1 + n_2 - 2$.

Now, if we wish to construct a 95 percent confidence interval for the carpenter and house painter data, we have

$$\Delta\bar{x} \pm t_{25;0.025}\hat{\sigma}_{\Delta x} = (6000 - 5400) + 2.06(269.74)$$

or

$$\$44.34 < \Delta\mu < \$1155.66.$$

Thus, the chances are 95 out of 100 that the true difference in annual average earings between carpenters and house painters ranges from \$44.34 to \$1155.66.

15-5 INFERENCES CONCERNING TWO POPULATION MEANS WITH DEPENDENT SAMPLES

So far, comparison between two expected values has been made under the assumption of independent samples. Two samples are said to be independent of each other if the elements in one are not related to those in the other in any significant or meaningful manner. On occasion we may be called upon to compare two population means on the basis of sample means computed from two dependent samples. Two samples are *dependent* if they are paired in the sense that each observation in one is associated with some particular observation in the other. Pairing may result when individuals on whom observations are made are matched by age, by sex, by political affiliation, or by some other criterion, or when the same individual element has a performance score and another characteristic in each sample. If two samples are dependent, they must contain the same elementary units, but different sets of paired observations are made on these units; that is, two measurements are made on each of the same experimental units.

To conduct a test with two dependent samples, we proceed as before by formulating the null hypothesis that the two population means are the same. We must now derive an appropriate testing statistic. To do this, we first take the difference between two matched observations for each element, d_i, obtaining a new variable D. We then reason that if the two samples are not different from each other, then D must average zero—that is, $\bar{d} = \sum d/n = 0$, where n is the number of pairs in the two samples. Next, we need the standard deviation for D for the estimation of the standard error of $\bar{d}$. The standard deviation for D can be computed in the usual manner as

$$\hat{s}_d = \sqrt{\frac{n \sum d^2 - (\sum d)^2}{n(n-1)}}.$$

With $\hat{s}_d$, the unbiased estimator of the standard error of $\bar{d}$ becomes $\hat{\sigma}_{\bar{d}} = \hat{s}_d/\sqrt{n}$. Finally, it can be shown that the ratio of $\bar{d}$ to its standard error is distributed as the Student's distribution, with $\delta = n - 1$. That is,

$$t_\delta = \frac{\bar{d}}{\hat{\sigma}_{\bar{d}}}, \tag{15–9}$$

which is used as the testing statistic in this case.

Suppose a methods engineer believes he has perfected a training program that can considerably shorten workers' assembly time for a certain mechanism. To check this belief, he plans to select ten workers at random and to make time and motion studies on them before and after they have gone through his training program. Then:

(1) HYPOTHESES: H_0: $\mu_1 = \mu_2$; H_1: $\mu_1 > \mu_2$.

(2) LEVEL OF SIGNIFICANCE: $\alpha = 0.01$.

(3) TESTING STATISTIC: $\bar{d}/\hat{\sigma}_{\bar{d}}$, which is distributed as t_{n-1}.

(4) DECISION RULE: Since $n = 10$ has already been determined, we have $P(t_9 > 2.821) = 0.01$. H_0 will be rejected if and only if $t \geq 2.821$.

(5) COMPUTATIONS: Suppose that results of the time and motion studies before and after the training program are as shown in the first three columns of the following table. We may then subtract the second observation from the first for each worker to get ds and d^2s, which are needed for this test.

Worker	*First study (minutes)*	*Second study (minutes)*	d	d^2
1	7	8	−1	1
2	8	8	0	0
3	10	7	3	9
4	11	6	5	25
5	18	10	8	64
6	16	9	7	49
7	12	9	3	9
8	12	8	4	16
9	6	7	−1	1
10	12	10	2	4
Total	112	82	+30	178

$$\bar{d} = \frac{\sum d}{n} = \frac{30}{10} = 3 \text{ minutes},$$

$$\hat{s}_d = \sqrt{\frac{n \sum d^2 - \sum d)^2}{n(n-1)}}$$

$$= \sqrt{\frac{10(178) - (30)^2}{10(10-1)}} = 3.1273 \text{ minutes},$$

$$\hat{\sigma}_{\bar{d}} = \frac{\hat{s}_d}{\sqrt{n}} = \frac{3.1269}{\sqrt{10}} = 0.9889 \text{ minutes},$$

and
$$t_9 = \frac{\bar{d}}{\hat{\sigma}_{\bar{d}}} = \frac{3}{0.9889} = 3.034.$$

(6) DECISIONS: Since the computed t value falls in the critical region, H_0 is rejected. The training program is evidently of some help in reducing the average assembly time.

We can also compute a confidence interval for the true difference between two population means with two dependent samples in the familiar way:

$$\bar{d} \pm t_{n-1,\alpha/2}\hat{\sigma}_{\bar{d}}. \tag{15-10}$$

15-6 F TEST OF THE EQUALITY BETWEEN VARIANCES

If we have two populations, X_1 and X_2, with variances σ_1^2 and σ_2^2, respectively, we naturally wish to ascertain whether these variances are equal or one is greater than the other. One convenient way to do this is to compute the difference between two sample variances. Another obvious procedure is to form a ratio of sample variances.

If the two samples are independent, then sample variances will be independent. Furthermore, if populations are normal (or approximately so) and the samples are large, $n_1 \geq 100$ and $n_2 \geq 100$, we can then construct a large-sample test for $\sigma_1^2 - \sigma_2^2$ on the basis of $\hat{s}_1^2 - \hat{s}_2^2$, which is an asymptotically normally distributed statistic.

Usually, when samples are small, though other assumptions mentioned above are fulfilled, we use the ratio of the sample variances as the basis for comparison of population variances. This procedure has two advantages. First, the ratio is independent of scale, provided that both populations are given in the same units. Second, the probability distribution of this ratio, under the null hypothesis that $\sigma_1^2 = \sigma_2^2$, is independent of population parameters. Let us see how this ratio is distributed.

When we have available two sample variances, $\hat{s}_1^2$ and $\hat{s}_2^2$, derived from two independent samples with n_1 and n_2 drawn from two normal populations with σ_1^2 and σ_2^2, then $\hat{s}_1^2$ would be an unbiased estimator of σ_1^2, and $(n_1 - 1)\hat{s}_1^2/\sigma_1^2$ would have a χ^2 distribution with $\delta_1 = n_1 - 1$. Similarly, $\hat{s}_2^2$ would be an unbiased estimator of σ_2^2 and $(n_2 - 1)\hat{s}_2^2/\sigma_2^2$ would be distributed as χ^2 with $\delta_2 = n_2 - 1$. Clearly, the ratio of these two estimators, $(\hat{s}_1^2/\sigma_1^2)/(\hat{s}_2^2/\sigma_2^2)$, would be distributed as $F_{\delta_1\delta_2}$. That is,

$$F_{\delta_1\delta_2} = \frac{\hat{s}_1^2/\sigma_1^2}{\hat{s}_2^2/\sigma_2^2}.$$

Since the null hypothesis is given as $\sigma_1^2 = \sigma_1^2$, this ratio becomes

$$F_{\delta_1\delta_2} = \frac{\hat{s}_1^2}{\hat{s}_2^2}. \tag{15–11}$$

To use (15–11) as the testing statistic for comparing two population variances, we state the null hypothesis either as $\sigma_1^2/\sigma_2^2 = 1$ or as $\sigma_1^2 = \sigma_2^2$. If the alternative is one-sided, it usually takes the form of an upper-tail test—namely, $\sigma_1^2 > \sigma_2^2$. Under this alternative, H_0 will be rejected if F is too large for the specified level of α. If the alternative is two-sided—that is, $\sigma_1^2 \neq \sigma_2^2$—$H_0$ will be rejected if F is either too small or too large. Note also that, with one-sided F tests, the large-sample variance should be used as the numerator in (15–11), so that only the upper tail of the F distribution need be considered.

Suppose that the purchasing agent for a truck company is considering the purchase of either make A or make B tires, which are known to have the same mean durability, though he suspects that make A has a greater variability than make B. He attempts to verify this by testing H_0: $\sigma_A^2 = \sigma_B^2$ against H_1: $\sigma_A^2 > \sigma_B^2$ at a level of significance of 0.10. To effect this test, he has tried 16 tires of make A and 21 tires of make B, obtaining the following results:

$$\hat{s}_A^2 = 38{,}990 \text{ miles}^2$$

and

$$\hat{s}_B^2 = 21{,}000 \text{ miles}^2.$$

With these data, we have

$$F_{15,20} = \frac{38{,}990}{21{,}000} = 1.86.$$

Since $P(F_{15,20} > 1.84) = 0.10$, the computed F value falls in the rejection region. Thus, the null hypothesis is rejected in favor of the alternative that

$\sigma_A^2 > \sigma_B^2$. If the decision is to be made on variability and a smaller variability is desired, make B tires should be bought.

Typically, the comparison between two variances is made with an upper-tail test rather than with a two-tail test, since it is more meaningful to ascertain if one variance is significantly greater than another, $\sigma_1^2 > \sigma_2^2$, than to determine whether two variances are significantly different from each other, $\sigma_1^2 \neq \sigma_2^2$. For example, in terms of the quality of products, we often wish to determine whether one product is less uniform (more variable and less desirable) than another. This type of information is more useful (more precise) in decision making than that the two products differ in uniformity.

If, for some reason, a two-tail test is desired, it can be made easily with the F test. For example, if we wish to test $\sigma_A^2 = \sigma_B^2$ against $\sigma_A^2 \neq \sigma_B^2$, using the data in the previous illustration, with $\alpha = 0.10$ as before, then H_0 will be rejected if the computed value of F is greater than 2.20 or if it is less than 1/2.33. These critical values are found by noting that $P(F_{15,20} > 2.20) = 0.05$ and that $P(F_{15,20} < 1/2.33) = 0.05$. This smaller rejection value is obtained by utilizing the reciprocal property of F distributions. Since we know that $P(F_{20,15} > 2.33) = 0.05$, then we know that $P(F_{15,20} < 1/2.33) = 0.05$. Note that the computed F value for this case is 1.86. It falls in the acceptance region $1/2.33 < F_{15,20} < 2.20$. Therefore, H_0 cannot be rejected at $\alpha = 0.10$ for the two-sided alternative. This illustrates once again that the same sample data may be significant for a single-tail test but insignificant for a two-tail test.

15–7 ANALYSIS OF VARIANCE

We have now learned how to make inferences about expectations, proportions, and variances of one or two populations from samples selected from these populations. We have also discovered that the chi-square tests provide a method of comparing more than two sample proportions. We naturally wish to know if there exists a general procedure by which more than two sample means can be evaluated. Such a procedure is important in statistical analysis, because often the researcher may wish to conduct an experiment for evaluating several designs, methods, products, and so on rather than for comparing a new theory or process with the standard one. For example, one may wish to compare the efficiency of three work designs for assembling a mechanism, or to evaluate the performance of four different makes of auto tires, or to study the relative productivity of k different kinds of fertilizers.

Clearly, it would be grossly inefficient to compare several samples by pairing them two at a time. If there were, say, five samples, we would have

$\binom{5}{2} = 10$ different tests to perform. Furthermore, even after making ten such tests, we still might not be able to generalize simultaneously about all the five populations from which the samples were drawn. Worse still, it is easy to show that such a tedious and time-consuming procedure actually has a high probability of leading to wrong conclusions pairwise. If we were to have ten t tests on the ten possible pairs of the five sample means, the probability of our arriving at a correct decision of no significant difference would be 0.95, given a 5 percent level of significance. Thus, the probability for our arriving at correct conclusions for all the ten t tests would be 0.95^{10}; the probability of obtaining at least one wrong conclusion would be as large as $1 - (0.95)^{10} = 0.401$. That is, more than 40 percent of the time we would make a type I error. It is easy to see that the larger the number of samples to be evaluated, the greater the probability of making wrong decisions. In short, the method of comparing only pairs of samples is at once uneconomical and inefficient for the ultimate purpose of comparing several populations.

An efficient procedure developed to compare several sample means is known as the *analysis of variance,* a method based on the comparison of variances estimated from various sources.

The analysis of variance originated in agrarian research, and its language is loaded with such agricultural terms as *blocks* (referring to land) and *treatments* (referring to populations, or samples, differentiated in terms of varieties of seed, fertilizer, or cultivation methods). Today, analysis-of-variance procedures are applied usefully in nearly every type of experimental design in both natural and social sciences. The analysis of variance is amazingly versatile: it can be readily adopted to furnish, within broad limits, a proper evaluation of data obtained from a large body of experiments that involve several continuous random variables. It can give us answers as to whether different sample data classified in terms of a single variable are meaningful. It can also provide us with meaningful comparisons of sample data that are classified according to two or more variables. In each case, the answer would involve significance, or lack of it, for each variable classification as well as possible joint effects of the variables-treatment combinations.

No attempt will be made here to give full coverage of this vast field of statistical techniques. Our goal is to introduce the reader to the simplest model that is frequently encountered in economic and business research. The model introduced here, called the *completely randomized, one-variable classification model,* is designed to test the equality among three or more means.

The Model

Suppose that we have c populations, $A_1, A_2, \ldots, A_c$, each normally distributed with mean μ_i and all possessing a common variance σ^2. These populations are often called *treatments*—representing c kinds of soybeans, c ways of teaching economics, or c makes of machines for producing a certain type of output. These c populations are assumed to possess a common variance, since there is reason to believe that different treatments used for the same purpose may differ in central tendencies but not in dispersion. Furthermore, we conceive that these populations together constitute a *grand* population with mean μ, called the *grand population mean*, defined as

$$\mu = \frac{1}{c} \sum_i \mu_i.$$

Under the preceding two assumptions, we wish to test the null hypothesis that all the treatment means are equal—that is,

$$\mu_1 = \mu_2 = \cdots = \mu_c.$$

If this H_0 is true, then we would expect $\mu_i = \mu$. However, if H_0 is false, then we would expect μ_i to deviate from μ by a quantity, say, α_i—that is,

$$\alpha_i = \mu_i - \mu, \qquad i = 1, 2, \ldots, c,$$

where α_i is called *treatment effect* for population A_i. Also, we know that

$$\sum \alpha_i = \sum (\mu_i - \mu) = 0.$$

For the previous arguments we can also see that

$$\mu_1 = \mu_2 = \cdots = \mu_c$$

and

$$\text{all } \alpha_i = 0$$

are equivalent statements, since when we say that the treatment means are equal, we imply that the effects due to treatments are nil.

Now, we assume that x_{ij}, the jth observation in the ith sample, is normally distributed and we state that $x_{ij} = \mu_i + e_{ij}$. Here, the e_{ij} are deviations of x_{ij} from μ_i due to chance fluctuations in random sampling. The e_{ij} are called the *error terms*, or *residuals*, and are assumed to be independent and distributed normally with zero mean and a variance identical with the common variance, σ^2, for the treatment populations.

Finally, noting that $\mu_i = \mu + \alpha_i$, we have as the basic model for the completely randomized, one-variable classification experiment the following linear form:

$$x_{ij} = \mu_i + e_{ij} = \mu + \alpha_i + e_{ij}. \qquad (15\text{–}12)$$

The Analysis

The core of the analysis of variance, irrespective of the type of model, lies in the partitioning of the total sum of squares into meaningful and distinct portions. To see how this is done for our present model, let us first arrange the data obtained from c samples, each with r observations, into a general form as shown by Table 15–1.

Table 15–1. Sample Data of the Linear Model

	Sample			
Observation	1	2	...	c
1	x_{11}	x_{21}	...	x_{c1}
2	x_{12}	x_{22}	...	x_{c2}
.	.	.	...	.
.	.	.	...	.
.	.	.	...	.
r	x_{1r}	x_{2r}	...	x_{cr}
Total	$\sum_j x_{1j}$	$\sum_j x_{2j}$	...	$\sum_j x_{cj}$
n_i	n_1	n_2	...	n_c

From the data in Table 15–1, two summary statistics can be computed.

(1) $\bar{x}_{i.}$ = the sample mean of the ith column

$$= \frac{1}{n_i} \sum_j x_{ij} = \frac{1}{n_i} (x_{i1} + x_{i2} + \cdots + x_{ir}),$$

which is an unbiased estimator of μ_i. Note that the dot (.) in the notation $\bar{x}_{i.}$ indicates that the column mean is obtained by summing on the index j.

(2) $\bar{\bar{x}}$ = the grand sample mean

$$= \frac{1}{N} \sum_i \sum_j x_{ij} = \frac{1}{N} (\sum_j x_{1j} + \sum_j x_{2j} + \cdots + \sum_j x_{cj}),$$

which is an unbiased estimator of μ. Here,

$$N = n_1 + n_2 + \cdots + n_c.$$

The variability of the set of i experimentally different samples can be partitioned into two distinct parts in terms of the sum of squared deviations about the grand sample mean. Each observation, x_{ij}, in sample i is expected to deviate from the grand sample mean by the amount $(x_{ij} - \bar{\bar{x}})$, which can be thought of as comprised of two parts,

$$(x_{ij} - \bar{\bar{x}}) = (x_{ij} - \bar{x}_{i.}) + (\bar{x}_{i.} - \bar{\bar{x}}).$$

The first component on the right of this expression is the deviation of x_{ij} from the mean of sample i. The second term is the deviation of the sample mean for sample i from the grand sample mean. Clearly, $(x_{ij} - \bar{\bar{x}})$ is an estimator of $(x_{ij} - \mu)$, and the two terms on the right are estimators of e_{ij} and α_i, respectively.

Now, if we square the deviation from $\bar{\bar{x}}$ for each observation in the sample and total these squared deviations over all individuals j in all sample groups i, we have

$$\sum_i \sum_j (x_{ij} - \bar{\bar{x}})^2 = \sum_i \sum_j [(x_{ij} - \bar{x}_{i.}) + (\bar{x}_{i.} - \bar{\bar{x}})]^2$$

$$= \sum_i \sum_j (x_{ij} - \bar{x}_{i.})^2 + \sum_i \sum_j (\bar{x}_{i.} - \bar{\bar{x}})^2$$

$$+ 2 \sum_i \sum_j (x_{ij} - \bar{x}_{i.})(\bar{x}_{i.} - \bar{\bar{x}}).$$

The last term in this expression is 0, since the value represented by the term $(x_{i.} - \bar{\bar{x}})$ is the same for all j in group i, and the sum of $(x_{ij} - \bar{x}_{i.})$ must be 0 when taken over all j in any group i. Thus, this expression can be written as

$$\sum_i \sum_j (x_{ij} - \bar{\bar{x}})^2 \qquad \text{(total variation: SST)},$$

$$= \sum_i \sum_j (x_{ij} - \bar{x}_{i.})^2 \qquad \text{[variation within (samples) or for error: SSE]},$$

$$+ \sum_i \sum_j (\bar{x}_{i.} - \bar{\bar{x}})^2 \qquad \text{[variation between (samples) or between (columns) : SSC]}.$$

This identity is often called the "partition of the sum of squares."

The logic of the partition of the sum of squares into two distinct parts can be put into intuitive terms quite easily. Observations in any sample will always show variability. Differences among individual observations here can come from two sources. Differences among individuals belonging to different treatment samples may result either from different treatments or from chance variation, or from both. The sum of squares between samples, SSC, reflects the contribution of both different treatments and chance to intersample variability. Individual observations in the same treatment sample, however, can differ from each other only because of chance variation, since each individual within the group receives exactly the same treatment. The sum of squares within samples, SSE, measures those intersample differences due to chance only. Therefore, in any group of i samples, it is possible to isolate the two kinds of variability: the sum of squares between groups, which reflects variation due to both treatments and chance, and the sum of squares within groups, which reflects chance variation alone.

The sums of squares are computed by expanding the binomials that define them, as follows:

$$\begin{aligned}\text{SST} &= \sum\sum (x_{ij} - \bar{\bar{x}})^2 \\ &= \sum\sum (x_{ij}^2 - 2x_{ij}\bar{\bar{x}} + \bar{\bar{x}}^2) \\ &= \sum\sum x_{ij}^2 - 2\bar{\bar{x}}\sum\sum x_{ij} + N\bar{\bar{x}}^2 \\ &= \sum\sum x_{ij}^2 - N\bar{\bar{x}}^2, \end{aligned} \tag{15–13}$$

$$\begin{aligned}\text{SSC} &= \sum\sum (\bar{x}_{i.} - \bar{\bar{x}})^2 \\ &= \sum n_i(\bar{x}_{i.} - \bar{\bar{x}})^2 \\ &= \sum n_i(\bar{x}_{i.}^2 - 2\bar{x}_{i.}\bar{\bar{x}} + \bar{\bar{x}}^2) \\ &= \sum n_i\bar{x}_{i.}^2 - 2\bar{\bar{x}}\sum n_i\bar{x}_{i.} + \bar{\bar{x}}^2\sum n_i \\ &= \sum n_i\bar{x}_{i.}^2 - 2N\bar{\bar{x}}^2 + N\bar{\bar{x}}^2 \\ &= \sum n_i\bar{x}_{i.}^2 - N\bar{\bar{x}}^2, \end{aligned} \tag{15–14}$$

$$\begin{aligned}\text{SSE} &= \sum\sum (x_{ij} - \bar{x}_{i.})^2 \\ &= \sum\sum (x_{ij}^2 - 2x_{ij}\bar{x}_{i.} + \bar{x}_{i.}^2) \\ &= \sum \left(\sum x_{ij}^2 - 2\bar{x}_{i.}\sum x_{ij} + n_i\bar{x}_{i.}^2\right) \\ &= \sum \left(\sum x_{ij}^2 - 2n_i\bar{x}_{i.}^2 + n_i\bar{x}_{i.}^2\right) \\ &= \sum\sum x_{ij}^2 - \sum n_i\bar{x}_{i}^2. \end{aligned} \tag{15–15}$$

As a check, we note that SST = SSC + SSE. Also, SSE can be obtained by subtraction; that is, SSE = SST − SSC, if we wish.

The preceding formulas for the sums of squares are applicable for both equal and unequal sample sizes. If the null hypothesis that treatment effects are nil is true, then each of these three sums of squares, divided by an appropriate number of degrees of freedom, is an unbiased estimator of the common population variance σ^2. The total number of degrees of freedom in this model is $N - 1$, since SST is computed with a total of N observations by using the sample grand mean as an estimator of μ: one degree of freedom is thus lost. To estimate σ^2 on the basis of SSC—variation between column means—the number of degrees of freedom is $c - 1$, since there are c column means to compare, and again the grand sample mean is used as an estimator of μ for the calculation of SSC. The number of degrees of freedom for estimating σ^2 based on SSE is $N - c$, since SSE is obtained from all the N observations with the c group means as estimators of the population mean. It is interesting to note that, just as in the case of sums of squares, the numbers of degrees of freedom are also additive. Namely, we have here $N - 1 = (c - 1) + (N - c)$.

Estimators for the common population variance are often called *mean squares* in the analysis of variance. For the present model, as we shall see, we are interested in comparing two mean squares: the mean square between column means, MSC = SSC/$(c - 1)$, and the mean square for error, MSE = SSE/$(N - c)$. Note that both MSC and MSE are distributed as chi-square variables with $c - 1$ and $N - c$ df, respectively.

It is customary to summarize calculations for sums of squares, together with their numbers of df and mean squares, in a table called "the analysis-of-variance table." For the present model, see Table 15–2.

Table 15–2. Analysis-of-Variance Table: Linear Model

Source of variation	*SS*	δ	*MS*
Between (columns)	SSC	$c - 1$	$\text{MSC} = \text{SSC}/(c - 1)$
Within; error	SSE	$N - c$	$\text{MSE} = \text{SSE}/(N - c)$
Total	SST	$N - 1$	

The Test

For the present model, the pair of hypotheses to be tested is

H_0: All treatment means are equal, or $\alpha_i = 0$;
H_1: Not all treatment means are equal, or not all α_i are zero.

The testing statistic to be used here is the F ratio of MSC to MSE. The rationale is quite simple. Under the assumption that e_{ij} are independent and distributed normally with 0 mean and a variance identical with σ^2, MSE must be an unbiased estimate of σ^2 whether or not H_0 is true, since MSE measures chance variation alone. On the other hand, MSC is an unbiased estimate of σ^2 if and only if H_0 is true. When the treatment effects are not nil, MSC actually estimates the quantity $\sigma^2 + a$, where a is a positive bias and measures the degree of differences in treatments. Under these circumstances, we would expect $F_{c-1,N-c} = \text{MSC}/\text{MSE} = 1$, or very close to 1, if H_0 is true, and we would expect this ratio to be greater than 1 if H_0 is false. Thus, the F test is of the upper-tail variety. When the computed value of F is, or is close to, 1, the null hypothesis, $\alpha_i = 0$, is accepted. When it is significantly greater than 1, we conclude that treatment effects are not nil.

First Example: Equal Sample Sizes

Suppose a manufacturing company has purchased three new machines of different makes and wishes to determine whether one of them is faster than the others in producing a certain output. Five hourly production figures are observed at random from each machine, and the results are presented in Table 15–3. What conclusions can be drawn at $\alpha = 0.05$? At $\alpha = 0.01$?

Following our general procedure of testing, the foregoing questions can be answered by the following test:

(1) Hypotheses: H_0: $\mu_1 = \mu_2 = \mu_3$; H_1: Not all three μ_i are equal.

(2) Level of Significance: $\alpha = 0.05, 0.01$.

(3) TESTING STATISTIC: $F_{2,12} = \text{MSC}/\text{MSE}$. (*Note*: For this F, $\delta_1 = c - 1 = 3 - 1 = 2$, and $\delta_2 = N - c = 15 - 3 = 12$.)

(4) DECISION RULE: Since $P(F_{2,12} > 3.89) = 0.05$ and $P(F_{2,12} > 6.93) = 0.01$, H_0 will be rejected at the 5 percent level if the observed F is greater than 3.89, and it will be rejected at the 1 percent level of significance if the observed F is greater than 6.93.

(5) COMPUTATIONS: Using the sample data from Table 15–3, we may compute first, quite conveniently, the column group means and the grand sample mean. These summary statistics we place at the bottom of the table that contains the original data.

Next we should compute the sums of squares. In doing so, we see that there are only three quantities, $\sum\sum x_{ij}^2$, $\sum n_i\bar{x}_{i.}^2$, and $N\bar{\bar{x}}^2$, in the equations for the three sums in this model; therefore it is advisable to compute them first before obtaining the sums of squares.

Table 15–3. Hourly Output Volumes of Three Machines for Five Hours

Observations	A_1	A_2	A_3	*Total*
1	25	31	24	
2	30	39	30	
3	36	38	28	
4	38	42	25	
5	31	35	28	
Total	160	185	135	480
n_i	5	5	5	
$\bar{x}_{i.}$	32	37	27	$\bar{\bar{x}} = 32$

$$\sum\sum x_{ij}^2 = 25^2 + 30^2 + \cdots + 28^2 = 15{,}810,$$

$$\sum n_i\bar{x}_{i.}^2 = n_1(\bar{x}_{1.}^2) + n_2(\bar{x}_{2.}^2) + n_3(\bar{x}_{3.}^2)$$

$$= 5(32^2) + 5(37^2) + 5(27^2) = 15{,}610,$$

$$N\bar{\bar{x}}^2 = 15(32^2) = 15{,}360.$$

Finally, the three sums of squares are as follows:

$$\begin{aligned}\text{SST} &= \text{total variation}\\ &= \sum\sum x_{ij}^2 - N\bar{\bar{x}}^2 = 15{,}810 - 15{,}360 = 450,\end{aligned}$$

$$\begin{aligned}\text{SSC} &= \text{variation between columns (machines)}\\ &= \sum n_i\bar{x}_{i.}^2 - N\bar{\bar{x}}^2 = 15{,}610 - 15{,}360 = 250,\end{aligned}$$

$$\begin{aligned}\text{SSE} &= \text{variation within columns (error)}\\ &= \sum\sum x_{ij}^2 - \sum n_i\bar{x}_{i.}^2 = 15{,}810 - 15{,}610 = 200.\end{aligned}$$

We summarize these sums of squares, along with their degrees of freedom and mean squares, in Table 15–4.

Table 15–4. Summary Table for Machine Data

Source of variation	*SS*	δ	*MS*
Machines (between columns)	250	$3 - 1 = 2$	125.00
Error (within columns)	200	$15 - 3 = 12$	16.67
Total	450	$15 - 1 = 14$	

$$F_{2,12} = \frac{\text{MSC}}{\text{MSE}} = \frac{125.00}{16.67} = 7.50.$$

(6) DECISIONS: Since the computed value is greater than both 3.89 and 6.93, H_0 is rejected at both the 5 percent and 1 percent levels of significance. The three new machines are significantly different in their mean speeds; or, the machine effects are significant.

Second Example: Unequal Sample Sizes

During the past four semesters, four different textbooks have been used in this author's class in mathematical economics, and the final grades for the students in these four classes are recorded in Table 15–5. In view of these data, can we conclude that students' scores are significantly affected by textbooks?

Table 15–5. Final Examination Scores

	Textbook				
	(1)	(2)	(3)	(4)	
	00	80	97	07	
	80	81	84	84	
	69	73	93	90	
	65	69	79	78	
		75	92	61	
		72			
Total	274	450	445	380	$\sum\sum x_{ij} = 1549$
n_i	4	6	5	5	$N = 20$
$\bar{x}_i$	68.5	75.0	89.0	76.0	$\bar{\bar{x}} = 77.45$

Computations for this example are as follows:

$$\sum \sum x_{ij}^2 = 60^2 + 80^2 + \cdots + 61^2 = 122{,}115,$$

$$\sum n_i \bar{x}_{i.}^2 = 4(68.5^2) + 6(75.0^2) + 5(89.0^2) + 5(76.0^2) = 121{,}004,$$

$$N(\bar{\bar{x}}^2) = 20(77.45^2) = 119{,}970,$$

$$\text{SST} = 122{,}115 - 119{,}970 = 2145,$$

$$\text{SSC} = 121{,}004 - 119{,}970 = 1034,$$

$$\text{SSE} = 122{,}115 - 121{,}004 = 1111.$$

Table 15–6. Summary Table for Textbook Data

Source of variation	*SS*	δ	*MS*
Textbooks (between columns)	1034	3	344.65
Error (within columns)	1111	16	69.44
Total	2145	19	

$$F_{3,16} = \frac{344.65}{69.44} = 4.96.$$

Since $P(F_{3,16} > 3.24) = 0.05$ and $P(F_{3,16} > 5.29) = 0.01$, there is a significant difference between textbooks at $\alpha = 0.05$; the null hypothesis that there is no textbook effect cannot be rejected at $\alpha = 0.01$. These conclusions are valid only if the four classes constitute four independent random samples.

15–8 THE LEAST SIGNIFICANT DIFFERENCE

The analysis of variance proper shows only whether or not there exists any significant difference among treatment means. It does not reveal where the difference is. When the F test indicates the existence of a significant difference, we naturally wish to know which of the means causes the difference. When sample sizes are equal, as in our first illustration in the last section, a simple procedure, called the *least significant difference*, can aid us in locating this source.

When $n_1 = n_2 = \cdots = n_c$ and when the F test leads to the rejection of $\alpha_i = 0$, the least significant difference, denoted as LSD, is defined as the smallest difference that could exist between two significantly different sample means. To derive a formula for LSD, we recall that a test for the difference between two sample means is made by the statistic

$$t = \frac{\bar{x}_1 - \bar{x}_2}{\hat{\sigma}_{\Delta\bar{x}}}.$$

If this ratio exceeds, say, $t_{\delta;0.05}$, we say the difference between the two sample means is significant at $\alpha = 0.05$. Thus, LSD can be considered as the difference between two sample means for which $t = t_{\delta;\alpha}$, where α is the level of significance:

$$\text{LSD} = t_{\delta;\alpha}\hat{\sigma}_{\Delta\bar{x}}.$$

Furthermore, if $n_1 = n_2 = n$ and $\sigma_1^2 = \sigma_2^2$ is assumed,

$$\hat{\sigma}_{\Delta\bar{x}} = \sqrt{\frac{\hat{\sigma}_1^2}{n_1} + \frac{\hat{\sigma}_2^2}{n_2}} = \sqrt{\frac{2\hat{\sigma}^2}{n}},$$

where, as it can be shown, $\hat{\sigma}^2 = \text{MSE}$, the mean square for error. We have, therefore,

$$\text{LSD} = t_{\delta;\alpha}\sqrt{\frac{2}{n}\,\text{MSE}},$$

where $\delta = N - c$. But, as will be shown later, when t has δ_2, then $t_{\delta_2}^2 = F_{1,\delta_2}$. The formula for the least significant difference becomes, finally

$$\text{LSD} = \sqrt{\frac{2}{n}\,(\text{MSE})F_{1,N-c,\alpha}}. \tag{15–16}$$

Any two treatment sample means in the analysis of variance are said to differ from each other significantly at a given level of α if their absolute difference is greater than LSD as defined by (15–16).

To apply (15–16) to our first illustration, we note $n_1 = n_2 = n_3 = 5$, $\text{MSE} = 16.67$, and $F_{1,12;0.01} = 9.33$. Thus

$$\text{LSD} = \sqrt{2/5\,(16.67)(9.33)} = 7.89.$$

From calculations in Table 15–3, we find that

$$|\bar{x}_{1.} - \bar{x}_{2.}| = |32 - 37| = 5 < 7.89 \qquad \text{(not significant)},$$

$$|\bar{x}_{1.} - \bar{x}_{3.}| = |32 - 27| = 5 < 7.89 \qquad \text{(not significant)},$$

$$|\bar{x}_{2.} - \bar{x}_{3.}| = |37 - 27| = 10 > 7.89 \qquad \text{(significant)}.$$

Thus, among the three pairs of column means, only one pair $(\bar{x}_{2.}, \bar{x}_{3.})$ differs significantly. Other things being equal, management should decide to use the second make of machines exclusively.

It is important to note that the LSD tests are also called *pairwise tests*, and they should be employed only when the F test has led to the rejection of H_0. When LSD is erroneously employed in a case where F value is insignificant, it is quite possible to find that some pair or pairs of sample means differ by an amount greater than the value of LSD. However, such differences are still explained by chance variations in random sampling from the same population or populations with identical means.

In a case where samples are not of equal size, there is no simple generalized procedure for determining which pair of samples cause the significant difference in the F test.

15–9 RELATIONSHIPS AMONG SAMPLING DISTRIBUTIONS

We have now covered the classical decision procedures with all the important sampling distributions: the binomial, normal, chi-square, t, and F probability models. Notice that these are not empirical but theoretical distributions. None is empirical in the sense that a decision maker has actually taken a large number of samples and discovered that the sample values occur according to the relative frequencies given by the function rule. Each is a theoretical distribution in the sense that a sampling distribution is deduced mathematically or by a priori reasoning. We have argued that, if simple random sampling of independent observations is made from certain kinds of populations, various sample statistics must possess distributions given by the several function rules. As theoretical distributions, the assumptions from which sampling distributions are derived are of great importance, since they are the justifications of our methods in making inferences.

In addition to the requirement of simple random sampling, the most basic assumption made in deriving sampling distributions is that the population must be a normal variable. All the three sampling distributions introduced in this and the last chapters rest on the premise of normality. The normal distribution may thus be thought of as the parent distribution of these others. We have seen that the chi-square distribution is a sum of squares of independent standard normal variables. We have also shown that the distribution of sample variance can be solved explicitly on the elementary level only for a normal population. Furthermore, the distribution of $\hat{s}^2$ depends on the chi-square distribution, which, in turn, is based on the assumption of a normal distribution of single observations. The F distribution, as a ratio of two independent chi-square variables, clearly also rests on the assumption of normal populations.

The close relationships among the normal, chi-square, F and t distributions can be established in a number of ways. We shall give a few that can be readily understood. First of all, the t ratio, for instance, may be considered as that of a standard normal variable Z to the square root of a variable that is independent of Z and that is distributed as χ_δ^2/δ. That is,

$$t = \frac{Z}{\sqrt{\chi_\delta^2/\delta}}.$$

To show that this definition of t is valid, we see that

$$t = \frac{\bar{x} - \mu}{\sqrt{\hat{s}^2/n}} = \frac{(\bar{x} - \mu)/\sigma}{\sqrt{\hat{s}^2/n\sigma^2}} = \frac{(\bar{x} - \mu)/(\sigma/\sqrt{n})}{\sqrt{\hat{s}^2/\sigma^2}}.$$

The numerator of this t ratio is clearly $n(0, 1)$ and is the normal sampling distribution of means. Moreover, we know that

$$\frac{(n-1)\hat{s}^2}{\sigma^2} = \chi_\delta^2$$

and

$$\frac{\hat{s}^2}{\sigma^2} = \frac{\chi_\delta^2}{\delta}.$$

Here $\delta = n - 1$. It follows, therefore, that

$$t = \frac{Z}{\sqrt{\chi_\delta^2/\delta}}.$$

The relationship between t and F distributions can be easily established by squaring both sides of the previous expression to obtain

$$t^2 = \frac{Z^2}{\chi_\delta^2/\delta}.$$

By definition, the numerator of t^2 is χ_1^2 and the denominator of t^2 is χ_δ^2. Hence, t^2 qualifies as $F_{1,n-1}$. In general,

$$t_\delta^2 = F_{1,\delta_2}, \quad \delta = \delta_2 \qquad \text{for } t^2.$$

That is, the square of t with δ is an F variable with 1 and δ_2 degrees of freedom.

Finally, we know that F is related to chi-square distributions directly. If $\chi_{\delta_1}^2$ and $\chi_{\delta_2}^2$ are two independent chi-square variables, we can state that

$$F = \frac{\chi_{\delta_1}^2/\delta_1}{\chi_{\delta_2}^2/\delta_2},$$

which clearly satisfies the definition that

$$F = \frac{\hat{s}_1^2}{\hat{s}_2^2} = \frac{\hat{s}_1^2/\sigma^2}{\hat{s}_2^2/\sigma^2} = \frac{\chi_{\delta_1}^2/\delta_1}{\chi_{\delta_2}^2/\delta_2},$$

We note that division of both the numerator and denominator by σ^2 in the above derivation to form chi-square distributions automatically specifies the null hypothesis $\sigma_1^2 = \sigma_2^2$ in comparing two population variances.

EXERCISES

15–1 The producer of a certain make of dry-cell battery claims its output has a mean life of more than 750 minutes. Suppose a sample of 15 of these batteries has been tested and yielded the following values: 730, 759, 725, 740, 754, 745, 750, 753, 730, 780, 725, 790, 719, 775, 700. From these data, can you conclude that the producer's claim is valid at a 1 percent level of significance?

15–2 A sample of ten measurements of the diameter of a sphere gives a mean of 4.08 inches and a standard deviation ($\hat{s}$) of 0.05 inch. Are these data consistent with the fact that the manufacturing process is adjusted to produce diameters with $\mu = 4$ inches, given $\alpha = 0.05$?

15–3 A cannery has just installed a new filling machine. A random sample of 20 filled cans yields a mean weight of the contents of 16.05 ounces with $\hat{s}$ of 1.5 ounces.

(a) Make an interval estimate of the true mean with $1 - \alpha = 0.95$.

(b) If the net contents of each can is supposed to be 16 ounces, is the machine properly adjusted?

15–4 Construct a 95 percent confidence interval for μ with the data in Exercise 15–2. In view of this interval, what can you say about the conclusion reached for Problem 15–2?

15–5 Two working designs are under consideration for adoption in a plant. A time and motion study shows that 12 workers using design A have a mean assembly time of 300 seconds with a standard deviation of 12 seconds, and that 15 workers using design B have a mean assemble time of 335 seconds with a standard deviation of 15 seconds. Is the difference in mean assembly time between the two working designs significant at a 1 percent level of significance?

15–6 Construct a 99 percent confidence interval for the true difference between mean assembly times of the working designs in the preceding problem.

15–7 Test the hypothesis that $\mu_1 = \mu_2$ against the one that $\mu_1 > \mu_2$ with $\alpha = 0.05$ for data as follows:

$$\bar{x}_1 = 72.50, \qquad \hat{s}_1^2 = 8.6, \qquad n_1 = 13;$$

$$\bar{x}_2 = 66.05, \qquad \hat{s}_2^2 = 6.1, \qquad n_2 = 7.$$

Construct a 95 percent confidence interval for $\Delta\mu$ with the same data.

15–8 To compare the efficiency of standard and electric typewriters, eight typists are chosen at random and thoroughly oriented to the operation of both kinds of typewriters.

Table 15–P1

	Average speed per minute	
Typist	*Electric*	*Standard*
1	75	79
2	89	62
3	79	54
4	85	67
5	102	81
6	115	78
7	97	66
8	69	73

Then they are asked to type on each kind of typewriter for ten minutes and their speeds, measured as average number of words typed per minute, are observed. The results of this experiment are shown in the last two columns of Table 15–P1. Can you conclude that the two types of typewriters are different in efficiency at $\alpha = 0.01$?

15–9 Suppose that a big corporation will decide to buy the standard typewriters unless the average speed of the electric typewriters is 30 words per minute greater with a probability of 0.99. In view of results obtained in the preceding problem, how would the corporation decide?

15–10 Ten students selected at random have the following final grades in physics and economics:

Student	1	2	3	4	5	6	7	8	9	10
Physics	66	72	50	81	62	73	55	90	77	85
Economics	75	70	65	88	59	85	60	97	82	90

Can you conclude that the mean grades in these two subjects are different? Construct a 95 percent confidence interval for the true mean difference for this illustration.

15–11 Two methods of performing a certain operation are compared. The results are assumed to be normal variables, with possible differences in expectations and variances. Suppose data obtained yield the following information:

$$\bar{x}_1 = 72.50, \qquad \hat{s}_1^2 = 8.6, \qquad n_1 = 13,$$

$$\bar{x}_2 = 66.05, \qquad \hat{s}_2^2 = 6.1, \qquad n_2 = 7.$$

Test the hypothesis that the two variances are equal at $\alpha = 0.10$.

15–12 With data in the preceding problem, can we accept H_0 if the alternative hypothesis is $\sigma_1^2 \neq \sigma_2^2$ at $\alpha = 0.10$?

15–13 Suppose there are two sources of raw materials under consideration. Both sources seem to have similar characteristics, but we are not sure about their respective uniformity. A sample of ten lots from source A yields a variance of 250, and a sample of eleven lots from source B yields a variance of 195. Is it likely that the variance of source A is greater than the variance of source B at $\alpha = 0.10$?

15–14 Would you reach the same conclusion for a two-tail test with data in the preceding problem?

15–15 A random sample consisting of ten colleges in New York State yields the following minimum salaries for instructors: 5000, 5400, 6000, 5200, 4700, 7200, 5400, 9000, 8000, and 7500. A random sample of eight colleges in California gives the following information on minimum salaries: 7400, 6000, 6500, 6200, 7500, 7800, 8000, and 8200. Test the hypothesis $\sigma_n^2 = \sigma_c^2$ against $\sigma_n^2 \neq \sigma_c^2$ at $\alpha = 0.10$.

15–16 An agricultural experimental station wishes to determine the yields of five varieties of corn. Each type of corn is planted on four plots of land with equal fertility, and the yields, in units of 100 bushels, are as follows:

Corn				
A	*B*	*C*	*D*	*E*
4	7	5	11	4
6	13	4	10	8
4	10	4	9	6
10	12	9	14	10

In view of these results, can we conclude that the mean yields of the five varieties of corn are the same at the 5 percent level of significance?

15–17 Apply the pairwise test to the results in the preceding problem.

15–18 Miss Smith, supervisor of a secretarial pool, has four typists under her supervision. She is concerned with the length of their coffee-breaks, so she times them. Her observations, recorded in minutes, for each girl are as follows. Can the differences in average time that the four typists spent on coffee-breaks be explained by chance variation?

Typist			
A	*B*	*C*	*D*
27	20	24	20
35	22	22	18
18	30	25	26
24	27	25	19
28	22	20	26
32	24	21	24
16	28	34	26
18	21	18	—
25	23	32	159
—	18	23	
223	30	22	
	32	—	
	—	266	
	297		

15–19 We wish to test whether there are any differences in the durability of three makes of computers. Samples of size $n_i = 5$ are selected from each make, and the frequency of repair during the first year of purchase is observed. The results are as follows:

Make		
A_1	A_2	A_3
10	6	5
11	7	3
9	8	4
9	6	2
8	5	1

In view of the data above, what conclusion can you draw? If a significant difference exists between A_i, which treatment mean seems to have produced the difference?

15–20 Four machines, A, B, C, and D, are used to produce a certain kind of cotton fabric. Samples of size $n_i = 4$, with each unit as 100 square yards, are selected from outputs of the machines at random, and the number of flaws in each 100 square yards are counted, with the following result :

Machine			
A	*B*	*C*	*D*
20	13	8	9
24	17	10	5
19	15	12	7
21	14	9	4

In view of the data above, can the null hypothesis $\alpha_i = 0$ be rejected at the 5 percent level of significance? At the 1 percent level of significance? Which pairs of means are different at $\alpha = 0.05$? at $\alpha = 0.01$?

15–21 The same examination is given to five classes in elementary statistics taught by five different instructors, A_i, and it is desired to test whether the average grades of the five classes are equal. Random samples of size $n_i = 5$ are drawn from each class, and the grades are shown below. What conclusion can you reach?

Instructor				
A_1	A_2	A_3	A_4	A_5
85	66	59	61	55
78	75	50	71	60
90	72	66	69	73
79	68	70	80	61
92	81	62	75	81

15–22 A random sample is selected from each of three makes of ropes, and their breaking strengths (in pounds) are measured, with the following results:

Source		
I	*II*	*III*
95	121	82
87	130	73
90	99	76
88	119	70
101	125	81
	127	71
	120	

Test whether there are any differences in the breaking strengths of the ropes.

16

BAYESIAN INFERENCE

16–1 INTRODUCTION

We have been concerned in the preceding four chapters with classical estimation and testing procedures. The classical analysis is purely empirical, since an inference is made on sample information alone. It suffers a serious limitation, also, in that the costs of wrong decisions are not explicitly evaluated. With interval estimation, we are given a specific level of confidence that the interval will include the parameter being estimated, but we are not required to consider what it will cost us if the interval does not contain the estimated parameter. With testing, a level of significance is arbitrarily selected as a measure of the maximum probability of committing the type I error. This level, as pointed out in Section 13–1, was thought to depend upon both our degree of belief in the null hypothesis and the costs associated with rejecting it when it is true. However, in the actual testing process, neither the prior beliefs nor the costs of wrong decisions were explicitly evaluated. Then, when determining the optimal sample size, we specify points on a power curve without giving precise reasons why these points are selected.

Despite these drawbacks, however, classical methods are still appropriate in a large class of decision situations where the monetary results cannot be measured accurately and are of negligible importance compared to non-economic consequences, as in many problems associated with scientific research. A scientist may be completely incapable of appraising realistically the losses from the wrong actions of either accepting or rejecting a research hypothesis. Neither can he evaluate the losses in estimating a parameter when the estimate may be used for a variety of purposes, some of which are unknown. In addition, in scientific research, one is often interested only in selecting an experiment whose results possess maximum information in some sense and not in arriving at an explicit terminal action. In such circumstances the classical approach seems much more sensible, since it has been designed for scientific or experimental work where the probability of error can be assessed more accurately than the consequences of a decision.

Bayesian theorists, however, make several rejoinders to the foregoing arguments. While ignoring loss functions explicitly, the classical theory does consider them implicitly. Otherwise, how can one decide upon the confidence level, significance level, and optimal sample size for a given decision situation? Furthermore, if it is desired to obtain information, should this requirement be formalized by introducing appropriate information measures as a part of loss functions? As the Bayesians are aware, such information measures are easier said than done! So the controversy persists.

In many business decision situations, objective probabilities are often unavailable and the costs of wrong decisions not only are of prime importance but also are subject to relatively accurate measurement. It is appropriate, then, for business decision makers to make explicit use of prior probabilities and estimated losses resulting from various decisions, as well as costs of sampling. The Bayesian approach has been designed to take all these aspects of a decision into account.

The Bayesian approach differs from the classical approach mainly in that prior information, often representing the decision maker's personal judgment, is formally utilized. That is, Bayesians desire to use any and all available information. Bayesian methods, too, can establish inferential procedures on the basis of combined information or in terms of decision-theoretic considerations. The latter approach, which bases inference on combined information and explicit loss functions, is the one we shall adopt. In other words, we shall present Bayesian inference from the standpoint of decision theory. In what follows, we shall first introduce the salient features of Bayesian estimation and testing and then point out the relationship between these two branches of inference via a discussion of the general form of Bayesian decision theory. From time to time we shall also compare Bayesian and classical methods. Finally, it will be pointed out that, under certain conditions, both methods may provide similar or even identical results but with a difference in interpretation.

16–2 ESTIMATION AS A BAYESIAN DECISION

When we approach inferential problems from the viewpoint of modern decision theory, an important consideration is the selection of a loss function, such as those introduced in Chapter 6, whose properties suit a given decision situation. We usually take into account the various consequences in terms of rewards and penalties for right and wrong decisions. Precisely, we are required to select a loss function that assigns a numerical value $L(a, \theta)$ to each pair of values of a and θ, where a and θ are the members of the action and state sets, respectively. With the loss function selected, we can apply the Bayesian strategy to select the act that has the minimum expected loss.

Turning to estimation problems, we note that the Bayesian approach can be associated with "many-state and many-action" decision situations, where the number of states may be infinite and where each act may be appropriate for one or more states. Regrets, or opportunity losses, are incurred if the decision maker *acts* as if the state were, say, $\hat{\theta}$, when the true value of the state is θ. The opportunity loss is usually assumed to be some function of the difference $\hat{\theta} - \theta$, and the loss function may vary, depending upon whether $\hat{\theta}$ underestimates or overestimates θ.

To bring out the salient features of the Bayesian estimation procedure, let us consider an example. Suppose the executives of a national TV network are reviewing the possible proportions of the total viewing audience for a proposed program. Such proportions are important for the purpose of deciding not only whether the program should be introduced at all but also at what hour of the day or evening it should be programmed. For the sake of simplicity, let us assume that only the following three states of nature are of interest:

$\theta_1 = 0.2$, viz., the program will have 20 percent of the audience;

$\theta_2 = 0.3$, viz., the program will have 30 percent of the audience;

$\theta_3 = 0.4$, viz., the program will have 40 percent of the audience.

Let us assume further that the executives' forecasts (predictions of possible sample results) are E_1, E_2, and E_3 for the three states of nature. According to past records, the executives have exhibited a "reliability" in their forecasts as summarized in Table 16–1. The reliability indices are the

Table 16–1. Reliability of Executives' Forecast

θ_j	$P(E_1 \mid \theta_j)$	$P(E_2 \mid \theta_j)$	$P(E_3 \mid \theta_j)$
θ_1	0.7	0.2	0.1
θ_2	0.3	0.6	0.1
θ_3	0.2	0.3	0.5

likelihoods of the forecasts (or possible sample results) given the various states of nature.

We shall denote the "acts" of forecasting (estimating) θ_1, θ_2, and θ_3 as $\hat{\theta}_1$, $\hat{\theta}_2$, and $\hat{\theta}_3$. Now, if we apply the classical maximum likelihood procedure to this problem, we estimate by $\hat{\theta}_j$ the state of nature θ_j that maximizes the likelihood, or conditional probability, $P(E_k \mid \theta_j)$. Thus, for our simple illustration, we need only go to Table 16–1 and find the largest likelihood for each possible forecast. We see that $\hat{\theta}_1$, $\hat{\theta}_2$, and $\hat{\theta}_3$ are the maximum likelihood estimators of θ_1, θ_2, and θ_3, respectively; that is,

If E_1 is observed, select "act" $\hat{\theta}_1$, since $P(E_1 \mid \theta_1) = 0.7 >$ others;

If E_2 is observed, select "act" $\hat{\theta}_2$, since $P(E_2 \mid \theta_2) = 0.6 >$ others;

If E_3 is observed, select "act" $\hat{\theta}_3$, since $P(E_3 \mid \theta_3) = 0.5 >$ others.

By the Bayesian approach, however, the optimal act (estimator) is selected by incorporating both prior distribution and opportunity losses into evaluation. Suppose the executives assume a uniform prior distribution for the state variable and are willing to assume a constant loss function of the form

$$L(\hat{\theta}, \theta) = \begin{cases} 0, & \text{for } \hat{\theta} = \theta; \\ 1, & \text{for } \hat{\theta} \neq \theta. \end{cases}$$

Then, the prior probabilities and losses of various actions can be summarized as in Table 16–2.

Table 16–2

		Opportunity losses		
θ_j	$P(\theta_j)$	$\hat{\theta}_1$	$\hat{\theta}_2$	$\hat{\theta}_3$
θ_1	1/3	0	1	1
θ_2	1/3	1	0	1
θ_3	1/3	1	1	0

Having established the prior probability distribution and opportunity losses, the Bayesian approach would then direct the decision maker to compute the posterior probabilities given each possible forecast (or sample outcome), to determine the expected losses for all possible acts (estimators), and then to select the act as the optimal estimator, for a given state, that has the lowest expected loss. These computations are done in Table 16–3 for our illustration.

Table 16–3. Calculation of Bayesian Estimators

		E_1			E_2			E_3	
Joint Prob.									
$P(\theta_1)P(E_k \mid \theta_1)$		$7/30$			$2/30$			$1/30$	
$P(\theta_2)P(E_k \mid \theta_2)$		$3/30$			$6/30$			$1/30$	
$P(\theta_3)P(E_k \mid \theta_3)$		$2/30$			$3/30$			$5/30$	
$P(E_k)$		$12/30$			$11/30$			$7/30$	
Posterior									
$P(\theta_1 \mid E_k)$		$7/12$			$2/11$			$1/7$	
$P(\theta_2 \mid E_k)$		$3/12$			$6/11$			$1/7$	
$P(\theta_3 \mid E_k)$		$2/12$			$3/11$			$5/7$	
Estimators	$\hat{\theta}_1$	$\hat{\theta}_2$	$\hat{\theta}_3$	$\hat{\theta}_1$	$\hat{\theta}_2$	$\hat{\theta}_3$	$\hat{\theta}_1$	$\hat{\theta}_2$	$\hat{\theta}_3$
$E[L(\hat{\theta}_j, E_k)]$*	$5/12$	$9/12$	$10/12$	$9/11$	$5/11$	$8/11$	$6/7$	$6/7$	$2/7$
Optimal est.		$\hat{\theta}_1$			$\hat{\theta}_2$			$\hat{\theta}_3$	

* For example, $E[L(\hat{\theta}_1, E_1)] = 0(7/12) + 1(3/12) + 1(2/12) = 5/12$.

If we trace through Table 16–3, we note that the calculations lead to the following:

If E_1 is observed, take "act" $\hat{\theta}_1$.

If E_2 is observed, take "act" $\hat{\theta}_2$.

If E_3 is observed, take "act" $\hat{\theta}_3$.

But these are exactly the same as the classical maximum likelihood estimators. Such results should not be surprising, since the particular example here, with a uniform prior and a constant loss function, is an equivalent state-action problem, and the special form of Bayesian strategy is to select as the optimal act that for which tho probability of the act given the state is a maximum. From this observation however, we may conclude that a classical statistician who uses MLE can get the same result as a Bayesian using the 0–1 constant loss function and an "equiprobable" prior. This conclusion seems to be a rather damaging description of the method of MLE, since neither this prior nor this loss function is easy to justify in cases such as the present one. However, that MLE is not nearly so restrictive in many situations. It can be shown that if the likelihood function (the conditional probability function) is unimodal and symmetric, such as a normal distribution, then MLE is equivalent to Bayesian estimation using any frequently encountered loss function, as introduced in Chapter 6.

We are interested now in establishing that the Bayesian approach may lead to different optimal estimators with different prior distributions and/or different loss functions. The loss function usually employed in estimation from the viewpoint of decision theory is the quadratic loss function. Suppose, for our example, according to the prior judgment of the executives, θ_1 and θ_2 are equally likely but θ_3 is only half as likely as either θ_1 or θ_2, and suppose the executives think the proper loss function should be of the quadratic form

$$L(\hat{\theta}, \theta) = 100(\hat{\theta} - \theta)^2;$$

then the prior probability distribution and opportunity losses would be as given in Table 16–4.

Table 16–4

		Opportunity loss		
State	*Prior*	$\hat{\theta}_1 = 0.2$	$\hat{\theta}_2 = 0.3$	$\hat{\theta}_3 = 0.4$
$\theta_1 = 0.2$	0.4	0	1	4
$\theta_2 = 0.3$	0.4	1	0	1
$\theta_3 = 0.4$	0.2	4	1	0

With this new information, as the computations in Table 16–5 show, the following optimal Bayes estimators are obtained:

If E_1 is observed, take "act" $\hat{\theta}_1$.

If E_2 is observed, take "act" $\hat{\theta}_2$.

If E_3 is observed, take "act" $\hat{\theta}_2$.

The fact that $\hat{\theta}_2$ is the optimal act whether E_2 or E_3 is observed clearly reveals that a rather low prior probability is assigned to the third state of nature, and reflects the relatively low reliability of $P(E_3 \mid \theta_3) = 0.5$.

We should note that the Bayesian method of estimation is usually more difficult and laborious than the maximum likelihood method, since it requires the specification of both a prior distribution and a loss function. This and the fact that most statisticians are strictly classical explains the wide application of maximum likelihood estimators in practice, even though the Bayesian method is more attractive from a theoretical point of view. Two observations are important here, however. First, although MLE has

Table 16–5

		E_1			E_2			E_3	
Joint*									
$P(\theta_1)P(E_k \mid \theta_1)$		0.28			0.08			0.04	
$P(\theta_2)P(E_k \mid \theta_2)$		0.12			0.24			0.04	
$P(\theta_3)P(E_k \mid \theta_3)$		0.04			0.06			0.10	
$P(E_k)$		0.44			0.38			0.18	
Posterior									
$P(\theta_1 \mid E_k)$		$28/44$			$8/38$			$4/18$	
$P(\theta_2 \mid E_k)$		$12/44$			$24/38$			$4/18$	
$P(\theta_3 \mid E_k)$		$4/44$			$6/38$			$10/18$	
Estimators	$\hat{\theta}_1$	$\hat{\theta}_2$	$\hat{\theta}_3$	$\hat{\theta}_1$	$\hat{\theta}_2$	$\hat{\theta}_3$	$\hat{\theta}_1$	$\hat{\theta}_2$	$\hat{\theta}_3$
$E[L(\theta_c, E_k)]$	$28/44$	$32/44$	$124/44$	$48/38$	$14/38$	$56/38$	$44/18$	$14/18$	$20/18$
Optimal est.		$\hat{\theta}_1$			$\hat{\theta}_2$			$\hat{\theta}_2$	

* The likelihoods used to compute these probabilities are taken from Table 16–1.

many attractive properties, as enumerated in Chapter 13, these are large-sample properties. When samples are small, MLE may lead to rather unbelievable results (results having extreme values), and it should, therefore, be used with great caution. (See Problem 16–5 for an example of this point.) Second, the Bayesian method of estimation can be considerably simplified if the prior density function can be reasonably approximated by certain known probability laws, since there are simple formulas that permit the revised probabilities to be calculated directly, given the sample information. For example, it was demonstrated in Chapter 7 that, if the probability density function over the states is normal, the expected opportunity loss can be determined directly by a simple formula. Furthermore, under the assumption of normal sampling, the resulting posterior distribution would also be normal and the optimal act—the estimator that minimizes the expected loss—would be the same as long as the loss function is symmetric, such as the constant, the quadratic, or the linear loss function.

We shall proceed to introduce Bayesian estimation for the normal case.

16–3 THE BAYESIAN ESTIMATOR OF μ

Consider a population X with variance σ^2 and an unknown mean μ that we wish to estimate. The distribution of X is assumed to be normal.

The Bayesian estimation procedure is to establish first a prior probability distribution of μ_X with mean μ_0 and variance σ_0^2. This distribution is assumed to be normal and to reflect the decision maker's subjective evaluation of the value of the parameter μ to be estimated. Precisely, μ_0 is considered as the prior estimator of μ. The confidence that the decision maker has in his judgment of the value of μ_0 can be revealed in his selection of σ_0^2. In general, the more vague the decision maker is about μ_0, the greater is the value he assigns to σ_0^2. Note that μ_X is the random variable in the prior distribution and its variability is measured by σ_0^2.

Next, a random sample of size n is drawn from the population under consideration and the sample mean is computed. Here, the sampling distribution of $\bar{x}$ is a likelihood function, $f(\bar{x} \mid \mu)$. This likelihood function, as the reader learned in our discussion of the theory of sampling, has expectation $E(\bar{x}) = \mu$ and variance $V(\bar{x}) = \sigma^2/n$. Furthermore, if X is normally distributed, then $f(\bar{x} \mid \mu)$ is normal irrespective of the size of n.

Finally, with the distribution of $\bar{x}$, the prior distribution can be revised to produce a subjective posterior distribution of the population mean, $f(\mu \mid \bar{x})$. Since both $f(\mu_X)$ and $f(\bar{x} \mid \mu)$ are normal, $f(\mu \mid \bar{x})$ will also be normal with mean and variance defined as

$$\mu_1 = \frac{(1/\sigma_0^2)\mu_0 + [1/V(\bar{x})]\bar{x}}{1/\sigma_0^2 + 1/V(\bar{x})}$$

$$= \frac{(\mu_0)V(\bar{x}) + \bar{x}(\sigma_0^2)}{\sigma_0^2 + V(\bar{x})} \tag{16-1}$$

and

$$\sigma_1^2 = \frac{1}{1/\sigma_0^2 + 1/V(\bar{x})}$$

$$= \frac{(\sigma_0^2)V(\bar{x})}{\sigma_0^2 + V(\bar{x})}. \tag{16-2}$$

Now, suppose we assume a quadratic loss function; then the posterior mean, μ_1, is the optimal estimator of the population mean, μ. Furthermore, we note that since the posterior distribution is normal in this case, all the three loss functions introduced before would lead to the same optimal estimator—because for a symmetrical distribution the mean, the mode, and the median coincide.

Note that the revised mean of the posterior distribution of μ_X is a weighted mean of the prior and the sample means. The weights are the reciprocals of the variances of the respective distributions. These weights are sometimes referred to as the *precision*, or *information*, *content* of the respective estimates. From (16-2) we see that the information content of the posterior distribution is the sum of the information contents of the prior distribution and the sampling distribution, since

$$\frac{1}{\sigma_1^2} = \frac{1}{\sigma_0^2} = \frac{1}{V(\bar{x})} \tag{16-3}$$

or

$$\frac{1}{\sigma_1^2} = \frac{1}{\sigma_0^2} + \frac{n}{\sigma^2}, \tag{16-4}$$

which indicates that the revised variance is smaller than that of either the prior distribution or the distribution of the sample mean. Thus, the posterior distribution, which combines estimates from both sources (judgment and sample), has more information than either source separately.

Also observe that to compute $V(\bar{x})$, we need the parent population variance. When it is unknown, we may use $\hat{s}^2$ in place of σ^2 to compute $V(\bar{x})$.

Let us consider an example. The manufacturer of a new machine designed to produce product X claims that its performance is characterized by a normal distribution of X with a mean of 25 units of output per hour and a variance of 6. That is, the manufacturer claims that $f(x) = n(25, 6)$.

Suppose the production manager of a certain firm would like to buy the machine if the mean output per hour is as claimed. After thorough study of the machine's design and construction, the production manager believes the chances are about 2 out of 3 the mean output would fall within the range of 22 ± 3 and the chances are about 96 out of 100 that the mean output would fall within the range of $22 \pm 2(3)$. In other words, according to his judgment the prior distribution of μ_X is normal with μ_0 (the prior estimator of μ) equal to 22 and a variance σ_0^2 equal to 9.

In order to revise his prior beliefs, the production manager observed four such machines in operation for an hour and obtained a sample mean $\bar{x} = 23$. Also, for $n = 4$, and assuming σ^2 is 6 as claimed by the manufacturer, we have

$$V(\bar{x}) = \frac{\sigma^2}{n} = \frac{6}{4} = 1.5.$$

Thus, the experiment yields a likelihood function $f(\bar{x} \mid \mu) = n(23, 1.5)$.

With the empirical results obtained, the revised, or posterior, mean and variance can now be computed with prior information as follows:

$$\begin{aligned}
\mu_1 &= \frac{(\mu_0)V(\bar{x}) + (\bar{x})\sigma_0^2}{\sigma_0^2 + V(\bar{x})} \\
&= \frac{22(1.5) + 23(9)}{9 + 1.5} \\
&= 22.86
\end{aligned}$$

and
$$\sigma_1^2 = \frac{(\sigma_0^2)V(\bar{x})}{\sigma_0^2 + V(\bar{x})}$$

$$= \frac{9(1.5)}{9 + 1.5}$$

$$= 1.29.$$

Hence, the posterior distribution becomes

$$f(\mu \mid \bar{x} = 21) = n(22.86, 1.29),$$

compared to the prior distribution,

$$f(\mu_X) = n(22, 9).$$

These results clearly indicate the Bayesian logic of adjusting a prior distribution by taking observed data into account, weighting the observed $\bar{x}$ according to the likelihood $P(\bar{x} \mid \mu)$. The end product is the posterior distribution, with mean increasing, 22.86, between the prior mean of 22 and the observed sample mean of 23. This should be expected in view of the sample result. In addition, the variance is reduced in the posterior distribution. This is a very attractive and useful feature of the Bayesian procedure.

Now, with the posterior distribution on hand, and assuming a quadratic loss function, the Bayesian optimal estimator for the population mean is the posterior mean (22.86). If a 95 percent confidence interval is required, we have

$$\begin{aligned} \mu_1 \pm 1.96\sigma_1 &= 22.86 \pm 1.96\sqrt{1.29} \\ &= 22.86 \pm 2.226, \end{aligned}$$

which does contain $\mu = 25$. If the production manager is willing to accept a 95 percent posterior interval estimate, then the machine should be purchased.

It is interesting to note that the classical maximum likelihood point estimator for μ in this example is $\bar{x} = 23$ and the corresponding 95 percent confidence interval is

$$\begin{aligned} \bar{x} \pm 1.96\sigma_{\bar{x}} &= 23 \pm 1.96\sqrt{1.5} \\ &= 23 \pm 2.400, \end{aligned}$$

which is less precise (wider) than the Bayesian interval, reflecting the fact that the classical procedure ignores the value of prior information.

In concluding, notice that the discussion in this section has been made under two important assumptions: a normal population distribution and a normal prior density function. If the first assumption holds, then a normal distribution of the sample mean will be generated; if the second assumption holds, then the revised density function for μ_X will also be normal. Suppose these assumptions are violated—would the conclusions here still hold? To answer this question, we first recall that if the population distribution is not normal, the sampling distribution of the mean is still normal according to the central limit theorem, if the sample is greater than 30. Next, it can be shown (as Robert Schlaifer has done in his *Introduction to Statistics for Business Decisions*) that if it is reasonable to assume (1) that the prior density function has the same mean and same variance (as assumed for a prior normal distribution) regardless of its shape and (2) that the prior density function is relatively smooth near the observed sample mean, then, as a rule of thumb, with a sample as large as 30, the revised density will be approximately normal and the pattern of the prior density function is irrelevant. That is, given $n > 30$, both the distribution of $\bar{x}$ and the prior density function can be assumed to be normal, and the equations of this section apply.

16–4 TESTING AS A BAYESIAN DECISION

The Bayesian approach to testing begins, as in the classical case, with the formulation of hypotheses. Again, assuming the two-action type of testing problem, we would have two competing hypotheses, H_0 and H_1, to be selected according to some decision procedure.

Having formulated the hypotheses, the decision maker next expresses his subjective feelings about the truth or falsity of each. These feelings must be expressed as probabilities forming the prior distribution of the states of nature. This distribution represents the decision maker's subjective betting odds on the hypotheses. For example, if we set $P(H_0) = 0.8$ and $P(H_1) = 0.20$, we mean the decision maker is 80 percent certain that H_0 is true and 20 percent certain that H_1 is true. In other words, he is willing to bet with 4-to-1 odds in favor of H_0.

Now, to decide between the two competing hypotheses, we need to know the costs or losses $L(a, \theta)$ of a wrong decision. Clearly, a_0 (action associated with accepting H_0) is optimal if H_0 is true while a_1 (action associated with rejecting H_0) is optimal if H_1 is true. For simplicity of notation, we shall denote $L(a_0, \theta)$ by L_0 and $L(a_1, \theta)$ by L_1.

With $P(H_i)$ and L_i determined, for $i = 0, 1$, a decision can already be made if sample data are not available. The prior Bayesian decision is simply this: select a_0 as the optimal act if and only if

$$E[L(a_0)] < E[L(a_1)] \tag{16–5}$$

or

$$L_0P(H_0) < L_1P(H_1).$$

To establish a posterior decision procedure, just as in the classical case, a criterion must be established for accepting or rejecting H_0 depending upon the outcomes of a random sample. Given such empirical evidence and if H_0 is true and it is accepted, no loss results. The probability of making this correct decision is $1 - \alpha$. However, if a true H_0 is rejected, we have committed a type I error with a loss of L_0. The probability of making this wrong decision is α—the level of significance. Thus, the average loss for a true H_0 can be stated as

$$\bar{L}_0 = 0(1 - \alpha) + L_0\alpha = L_0\alpha. \tag{16–6}$$

Similarly, if H_1 is true and the sample result leads to its acceptance, no loss will occur. The probability of making this correct decision is the power of test—that is, $1 - \beta$. However, if a true H_1 is rejected, we have committed a type II error with a loss L_1. The probability for this wrong decision is β. Thus, the average loss for a true H_1 becomes

$$\bar{L}_1 = 0(1 - \beta) + L_1\beta = L_1\beta. \tag{16–7}$$

Clearly, the overall measure of worth of the decision procedure can be measured by the expected value of the two preceding average losses:

$$E(\bar{L}) = L_0\alpha P(H_0) + L_1\beta P(H_1). \tag{16–8}$$

The Bayesian approach to testing is to find a decision procedure that will minimize this expected value of average losses. It turns out that $E(\bar{L})$ is a function of K—the critical value of the testing statistic. Given the sample size n, each value of K determines a particular decision procedure. Let us see how the optimal K is found under some simplifying conditions.

Suppose X is normally distributed with an unknown mean μ and a known variance σ^2. Furthermore, suppose the decision situation is to select between the hypotheses $\mu = \mu_0$ and $\mu = \mu_1$ (with $\mu_0 < \mu_1$), and suppose we adopt the decision procedure of accepting $\mu = \mu_0$ if and only if $\bar{x} < K$. Then it can be shown that, given $P(H_i)$ and L_i, the value of K that would minimize $E(\bar{L})$ is given by the expression

$$K = \frac{1}{\mu_1 - \mu_0}\left(\frac{\sigma^2}{n}\right)\ln\left[\frac{L_0P(H_0)}{L_1P(H_1)}\right] + \frac{\mu_0 + \mu_1}{2}, \tag{16–9}$$

where ln refers to natural logarithms to the base e. The common logarithms tabulated in Appendix B can be converted to natural logarithms by multiplying their values by 2.30.

It is interesting to note that the equation for K is independent of the sample result $\bar{x}$, as in all Bayesian testing problems that can be evaluated before $\bar{x}$ is actually observed. However, the value of K does depend upon prior information $P(H_i)$ and the loss functions. It can also be seen that if $L_0 = L_1$ and if the two competing hypotheses are equiprobable [i.e., if

$P(H_0) = P(H_1)$], then the ln term disappears and the equation of K is reduced to a special and intuitively obvious decision rule: accept H_0 if and only if the observed sample mean is closer to μ_0 than to μ_1. Finally, given the population variance σ^2, the value of K varies inversely with the sample size n. The larger n becomes, the smaller K becomes, and so a smaller value of $\bar{x}$ is required for acceptance of H_0. This also reinforces the observation that the larger the sample size is, the more powerful the test is. Alternatively, we say that the greater σ is, the greater K is and the less sensitive the test becomes for a fixed sample size, since the value of $\bar{x}$ can be much larger for acceptance of H_0.

An illustrative example is in order. A critical dimension of a certain product is normally distributed with specifications of $\mu = 30$ and $\sigma^2 = 2$. These specifications are met when the production process is in control. When it is out of control, it tends to have a mean of 32. According to past experience and the production manager's judgment, the process is in control 70 percent of the time and out of control 30 percent of the time. If the process is in control but it is judged otherwise, a loss of \$300 will result, representing the costs of unnecessarily adjusting the process. If the process is out of control but it is judged not to be, a loss of \$800 will occur, representing the rework costs and/or economic wastes of scrapping the useless units produced. A decision must be made continuously between the competing hypotheses:

$$H_0\colon \mu_0 = 30 \text{ or the process is in control;}$$

$$H_1\colon \mu_1 = 32 \text{ or the process is out of control.}$$

Now, if this decision is to be made on prior information only, we will always accept H_0, since the expected loss for a_0,

$$\bar{L}_0 = L_0 P(H_0) = 300(0.70) = \$210,$$

is smaller than the expected loss for a_1,

$$\bar{L}_1 = L_1 P(H_1) = 800(0.3) = \$240.$$

In quality control, as exemplified by this problem, the decision is usually on the basis of both prior judgment and sample information at regular time intervals, since sampling is quite convenient and relatively inexpensive in this branch of statistical work.

Now, suppose a simple random sample of say, $n = 9$, is to be observed from time to time in order to decide whether the process is in control. The general form of a decision procedure based on the sample mean becomes:

$$\text{Accept } H_0 \text{ if } \bar{x} < K;$$

$$\text{Reject } H_0 \text{ if } \bar{x} \geq K.$$

By (16–9), and for the current example,

$$K = \frac{1}{\mu_1 - \mu_2}\left(\frac{\sigma^2}{n}\right)\ln\left(\frac{L_0P(H_0)}{L_1P(H_1)}\right) + \frac{\mu_0 + \mu_1}{2}$$

$$= \frac{1}{32 - 30}\left(\frac{2}{9}\right)2.3\left[\log\left(\frac{210}{240}\right)\right] + \frac{30 + 32}{2}$$

$$= \left(\frac{2}{18}\right)2.3(-0.058) + 31$$

$$= 30.985.$$

Thus, if we establish the decision rule that the process is in control if and only if $\bar{x} < 30.985$, then the expected value of average losses will be a minimum. To calculate $E(\bar{L})$, we first compute

$$\alpha = P(\bar{x} \geq K \mid \mu_0 = 30)$$

$$= 1 - N\left(\frac{30.985 - 30}{2/\sqrt{9}}\right)$$

$$= 1 - N(1.48)$$
$$= 1 - 0.9304$$
$$= 0.0696$$

and

$$\beta = P(\bar{x} < K \mid \mu_1 = 32)$$

$$= N\left(\frac{30.985 - 32}{2/\sqrt{9}}\right)$$

$$= N(-1.52)$$
$$= 0.0643.$$

With these values, we have the minimum expected average losses as

$$E(\bar{L}) = L_0\alpha P(H_0) + L_1\beta P(H_1)$$
$$= 300(0.0696)(0.70) + 800(0.0643)(0.30)$$
$$= \$30.05.$$

To verify that this minimum $E(\bar{L})$ occurs at $K = 30.985$, the reader may repeat the process above to compute $E(\bar{L})$ for a few selected values of K, such as 29.5, 30, 30.5, 31, 31.5, and 32. He will find, for each of these K values, that $E(\bar{L})$ will be greater than \$30.05.

Note that decision making with information involves a cost of sampling, which must be added to the minimum expected average losses determined by the foregoing procedure. If, for example, the total cost of sampling in our illustrative case is \$20, then the total average expected loss for the previous posterior optimal decision is \$30.05 + \$20.00 = \$50.05, which is much less than the expected loss of \$210 for the optimal decision on prior information alone. Thus, decision with information is favored in this case.

16–5 BAYESIAN INFERENCE WITH NONEQUIVALENT STATE-ACTION PROBLEMS

The Bayesian methods of estimation and testing presented in the previous two sections are actually nonequivalent state-action problems, already introducted in Section 7–2. Typically, in such problems, the number of states exceeds the number of available actions. A relatively simple but quite useful case involves only two actions and a continuous state variable under the assumption of linear payoff functions. For such decision situations, the Bayesian strategy simplifies, and the expected opportunity loss can be computed by a simple formula. In many cases both estimation and testing satisfy these requirements. For example, the new-product problem introduced in Chapter 7 can be solved either as an estimation problem or as a testing problem. In the setting of estimation, we try to estimate whether the actual number of sales is equal to or less than the break-even sales, or if it is above the break-even sales. This implies that choice of one action if the first is true and of another action if the second is true. In the setting of testing, this problem can be solved by testing H_0: the sales volume is equal to or less than x_b against H_1: the sales volume is greater than x_b. Of course, the terminal decision would be the same, given the same prior density for the state variable and the same sample result. In such situations an inferential process can be conducted directly, without specifying whether it is an estimation or testing problem, with the special form of the Bayesian strategy for nonequivalent state-action problems with linear payoffs. Let us see how this is done with the new-product example introduced in Chapter 7.

Recall that the manufacturer in the new-product example is confronted with the following linear payoff function:

$$R(a_i, X) = -\$4{,}000{,}000 + 0.8x,$$

which gives a break-even sales volume of $x_b = 5{,}000{,}000$ units. Furthermore, according to the manufacturer's judgment, a prior density of sales was given as $n(\mu_0 = 6{,}000{,}000; \sigma_0 = 2{,}000{,}000)$. With this information, the prior expected payoff for the optimal act, a_1 = introducing the new product, is given as

$$\begin{aligned} EP(a_1) &= -\$4{,}000{,}000 + \$0.8(6{,}000{,}000) \\ &= \$800{,}000. \end{aligned}$$

Finally, owing to the assumption of a normal state variable and a linear payoff function, the expected opportunity loss, EOL, or the expected value with perfect information EVPI, can be computed directly:

$$\begin{aligned} \text{EOL} = \text{EVPI} &= b\sigma L_n(D) \\ &= \$316{,}480. \end{aligned}$$

(See Section 7–2.) This indicates the terminal decision might be made with additional information.

Prior Analysis in Terms of Means

For the sake of compactness and simplicity, before a posterior analysis is applied to this illustration, let us convert the manufacturer's prior distribution for the volume of sales into one of mean sales by assuming that he has 50,000 retailing outlets as his regular customers. That is, we assume the prior random variable to be μ_X with $\mu_0 = 6{,}000{,}000/50{,}000 = 120$ and $\sigma_0 = 2{,}000{,}000/50{,}000 = 40$. The break-even value now becomes $\mu_b = 5{,}000{,}000/50{,}000 = 100$. Since $\mu_b < \mu_0$, as before, the optimal act with prior information should be to introduce the new product with

$$\begin{aligned} EP(a_1) &= -\$4{,}000{,}000 + (0.8)(50{,}000)\mu_0 \\ &= -\$4{,}000{,}000 + \$40{,}000(120) \\ &= \$800{,}000 \end{aligned}$$

and

$$\begin{aligned} \text{EVPI} &= b\sigma_0 L_n(D_0) \\ &= 40{,}000(40)(0.1978) \\ &= \$316{,}480.00, \end{aligned} \tag{16–10}$$

where

$$D_0 = \left| \frac{\mu_b - \mu_0}{\sigma_0} \right| = 0.50; \qquad L_n(D_0) = 0.1978. \tag{16–11}$$

Note that (16–10) and (16–11) are identical with (7–2) and (7–3) introduced before, except that now we have added subscripts "0" to indicate prior values, and the distributions are given in terms of means.

Posterior Analysis

Suppose the manufacturer decides to obtain additional information by sending model units of the new product to 64 randomly selected retailing outlets; and suppose, after a period of time, their reports indicate that

$$\bar{x} = 98 \quad \text{and} \quad \hat{s} = 160.$$

Since the population standard deviation is unknown, we shall use $\hat{s}$ as its estimator; thus,

$$\hat{\sigma}_{\bar{x}} = \frac{\hat{s}}{\sqrt{n}} = \frac{160}{\sqrt{64}} = 20.$$

Note that the sample mean is less than the break-even mean sales. Had we obtained a sample mean greater than μ_b, then we would know immediately that the terminal decision would also select a_1 as in the prior analysis. In view of the present result, we must have the posterior estimate of μ before we can know for sure which act—marketing or abandoning the product—is optimal. With the sample information, the posterior distribution is defined by

$$\begin{aligned}
\mu_1 &= \frac{(\mu_0)V(\bar{x}) + (\bar{x})\sigma_0^2}{\sigma_0^2 + V(\bar{x})} \\
&= \frac{120(20^2) + 98(40^2)}{40^2 + 20^2} \\
&= 102.4; \\
\sigma_1^2 &= \frac{(\sigma_0^2)V(\bar{x})}{\sigma_0^2 + V(\bar{x})} \\
&= \frac{(40^2)(20^2)}{40^2 + 20^2} \\
&= 320; \\
\sigma_1 &= \sqrt{320} \\
&= 17.8885.
\end{aligned}$$

With these results, the posterior estimate of μ_X is greater than μ_b and the optimal act is still a_1. Furthermore, for this optimal act,

$$\begin{aligned}
\text{posterior } EP(a_1) &= -4{,}000{,}000 + 40{,}000\mu_1 \\
&= -4{,}000{,}000 + 40{,}000(102.4) \\
&= \$96{,}000.00,
\end{aligned}$$

$$\begin{aligned}
\text{posterior EVPI} &= b\sigma_1 L_n(D_1) \qquad (16\text{–}12) \\
&= 40{,}000(17.8885)(0.3354) \\
&= \$239{,}992.12,
\end{aligned}$$

where

$$\begin{aligned}
D_1 = D &= \left|\frac{\mu_b - \mu_1}{\sigma_1}\right| \qquad (16\text{–}13) \\
&= \left|\frac{100 - 102.4}{17.8885}\right| \\
&= 0.134; \\
L_n(D_1) &= 0.3354.
\end{aligned}$$

It is interesting to note that both the expected payoff and the expected opportunity loss in the posterior analysis are much smaller than in the prior analysis. The great reduction in the posterior expected payoff for the optimal act is clearly produced by the great reduction in μ_1 from μ_0 due to the samples' yielding a sample mean below $\mu_0 = 120$. The reduction in the expected opportunity loss is the result of the reduction in σ_1 from σ_0. In other words, some of the uncertainty attached to the prior information has now been reduced by sample information. The behavior of the expected opportunity loss is similar to that of the expected value of sample information, which we shall now discuss in more detail.

Expected Value of Sample Information and Optimal Sample Size

The previous subjective posterior distribution was generated by a particular sample with $n = 64$, $\bar{x} = 98$, and $\hat{s} = 160$; a different sample would give a different posterior distribution. Thus, when it is worthwhile to obtain additional information, it is desirable to estimate the change that will take place in the prior betting distribution. That is, we can conduct a preposterior analysis to give some idea of the posterior distribution before actually taking a sample and revising the prior distribution. In the process of evaluating the expected posterior distribution, we are also able to determine the economic value of sample information of a given sample size and the optimal sample size.

The determination of the expected value of sample information is a very simple matter when either the population standard deviation σ is known or an estimator of σ is made available from, say, a pilot sample. Then the standard error of $\bar{x}$ can be computed and the variability in the posterior distribution can be estimated. Let us note first that sample information has economic value, because it reduces the posterior expected loss by supplying us with additional information that tends to reduce the variability (uncertainty) in the prior distribution. Thus, to find the expected value of sample information we need a measure that will indicate the amount of reduction in the variance from the prior to the posterior distribution. This measure—let us denote it by σ_*^2—is the difference between prior and posterior variances. That is,

$$\sigma_*^2 = \sigma_0^2 - \sigma_1^2$$

$$= \sigma_0^2 - \left[\frac{(\sigma_0^2)V(\bar{x})}{\sigma_0^2 + V(\bar{x})}\right]$$

$$= \frac{(\sigma_0^2)^2}{\sigma_0^2 + V(\bar{x})}. \qquad (16\text{–}14)$$

With σ_*^2, we can determine the expected value of the sample information, EVSI, under the assumption of a normal state and a linear payoff function, in the same manner as for EVPI:

$$\text{EVSI} = b\sigma_* L_n(D_*), \tag{16–15}$$

where

$$D_* = \left| \frac{\mu_b - \mu_0}{\sigma_*} \right|. \tag{16–16}$$

For the sample results in the previous section, recall that $\sigma_0^2 = 1600$ and $\sigma^2\bar{x} = 400$. We have then

$$\begin{aligned} \sigma_* &= \frac{1600}{\sqrt{1600 + 400}} \\ &= 35.7771, \\ D_* &= \left| \frac{100 - 120}{35.7771} \right| \\ &= 0.56, \\ L_n(D_*) &= 0.1799. \end{aligned}$$

These values yield

$$\begin{aligned} \text{EVSI} &= 40{,}000(35.7771)(0.1799) \\ &= \$257{,}452. \end{aligned}$$

Note, with $n = 64$, the expected value of sample information is already more than 81 percent of the expected value of perfect information (\$316,480) computed before.

Before proceeding further, let us observe the possible effects of the determinants of EVSI on the size of EVSI by referring to equations (16–15) and (16–16). First, the smaller the difference is between μ_0 and μ_b, the more difficult it is to make the correct decision and, therefore, the greater would be the EVSI. Second, other things being equal, EVSI would also be larger, the larger is the value of σ_0. The reason is that the greater σ_0 becomes, the more uncertain we are about the true μ. This greater uncertainty would increase the likelihood of making the wrong decision; or, sample information would become more valuable. Third, given other determinants, the smaller the population variance σ^2 or its estimator $\hat{s}^2$ is, the greater would be the expected value of sample information. One reason is that $V(\bar{x})$ would be smaller, and a more concentrated sampling distribution of $\bar{x}$ would emerge for a less dispersed population; another is that D_* would increase, making a loss less likely. Finally, it is obvious that EVSI increases with the increase in n until it reaches EVPI as a limit.

Turning now to the question of optimal sample size, we observe that sampling involves a cost that is usually of this linear form:

$$C(n) = F + v(n), \qquad (16\text{–}17)$$

where F represents the total fixed cost of sampling (designing, administering, analyzing, reporting, and so on) that will be incurred if a sample of any size is taken at all, and v represents the unit variable cost of sampling under the assumption that the incremental cost of each additional item is identical.

Next, we observe that the difference between EVSI and $C(n)$ is defined as the expected net gain from sampling, ENGS. That is,

$$\text{ENGS} = \text{EVSI} - C(n). \qquad (16\text{–}18)$$

The optimal sample size, n^*, is that size of sample for which the expected net gain from sampling is a maximum.

Note that EVSI increases continuously with increasing sample size, but it increases with a decreasing rate of increase with the EVPI as its limit. Note also that $C(n)$ is a linear function. Owing to the behavior of these two functions, the ENGS function would increase at the beginning, reach a maximum, and eventually decline with continuous increase in n. The sample size corresponding to the maximum of the ENGS is the optimum sample size. See Figure 16–1.

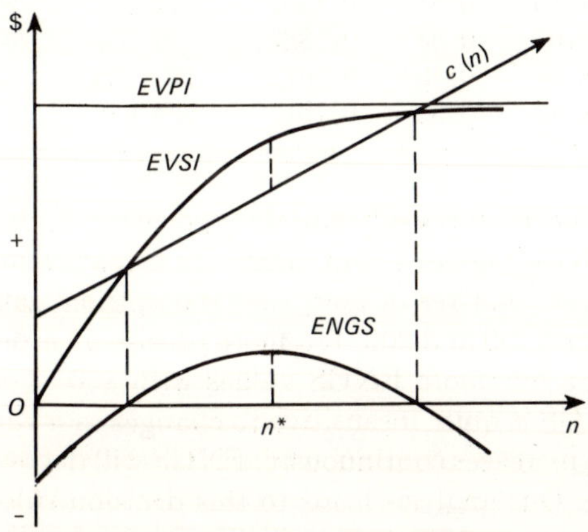

Figure 16–1 Graphic Illustration for the Determination of Optimal Sample Size n^*

Under certain conditions, a mathematical expression can be formulated to determine n^* (as in Chapter 35 of Schlaifer's *Introduction to Statistics for Business Decisions*). At an elementary level, n^* may be found by trial and error—by the so-called iternative procedure.

Returning to our illustrative case, suppose the sampling cost is given by

$$C(n) = \$5000 + \$200(n).$$

Thus, for $n = 64$, we have

$$\begin{aligned}\text{ENGS} &= \text{EVSI} - C(n)\\ &= 257{,}452 - 17{,}800\\ &= \$239{,}652.\end{aligned}$$

This large positive value of ENGS indicates sampling is quite worthwhile. But 64 is not necessarily the optimal sample size. To approximate the optimal sample size, we may compute ENGS with a few values on either side of 64 and converging on the best n until we approximately maximize ENGS. These calculations are summarized in Table 16–6. For these calculations, we have assumed $\sigma^2 = \hat{s}^2 = 160^2$ throughout.

Table 16–6. Approximation of Optimal Sample Size for the New-Product Example

n	$V(\bar{x})$	σ_*	D_*	$L_n(D_*)$	EVSI	$C(n)$	ENGS
50	512	34.80	0.57	0.1772	\$246,592	\$15,000	\$231,592
60	427	35.50	0.56	0.1799	255,458	17,000	238,458
80	320	36.47	0.55	0.1828	266,699	21,000	245,669
100	256	37.15	0.54	0.1857	275,950	25,000	250,950
150	171	38.07	0.53	0.1887	287,352	35,000	252,352
170	151	38.21	0.52	0.1917	292,994	39,000	253,994
200	128	38.47	0.52	0.1917	294,988	45,000	249,988

Results in Table 16–6 reveal that the approximate optimal sample size is 170, since ENGS increases continously to a maximum at $n = 170$ and then decreases. As a matter of fact, the true optimal sample size must be somewhere between 150 and 200. If a more precise n^* is desired, we can find it by computing a few more ENGS values with $150 < n < 200$. Observe also that the ENGS is quite insensitive to changes in n with $n > 150$ in this case, but when n increases continuously, ENGS will decrease and eventually become negative. Our analysis leads to this decision rule for our example: Take a sample of $n = 170$ and revise the prior distribution; if $\mu_1 > \mu_b$, take a_1, and otherwise, take a_2.

16–6 BAYESIAN VERSUS CLASSICAL INFERENTIAL PROCEDURES

From the foregoing presentation, we can see that the Bayesian estimation procedure is the optimal method of minimizing the loss of wrong decisions provided that a prior distribution and a loss function can be specified.

In contrast to the classical approach, Bayesian methods often provide more reasonable point estimates when the sample results happen to be the extreme values (see, for example, Exercise 16–5) and more precise interval estimates. This advantage is especially pronounced when the sample is small and when the prior information is not very vague. If, as can be seen from equations (16–1) and (16–2), either n or σ_0^2 approaches infinity, then the posterior mean μ_1 would approach the sample mean $\bar{x}$ and the posterior variance σ_1^2 would approach the variance of the sample mean $V(\bar{x})$. In other words, Bayesian estimators would approach the classical estimators if either perfect information could be made available, $n \to \infty$, or the prior information were completely useless, $\sigma_0^2 \to \infty$.

Another fundamental difference between the Bayesian and classical estimation procedures is that the former treats the parameter being estimated as a random variable while the latter considers it as an unknown fixed value. Undoubtedly, there is often some difficulty in interpreting a parameter as a random variable, as if it were drawn at random from a deck of cards. We have found, however, that it is useful to think of a parameter as having a subjective probability distribution that can be used to express the odds given by a decision maker in betting on the parameter's value. Hence, the decision maker can narrow down his prior knowledge of a parameter into a subjective distribution; then the posterior subjective distribution of the parameter will reveal how the empirical evidence changed the initial betting odds.

It should be noted also that the Bayesian estimation procedure requires at least rough-and-ready specifications of the prior and loss functions, which are often unknown; subjective judgment is always involved. This has been a major criticism of Bayesian statistics in general. Incidentally, the classical technique, which requires no explicit specification of the loss function, is by no means free of the same "sin" of subjectivity. Its arbitrary selection of the confidence coefficient, for example, involves the decision maker's subjective judgment. Indeed, one of the most important contributions of Bayesian statistics is to make explicit the assumptions implied by the classical methods.

Some important differences between the Bayesian and classical testing procedures must also be recognized.

First, the Bayesian approach explicitly regards the parameter under test as a random variable in its specification of the prior distribution of H_i, whereas the classical approach implicitly treats the parameter as a random variable in giving consideration to the selection of the level of significance.

Next, the two approaches differ in the process of reasoning. Bayesians ask: "What is the probability of the hypothesis given the sample result?" Classicists inquire: "What is the probability of the sample result given the hypothesis?"

Third, Bayesian testing seeks to specify more precisely the decision maker's beliefs in H_i and the costs of wrong decisions. In other words, it explicitly includes the two criteria of prior probabilities and costs in the analysis for the determination of α and β. Here the Bayesian approach obviously is more satisfactory, even though it requires more effort.

Finally, classical decision procedures are based on empirical observations alone and are confined to estimation and testing. On the other hand, Bayesian statistics embraces decision procedures without sample information when the decision maker does not have the option of obtaining it, and procedures with both subjective judgment and sample results. Furthermore, the Bayesian approach offers criteria of decision making for a wide variety of problems—not just inferential methods. Classical inferences, however, remain useful under certain circumstances.

EXERCISES

16–1 For the example in Section 16–2, what would be the optimal Bayesian esti mators for the various states if E_1, E_2, or E_3 were observed, assuming uniform prior distribution and a quadratic loss function as given in the text?

16–2 Consider again the example in Section 16–2. If the prior probabilities are $P(\theta_1) = 0.6$, $P(\theta_2) = 0.3$, and $P(\theta_3) = 0.1$, and if the same quadratic loss function is assumed, what are the optimal Bayesian estimators for the various states now, given each possible sample result, E_1, E_2, or E_3?

16–3 An important theorem of Bayesian statistics states that if the prior distribution is $f_0(\theta) = n(\theta_0, \sigma_0^2)$ and if the empirical evidence of a single observation is $f(x \mid \theta) = n(\theta, \sigma^2)$, then it can be shown that the posterior distribution would be $f_1(\theta \mid x) = n(ab, a)$, where

$$\frac{1}{a} = \frac{1}{\sigma_0^2} + \frac{1}{\sigma^2} \tag{16–19}$$

and

$$b = \frac{\theta_0}{\sigma_0^2} + \frac{x}{\sigma^2}. \tag{16–20}$$

Can you explain the differences between this set of equations and equations (16–1) and (16–2)?

16–4 An incoming shipment of steel wires has an unknown mean but a known standard deviation of $\sigma = 25$ pounds. A random observation of a single steel wire yields $x = 1000$ pounds.

(a) What is the classical 95 percent confidence interval for μ?

(b) Suppose an engineer believes that this make of steel wires has a prior normal distribution with $\mu_0 = 1200$ pounds and $\sigma_0^2 = 900$; how can this information be used to define a posterior distribution of μ?

(c) With the result in (b) and assuming a quadratic loss function, what is the Bayesian optimal estimator of μ? The Bayesian 95 percent confidence interval?

16–5 As you know, the classical unbiased estimator of π is $p = x/n$. It can be shown that, assuming a quadratic loss function, the Bayesian optimal estimator of π is

$$\hat{\pi} = \frac{x + 1}{n + 2} \tag{16–21}$$

if the subjective prior distribution of π is uniform. Suppose now that a random sample of $n = 10$ is drawn from the student population of a certain college in order to estimate the true proportion of students who oppose President Nixon's domestic policies, and it is found that $x = 2$.

(a) What is the classical estimate of π?

(b) What is the Bayesian estimate of π assuming a prior uniform distribution?

(c) Which estimate seems to be more credible according to your judgment? Explain.

16–6 Given that wrong decisions entail losses as shown in the following table.

	State	
Act	$P(H_0) = 0.80$	$P(H_1) = 0.20$
a_0: Accept H_0	10	120
a_1: Accept H_1	30	20

Then we have

$$\bar{L}_0 = P(H_0)l_{00} + P(H_1)l_{01}$$

and

$$\bar{L}_1 = P(H_0)l_{10} + P(H_1)l_{11},$$

where l_{ij} stands for the loss by taking the ith action corresponding to the jth state.

(a) What are the values of $\bar{L}_0$ and $\bar{L}_1$, respectively?

(b) Which is the optimal act on the prior information only?

16–7 We wish to test $\mu_0 = 75$ against $\mu_1 = 77$, with

$$P(H_0) = 0.8, \qquad P(H_1) = 0.20,$$

$$\sigma = 3, \qquad n = 4,$$

$$L_0 = 300, \qquad L_1 = 600.$$

(a) Show that the critical value of $\bar{x}$ is $K = 76.78$.

(b) Show that for this rule $\alpha = 0.12$ and $\beta = 0.44$, with $E(\bar{L}_1) = \$81.60$

16–8 An automobile manufacturing firm buys batteries in large quantities at regular intervals. The mean battery life varies from one shipment to another. The two hypotheses are

$$H_0\colon \mu_0 = 35 \text{ months};$$
$$H_1\colon \mu_1 = 40 \text{ months}.$$

If H_0 is true, the manufacturer wants to return the shipment to the supplier. Returning the shipment when $\mu = 40$ is an error with a loss of \$2000. If H_1 is true, the manufacturer wants to receive the shipment and pay for it. To receive it when $\mu = 35$ is an error with a loss of \$10,000. The purchasing agent of the manufacturer is 15 percent sure that $\mu = 35$ and 85 percent sure that $\mu = 40$.

(a) What is the optimal act based on prior considerations alone?

(b) Suppose it is known that distribution of battery life is normal with $\sigma = 3$, and the purchasing agent adopts the following decision rule: take a random sample of 25 and

$$\text{accept } H_0 \text{ if } \bar{x} < K; \qquad \text{accept } H_1 \text{ if } \bar{x} \geq K.$$

What would be the critical value of K and the expected value of average losses?

16–9 Rework the preceding problem, assuming that $n = 16$.

16–10 For the preceding two problems if the cost of sampling is $C(n) = \$100 + \$20(n)$, which sample size would have lower terminal expected value of average losses?

16–11 A process is considered in control when it yields a mean of 22 and a variance of 16. When it is out of control, the mean becomes 20. Probabilities of these two states and costs of wrong decisions are summarized below:

Act	State probability	
	$P(\mu_0 = 20) = 0.20$	$P(\mu_1 = 22) = 0.80$
a_0: Accept H_0	50	100
a_1: Accept H_1	200	80

(a) What are the expected losses for a_0 and a_1, respectively?

(b) What is the optimal act based on prior information alone?

(c) If a sample $n = 16$ is observed before a terminal decision is made, what is the critical value of $\bar{x}$? What is the minimum expected value of average losses with this rule?

(d) What is the difference between this problem and the illustrative example of Bayesian testing procedure in the text so far as the designations of H_0 and H_1 is concerned? What is the significance of this difference in connection with the requirement in the text that $\mu_0 < \mu_1$ in determining K?

16–12 A production process is adjusted to yield a mean weight of the output as 100 and a standard deviation of 10, as shown in column (1) of Table 16–P1. The prior distribution of the mean is judged to be approximately normal with a standard deviation of 12.28, as shown in column (2). Losses involved in erroneously accepting that the process is in control (i.e., $\mu = 100$), L_0, for various states are given in column (3). Losses involved in erroneously rejecting that the process is in control, L_1, for various states are given in column (4).

Table 16–P1

(1) μ*	(2) $P(\mu)$	(3) L_0	(4) L_1
70	0.016	$90	$0
80	0.094	60	20
90	0.234	40	30
100	0.312	0	50
110	0.234	40	30
120	0.094	60	20
130	0.016	90	0
	1.000		

* This column contains the midpoints of the classes for a continuous random variable.

(a) In terms of prior information alone, is a_0 (always accept that the process is in control) or a_1 (always reject that the process is in control) the optimal act?

(b) Suppose it is decided to set the quality standards, or control limits, as $\bar{x} \pm 2.0\sigma_{\bar{x}}$, and a decision is to be made before each production run with $n = 16$; what are the critical values for rejecting that the process is in control? Also, assume $C(n) = \$0.10n$; what are the net expected average losses of the optimal act?

(c) In view of results obtained in (a) and (b), which decision rule is better—prior or posterior analysis?

16–13 With the previous problem, still assuming $n = 16$, what would be the expected average losses if the control limits are changed to $\bar{x} \pm 2.5\sigma_{\bar{x}}$ or to $\bar{x} \pm 3.0\sigma_{\bar{x}}$?

16–14 Suppose, for the preceding two problems, that the control limits were $\bar{x} \pm 2.5\sigma_{\bar{x}}$; what would be the expected average losses with $n = 9$ or 25, respectively? Given the three sample sizes and the three control limits, what is the best decision rule if both the sample size and control limits are taken into consideration?

16–15 In a decision situation, assume (1) the population is normally distributed with an unknown mean and a standard deviation of $\sigma = 25$, (2) a prior distribution $n(\mu_0 = 22, \sigma_0 = 10)$, (3) $C(n) = 15 + 2n$, (4) $b = \$90$, and (5) $\mu_b = 17$.

(a) Compute EVPI with prior information.

(b) Compute EVSI for $n = 16$, $n = 25$, and $n = 36$. Which size among the three computed above is the approximate optimal sample size?

16–16 For data in the previous problem and for $n = 36$, what is the EVSI if

(a) $\mu_b = 20$ while other values remain the same?

(b) $\sigma_0 = 15$ while other values remain the same?

(c) What can you say, in view of results in this and the preceding problem, about the effects of n, σ_0, and μ_b on the value of EVSI?

16–17 The Sigma Company is considering the introduction of a new product. Sigma can distribute its products through 10,000 retail stores. The demand for the new product is thought to be a random variable with an unknown mean but a standard deviation of 30 units per store. The marketing department is able to estimate subjectively that mean sales can be characterized as $f(\mu_{\bar{x}}) = n(45, 12)$. The introduction of the new product would require a total fixed cost of \$120,000 with an estimated incremental contribution rate (price less average variable cost) of \$0.25.

(a) What is the optimal decision with the data above—to market or not to market the new product? What is the EVPI?

(b) Suppose a random sample of 81 retail stores has been taken and a sample mean $\bar{x} = 55$ has been observed. What action should be taken posterior to the sample? What is the posterior EVPI?

16–18 Suppose, for the preceding problem, a preposterior analysis is to be conducted with $n = 64$, $n = 81$, $n = 121$, and $n = 169$, and suppose $C(n) = \$250 + \$20(n)$. What is the EVSI for each sample size? What is the approximate optimal sample size?

16–19 Suppose you would like to retail women's clothing and you are told that a dress shop, with an inventory of 6000 different items, is up for sale for \$250,000. You hire a women's clothing expert to evaluate this proposition, and he informs you that in his prior judgment the mean value of the inventory is about \$45 per item with a standard deviation of \$10; that is, $f(\mu_{\bar{x}}) = n(45, 10)$.

(a) Would you buy the shop on the basis of prior information alone?

(b) Suppose you wished to take an actual sample before making a final decision and suppose $C(n) = \$300 + \$5(n)$; what should be the approximate size of the sample?

17

SIMPLE LINEAR ASSOCIATION ANALYSIS

17–1 ASSOCIATION BETWEEN VARIABLES

So far we have been concerned with analysis of *univariate data*. That is, a single observation is made on each elementary unit of the sample, with a sample of values for a single variable; the computed sample statistics are then used to make inferences about the corresponding parameters of the parent *univariate population*. In this chapter we shall study *bivariate populations*—those that involve two variables. For example, we may observe the height and weight of each individual in a population of adult men. All the measured heights would form a variable, say, X; all the weights, another variable, say, Y.

The main problem in analyzing bivariate data is to discover and measure the association between the variables and to determine how they vary together. Thus, we may observe that taller men are usually heavier than shorter men. This is, of course, only true on the average, since there are exceptions, such as short heavy men or tall light men. If, however, this average relationship can be mathematically established in some functional form, we will be in a position to estimate quite accurately, on the average, the weight of a man on the basis of his height. Such a procedure is called *estimation by association*.

The technique of estimation by association is far more sophisticated and useful than our homely example of heights and weights. This procedure is actually one of prediction, which is the central function of science. The main task of any scientific study is to discover the general relationships between observed variables and to state the nature of such relationships precisely in mathematical terms, so that the value of one variable can be predicted on the basis of the value of another. A few examples may show the important place that decision by association occupies in business and economic statistics.

Many direct tests of product quality are destructive. However, if it can be found that one property of a product is related to another, the former can then be evaluated in terms of the latter. Thus, the breaking strength of metal parts manufactured from compressed metal powder can be determined by density as measured in weight; the shearing strength of joints on airplane wings can be tested by observing the diameter of welds; and so forth. These examples illustrate that decision by association often reduces costs in decision making.

We know that average consumer expenditures are associated with a variety of factors: disposable income, interest rate, population size, age composition, size and distribution of liquid assets, and so on. However, to relate all these variables to consumption is extremely cumbersome. Moreover, these variables may not be operating on consumption at the same time, or to the same degree. Under these circumstances, the most fruitful tactic is to relate consumption to the most important single factor. This is determined by studying the nature and degree of association of each factor to consumption. If we find, say, that disposable income exerts a clear-cut and most powerful influence upon consumption, we may then predict consumption expenditures on the basis of disposable income alone. In this way sometimes we can find a sufficiently good explanatory variable when restricting our investigation to a bivariate analysis.

To estimate the selling expenses of each branch office of a chain retailing organization on the basis of mean selling expenses may be infeasible because of an extremely large standard deviation. Since there is reason to believe that selling expenses are usually associated with the volume of sales, more accurate estimates of the former can often be made on the basis of the latter. In this way association analysis may be able to increase precision in prediction.

The hiring of personnel on probation is usually time-consuming and expensive. If it can be established that the performance of workers is associated with the results of a certain test, then time and money can be saved by predicting an applicant's job performance on the basis of the related variable, test score.

There are two related but distinct aspects of the study of association between variables. The first, *regression analysis*, attempts to establish the "nature of relationship" between variables; that is, in regression analysis, we study the functional relationship between the variables so that we may be able to predict the value of one on the basis of another. Conventionally, the variable that is the basis of prediction is called the *independent variable*, and the variable that is predicted is called the *dependent variable*.

The second aspect of association analysis, referred to as *correlation analysis*, is concerned with determining the "degree of relationship" between variables. Here the designation of dependent and independent variables is purely a personal choice and has no practical significance.

Association analysis may also be distinguished as simple and multiple. *Simple association analysis* refers to a situation in which there is only one independent variable and one dependent variable—that is, bivariate data. *Multiple association analysis* studies the relationship among three or more variables (one dependent and two or more independent variables.) Association analysis can also be differentiated as *linear* and *nonlinear*, according to the type of relationship the variables have. In this chapter we shall be concerned with simple linear association models. A brief discussion on multiple linear and nonlinear association analysis will be given in Chapter 20.

17–2 THE LINEAR BIVARIATE REGRESSION MODEL

In regression analysis, as in other types of statistical studies, we usually proceed by observing the sample data and using the results obtained as estimates of the corresponding population relationship. To make valid inferences, we must assume some population model. For a bivariate population, many possible models can be constructed to describe the mutual variations of the two variables. The on that interests us at this time is called the *simple linear regression model*, which is constructed under the following set of assumptions:

(1) In the simple linear regression model, the values of the independent variable are assumed to be fixed quantities that are selected and controlled by the experimenter. Associated with each fixed value of X is a subpopulation of Y, the dependent variable, which is assumed to be a random quantity. The requirement that the independent variable assume fixed values, however, is not a critical one. Useful results can still be obtained by regression analysis in the case where both Y and X are random variables.

(2) The average relationship between Y and X can be adequately described by a linear equation whose geometrical presentation is a straight line, $Y = A + Bx$. The height of the line tells the average value of Y at a fixed value of X. When $X = 0$, the average value of Y is equal to A. The value of A is called the Y *intercept*, since it is the point at which the straight line crosses the Y axis. The slope of the line is measured by B, which gives the average amount of change of Y per unit change in the value of X. The sign of B also indicates the type of relationship between Y and X.

(3) Associated with each value of X is a subpopulation of Y. The distribution of each Y may be assumed to be normal or nonspecified in the sense that it is unknown. In any event, the distribution of each subpopulation Y is conditional on the value of X.

(4) The mean of each subpopulation Y is called the expected value of Y for a given X; $E(Y \mid X) = \mu_{YX}$. Furthermore, under the assumption of a linear relationship between X and Y, all values of $E(Y \mid X)$, or μ_{YX}, must fall on a straight line. That is,

$$E(Y \mid X) = \mu_{YX} = A + Bx, \tag{17-1}$$

which is the *population regression equation* for our bivariate linear model. In this equation, A and B are called the population regression coefficients. The foregoing assumptions are portrayed by Figure 17–1.

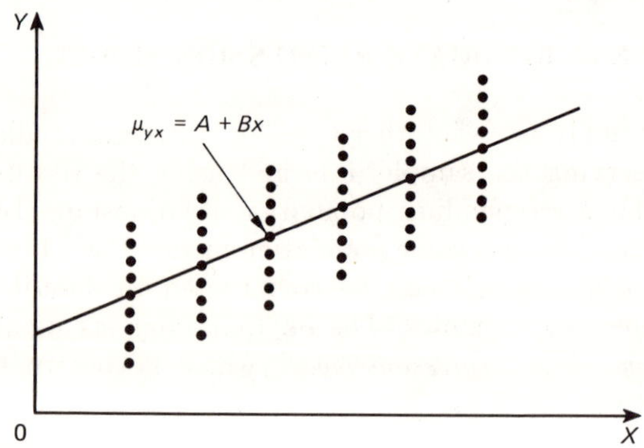

Figure 17–1 Distribution of Y's Associated with X's

(5) An individual value in each subpopulation Y, as indicated by Figure 17–2, may be expressed as

$$y = E(Y \mid X) + e, \tag{17-2}$$

where e, as in the linear model of the analysis of variance introduced in Chapter 15, is called the *error term* or the *stochastic term*. The errors, like the Y's, are assumed to be independent random variables. The expectation of these errors is zero; $E(e) = 0$. Furthermore, if the error term can be assumed to be normal, then the subpopulations, Y's, can also be assumed to be normal.

(6) Finally, we assume that variances of all subpopulations, called *variances of the regression*, are identical. That is,

$$\sigma_1^2 = \sigma_2^2 = \cdots = \sigma_n^2 = \sigma^2 = \sigma_{YX}^2 = E[Y - E(Y \mid X)]^2.$$

It is interesting to note that the common variance of the regression is

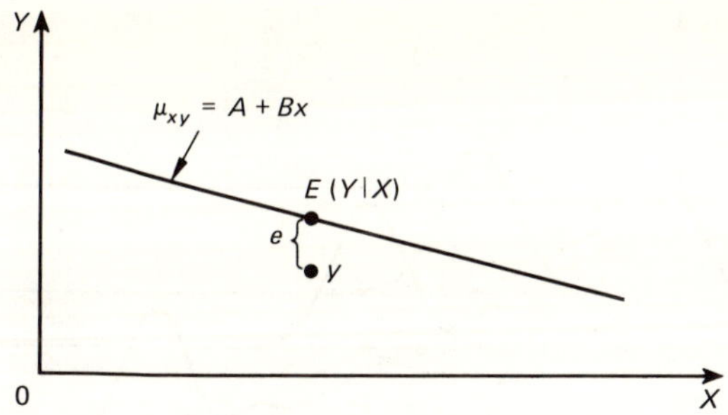

Figure 17–2 Deviation of an Observation of *Y* from the Regression

also identical with the variance of the error term:

$$V(Y) = V(e) = \sigma_{YX}^2,$$

since $E(e) = 0$, and thus

$$\sigma_{YX}^2 = [Y - E(Y \mid X)]^2 = E(e)^2 = E[e - E(e)]^2 = V(e).$$

Because e shows the error of Y with respect to μ_{YX}, σ_{YX}^2 is often called the *residual variance.* The definitional equation for this value is

$$\sigma_{YX}^2 = E[Y - E(Y \mid X)]^2. \tag{17–3}$$

The square root of σ_{YX}^2 is called the *standard deviation of the regression,* denoted by σ_{YX}.

To summarize: The simple linear regression model consists of the following three equations:

$$\mu_{YX} = A + Bx,$$

$$y = E(Y \mid X) + e,$$

$$\sigma_{YX}^2 = E[Y - E(Y \mid X)]^2.$$

These equations describe and summarize regression behavior of bivariate population data. Furthermore, if the subpopulations are normally distributed, this model can be portrayed by Figure 17–3.

17–3 ESTIMATING THE POPULATION REGRESSION EQUATION

With linear regression analysis, the main task is to estimate the population regression coefficients A and B in equation (17–1) on the basis of n pairs of sample observation: $(y_1\ x_1)$, $(y_2\ x_2)$, . . . , (y_n, x_n). Let us denote

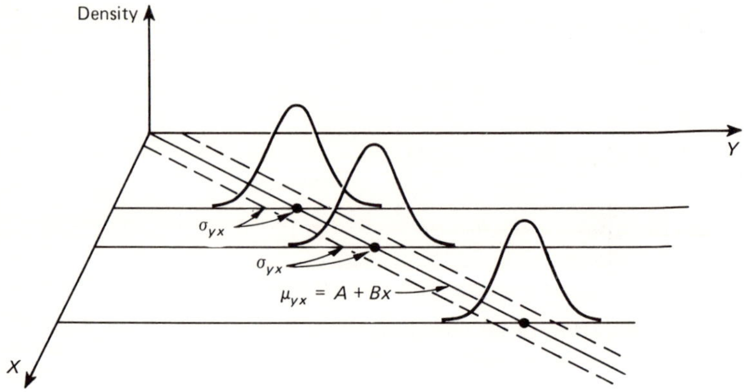

Figure 17–3 The Linear Regression Model

the estimators of A and B by a and b, respectively; then the *sample regression equation* becomes

$$\bar{y}_c = a + bx. \tag{17–4}$$

Since $\mu_{YX} = A + Bx$, then $\bar{y}_c$, or $(a + bx)$, is in effect an estimator of μ_{YX}. Here $(a + bx)$ is called a *linear estimator* because it represents a linear relationship.

From our discussion on the properties of estimators in Chapter 12, we can say that $(a + bx)$ is an unbiased estimator of μ_{YX} if

$$E(a + bx) = A + Bx,$$

which implies that

$$E(a) = A \quad \text{and} \quad E(b) = B.$$

Furthermore, if many procedures are available to estimate $(A + Bx)$, $(a + bx)$ is said to be the minimum variance, or the best, estimator of μ_{YX} if the scatter of $(a + bx)$ around $(A + Bx)$ is less than the dispersion of any other estimator obtained by any other procedure. If the distributions of subpopulations, Y's, are not specified, but if they possess equal variances corresponding to fixed X values, then, according to the Gauss-Markov theorem, the best linear estimator of $(A + Bx)$ can be obtained by the method of least squares. On the other hand, if the subpopulation distributions have been known or assumed to be normal, then we may apply the maximum likelihood method to obtain an unbiased estimator of $(A + Bx)$ because we now know the form of the density function of the Y's.

Let us see how the best estimator of μ_{YX} can be obtained by the least squares method. First, define the estimator of e as

$$\hat{e} = y - \bar{y}_c = y - a - bx.$$

That is, $\hat{e}$ is the residual of y from $\bar{y}_c$. Then the procedure for estimating the population regression coefficients by the least squares method is to find a and b that will minimize the sum of squared deviations, $\hat{e}$. That is, a and b must be found such that

$$\sum \hat{e}^2 = \sum (y - a - bx)^2 = \text{minimum.}$$

This requirement is met, according to the Gauss-Markov theorem, if a and b are determined by solving simultaneously the following set of normal equations:

$$(1) \qquad \sum y = na + b \sum x;$$

$$(2) \qquad \sum xy = a \sum x + b \sum x^2.$$

It turns out that the estimation of $(A + Bx)$ by the maximum likelihood method also requires the solution of the same set of normal equations for a and b.

All quantities in the normal equations above can be computed from sample data. For the sake of simplicity, however, we usually solve for a and b first in terms of other quantities in the normal equations. Then values of a and b can be computed directly from sample data. To solve for b, we multiply the first equation by $\sum x$ and the second by n, and then subtract the first result from the second so that a can be eliminated. As a result, we have

$$b = \frac{n \sum yx - \sum y \sum x}{n \sum x^2 - (\sum x)^2}. \qquad (17\text{–}5).$$

Substituting the right side of the foregoing expression for b in either of the two normal equations yields

$$a = \frac{\sum y}{n} - b\frac{\sum x}{n} = \bar{y} - b\bar{x}, \qquad (17\text{–}6)$$

where $\bar{y}$ and $\bar{x}$ are the means of Y and X, respectively.

To illustrate, let us suppose that a manufacturer of spot welds wishes to predict shearing strength by the diameters of spot welds instead of destroying the product with a direct test. A sample of ten welds, taken to establish the relationship between the two variables, yields the result shown in Table 17–1.

The first step in regression analysis is to construct a diagram of the sample data to see whether or not the dependent variable does depend to some extent upon the independent variable, and whether the average relationship between them can indeed be explained by a straight line. In

Table 17–1. Diameters and Shearing Strength of Spot Welds of Aluminum

Shearing strength (in thousands of pounds) Y	*Weld diameter (in inches)* X
7.0	2.4
5.3	1.8
4.2	1.6
3.3	1.0
3.8	1.2
6.6	1.1
8.5	2.8
6.6	1.6
4.5	1.5
8.8	2.3

this *scatter diagram* each point corresponds to a pair of values, (y_i, x_i). The scatter diagram for our illustrative data is shown in Figure 17–4.

Examining the scatter diagram, we can obtain some distinct impressions. First, there is a clear positive relationship betweeen diameter and shearing strength of the welds; the former, on the average, seems to increase with an increase in the latter. Second, the points seem to scatter considerably from each other, but they also seem to have roughly a linear relationship; that is, their average relationship can be described approximately by a straight line. These observations strengthen our brief that linear regression analysis can be fruitfully applied to the sample data on hand.

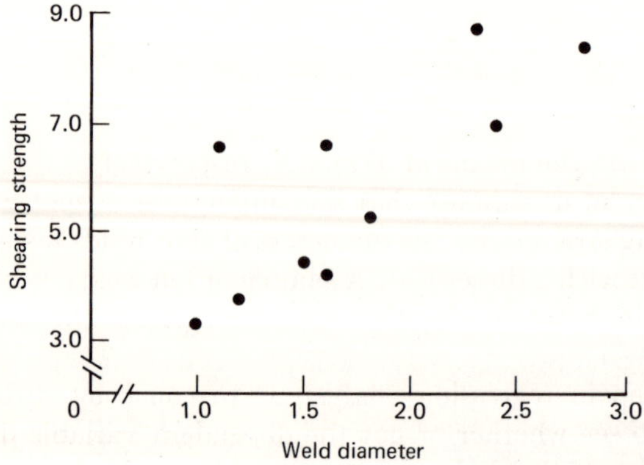

Figure 17-4 Scatter Diagram for Weld Diameters and Shearing Strength

The next step is to fit a straight line through the points in the scatter diagram. This we do by computing a and b, the sample regression coefficients, with formulas (17–5) and (17–6). To compute these values we need to accumulate $\sum x$, $\sum y$, $\sum xy$, and $\sum x^2$. For later calculations we also need $\sum y^2$. All of these sums can be arranged in one table—for our example, Table 17–2. From the sums in Table 17–2 we have

$$b = \frac{(10)(109.53) - (17.3)(58.6)}{(10)(33.15) - (17.3)^2}$$

$$= 2.531 \text{ thousand pounds per inch}$$

and

$$a = 5.86 - (2.531)(1.73)$$
$$= 1.481 \text{ thousand pounds.}$$

Our estimate of the population regression equation is therefore

$$\bar{y}_c = 1.481 + 2.531x.$$

Table 17–2. Sums Required for Regression Computations

y	x	yx	y^2	x^2
7.0	2.4	16.80	49.00	5.76
5.3	1.8	9.54	28.09	3.24
4.2	1.6	6.72	17.64	2.56
3.3	1.0	3.30	10.89	1.00
3.8	1.2	4.56	14.44	1.44
6.6	1.1	7.26	43.56	1.21
8.5	2.8	23.80	72.25	7.84
6.6	1.6	10.56	43.56	2.56
4.5	1.5	6.75	20.25	2.25
8.8	2.3	20.24	77.44	5.29
58.6	17.3	109.53	377.12	33.15

$\bar{y} = 5.85, \quad \bar{x} = 1.73$

The estimate of A, as in this instance, has no practical significance. In general, the a value has only the mathematical significance that it, together with b, locates the position of the regression line. The b value, however, is always of importance. In the present case it is an estimate of the average amount that the shearing strength is increased by each additional inch in the diameter of welds. Specifically, the value 2.531 means that the shearing strength is estimated to increase by an average amount of 2.531 thousand pounds per inch increase in diameter.

17–4 COMPUTATIONS WITH CODED DATA AND SIMPLIFIED NOTATIONS

When values of X and Y are large, it becomes quite time-consuming to compute the sample regression coefficients. There are two ways to save time and effort in computing a and b. The first is referred to as a *shift of the origin*. This is done by deducting a constant, say c, from each value of X, which results in a new variable, X', and by deducting a constant, say k, from each Y, which results in a new variable, Y'. Then calculations are made with X' and Y'.

As shown by Figure 17–5, the shift of the origin from $(0, 0)$ to (c, k)

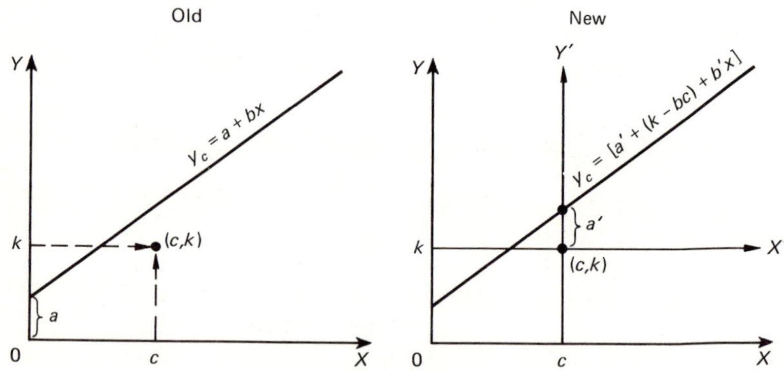

Figure 17–5 Effects of Shifting Origin from (0, 0) to (c, k)

does not affect the slope but it does affect the Y intercept. Thus, while

$$b = b' = \frac{n \sum y'x' - \sum y' \sum x'}{n \sum x'^2 - (\sum x')^2} \tag{17–7}$$

without further adjustment, a should be computed as

$$\begin{aligned} a &= (\bar{y}' + k) - b(\bar{x}' + c) = \bar{y}' - b\bar{x}' + (k - bc) \\ &= a' + (k - bc), \end{aligned} \tag{17–8}$$

since $X' = X - c$, and $Y' = Y - k$.

The second short-cut is called *reducing the scale*. According to this procedure, both X and Y are divided by a constant, say g, to obtain the new variables, X' and Y'. Again, calculations with X' and Y' will have no effect upon b (unless a different constant is used for X and Y), and again a is affected and must be converted back to the original scale by multiplying a' by g. That is,

$$a = a'g. \tag{17–9}$$

As an extension of reducing the scale, the scale can also be *inflated* before calculations are made. In this case both X and Y are multiplied by a constant to yield X' and Y'. This procedure is especially appropriate when we wish to get rid of the decimal points in the original data. For example, in our illustrative data, we may multiply both X and Y by 10 before a and b are computed. Here again, $b = b'$ but $a = a'/m$, where m is the common multiplier.

Shifting the origin and reducing the scale can be combined in order to simplify computations of the estimators of population regression coefficients. For instance, we may let $X' = (X/g) - c$ and $Y' = (Y/g) - k$, or we may let $X' = mX - c$ and $Y' = mY - k$ for computational purposes.

Another way to facilitate computations and to simplify various formulas is via simplified notations for those quantities that are used frequently in regression and correlation analysis. If $d_1 = y - \bar{y}$ and $d_2 = x - \bar{x}$, then it is easy to show that

$$\begin{aligned} n \sum d_1^2 &= n \sum y^2 - \left(\sum y\right)^2, \\ n \sum d_2^2 &= n \sum x^2 - \left(\sum x\right)^2, \\ n \sum d_1 d_2 &= n \sum yx - \sum y \sum x. \end{aligned} \tag{17-10}$$

Using these forms, for example, we can write the equation for b as

$$b = \frac{n \sum d_1 d_2}{n \sum d_2^2} = \frac{\sum d_1 d_2}{\sum d_2^2}. \tag{17-11}$$

With our illustrative data, we have

$$\begin{aligned} n \sum d_1^2 &= 10(377.12) - (58.6)^2 = 337.24, \\ n \sum d_2^2 &= 10(33.15) - (17.3)^2 = 32.21, \\ n \sum d_1 d_2 &= 10(109.53) - (58.6)(17.3) = 81.52, \end{aligned}$$

and

$$b = \frac{81.52}{32.21} = \frac{8.152}{3.221} = 2.531,$$

as obtained before.

17-5 ESTIMATING THE VARIANCE AND THE STANDARD DEVIATION OF THE REGRESSION

The sample regression equation, $\bar{y}_c = a + bx$, is called the *predictive equation* because its main function is to predict the mean value of Y, or the value of an individual observation of Y, associated with a given value of X. But how good is it as a predictive device? The answer evidently rests upon the variability of the individual Y values from the computed values (averages) of Y associated with the X values. One way to visualize such

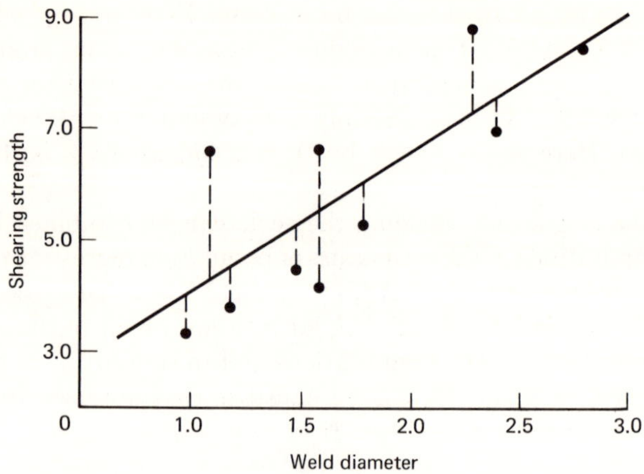

Figure 17–6 Deviations from the Regression

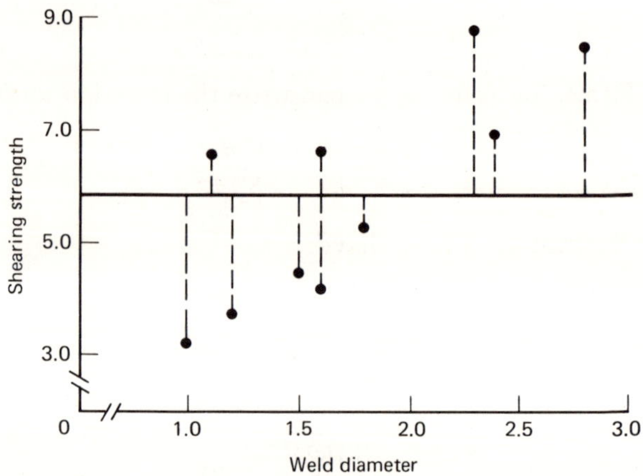

Figure 17–7 Deviations from the Mean

dispersion is to draw a regression line through the points in the scatter diagram and to connect this line with the points by vertical lines, as depicted in Figure 17–6. To draw the regression line, we need only plot two $\bar{y}_c$ values with, perhaps, a third as a check. The regression line, of course, represents estimated values of μ_{YX}.

Examining Figure 17–6, we find that the line seems to fit the points fairly well, but the points still deviate somewhat from the line. The numerical measure of such deviations is the estimate of the population variance, or the estimate of the population standard error, of the regression. The σ_{YX}^2 is estimated by using the minimum variance, or maximum likelihood, estimators of a and b as follows:

$$\hat{\sigma}_{YX}^2 = \frac{\sum_i (y_i - \bar{y}_c)^2}{n - 2} = \frac{\sum_i (y_i - a - bx_i)^2}{n - 2}. \tag{17-12}$$

This is an unbiased estimator of σ_{YX}^2. Here, $(n - 2)$ degrees of freedom is used as the denominator. The two degrees of freedom lost correspond to the number of regression coefficients. The positive square root of the estimator $\hat{\sigma}_{YX}^2$ is called the estimator of the population standard deviation of the regression, or the *sample standard deviation of regression.* It is also sometimes, rather misleadingly, named the *standard error of estimate.*

Quantities required to compute $\hat{\sigma}_{YX}^2$ are given in Table 17-3. From this

Table 17-3. Computation of the Estimate of Standard Deviation of Regression

Observed values				
Shearing strength y (1)	*Diameter* x (2)	*Predicted shearing strength* y_c (3)	$y - y_c$ (4)	$(y - y_c)^2$ (5)
7.0	2.4	7.56	−0.56	0.3136
5.3	1.8	6.04	−0.74	0.5476
4.2	1.6	5.53	−1.33	1.7689
3.3	1.0	4.01	−0.71	0.5041
3.8	1.2	4.52	−0.72	0.5184
6.6	1.1	4.27	+2.34	5.4756
8.5	2.8	8.57	−0.07	0.0049
6.6	1.6	5.53	+1.07	1.1449
4.5	1.5	5.28	−0.78	0.6084
8.8	2.3	7.30	+1.50	2.2500
58.6	17.3	58.60	0	13.1364

table, and by (17-12), we have for our illustrative example

$$\hat{\sigma}_{YX}^2 = \frac{13.1364}{10 - 2} = 1.64$$

and

$$\hat{\sigma}_{YX} = \sqrt{1.64} = 1.28 \text{ thousand pounds.}$$

Equation (17-12), the theoretical formula for $\hat{\sigma}_{YX}^2$, involves a great deal of work, especially when samples are large. For practical work, this measure is usually obtained by the following expression:

$$\hat{\sigma}_{YX}^2 = \frac{\sum y^2 - a \sum y - b \sum yx}{n - 2}, \tag{17–13a}$$

which, in terms of deviations from means, can be written as

$$\hat{\sigma}_{YX}^2 = \frac{[\sum d_1^2 - (\sum d_1 d_2)^2/\sum d_2^2]}{n - 2}. \tag{17–13b}$$

Applying it to our example, we have

$$\hat{\sigma}_{YX}^2 = \frac{33.724 - (8.152)^2/3.221}{10 - 2}$$

$$= \frac{13.0922}{8} = 1.64,$$

as before.

A precise interpretation of $\hat{\sigma}_{YX}^2$ or $\hat{\sigma}_{YX}$ will be given later in connection with our discussion on the "coefficient of determination." For the time being, we may consider the sample variance, or the sample standard deviation, of the regression as a measure of the "closeness" of the relationship between the two variables. The smaller either measure is, the closer are the Y values to the regression line, and thus the more accurate is the regression equation as a predictive device. We should also note that larger samples will aid in locating the regression more accurately but will not reduce the standard deviation of regression very much.

An interesting indication of the utility of regression analysis may be seen here in a comparison between the standard deviation of the regression and the standard deviation of Y computed with $\bar{y}$. In our illustration, $\hat{\sigma}_Y$ of shearing strength does not take diameter into consideration. Let us compare Figures 17–6 and 17–7, both plotted with the data of our present example. Figure 17–6 shows the vertical deviations of y's from the regression line, and Figure 17–7 shows the vertical deviations of y's from their mean $\bar{y}$. It is clear from these two diagrams that $\sum (y_i - \bar{y}_c)^2$ is much smaller than $\sum (y_i - \bar{y})^2$. To put this difference in numerical terms, we find that

$$\hat{\sigma}_Y = \sqrt{\frac{\sum (y - \bar{y})^2}{n - 1}} = \sqrt{\frac{\sum d_1^2}{n - 1}}$$

$$= \sqrt{\frac{33.724}{9}} = 1.94 \text{ thousand pounds.}$$

which is 0.66 (= 1.94 − 1.28) thousand pounds greater than $\hat{\sigma}_{YX}$. This rather large reduction in variation suggests that knowledge of the diameter helps us considerably in estimating shearing strength. In other words, it suggests that the relationship between shearing strength and diameter is rather close, and that the average shearing strength may vary to a large extent with diameter. We shall return to this point later.

17–6 INFERENCE ABOUT POPULATION REGRESSION COEFFICIENTS

Having obtained the sample regression equation, which shows the existence of a certain functional relationship, and having concluded that the regression equation may be a useful one on the basis of the sample standard deviation of the regression, the reader may jump to the conclusion that now the sample regression equation can be used to predict Y values on the basis of X values.

But in fact, even if the sample regression equation is identical with the true regression equation, prediction may still contain some error because the relationship between two variables in a bivariate population may not be perfect. The size of this error is measured by the population standard deviation of regression. This type of error is not due to sampling, and there is no way to evaluate it because the population regression coefficients and standard deviation of the regression are unknown. In predicting with the sample regression equation, moreover, there is an additional error due to variability in sampling. For repeated sampling with the same sample size, the estimates of A and B would tend to vary from sample to sample. Sampling errors of a and b, as usual, are evaluated in terms of their respective sampling distributions.

Sampling Distribution of *a*

The sampling distribution of a (we state here without proof) follows a t distribution with $\delta = n - 2$. Its expectation is $E(a) = A$ and its variance,

$$V(a) = \sigma_a{}^2 = \frac{(\sigma_{YX}{}^2)(\sum x^2)}{n \sum (x - \bar{x})^2},$$

where the sum, $\sum$, is taken over the sample. Since a is often of little practical importance, we shall not discuss inference about A on the basis of a, except to note that tests and estimations can be made in a fashion similar to that for B, discussed below.

Sampling Distribution of *b*

The estimated slope, b, is a crucial value in making predictions. It may be positive or negative, suggesting a positive or negative relationship between the two variables. But b, being a sample statistic, is subject to sampling variability. That is, because of chance variations, b may assume positive or negative values but the population slope may actually be zero. If this is true, then the sample regression equation is of no value whatsoever as a predictive device. Thus, we must know the nature of the sampling error in b and its size before we can use the sample regression equation with confidence.

The sampling distribution of b (we again state without proof) is normal if $n > 30$ and is distributed as Student's t with $n - 2$ degrees of freedom, otherwise. Furthermore, in both cases, the distribution of b has as its expectation $E(b) = B$, and as its variance

$$V(b) = \sigma_b^2 = \frac{\sigma_{YX}^2}{\sum (x - \bar{x})^2}.$$

The unbiased estimator of $V(b)$ is

$$\hat{\sigma}_b^2 = \frac{\hat{\sigma}_{YX}^2}{\sum (x - \bar{x})^2} = \frac{\hat{\sigma}_{YX}^2}{\sum d_2^2}. \tag{17–14}$$

The square root of the unbiased estimator of $V(b)$ is the estimator of the standard error of b and is denoted by $\hat{\sigma}_b$.

Test about, and Confidence Interval for, *B*

Conventionally, to test the significance of b we formulate the null hypothesis that the population slope is zero, that is, $B = 0$; and the alternative hypothesis that the population slope is not zero, that is, $B \neq 0$. When samples are large, we use as the testing statistic

$$Z = \frac{b - B_0}{\hat{\sigma}_b}. \tag{17–15}$$

When samples are small, we use

$$t = \frac{b - B_0}{\hat{\sigma}_b} \tag{17–16}$$

as the testing statistic. Here, for t, $\delta = n - 2$.

Suppose that we wish to test $B = 0$ against $B \neq 0$ with our illustrative data at, say, $\alpha = 0.05$, with $\delta = 10 - 2 = 8$; then the critical values would be ± 2.306. But

$$t = \frac{2.531 - 0}{\sqrt{\dfrac{1.64}{3.221}}}$$

$$= \frac{2.531}{0.714} = 3.545,$$

which is greater than 2.306. Thus we conclude that the difference between the estimate of slope and the hypothetical population slope cannot be explained by chance variation alone, and the null hypothesis is rejected at the 5 percent level of significance. This conclusion suggests that H_1 is true and that the sample regression equation should not be discarded. It can be used for predictive purposes.

The corresponding 95 percent confidence interval for B can be computed in the usual fashion:

$$b \pm 2.306\hat{\sigma}_b = 2.531 \pm 2.306(0.714) = 0.885 \text{ to } 4.177,$$

or

$$0.885 < B < 4.177.$$

The positive value of the lower confidence limit here indicates that we have established, at the 95 percent level of confidence, that the true slope is positive.

17–7 CONFIDENCE LIMITS OF PREDICTION

After we have decided that the sample regression equation should be retained, we may then use it to predict the mean of Y on X (μ_{YX}), or a new observation of Y (y), given the value of X. For this, we must be able to answer the question: how reliable is our prediction? The reply here again is made with our understanding of the distribution of $\bar{y}_c$ and y.

Confidence Interval for μ_{YX}

The estimator of μ_{YX} is $\bar{y}_c$. It can be shown that the sampling distribution of $\bar{y}_c$ is a t distribution, with $\delta = n - 2$. The expectation of $\bar{y}_c$ is μ_{YX}. Thus, $\bar{y}_c$ is an unbiased estimator of μ_{YX}. The variance of $\bar{y}_c$ can be shown to be

$$V(\bar{y}_c) = \sigma_{YX}^2\left[\frac{1}{n} + \frac{(x_0 - \bar{x})^2}{\sum (x - \bar{x})^2}\right],$$

where x_0 is a given fixed X value.

The square root of $V(\bar{y}_c)$ is the standard error of $\bar{y}_c$ and is denoted by $\sigma_{\bar{y}_c}$. The unbiased estimator of $V(\bar{y}_c)$ is

$$\hat{\sigma}_{\bar{y}_c}{}^2 = \hat{\sigma}_{YX}{}^2\left[\frac{1}{n} + \frac{(x_0 - \bar{x})^2}{\sum (x - \bar{x})^2}\right]$$

$$= \hat{\sigma}_{YX}{}^2\left[\frac{1}{n} + \frac{(x_0 - \bar{x})^2}{\sum d_x{}^2}\right] \qquad (17\text{–}17)$$

and the positive square root of this quantity is the estimator of the standard error of $\bar{y}_c$, denoted by $\hat{\sigma}_{\bar{y}_c}$.

Examination of (17–17) shows that the variance or standard error for $\bar{y}_c$ depends upon the given value of X. If $x_0 = \bar{x}$, it has its minimum value, since then $V(\bar{y}_c) = \hat{\sigma}_{YX}{}^2/n$. The farther x_0 is from $\bar{x}$, the greater is the variance or the standard error. Consequently, predictions of μ_{YX} are more precise the closer the fixed values of X are to $\bar{x}$. We become less and less sure about the relationship between the two variables the more extreme the X values become.

With the standard error of $\bar{y}_c$ estimated, a confidence interval for μ_{YX}, given the value of X, can be constructed in the usual way. For example, if we wish to construct a 95 percent confidence interval for our illustrative data, given that $x_0 = 1.48$, we have first

$$\hat{\sigma}_{\bar{y}_c} = 1.28\sqrt{\frac{1}{10} + \frac{(1.48 - 1.73)^2}{3.221}} = 0.442.$$

Next we note, for $x_0 = 1.48$,

$$\bar{y}_c = 1.481 + 2.531(1.48) = 5.23.$$

Finally,

$$\begin{aligned}\bar{y}_c \pm t_{n-2;\alpha/2}\hat{\sigma}_{\bar{y}_c} &= 5.23 \pm 2.306(0.442)\\ &= 4.21 \text{ to } 6.25\end{aligned}$$

or

$$4.21 < \mu_{YX} < 6.25.$$

Confidence intervals constructed for μ_{YX} at various fixed values of X are called a *confidence band for* μ_{YX}. A 95 percent confidence band is computed in Table 17–4 and the results are plotted in Figure 17–8. To speed up

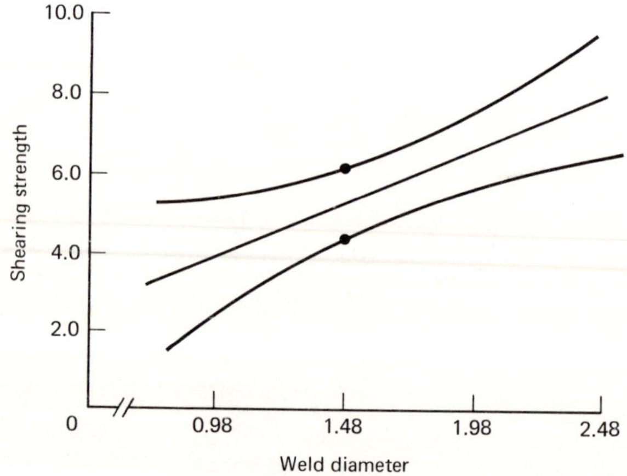

Figure 17–8 Confidence Interval for the True Regression Line

calculations, we may note that the values of X are selected to be symmetrical with respect to $\bar{x}$. Thus, the calculations of the second half are duplicates of those in the first half. Note how the general shape of the confidence band follows previous comments on the standard error of $\bar{y}_c$. It should be clear that this confidence band is for the true population regression line, not for the individual observations of Y, the y's. The result here should be interpreted as follows: we are 95 percent certain that the population regression line will fall within the confidence band, as shown by Figure 17–8. It is interesting to observe also that our confidence band is rather wide, but still not wide enough to contain a regression line of zero slope. This reinforces our confidence that the true regression is to some extent positively sloped.

Table 17–4. Computation of 95 Percent Confidence Band for True Regression Line

x_0	$(x_0 - \bar{x})$	$(x_0 - \bar{x})^2$	$\hat{\sigma}_{\bar{y}_c}$	$\bar{y}_c$	$\bar{y}_c \pm 2.306(\hat{\sigma}_{\bar{y}_c})$
0.98	−0.75	0.5625	0.671	3.96	2.41–5.51
1.23	−0.50	0.2500	0.539	4.59	3.34–5.84
1.48	−0.25	0.0625	0.442	5.23	4.21–6.25
1.73	0.00	0.0000	0.405	5.86	4.93–6.79
1.98	+0.25	0.0625	0.442	6.49	5.47–7.51
2.23	+0.50	0.2500	0.539	7.13	5.88–8.38
2.48	+0.75	0.5625	0.671	7.76	6.21–9.31

Confidence Interval for an Individual Observation *y*

Sometimes we are interested in predicting the value of an individual observation of Y instead of the average value of all Y values associated with a given value of X. For example, given a student whose IQ is 105, what is his predicted performance at college? Here, we are not asking for the average performance of all students whose IQ is 105. The problem of predicting an individual value is dealt with by constructing a confidence interval for y.

For the construction of a confidence interval for y, we start with an evaluation of the distribution of the difference $\bar{y}_c - y$. Given the value of X, the expected value of this difference is

$$E(\bar{y}_c - y) = E(\bar{y}_c) - E(y) = \mu_{YX} - \mu_{YX} = 0,$$

and the variance of the difference is

$$V(\bar{y}_c - y) = \hat{\sigma}_{YX}^2 \left[1 + \frac{1}{n} + \frac{(x_0 - \bar{x})^2}{\sum (x - \bar{x})^2}\right].$$

Again, using $\hat{\sigma}_{YX}^2$ as an estimator of σ_{YX}^2, we have the unbiased estimator of $V(\bar{y}_c - y)$ as follows:

$$\hat{\sigma}_{\bar{y}_c - y}^2 = \hat{\sigma}_{YX}^2 \left[1 + \frac{1}{n} + \frac{(x_0 - \bar{x})^2}{\sum (x - \bar{x})^2}\right]$$

$$= \hat{\sigma}_{YX}^2 \left[1 + \frac{1}{n} + \frac{(x_0 - \bar{x})^2}{\sum d_2^2}\right]. \qquad (17\text{–}18)$$

Note that $V(\bar{y}_c - y)$ is greater than $V(\bar{y}_c)$. This implies that the prediction of y is less precise than that of μ_{YX}. Again the positive square root of (17–18) is an unbiased estimator of the standard error of $(\bar{y}_c - y)$ and is denoted by $\hat{\sigma}_{\bar{y}_c - y}$.

It can be shown that the ratio $(\bar{y}_c - y)/\hat{\sigma}_{\bar{y}_c - y}$ is distributed as t_{n-2}. From this fact, we can easily derive a confidence interval for y as follows:

$$P\left(-t_{n-2;\,\alpha/2} < \frac{\bar{y}_c - y}{\hat{\sigma}_{\bar{y}_c - y}} < t_{n-2;\,\alpha/2}\right) = 1 - \alpha,$$

which, by rearranging terms, becomes

$$\bar{y}_c - t_{n-2;\,\alpha/2}\hat{\sigma}_{\bar{y}_c - y} < y < \bar{y}_c + t_{n-2;\,\alpha/2}\hat{\sigma}_{\bar{y}_c}.$$

Again, as before, given $x_0 = 1.48$, we have as our point estimate of an individual or new observation y

$$y = 1.481 + 2.531(1.48) = 5.23,$$

for which the standard error is

$$\hat{\sigma}_{\bar{y}_c - y} = 1.28\sqrt{1 + \frac{1}{10} + \frac{(1.48 - 1.73)^2}{3.221}} = 1.35.$$

With these values, a 95 percent confidence interval for y with $x_0 = 1.48$ becomes

$$5.23 - 2.306(1.35) < y < 5.23 + (2.306)(1.35)$$

or

$$2.12 < y < 8.34.$$

This is quite wide, owing to small n.

A confidence band can also be computed for y at various values of X. The interpretation here is the same as that for the regression line, except that, for obvious reasons, the confidence band for y's is much wider.

To conclude this section, we should note that in the prediction of either μ_{YX} or y, we should not fix the value of X outside the range of x's employed to determine the sample regression equation. The reason is that the relationship between the two variables below or above the range of the X values may not be the same as that which has been observed within the range. In general, the narrower the prediction limits, the more useful is the independent variable as the basis for predicting the values of the dependent variable. The factors that influence the width of prediction limits for both the average value of y's and individual new observations are the same and are indicated in equations for their respective variances. These factors are summarized below:

(1) The prediction intervals vary directly with the estimated standard deviation of the regression. The size of sample standard deviation of regression, however, does not depend very much upon the sample size; it is inherent in the particular bivariate population.

(2) The prediction interval varies inversely with n for two reasons: the larger n is, the larger is the quantity $\sum (x - \bar{x})^2$ and, therefore, the smaller is the standard error; the larger n is, the smaller is the critical value of $t_{n-1;\,\alpha/2}$ and, therefore, the narrower is the confidence interval.

(3) The greater the variability of the independent variable is, the narrower will be the interval, since the quantity $\sum (x - \bar{x})^2$ will be large.

(4) As has already been pointed out, given the first three factors, the prediction limits become wider as the fixed X value departs farther from $\bar{x}$. This is so because the quantity $(x_0 - \bar{x})^2$ will increase as the difference between x_0 and $\bar{x}$ increases, therefore increasing the standard error.

17-8 POPULATION CORRELATION COEFFICIENT

Up to now, we have discussed the *type* of relationship that exists between two variables. In some situations, however, we may be interested in the *degree* of relationship between two variables. For example, the federal government, in deciding on educational grants to a state or city government, may consider per capita income as a measure of the latter's ability to finance its own schools. Thus, the federal government may decide that per capita income is an appropriate criterion for the decision, because there is a high degree of relationship between per capita income and expenditures per student in each locality. Otherwise, another decision criterion should be chosen.

The measure for the degree of relationship between two variables is called the *correlation coefficient*, universally denoted as ρ, the Greek letter *rho*. The assumptions that constitute a bivariate linear correlation population model are briefly noted below:

(1) Both Y and X are random variables. As such, they need not be designated as dependent and independent; any designation will yield the same result.
(2) The bivariate population is normal. A bivariate normal population is one in which both Y and X are normally distributed. The expectation and variance of X are $E(X) = \mu_X$ and $V(X) = \sigma_X^2$, respectively, and the expectation and variance of Y are $E(Y) = \mu_Y$ and $V(Y) = \sigma_Y^2$, respectively.
(3) The relationship between Y and X is linear. This assumption implies that all the means of Y's associated with each and every X value, μ_{YX}, fall on the regression line of Y on X: $\mu_{YX} = A + Bx$. Also, all the means of X's associated with each and every Y value, μ_{XY}, fall on the regression line of X on Y: $\mu_{XY} = A' + B'y$. Furthermore, the population regression lines in the two equations are the same if and only if the relationship between Y and X is perfect. Otherwise, with Y dependent, intercepts and slope will differ from the regression equation with X dependent.

Under the foregoing assumptions, the population correlation coefficient is defined in terms of the covariance between Y and X as

$$\rho = \frac{\text{cov }(Y, X)}{\sigma_Y \sigma_X} = \frac{E[(Y - \mu_Y)(X - \mu_X)]}{\sqrt{E(Y - \mu_Y)^2}\sqrt{E(X - \mu_X)^2}}. \tag{17–19}$$

Note that cov (Y, X) is a measure indicating the systematic association between two jointly distributed random variables Y and X.

A number of things may be observed from the definition above. First, it is a function with five parameters: μ_X, σ_X, μ_Y, σ_Y, and ρ. The last, as mentioned before, is the correlation coefficient for the bivariate normal population. Second, ρ is symmetric with respect to Y and X. Finally, ρ is independent of units of measurement, since it is the covariance between two standard normal variables. In other words, ρ is a pure number, because it is defined as the ratio of the covariance between Y and X to the product of their respective standard deviations. As such, when cov $(Y, X) = 0$, ρ would be 0, indicating that there is no relationship between the two variables. When there is perfect covariability between Y and X, and Y and X vary in the same direction, $\rho = 1$. Similarly, when there is perfect covariability but Y and X vary in opposite directions, $\rho = -1$. When some degree of covariability exists between Y and X, we have $-1 < \rho < 0$ or $0 < \rho < 1$.

It is convenient to think of the bivariate normal distribution as a three-dimensional surface instead of a two-dimensional frequency curve. Figure 17–9 is a generalized portrait of the density of a joint normal distribution.

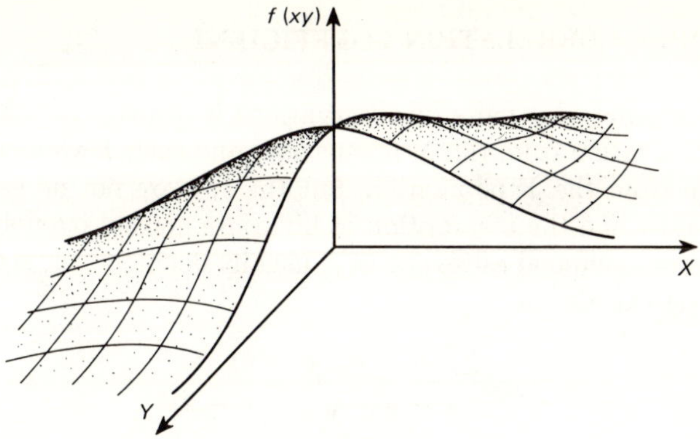

Figure 17–9 Bivariate Normal Distribution

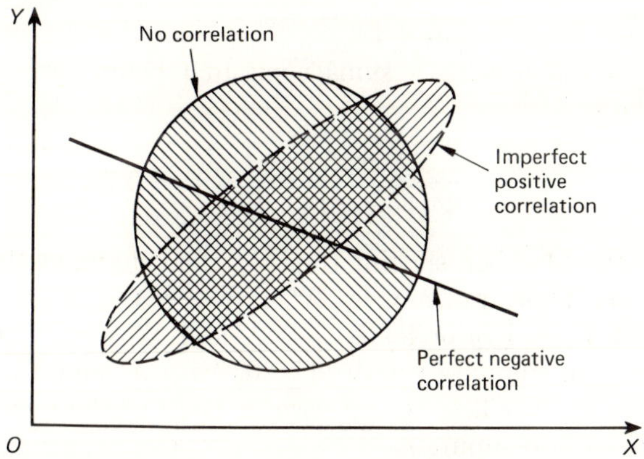

Figure 17–10 Ellipses from Correlation Surfaces

This figure reveals that our particular bivariate normal surface has some degree of covariability, since the mound-shaped surface is not parallel to either the Y or the X axis. If we slice this population surface at a given X value, we have a frequency curve giving the frequency of Y at that X value. That is, we have the distribution of the subpopulation Y that corresponds to the selected X value. Similarly, if we slice the population surface at a selected Y value, a subpopulation of X associated with this particular Y emerges. Now, if we take any cross section of the surface parallel to the XY plane, this cross section will be an ellipse. Furthermore, this cross section will approach a circle when $\rho = 0$ (indicating no correlation), a straight line when $\rho = \pm 1$ (indicating perfect correlation), and some in-between shape when imperfect correlation exists. These cases are illustrated by the shapes in Figure 17–10.

17–9 SAMPLE CORRELATION COEFFICIENT

When a sample of n pairs of observations is drawn, each X value is a random observation from the X population, and each Y value is a random observation from the Y population, but the two are not necessarily independent; and when the assumption of bivariate normal population is met, the maximum likelihood estimator of ρ, usually denoted as r, is obtained by the following expression:

$$\hat{\rho} = r = \frac{\sum (y - \bar{y})(x - \bar{x})}{\sqrt{\sum (y - \bar{y})^2}\sqrt{\sum (x - \bar{x})^2}}$$

$$= \frac{n \sum d_1 d_2}{\sqrt{(n \sum d_1^2)(n \sum d_2^2)}} = \frac{\sum d_1 d_2}{\sqrt{\sum d_1^2 \sum d_2^2}}, \tag{17–20}$$

which may vary from -1 to $+1$.

Referring to the necessary sums computed before, we have, for our illustrative data,

$$r = \frac{8.152}{\sqrt{(33.724)(3.221)}} = +0.78.$$

Since r is a pure number, shifting the origin and changing the scale do not affect its value. Thus, we may let $X' = X - c$ and $Y' = Y - k$, or let $X' = X/g$ and $Y' = Y/g$ or let $X' = mX$ and $Y' = mY$, and in each of these cases we can compute r with (17–20) with X' and Y' without additional adjustment. Needless to say, two or more of these coding methods can be employed to compute r.

17–10 INFERENCE ABOUT ρ

The estimator r, like other statistics, is subject to sampling variations. When we obtain a positive or negative r, we may not be at all sure, because of sampling error, that the corresponding value of ρ is also positive or negative. This situation calls for a test of significance.

For the purpose of testing, we note that the sampling distribution of r varies with the value of ρ and the sample size. Under the assumption that the sample is drawn from a normal bivariate population, r tends to be distributed normally as the sample size increases. This tendency is much stronger for values of ρ that are close to 0 than for values close to -1 or $+1$. When $\rho = 0$ and n is large, in other words, the distribution of r closely resembles a normal distribution. Figure 17–11 illustrates this point.

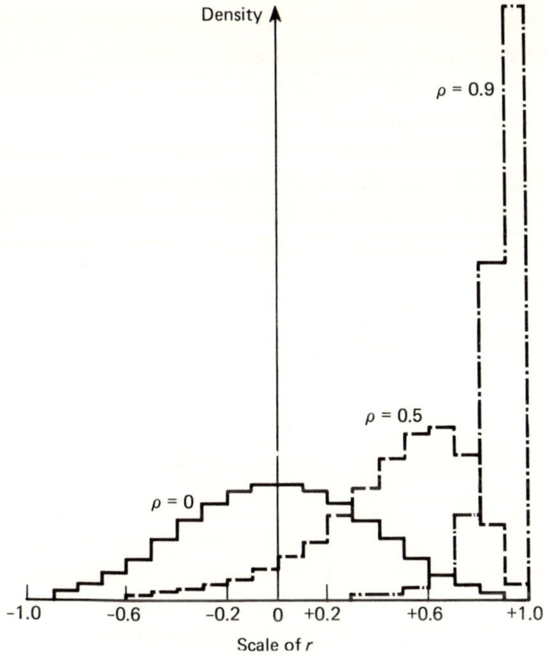

Figure 17–11 Sampling Distributions of Coefficient of Correlation Samples of Size 10, $\rho = 0$, $\rho = 0.5$, and $\rho = 0.9$

The property that, when $\rho = 0$ and n is large, r is approximately normal, enables us to conduct the simplest and perhaps the most useful test on the null hypothesis that there is no correlation in the bivariate population, $\rho = 0$, against the alternative that $\rho \neq 0$. When $n \leq 30$, the statistic used to test this H_0 is

$$t = r\sqrt{\frac{n-2}{1-r^2}}, \tag{17–21}$$

which is distributed as t_{n-2} if the null hypothesis is true.

Applying such a test to our illustrative data at $\alpha = 0.05$, we would accept H_0: $\rho = 0$ if $-2.306 < t_8 < 2.306$. For the value of r obtained in the last section, however,

$$t = 0.78\sqrt{\frac{10-2}{1-(0.78)^2}} = 3.523.$$

Consequently, the null hypothesis that there is no correlation in the population is rejected at the specified level of significance. This means that a value of r as large as 0.78 with $\delta = 8$ cannot be the result of chance variations. It is significant.

Note that the null hypotheses $\rho = 0$ and $B = 0$ are equivalent, since, as can be shown, $B = \rho(\sigma_Y/\sigma_X)$. As a result, the t tests in both cases should yield the same value. The t values are different in our example owing to rounding.

With the t test just introduced, we cannot state explicitly the extent of the relationship between the two variables when the null hypothesis is rejected. If we desire to test a hypothesis that ρ has some value other than 0, or if we desire to construct a confidence interval for ρ, we may employ what is called the z transformation. That is, we may make a transformation of the asymmetric distribution of r into an approximately normal distribution as follows:

$$z_r = \frac{1}{2} \ln \frac{1 + r}{1 - r}, \tag{17–22}$$

where ln is a natural logarithm and is approximately equal to 2.3026 log. Here, log stands for the common logarithm of base 10. It can be shown that z_r is approximately normally distributed with $E(z_r) = z_\rho$, and the estimated standard error is

$$\hat{\sigma}_z = \frac{1}{\sqrt{n - 3}}. \tag{17–23}$$

To test a hypothesis about ρ by r, we have now the testing statistic

$$Z = \frac{z_r - z_\rho}{\hat{\sigma}_z}, \tag{17–24}$$

which is approximately $n(0, 1)$.

To avoid the computations with logarithms, we employ a table of values for z_r corresponding to various values of r as given by Table B–12.

Now, suppose that we wish to test H_0: $\rho = 0.85$ against H_1: $\rho \neq 0.85$ for our illustrative example; we refer to Table B–12 and find

$$z_\rho = 1.25615 \qquad \text{for } \rho = 0.85,$$

$$z_r = 1.04537 \qquad \text{for } r = 0.78.$$

Hence,

$$Z = \frac{1.04537 - 1.25615}{1/\sqrt{10 - 3}} = -0.56.$$

Since the critical value is greater than -1.96, H_0 is accepted at $\alpha = 0.05$.

Confidence limits may be computed for z_ρ as follows:

$$P(z_r - z_{\alpha/2}\sigma_z < z_\rho < z_r + z_{\alpha/2}\sigma_z) = 1 - \alpha. \tag{17–25}$$

We can convert z_r into r in this expression by looking at the entries in Table B–12 for the value of z_ρ and then checking the stub and column headings to find the corresponding value of r. Owing to rounding and/or limited entries in the table, this conversion may be only an approximation.

To illustrate the application of (17–25), we shall now make a 95 percent confidence interval for the problem on hand, by noting that $\hat{\sigma}_z = 0.378$,

$$1.04537 - 1.96(0.378) < z_\rho < 1.04537 + 1.96(0.378),$$

or

$$0.30449 < z_\rho < 1.78625.$$

From Table B–12, we find

z_ρ	r
0.30	0.29
1.79	0.95

Hence, the 95 percent confidence interval becomes

$$0.29 < \rho < 0.95,$$

which includes $\rho = 0.85$.

Also, the significance of a difference between two sample correlation coefficients obtained from two independent samples and their difference can be tested by the statistic

$$Z = \frac{z_{r1} - z_{r2}}{\sqrt{\hat{\sigma}_{z_1}{}^2 + \hat{\sigma}_{z_2}{}^2}}, \tag{17–26}$$

where

$$\hat{\sigma}_{z1} = \frac{1}{\sqrt{n_1 - 3}} \quad \text{and} \quad \hat{\sigma}_{z2} = \frac{1}{\sqrt{n_2 - 3}}.$$

17–11 POPULATION COEFFICIENT OF DETERMINATION

We have defined the correlation coefficient of bivariate normal population as the degree of convariability between Y and X. This definition led us to compute r as the maximum likelihood estimator of ρ. Suppose that we are now concerned with a bivariate population whose distribution is unspecified. We can then define the population correlation coefficient as a measure of closeness-of-fit of the regression line. This new definition is based on the concept of coefficient of determination, which we are about to introduce with the aid of Figure 17–12. This figure attempts to show the relationship between an individual value of Y, y, and the population regression line of Y on X, as well as the relationship between y and the mean of Y, μ_Y. We

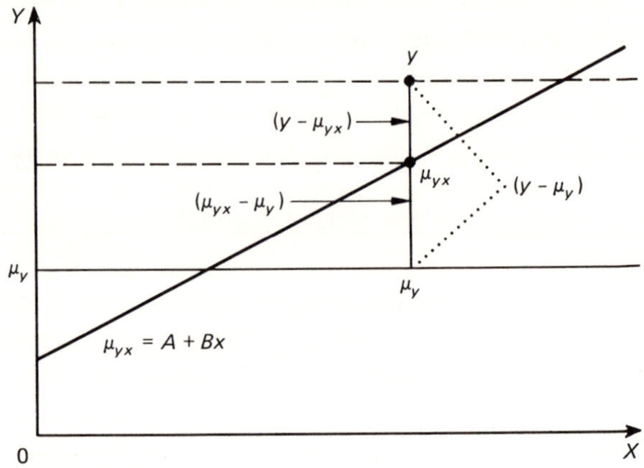

Figure 17–12 Partition of $(y - \mu_y)$

see that the following relationship holds:

$$\text{total error} = \text{unexplained error} + \text{explained error}$$

$$(1) \qquad (y - \mu_Y) = [(y - \mu_{YX}) \qquad + (\mu_{YX} - \mu_Y)]$$

In this expression, the deviation of y from μ_Y is the *total error*. The second term on the right side of (1), $\mu_{YX} - \mu_Y$, is called the *explained error* because its pattern is definite, in the sense that it measures the amount of error that has been removed after the regression line has been fitted. The term $(y - \mu_{YX})$ is called the *unexplained error* because it is a random quantity and is unpredictable; it measures the part of the total error that still remains or is unexplained after the regression line has been fitted.

Suppose we decompose each and every point in the scatter diagram as indicated by (1): squaring both sides of the equation, and summing over all points in the population, we would then obtain

$$(2) \quad \sum (y - \mu_Y)^2 = \sum (y - \mu_{YX})^2 + \sum (\mu_{YX} - \mu_Y)^2 - 2\sum (y - \mu_{YX})(\mu_{YX} - \mu_Y),$$

in which the last sum is zero, since

$$\begin{aligned}\sum (y - \mu_{YX})(\mu_{YX} - \mu_Y) &= \sum (y - A - Bx)(A + Bx - \mu_Y) \\ &= A\sum (y - A - Bx) + A\sum x(y - A - Bx) \\ &\quad - \mu_Y \sum (y - A - Bx) = 0.\end{aligned}$$

Thus, (2) becomes

$$\sum (y - \mu_Y)^2 = \sum (y - \mu_{YX})^2 + \sum (\mu_{YX} - \mu_Y)^2, \qquad (17\text{–}27)$$

which says that the *total sum of squares* $\sum (y - \mu_Y)^2$ is partitioned into two independent parts: the *unexplained sum of squares* $\sum (y - \mu_{YX})^2$, and the *explained sum of squares* $(\mu_{YX} - \mu_Y)^2$. [Compare these results with (15–13) to (15–15) inclusive.]

Our interest in decomposing the total sum of squares rests on the fact that the ratio of the explained sum of squares to the total sum of squares gives us a parameter of a bivariate population called the *coefficient of determination*. Denoting this parameter as ρ^2, we have

$$\rho^2 = \frac{\sum (\mu_{YX} - \mu_Y)^2}{\sum (y - \mu_Y)^2}. \tag{17–28}$$

A number of properties of the population coefficient of determination are of interest to us. First, the value of ρ^2 ranges from 0 to 1. Second, we see from equations (17–27) and (17–28) that if the unexplained error is zero, the total error is equal to the explained error and, therefore, $\rho^2 = 1$. Thus, ρ^2 is a measure that shows the relative reduction of the total error when a regression line is fitted. For example, if $\rho^2 = 0.8$, it means that 80 percent of the variation in Y is "explained" by the relationship between Y and X. When $\rho^2 = 1$, then we know that $y = \mu_{YX}$ and $e = 0$, indicating that all points fall on the regression line. When $\rho^2 = 0$, then the explained error equals zero and $\mu_{YX} = \mu_Y$. Graphically, this means that the regression line is parallel to and coincides with the line for μ_Y. See Figure 17–13(a) and (b). It is important to note, in this connection, that the coefficient of determination is a measure of the improvement in reducing the total error. It is a measure of the linear relationship between Y and X.

Third, when ρ^2 is close to 1, the individual points will be scattered closely and give the appearance of a straight line. When ρ^2 is close to 0, the points will be widely dispersed and will not resemble a straight line. Thus, ρ^2 can also be considered as a measure of *linearity* of points. In the case in which Y is a constant variable (that is, each individual Y value is equal to, say, c), for each value of X, $y = c$ and $\bar{y} = c$. Then, as shown by Figure 17–14, all the points would fall on the regression line, which is identical with the horizontal line for μ_Y. However, now the y's do not have a distribution and $B = 0 = \sum (y - \mu_Y)^2$. As a result, the formula defining the coefficient of determination becomes meaningless. There is, in other words, no regression problem at all.

Finally, as we have pointed out before, the population standard deviation of regression is a measure of the scatter of y's around μ_{YX}. However, if the subpopulations, Y's, do not possess any specific form of distribution, an explicit evaluation of σ_{YX} cannot be made. In this case, ρ^2 is a much more obvious measure of the scatter or of the goodness-of-fit. It is easy to show, fortunately, that σ_{YX} and ρ^2 are closely related. To bring out this relationship, let us divide both sides of the equation for the sums of squares

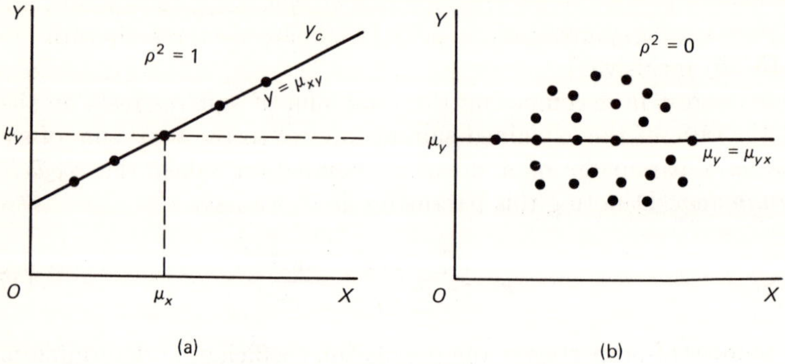

Figure 17–13 ρ^2 as a Measure of Reducing Total Error

by $\sum (y - \mu_Y)^2$, obtaining

$$1 = \frac{\sum (y - \mu_{YX})^2}{\sum (y - \mu_Y)^2} + \frac{\sum (\mu_{YX} - \mu_Y)^2}{\sum (y - \mu_Y)^2}.$$

The second term on the right side of this expression is our definition for ρ^2; thus

$$\frac{\sum (y - \mu_{YX})^2}{\sum (y - \mu_Y)^2} = 1 - \rho^2.$$

Note that, from a computational viewpoint,

$$\sigma_{YX}{}^2 = \frac{1}{N} \sum_i (y - \mu_{YX})^2$$

and

$$\sigma_Y{}^2 = \frac{1}{N} \sum_i (y - \mu_Y)^2,$$

where N is the population size. We have, therefore,

$$\sigma_{YX}{}^2 = (1 - \rho^2)\sigma_Y{}^2. \tag{17–29}$$

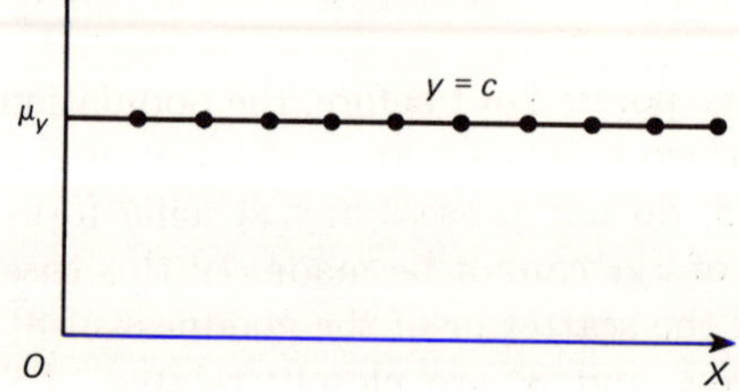

Figure 17–14 Graphic Illustration of the Absence of Regression Problem

This result states that, after the regression line has been fitted, the variance of Y has been reduced by $(100)\rho^2$ percent, and a residual of $100(1 - \rho^2)$ percent of $\sigma_Y{}^2$ remains unexplained. When $\rho^2 = 1$, all of $\sigma_Y{}^2$ has been removed, or all the variations in Y have been explained by the variations in X. When $\rho^2 = 0$, there has been no improvement at all by fitting the regression line, and $\sigma_{YX}{}^2 = \sigma_Y{}^2$.

In concluding this section, let us note that the square root of the population coefficient of determination is the population correlation coefficient:

$$\sqrt{\rho^2} = \pm\rho, \tag{17–30}$$

where the sign of the correlation coefficient is the same as that for B, the population regression slope. When we have a bivariate population, ρ or ρ^2 can be interpreted as a measure of covariability as well as of closeness-of-fit. When X's are fixed values, however, cov (Y, X) may be undefined, and ρ or ρ^2 can be interpreted only as a measure of closeness-of-fit.

17–12 SAMPLE COEFFICIENT OF DETERMINATION

Given the minimum variance estimator of the population regression line, $\bar{y}_c = a + bx$, we have the following valid relationship:

$$\underset{\text{total error}}{\sum (y - \bar{y})^2} = \underset{\text{unexplained error}}{\sum (y - \bar{y}_c)^2} + \underset{\text{explained error}}{\sum (\bar{y}_c - \bar{y})^2}. \tag{17–31}$$

From this, we define the *sample coefficient of determination* as

$$r^2 = \hat{\rho}^2 = \frac{\sum (\bar{y}_c - \bar{y})^2}{\sum (y - \bar{y})^2}. \tag{17–32}$$

The sample coefficient of determination, as an estimator for ρ^2, measures the closeness-of-fit of the sample regression line to sample observations. It can be shown that r^2 is a consistent estimator of ρ^2 as n becomes large. We may also note that the square root of r^2 is the sample correlation coefficient,

$$\sqrt{r^2} = \pm r, \tag{17–33}$$

where the sign is the same as that for b.

For data below, X stands for the ages of five trees in years and Y stands for the heights of the trees in feet:

Y: 2 2 4 3 5

X: 1 2 3 4 5

For these data the sample regression equation is

$$\bar{y}_c - 1.10 + 0.7x.$$

Now, to check the relationship between the sums of squares for error and compute the sample coefficient of determination, we accumulate the necessary sums in Table 17–5. From quantities therein, we find that

$$\sum (y - \bar{y})^2 = \sum (y - \bar{y}_c)^2 + \sum (\bar{y}_c - \bar{y})^2$$
$$6.8 \quad = \quad 1.9 \quad + \quad 4.9$$

and that

$$r^2 = \frac{\sum (y_c - \bar{y})^2}{\sum (y - \bar{y})^2} = \frac{4.90}{6.80} = 0.72.$$

Table 17–5

y	$(y - \bar{y})$	$(y - \bar{y})^2$	y_c	$(y - y_c)$	$(y - y_c)^2$	$(y_c - \bar{y})$	$(y_c - \bar{y})^2$
2	−1.2	1.44	1.8	+0.2	0.04	−1.4	1.96
2	−1.2	1.44	2.5	−0.5	0.25	−0.7	0.49
4	+0.8	0.64	3.2	+0.8	0.64	0.0	0.00
3	−0.2	0.04	3.9	−0.9	0.81	+0.7	0.49
5	+1.8	3.24	4.6	+0.4	0.16	+1.4	1.96
	0.0	6.80		0.0	1.90	0.0	4.90

The value of r^2 indicates that 72 percent of the changes in the height of the trees are associated with the changes in the age of the trees. There remains $100 - 72 = 28$ percent variability in the height of the trees that is unexplained by the regression. Also, since b is positive in this case, we have

$$r = \sqrt{0.72} = +0.85.$$

Here, X values are fixed; the value of r^2 or r is thought of as a measure of closeness-of-fit.

Using the data in Table 17–3 concerning spot welds of aluminum, we have

$$r^2 = \frac{20.59}{33.72} = 0.61.$$

In this case, where both X and Y are random variables and we have a bivariate distribution, r^2 may be thought of as a measure of closeness-of-fit as well as of the degree of covariability.

17–13 RANK-CORRELATION COEFFICIENT

When a random sample of n pairs is drawn from a bivariate population, where observations are in ordinal measurements instead of cardinal numbers, the degree of relationship between Y and X cannot be determined by the technique of correlation coefficient presented before. In such a situation, the covariability or the lack of it between Y and X is established by what is called the *rank-correlation coefficient* developed by C. Spearman in 1904. This measure is especially useful when quantitative measures for certain factors (such as in the evaluation of leadership ability or the judgment of feminine beauty) cannot be fixed, but the individuals in the group can be arranged in order, thereby obtaining for each individual a number indicating his (her) rank in the group. In any event, the rank-correlation coefficient is applied to a set of ordinal rank numbers, with 1 for the individual ranked first in quantity or quality, and so on, to n for the individual ranked last in a group of n individuals (or n pairs of individuals).

Let us symbolize the rank of individual i on variable X by x_i and the rank of i on variable Y by y_i; then we define d_i as $(x_i - y_i)$. Furthermore, the x_i and y_i may represent two sets of ranks assigned to the same set of objects by two different judges. With these notations, the Spearman's rank-correlation coefficient is defined as

$$r_s = 1 - \frac{6 \sum_i d_i^2}{n(n^2 - 1)}. \tag{17-34}$$

From this definition we see that $-1 \leq r_s < +1$. If the rank orders agree perfectly, the ranks assigned to individuals should have perfect positive correlation. If the ranks assigned to one set, x_i, are exactly the opposite of the ranks in the other set, y_i, then r_s should reflect perfect negative correlation. A zero r_s clearly represents the absence of any relationship between the two sets of ranks.

It is important to note that formula (17–34) has been derived under the assumption that there are no ties in either or both rankings. When ties do occur, they may be treated in any of the following three ways. First, we may assign mean ranks to sets of tied individuals; that is, when two or more individuals are tied in order, each is assigned the mean of the ranks they would otherwise take. Second, we may compute an ordinary correlation coefficient, r, with the ranks as if they were simply numerical scores; the result may be regarded as a Spearman's correlation adjusted for ties. Finally, if the main aim of our analysis is to test the significance of r_s, we should adopt the conservative procedure of breaking the ties in such a way that the absolute value of r_s is as small as possible.

When no ties in rank exist, the exact sampling distribution of r_s can be worked out. Tables for distributions of r_s are available for small n. When $n \geq 20$, we may test the hypothesis $r_s = 0$ against an appropriate alternative by using the statistic

$$t = \frac{r_s\sqrt{n - 2}}{\sqrt{1 - r_s^2}}, \tag{17-35}$$

which is approximately distributed as Student's t with $n - 2$ degrees of freedom. For values $8 \leq n \leq 20$, the t test can still be employed to yield satisfactory results if the t ratio is computed with an r_s that is calculated with a correction for continuity. Let us denote the corrected rank-correlation coefficient as r_s'; then

$$r_s' = 1 - \frac{\sum_i d_i^2}{\frac{1}{6}[n(n^2 - 1)] + 1}. \tag{17–36}$$

With this measure,

$$t_\delta = \frac{r_s'\sqrt{n - 2}}{\sqrt{1 - r_s'^2}} \tag{17–37}$$

is again approximately distributed at t_{n-2}.

Let us consider an example. In a beauty contest, two judges each ranked ten girls in the order of their "beautiful features." The results are shown in Table 17–6. Did the judges tend to agree at $\alpha = 0.05$?

$$r_s = 1 - \frac{6(24)}{10(10^2 - 1)} = 0.8545.$$

Since n is rather small in our illustration, let us compute

$$r_s' = 1 - \frac{24}{\frac{1}{6}[10(10^2 - 1)] + 1} = 0.8554.$$

To determine whether this result is significant at $\alpha = 0.05$, we obtain

$$t_8 = \frac{0.8554\sqrt{10 - 2}}{\sqrt{1 - (0.8554)^2}} = 4.67.$$

Table 17–6

Contestant	*Judge I*	*Judge II*	d_i	d_i^2
1	8	9	−1	1
2	3	5	−2	4
3	9	10	−1	1
4	2	1	1	1
5	7	8	−1	1
6	10	7	3	9
7	4	3	1	1
8	6	4	2	4
9	1	2	−1	1
10	5	6	−1	1
			0	24

Since $P(t_8 > 1.860) = 0.05$, the observed value of rank-correlation is significant. The null hypothesis $\rho_s = 0$ is rejected in favor of the alternative, $\rho_s > 0$. The two judges have a high degree of agreement.

17-14 CONCLUDING REMARKS

In this chapter we have discussed, in a rather detailed fashion, the simple linear regression and correlation models by presenting the measure of coefficient of determination as the link between the two. Because these two models are constructed under different assumptions, they furnish different types of information, and it is not always clear to a beginning student which measure should be used in given problem situations. It may be useful, then, to compare the two models again here.

First, the same type of relationship holds for both regression analysis and correlation analysis, as indicated by the fact that r takes the same sign as b. Also, if the value of b is significant at a given level of significance, r is also significant at that level.

Second, r, via r^2, may be considered as a measure of the closeness-of-fit of the regression line. In general, the greater the value of r^2, the better is the fit and the more useful the regression equation as a predictive device. (Note that the greater the value of r^2 is, the closer r is to $+1$ or -1.)

Given the first two observations, we must realize that the main objective of regression analysis is to establish a functional relationship between the dependent and independent variables so that the former can be predicted by the latter; however, the correlation coefficient is a measure of the closeness-of-fit of the regression line and/or a measure of the degree of covariability between Y and X.

Third, a given value of the correlation coefficient is consistent with an infinite number of regression lines. For instance, in the case of perfect correlation, when all points are shifted upward or downward by an equal amount, the value of the correlation coefficient will remain as 1 but a different Y intercept will result. Similarly, if all points are rotated at a given point so that they all fall precisely on a different straight line, the slope, and perhaps the Y intercept, will change while the correlation coefficient will still assume the value of $+1$ or -1.

Finally, the main difference in assumptions between the two models is that Y is assumed to be a random variable and X is assumed to take fixed values in the regression analysis, while both Y and X are hypothesized as random variables in the correlation analysis.

Theoretically, therefore, regression is a directional method but correlation is not. The former should be used if one variable is clearly dependent upon the other, or one is measured subsequent to the other. The latter is appropriate when the two variables have a joint distribution, neither of which can be considered as being subsequent to, or as being a consequence of, the other.

Practically, on the other hand, the purpose of the investigation is perhaps the most important consideration in selecting between regression and correlation techniques. The choice depends, in other words, on whether we want to have a predictive equation or merely wish to determine the degree of association. As a matter of fact, in practical situations we are often more interested in finding out the nature of the relationship for the objective of prediction. In this connection, the reason for our interest in the degree of relationship may be only to select the most useful independent variable, among several possible independent variables, for regression analysis with the dependent variable that has been chosen.

Association analysis, while a very useful device, often opens up many opportunities for misinterpretation. The most unwarranted interpretation stems from the confusion between association and causation, arising chiefly from the conventional designation of dependent and independent variables. Many people take the independent variable to be the cause and the dependent variable to be the effect in a given situation, but this need not be so.

The link between association and causation can be stated thus: the presence of association does not imply causation; but the existence of causation always implies association. Statistical evidence can only establish the presence or absence of association between variables. Whether causation exists or not depends purely on reasoning. For example, there is reason to believe that high family income "causes" high family consumption expenditures; hence these two variables should be positively associated. There is also reason to support the premise that the high price of prime cuts of beef "causes" low quantities of this beef to be sold; hence these two variables should be negatively associated. That good weather is associated with high death rates in areas favored by retired people clearly does not mean that good weather "causes" high death rates.

Why is it that association need not imply causation? One reason is that the association between two variables may be the result of pure chance, such as soil erosion in Alaska and the amount of alcohol consumed in South America. Another reason is that association between two variables may be due to the influence of a third common factor. Since 1945, for example, it has been found that there is a close relationship between teachers' salaries and liquor consumption in the United States. Can we say that the increase in liquor consumption is the effect of the increase in teachers' salaries? Certainly not. Indeed, a more plausible explanation is that both variables have been influenced by the continuous increase in national income during the same period. A third possibility is that in a real relationship we may be unable to determine which variable is the cause and which is the effect; we often find that both variables may be the cause and the effect at the same time. It may also happen that cause and effect change places from time to time. The association between per capita income and per capita expenditures for education, for example, may be of this kind. The higher the per capita income is in a state, the more money can be spent for each student; the more money the state spends for education, the higher will be the per capita income in that state; we cannot be sure, however, that one has produced the other or vice versa.

Association observed between variables that could not conceivably be causally related are called *spurious*, or *nonsense, correlations*. More appropriately, however, we should remember that it is the *interpretation* of the degree of association that is spurious, not the degree of association itself.

A last word of warning: Errors in association analysis include not only reading causation into spurious associations but also interpreting spuriously a perfectly valid association.

EXERCISES

(*Note*: The student should save calculations for Exercises 17–1 to 17–6 to use in Exercises 17–7 to 17–12. Because of rounding differences, the student's answers to the exercises in this chapter may not agree exactly as given.)

17–1 The number of defective items produced per unit of time, Y, by a certain machine is thought to vary directly with the speed of machine X, measured in rpm. Observations for twelve time periods yield the following results:

X	Y	X	Y
13.2	9.4	10.9	5.7
14.9	12.2	17.4	12.3
8.1	6.0	13.8	9.2
16.4	11.4	10.2	7.0
13.1	9.6	15.8	9.0
10.8	7.5	12.0	7.0

(a) Construct a scatter diagram Y on X for the data above.

(b) Verify that this set of data yields an estimate of the population regression equation as follows:

$$\bar{y}_c = -0.5967 + 0.7245x.$$

(c) Give a verbal interpretation to a and b just obtained and indicate whether a is of practical importance here.

(d) Is b obtained significant at $\alpha = 0.05$? $\alpha = 0.01$?
(e) Construct a 95 percent confidence interval for B.

17–2 The following sample contains the price and the quantity in relation to the supply of a commodity. Use quantity as the dependent variable in this case.

Price	*Quantity*	*Price*	*Quantity*
25	60	15	40
20	85	20	55
35	110	30	90
40	95	40	115
60	140	50	120
55	160	70	180
45	80	45	95

(a) Construct a scatter diagram of Q on P.
(b) Verify that the estimated supply function is

$$\bar{q}_s = 15.71 + 2.191p.$$

(c) Give verbal interpretations to a and b just obtained, and indicate whether a has any practical meaning here.
(d) Is b significant at $\alpha = 0.05$? $\alpha = 0.01$?
(e) Construct a 95 percent confidence interval for B.

17–3 The following sample contains the price and the quantity in relation to the demand for a commodity. Use Q as the dependent variable.

Price	*Quantity*	*Price*	*Quantity*
\$6.10	65	\$6.80	53
7.40	25	2.50	44
6.40	41	7.00	48
3.00	66	2.30	87
1.00	78	2.00	60
4.00	60	6.00	18
7.10	32	3.80	34
4.50	28	8.00	10

(a) Construct a scatter diagram of Q on P.
(b) Verify that the estimated demand function is

$$\bar{q}_d = 78.36 - 6.48p.$$

(c) Give verbal interpretations to a and b and indicate whether a is of practical importance.
(d) Is b significant at $\alpha = 0.05$? $\alpha = 0.01$?
(e) Construct a 95 percent confidence interval for B.

17–4 The annual disposable income and consumption expenditure of twelve families selected at random from a certain section of New York City are as follows:

Income	*Consumption*	*Income*	*Consumption*
\$8,000	\$7,000	\$8,000	\$8,000
15,000	12,000	13,000	11,000
20,000	18,000	7,000	8,000
35,000	30,000	6,000	6,000
28,000	20,000	12,000	10,000
25,000	24,000	15,000	18,000

[*Hint*: Use consumption expenditure, C, as the dependent, and disposable income, Y, as the independent, variable. Do your calculations with $Y' = Y/1000$ and $C' = C/1000$.]
(a) Construct a scatter diagram for C on Y.
(b) Verify that the estimated consumption function is

$$\bar{C}_c = 1660 + 0.7921y.$$

(c) Give verbal interpretations to a and b and indicate whether a is of practical importance.
(d) Is b significant at $\alpha = 0.05$? $\alpha = 0.01$?
(e) Construct a 95 percent confidence interval for B.

17–5 The president of a chain retailing organization believes there is a positive relationship between the sales of his company's product and per capita income in the year just past. He decides to associate 1967 sales with 1966 per capita income in the 15 cities where his company has branches. The data are as follows, where X is 1966 per capita income in thousands of dollars and Y is 1967 per capita sales in dollars.

X	Y	X	Y	X	Y
2.0	15	2.2	23	1.8	12
2.1	25	2.5	25	2.3	22
2.0	17	2.2	16	2.2	18
2.3	23	2.1	18	2.0	18
2.7	12	1.6	11	1.5	10

(a) Construct a scatter diagram.
(b) Verify that the estimated regression of Y on X is

$$\bar{y}_c = -0.55 + 8.6765x.$$

(c) Interpret a and b. Is a of any practical meaning here?
(d) Is b significant at $\alpha = 0.05$? $\alpha = 0.01$?
(e) Construct a 95 percent confidence interval for B.

17–6 The following random sample gives the number of hours of study for, X, and grades of, Y, an examination in elementary statistics for twelve students:

X	3	3	3	4	4	5	5	5	6	6	7	8
Y	45	60	55	60	75	70	80	75	90	80	75	85

(a) Construct a scatter diagram.
(b) Verify that the sample regression is

$$\bar{y}_c = 38.36 + 6.6x.$$

(c) Interpret a and b. Does a have a practical meaning here?
(d) Is b significant at $\alpha = 0.05$? $\alpha = 0.01$?
(e) Construct a 95 percent confidence interval for B.

17–7 For data in Problem 17–1, given $x_0 = 10$:
(a) Show that $\hat{\sigma}_{\bar{y}_c} = 0.4947$ and $\hat{\sigma}_{\bar{y}_c - y} = 1.2283$.
(b) Construct 95 percent predictive intervals for μ_{YX} and y with $x_0 = 10$.

17–8 For data in Problem 17–2:
(a) Show that the 95 percent confidence predictive intervals for μ_{qp} and q, at $p_0 = 42$, are respectively

$$97.19 < \mu_{q42} < 118.28$$

and

$$67.49 < q_{42} < 148.00.$$

17–9 For data in Problem 17–3, show that, given $p_0 = 4.2$ and $1 - \alpha = 0.95$, we have

$$41.75 < \mu_{q4.2} < 60.56$$

and

$$13.97 < q_{4.2} < 88.34.$$

17–10 For data in Problem 17–4, show that, given $y_0 = 11{,}500$ and $1 - \alpha = 0.95$, we have

$$9186 < \mu_{C_{11,500}} < 12{,}352$$

and

$$1865 < C_{11,500} < 19{,}863.$$

17–11 Construct 95 percent predictive intervals with data in Problem 17–5 for μ_{YX} and y given that $x_0 = 2.105$.

17–12 Construct 95 percent predictive intervals for μ_{YX} and y with data in Problem 17–6, given $x_0 = 4.5$.

Instructions for Exercises 17–13 to 17–16:

(a) Compute the sample correlation coefficient.

(b) Test $\rho = 0$ against $\rho \neq 0$ at $\alpha = 0.05$.

(c) Test whether ρ is equal to the value specified in H_0 and construct a corresponding confidence interval.

(d) Explain whether r is real or spurious.

17–13 The following data refer to the entrance test scores, X, and the labor efficiency test, Y, of 20 machine operators. H_0: $\rho = 0.90$.

Worker	*X*	*Y*	*Worker*	*X*	*Y*
1	35	75	11	20	30
2	50	100	12	50	60
3	55	95	13	60	105
4	50	80	14	30	60
5	60	80	15	50	90
6	40	45	16	35	75
7	60	80	17	35	80
8	40	80	18	55	60
9	45	75	19	60	80
10	55	90	20	35	60

17–14 Use data in Problem 17–1. H_0: $\rho = 0.95$.

17–15 Use data in Problem 17–4. H_0: $\rho = 0.95$.

17–16 A researcher has gathered data for annual per capita consumption of chewing tobacco, X, and the number of thefts reported in a sample of urban areas, Y, for the period 1948 to 1955 as shown below. H_0: $\rho = 0$.

Year	*X* (*pounds*)	*Y* (*thousands*)
1948	0.52	73
1949	0.49	71
1950	0.49	77
1951	0.45	89
1952	0.43	97
1953	0.42	102
1954	0.41	97
1955	0.40	104

Instructions for Exercises 17–17 and 17–18:

(a) Show that total error = unexplained error + explained error.
(b) Compute r^2. What does r^2 measure in each case?

17–17 A sample of five adult men for whom heights and weights are observed gives the following results:

Subject	*Height*: X	*Weight*: Y (lb)
0	5′4″	130
1	5′5″	145
2	5′6″	155
3	5′7″	165
4	5′8″	170

17–18 The following sample data give the amount of fertilizer, X, and the yield of potatoes, Y.

X (lb)	1.5	2.0	2.5	3.0	3.5	4.0	4.5	5.0	5.5
Y (lb)	10	9	12	14	13	15	17	14	14

17–19 The following are two sets of ranks assigned by two professors of management to ten characteristics for business leadership:

Characteristic	*Prof. I*	*Prof. II*
A	10	6
B	1	9
C	2	3
D	9	10
E	8	2
F	6	7
G	7	5
H	5	1
I	4	8
J	3	4

(a) Compute r_s and r_s'.
(b) In view of these results, can we conclude that there is agreement between the two professors?

17–20 The data below are costs and sales, in thousands of dollars, for twelve drug stores:

Store	*Costs*	*Sales*
1	11	19
2	10	15
3	14	20
4	13	14
5	12	16
6	20	33
7	21	32
8	15	18
9	22	29
10	18	22
11	19	23
12	16	20

(a) Compute the rank-correlation coefficient for these data.
(b) Is the result significant at $\alpha = 0.05$? (State your null and alternative hypotheses.)
(c) Assuming a normal bivariate population, compute the correlation coefficient r.
(d) Is r significant at $\alpha = 0.05$?

18

MEASURES OF SECULAR TRENDS

18–1 INTRODUCTION TO TIME-SERIES ANALYSIS

Reasons for Time-Series Analysis

A time series may be defined as a collection of magnitudes, belonging to different time periods, of some variable or composite of variables, such as production of steel, per capita income, gross national product, price of tobacco, or index of industrial production. A time series, then, portrays the variations of a variable through time. Like all kinds of social behavior, the movements of time series are generated by the systematic and stochastic forces of the society.

Time series are sometimes investigated merely because of an interest in history. Sometimes the interest is in the analysis of relationships between variables through time. Often, however, the ultimate interest is the future: statistical forecasting. In this connection, for example, if we find that a certain underlying and persistent trend of a series has continued for decades, we would be unwise to ignore the possibility that it will continue. The belief that past behavior of a series may continue into the future forms a rational basis for statistical forecasting.

At the very outset we should note that a time series is also a sample taken from a population, but it is not selected at random. For such samples no tried-and-proven body of procedures for inference is available. Hence, the help statistics can offer in dealing with time series is mainly descriptive.

Types of Economic Change

The actual movements of a time series are caused by a variety of factors—some economic, some natural, and some institutional. Some tend to affect

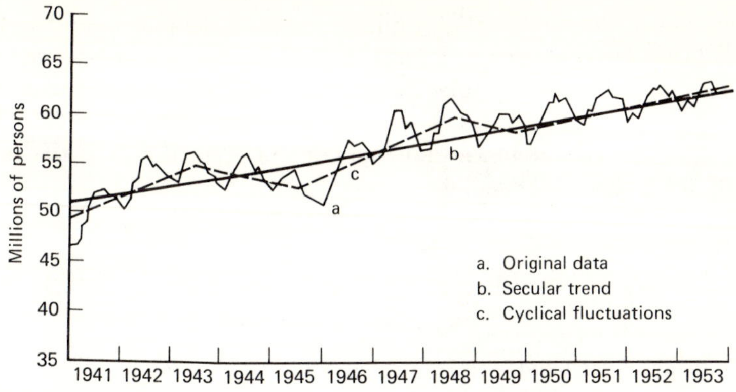

Figure 18–1 Employment in the United States, Bureau of Census Estimates, 1941–1953, by Month (SOURCE: *Federal Reserve Charts on Bank Credits, Money Rates, and Business,* November 1953)

only the long-run movement of a series, others to produce its short-run fluctuations. Thus, a time series includes different types of change. Looking now at Figure 18–1, we observe first that even though the movements of the original series seem quite irregular, on the average they tend to move continously upward through time. This general movement, persisting over a long period, is called a *secular trend.* It is represented by the diagonal line drawn through the irregular curve.

Next, if we study the irregular curve year by year, we see that the curve starts with a low, reaches a high about the middle of the year, and then decreases again. This type of fluctuation, which completes a whole sequence within the span of a year and has about the same pattern year after year, is called *seasonal variation.*

Furthermore, looking at the broken curve superimposed on the original irregular curve, we find pronounced fluctuations moving up and down every few years throughout the length of the chart. These are known as *business cycles,* or *cyclical fluctuations*—so called because they comprise a series of repeated sequences, just as a wheel goes round and round.

Finally, the little sawtooth irregularities on the original curve represent what are referred to as *irregular movements.*

Time-Series Models and Methods of Analysis

Analysis of a time series, either to study one component in its own right or to eliminate one or more components from the original series, calls for decomposition of the series. To decompose a series, we must assume that some type of relationship exists among its four components. Usually we proceed on the assumption that a time series is made up of several additive or multiplicative components. Trends, cycles, and seaonals are in some sense considered as rather stable functions of time; the irregular movements are not.

The *additive model* assumes that the value of original data is the sum of the four components. Thus, let

Y = value of the original time series,
T = value of trend,
C = value of cycle,
I = value of irregular, and
S = value of seasonal;

then the additive model can be expressed as

$$Y = T + S + C + I. \tag{18-1}$$

The *multiplicative model* assumes that the value of original data is the product of the values of the four components. That is,

$$Y = TSCI. \tag{18-2}$$

Note that the additive model assumes the four components to be independent of each other. This means that the individual components are the outcomes of four independent sources of causes. Thus, for example, no matter how high or how low the trend value is, it will have no effect on the seasonal variation or the cyclical fluctuation. The trend of the level of employment since 1900 has been increasing continuously; the additive assumption implies that the seasonal variation in employment from year to year is in no way affected by the continuous growth in employment through time.

The multiplicative model is the one we shall use in our later discussion. This model assumes that the four components are due to different causes but also that they are related to each other. This assumption allows convenient isolation of the components by a series of divisions. It can also be argued that the outcomes of the four components are not due to different causes and cannot be isolated. Particularly, one may argue that both trend and cycles are often subjected to many common factors, such as national income, population, and changes in consumers' preferences.

There are available two common approaches to analysis of time series. One was offered by Warren M. Persons; the other by the National Bureau of Economic Research. The Persons method is in common use, and we shall employ it in this text. Persons, in 1919, generalized and described a method of distinguishing trends from cycles that had long been extensively used. The idea is a very simple one.

First of all, a simple type of linear function is often selected to represent the average long-run underlying movement of a time series. The main technique used to fit the trend line is that of least squares. Next, an effort is made to discover if the data contain seasonal variations. If so, we eliminate them by dividing the original data by a seasonal index constructed for the purpose. Of course, if annual data are used, seasonals do not show up, and this step is unnecessary. Finally, the deseasonalized data are expressed as deviations, usually in percentage form, from the trend line. These relative deviations represent the cyclical fluctuations in the series. From these, the turning points and the amplitude of cycles in the series may then be examined.

The National Bureau of Economic Research has engaged in the study of time series for more than half a century under the leadership of Wesley C. Mitchell and Arthur F. Burns. It has developed a method of its own for analyzing time series in order to study the behavior of cyclical fluctuations only. This method, for various reasons, has been used by few investigators outside the Bureau. Nevertheless, the Bureau has made invaluable contributions to our understanding of the behavior of business cycles. Its findings are frequently quoted in writings on economic fluctuations and forecasting. They are also often used by firms in their planning and are a basis of many government policies. Thus, the student should gain at least a nodding acquaintance with the Bureau's method by reading Burns and Mitchell: *Measuring Business Cycles.*

18–2 TYPES OF SECULAR TREND AND METHODS OF TREND FITTING

In this chapter we are concerned with the measurement of secular trends—the long growth or decline of time series. Different time-series data can assume a variety of patterns of long-run behavior as represented in Figure 18–2. Among these trend equations, the arithmetic straight-line, the second-degree polynomial, and the exponential, or geometric straight-line, trends are by far the most important because they can adequately characterize the growth or decline of many time series. We shall concentrate here on these three trends.

Many methods are available for fitting trends to time-series data. In general, when we fit a line to a set of data, we want the fit to be close, by which we mean that the deviations of the observations from the line will be small. There are several ways to accomplish this. One is to minimize the sum of the absolute deviations. The other is to minimize the sum of squares of the deviations, which leads to the method of least squares—the device

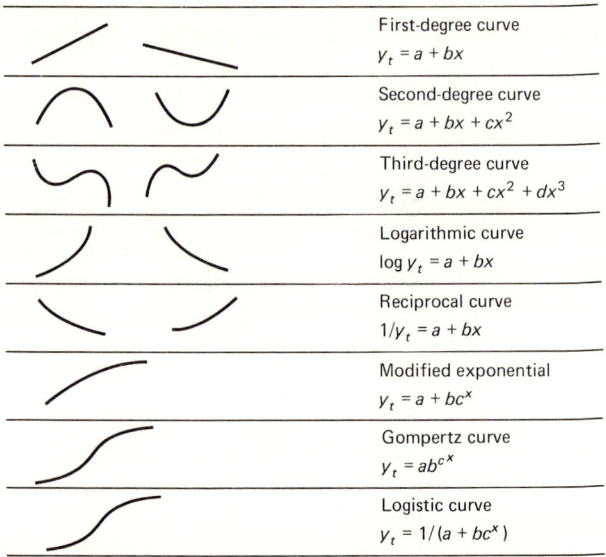

Figure 18–2 Illustration of Some Typical Forms of Trend Equations

most widely employed. The reader is already familiar with this method in connection with our discussion on regression analysis. Note that this method is a device of estimating parameters and may be applied to linear as well as curvilinear equations.

18–3 ARITHMETIC STRAIGHT-LINE TREND

When theoretical knowledge or practical experience suggests that a given time series may tend to grow or decay by a constant amount per unit of time, then the trend of this series can be appropriately described by a first-degree polynomial, or arithmetic straight-line equation, as follows:

$$y_t = a + bx, \tag{18–3}$$

where

y_t = the computed trend value,
a = the trend value (in original unit of Y) at the origin,
b = the amount of change in trend value per time unit, and
x = the value of the independent variable time.

As you have already learned from our discussion of regression, to derive the values of constants a and b in a linear equation by the least squares method we must solve the following two normal equations simultaneously:

$$\sum y = na + b \sum x,$$

$$\sum xy = a \sum x + b \sum x^2.$$

The solution of a and b from these equations in the case of time-series analysis is greatly simplified if we use the middle of the series as the origin. Since the time units in a series are usually of uniform duration and are consecutive numbers, when the middle period is taken as the origin, not only the mean of the time variable, $\bar{x}$, but also the sum of time units, $\sum x$, will be zero. As a result, the two normal equations become

$$\sum y = na,$$

$$\sum xy = b \sum x^2.$$

From these we can then derive

$$a = \frac{\sum y}{n} \tag{18–4}$$

and

$$b = \frac{\sum xy}{\sum x^2}. \tag{18–5}$$

It will be recognized that equation (18–4) is that for the arithmetic mean of Y. Thus, a, the value of Y at the origin, is the arithmetic mean of the Y variable. The value of b, of course, is the average amount of change in the trend values per unit of time.

We have noted that in computing trend it is convenient to use the middle of the series as the origin. Now, we may note that if the series contains an odd number of periods, the origin is the middle of the middle period. Thus, if the series extends from 1956 to 1968, in years, the origin is taken at the end of June, 1962. If an even number of periods is involved, the origin is set between the two middle periods. Thus, if the series covers a period from 1956 to 1967, again in years, the origin then falls between 1961 and 1962—that is, midnight, December 31, 1961.

Now, let us apply this method by using data in Table 18–1, which contains an odd number of years. From the sums accumulated in this table, we have

$$a = \frac{\sum y}{n} = \frac{23{,}146}{15} = 1543.07$$

and

$$b = \frac{\sum xy}{\sum x^2} = \frac{-41{,}451}{280} = -148.04.$$

Table 18–1. Computation of Least Squares Arithmetic Straight-Line Trend for Shipments of Radiators and Convectors, 1954–1968

(Monthly average in thousands of square feet of radiation)

Year (1)	x (2)	x^2 (3)	y (4)	xy (5)	y_t (6)
1954	−7	49	2412	−16,884	2579
1955	−6	36	2572	−15,432	2431
1956	−5	25	2464	−12,320	2283
1957	−4	16	2074	−8,296	2135
1958	−3	9	1869	−5,607	1987
1959	−2	4	1965	−3,930	1839
1960	−1	1	1715	−1,715	1691
1961	0	0	1450	0	1543
1962	1	1	1309	1,309	1395
1963	2	4	1324	2,648	1247
1964	3	9	1050	3,150	1099
1965	4	16	967	3,868	951
1966	5	25	625	3,125	803
1967	6	36	817	4,902	655
1968	7	49	533	3,731	507
	0	280	23,146	−41,451	

SOURCE: *Business Statistics*, 1967, 16th Biennial Edition, Department of Commerce; *Survey of Current Business*, October 1969, Vol. 49, No. 10, Department of Commerce (1967–1968 figures).

With these results, the trend equation can now be written as

$$y_t = 1543.07 - 148.04x.$$

(Origin, 1961; time unit, 1 year; Y, monthly average shipments of radiators and convectors, 1954–1968, in thousands of square feet of radiation.)

Trend values can be obtained by substituting the X values for each year in the trend equation. For instance, in our example, the trend value for 1954 is

$$y_{1954} = 1543.07 - 148.04(-7) = 2579.$$

The trend fitted to the original data is shown by Figure 18–3.

As an example of fitting a straight-line trend to a series with an even number of years, let us consider the data in Table 18–2, covering 16 years. If we decide that the origin falls at the middle of the series, it falls between

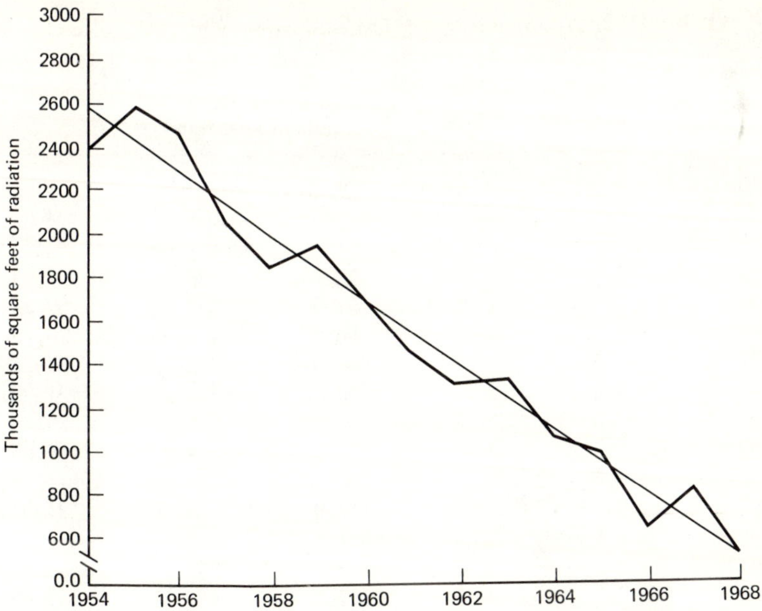

Figure 18–3 Least Squares Arithmetic Straight-Line Trend for Shipments of Radiators and Convectors, 1954–1968 (SOURCE: Table 18–1)

1960 and 1961. Thus, the middle of 1960 is six months to the left of the origin and the middle of 1961 is six months to the right of the origin. Furthermore, if we wish to keep the time unit at 1 year, then clearly the X value for 1960 is -0.5—that is, a half unit before the origin. The middle of 1959 is one unit and a half before the origin and thus has a value of -1.5, and so on. Similarly, consecutive positive values of 0.5, 1.5, 2.5, and so on, should be assigned to every year more recent than the origin. An alternative method is to consider the time unit as six months. Then each year has two time units, and for each year prior to the origin, consecutive negative values of -1, -3, -5, and so on should be assigned. Likewise, for more recent years, the X values are 1, 3, 5, and so forth.

Table 18–2 and Figure 18–4 illustrate the computation of a least squares arithmetic straight-line trend for data that have an even number of years. For the data in this table we have the values of the Y intercept and the slope of the trend line as

$$a = \frac{\sum y}{n} = \frac{1890.3}{16} = 118.144,$$

and
$$b = \frac{\sum xy}{\sum x^2} = \frac{2559.25}{340} = 7.527.$$

Table 18–2. Computation of Least Squares Arithmetic Straight-Line Trend for Synthetic Rubber Production in the United States, 1953–1968

(Monthly averages in thousands of long tons)

Year (1)	x (2)	x^2 (3)	y (4)	xy (5)	y_t (6)
1953	−7.5	56.25	70.7	−530.25	61.69
1954	−6.5	42.25	51.9	−337.35	69.22
1955	−5.5	30.25	80.9	−444.95	76.75
1956	−4.5	20.25	90.0	−405.00	84.27
1957	−3.5	12.25	93.2	−326.20	91.80
1958	−2.5	6.25	87.9	−219.75	99.33
1959	−1.5	2.25	114.9	−172.35	106.85
1960	−0.5	0.25	119.5	−59.75	114.38
1961	0.5	0.25	117.0	58.50	121.91
1962	1.5	2.25	131.1	196.65	129.43
1963	2.5	6.25	134.1	335.25	136.96
1964	3.5	12.25	147.0	514.50	144.49
1965	4.5	20.25	151.1	679.95	152.02
1966	5.5	30.25	164.1	902.55	159.54
1967	6.5	42.25	159.3	1035.45	167.07
1968	7.5	56.25	177.6	1332.00	174.60
Total		340.00	1890.3	2559.25	

SOURCE: *Business Statistics*, 1967, 16th Biennial Edition, Department of Commerce; *Survey of Current Business*, October 1969, Vol. 49, No. 10, Department of Commerce (1967–1968 figures).

Thus, the trend equation becomes

$$y_t = 118.144 + 7.527x.$$

(Origin, 1960–1961; time unit, 1 year; Y, monthly average production of synthetic rubber, 1953–1968, in thousands of long tons.)

As previously pointed out, our interest in trend analysis is not confined to determining the long-run pattern of a series in the past. We are also concerned with forecasting trend values. For this we use the same technique that we use to compute the trend values for periods covered by the series. For instance, if we are interested in forecasting the trend value of production of synthetic rubber for 1970, we simply observe that 1970 is 9.5 years ahead of the origin for the trend equation we have just established, and thus

$$Y_{1970} = 118.144 + 7.527(9.5) = 189.65.$$

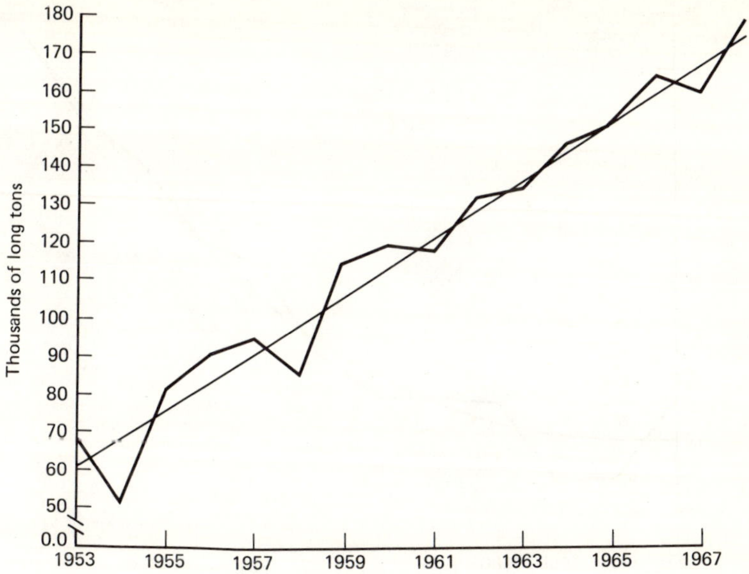

Figure 18–4 Least Squares Arithmetic Straight-Line Trend for Synthetic Rubber Production in the United States, 1953–1968 (SOURCE: Table 18–2)

We may then state that, under the assumption that the same trend factors that produce the trend equation for 1953 and 1968 will remain operative, the average monthly production of synthetic rubber will have a trend value of 189.66 thousands of long tons in 1970.

18–4 SECOND-DEGREE POLYNOMIAL TRENDS

Polynomial trends are of the general form:

$$y_t = a + bx + cx^2 + dx^3 + \cdots + rx^k.$$

The arithmetic straight-line trend is a special case, having only the first two terms on the right side of this expression. With three terms on the right side we would have a parabola. With four terms on the right, we would have a cubic equation, and so forth. One can always force a polynomial to fit the data quite closely by adding enough terms.

All polynomials can be fitted by the method of least squares. In particular, if we wish to fit a second-degree parabolic trend, we will have the trend equation

$$y_t = a + bx + cx^2, \tag{18–6}$$

where a is still the Y intercept, b is the slope of the curve at the origin, and

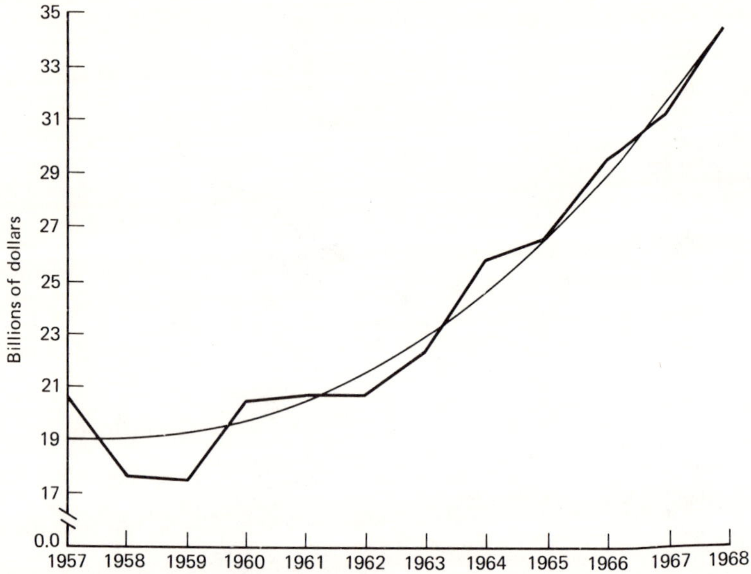

Figure 18–5 Second-Degree Parabolic Trend for Volume of United States Merchandise Exports, 1957–1968 (SOURCE: Table 18–3)

c is the rate of change in the slope. It should be noted that, just as b is a constant in the first-degree curve, c is a constant in the second-degree curve. Here the c value determines whether the curve is concave or convex and the extent to which it departs from linearity. The second-degree curve, as shown in Figure 18–5, has curvilinearity without change of inflection, and under these conditions c determines the existence or nonexistence of curvilinearity.

Since there are three constants in the second-degree curve, if it is to be fitted by the method of least squares to a series, three normal equations, as given below, are required:

$$\sum y = na + b\sum x + c\sum x^2,$$
$$\sum xy = a\sum x + b\sum x^2 + c\sum x^3;$$
$$\sum x^2y = a\sum x^2 + b\sum x^3 + c\sum x^4.$$

When the middle of the time series is taken as the origin, then all the sums of the odd powers of x become zero, and the three normal equations are reduced to

$$\sum y = na + c\sum x^2,$$
$$\sum xy = b\sum x^2,$$
$$\sum x^2y = a\sum x^2 + c\sum x^4.$$

A little algebraic manipulation with these equations will show that

$$c = \frac{n \sum x^2y - \sum x^2 \sum y}{n \sum x^4 - (\sum x^2)^2}, \tag{18-7}$$

$$a = \frac{\sum y - c \sum x^2}{n}, \tag{18-8}$$

and

$$b = \frac{\sum xy}{\sum x^2}. \tag{18-9}$$

To obtain the coefficients for a second-degree parabolic trend, therefore, we need five sums, which are $\sum x^2$, $\sum x^4$, $\sum y$, $\sum xy$, $\sum x^2y$. Table 18-3 illustrates the procedure for obtaining these totals. For this series, we have

$$c = \frac{(12)(3629.675) - (143)(287.5)}{(12)(3038.75) - (143)^2} = 0.1526,$$

$$a = \frac{287.5 - (0.1526)(143)}{12} = 22.14,$$

Table 18-3. Computation of Second-Degree Parabolic Trend for Annual Total Volume of United States Merchandise Exports, 1957–1968

(annual totals in billions of dollars)

Year (1)	x (2)	x^2 (3)	x^4 (4)	y (5)	xy (6)	x^2y (7)	y_t (8)
1957	−5.5	30.25	915.0625	20.7	−113.85	626.175	19.14
1958	−4.5	20.25	410.0625	17.7	−79.65	358.425	19.00
1959	−3.5	12.25	150.0625	17.6	−61.60	215.600	19.16
1960	−2.5	6.25	39.0625	20.5	−51.25	128.125	19.63
1961	−1.5	2.25	5.0625	20.9	−31.35	47.025	20.41
1962	−0.5	0.25	0.0625	20.9	−10.45	5.225	21.49
1963	0.5	0.25	0.0625	22.4	11.20	5.600	22.87
1964	1.5	2.25	5.0625	25.7	38.55	57.825	24.56
1965	2.5	6.25	39.0625	26.6	66.50	166.250	26.56
1966	3.5	12.25	150.0625	29.4	102.90	360.150	28.86
1967	4.5	20.25	410.0625	31.0	139.50	627.750	31.46
1968	5.5	30.25	915.0625	34.1	187.55	1031.525	34.37
Total		143.00	3038.7500	287.5	198.05	3629.675	

SOURCE: Adopted from various issues of *Federal Reserve Bulletin*; 1962 figures significantly affected by strikes.

and $$b = \frac{198.05}{143} = 1.385.$$

With these values, the trend equation becomes:

$$y_t = 22.14 + 1.385x + 0.1526x^2.$$

(Origin; 1962–1963; X = 1 year; Y = annual total U.S. merchandise exports, in billions of dollars.)

The trend values are obtained by substituting the x's and x^2's into the trend equation. To draw curvilinear trends, it is advisable to plot all the trend values on the chart. The trend drawn through the original data for our example is shown in Figure 18–5.

It may be pointed out that the value of b in a second-degree curve is identical with that in the straight-line trend for the same series if the second-degree curve is fitted from the middle of the series. In our present situation, furthermore, the positive value for b is responsible for the general upward drift, and the positive value of c makes the curve convex to the X axis. Conversely, negative values of b and c will give the curve a downward drift concave to the base. A second-degree curve changes its direction only once.

When the value of x^2 becomes larger at the extremes of the series, the influence of c upon the trend increases, and the departure from linearity becomes quite marked. This limits somewhat the usefulness of this type of trend for the forecasting of trend values.

18–5 THE EXPONENTIAL, OR GEOMETRIC STRAIGHT-LINE, TREND

Fitting trends after transformation of data is a common practice. An arithmetic straight line that will not fit the original data may fit, for example, the logarithms of the data as typified by the exponential trend,

$$y_t = ab^x. \tag{18–10}$$

This expression is sometimes called a growth equation, since it is often used to explain the phenomenon of growth through time. For example, the reader may be familiar with the compound-interest formula

$$P_n = P_o(1 + i)^n,$$

where P_0 is the initial capital, i is the rate of interest, and P_n is the capital value after n years. Now, if we set $P_n = y_t$, $P_0 = a$, $1 + i = b$, and $n = x$, then the equation is identical with the exponential trend equation (18–10).

In any event, the exponential trend is an appropriate explanation for the underlying long-run drift of a time series if its growth is approximately at a constant rate per unit of time, the rate of growth being $(b - 1)$, such as national product, population, or production of an industry.

Now, returning to the problem of fitting an exponential curve to data, we note that if we take logarithms on both sides of (18–10), it becomes a linear function:

$$\log y_t = \log a + x(\log b), \tag{18–11}$$

which has been named as a *geometric straight-line* trend. To obtain the constants a and b in the expression by the method of least squares, we need to solve the following set of equations:

$$\sum \log y = n(\log a) + \log b(x),$$

$$\sum x(\log y) = \log a(\sum x) + \log b(\sum x^2).$$

Again, by setting the origin at the middle of the series, we reduce these equations to

$$\sum \log x = n(\log a),$$

$$\sum x(\log y) = \log b(\sum x^2).$$

From these we have Y intercept and slope as

$$\log a = \frac{\sum \log y}{n} \tag{18–12}$$

and

$$\log b = \frac{\sum x(\log y)}{\sum x^2}. \tag{18–13}$$

As an example of fitting a geometric straight-line trend, we use the data of annual GNP of the United States from 1954 to 1968. The original data and the various computations for the trend are presented in Table 18–4. From these data we obtain

$$\log a = \frac{\sum \log y}{n} = \frac{40.78046}{15} = 2.71870,$$

$$a = 523.2;$$

$$\log b = \frac{\sum x(\log y)}{\sum x^2} = \frac{4.66528}{280} = 0.01666,$$

$$b = 1.04.$$

(It may be noted that a is the logarithmic value of trend at the origin and that b is the logarithm of the rate of change per unit of time stated as a ratio.)

Table 18–4. Computation of Geometric Straight-Line Trend for Gross National Product of the United States in Constant Dollars, 1954–1968

(annual totals in billions of 1958 dollars)

Year (1)	x (2)	x^2 (3)	y (4)	$\log y$ (5)	$x \log y$ (6)	$\log y_t$ (7)	y_t (8)
1954	−7	49	407.0	2.60959	−18.26713	2.60207	400.0
1955	−6	36	438.0	2.64147	−15.84882	2.61873	415.6
1956	−5	25	446.1	2.64943	−13.24715	2.63539	431.9
1957	−4	16	452.5	2.65562	−10.62248	2.65205	448.8
1958	−3	9	447.3	2.65060	−7.95180	2.66871	466.3
1959	−2	4	475.9	2.67752	−5.35504	2.68537	484.6
1960	−1	1	487.7	2.68815	−2.68815	2.70203	503.5
1961	0	0	497.2	2.69653	0.0	2.71869	523.2
1962	1	1	529.8	2.72411	2.72411	2.73535	543.7
1963	2	4	551.0	2.74115	5.48230	2.75201	564.9
1964	3	9	581.1	2.76425	8.29275	2.76867	587.0
1965	4	16	616.7	2.79007	11.16028	2.78533	610.0
1966	5	25	652.6	2.81465	14.07325	2.80199	633.8
1967	6	36	673.1	2.82808	16.96848	2.81865	658.6
1968	7	49	706.7	2.84924	19.94468	2.83531	684.4
	0	280	7962.7	40.78046	4.66528		

SOURCE: *Business Statistics*, 1967, 16th Biennial Edition, Department of Commerce; *Survey of Current Business*, October 1969, Vol. 49, No. 10, Department of Commerce (1967–1968 figures).

The trend equation in logarithmic terms, for our example, is

$$\log y_t = 2.71870 + 0.01666x.$$

(Origin, 1961; $X = 1$ year; $Y =$ GNP of United States, in billions of 1958 dollars.)

We may also write this result in the exponential form as

$$y_t = (523.2)1.04^x.$$

With the exponential trend, we usually find the trend values in logarithmic terms first by substituting the X values in the geometric straight-line trend equation. We then obtain natural values by finding the antilogs of the logarithmic trend values.

Note that Figure 18–6 is constructed by plotting the natural values on semilogarithmic paper, with the logarithmic scale extending along the Y axis.

In fitting the trend with this procedure, we satisfy the condition that the sum of the squares of the logarthmic deviations from the trend is a minimum.

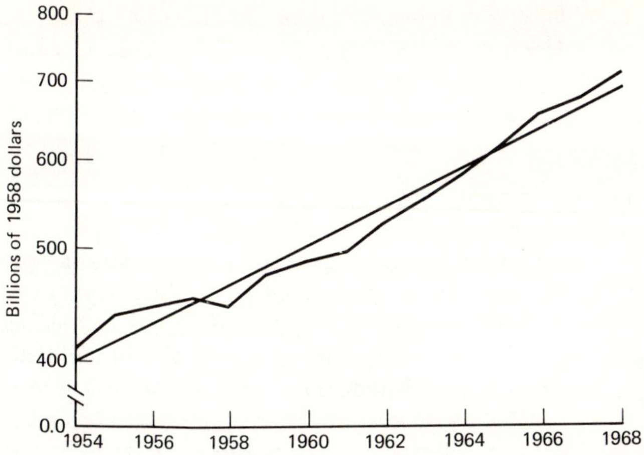

Figure 18-6 Geometric Straight-Line Trend for Gross National Product of the United States in Constant Dollars, 1954–1968 (SOURCE: Table 18-4)

That is, the deviations to which this condition relates are the differences between logarithms of the original values and the logarithms of the trend values. The geometric straight line, therefore, should not be thought of as the same as the least squares arithmetic straight line, for which the sum of the squares of natural-number deviations is a minimum.

When we use this method, it should also be remembered that both the Y intercept and the slope are logarithmic values. It can be seen, from equation (18-12), that log a is the logarithm of the geometric mean of the series. The antilog of this value is the trend value in natural numbers at the origin. In our illustrative problem, the antilog of 2.71870 is 523.2, which is, therefore, the trend value in 1961.

Log b is the logarithm of the ratio of each time unit to the preceding time unit. In our example the value of b is 0.01666, which is the logarithm of 1.04. This value of 1.04 is the ratio of the trend value of each year to the preceding year. To facilitate the interpretation of this value, we usually express it as a percentage and then deduct 100 from it. The result is the constant rate of change in Y per unit of X. For the series under consideration we have $104 - 100 = 4$ percent. Thus we state: the annual real GNP in the United States, on the average, increases at a constant rate of 4 percent per year in real terms.

The property of b for a geometric straight line is of considerable significance. When it has been transformed into a pure percentage it defines the average annual rate of growth or decline of the series. Being an abstract measure, it permits comparison of trends, all described by straight lines on ratio (semi-log) paper, of series with different original units. Series, such as production of all kinds, population, or national income, become immediately comparable, and conclusions about the direction and magnitude of economic activities can readily be drawn. This measure, consequently, provides an effective device for the study of the socioeconomic changes in a nation.

18–6 METHOD OF MOVING AVERAGES

A moving average may be considered as an artificially constructed time series in which each periodic figure is replaced by the mean of the value of that period and those of a number of the preceding and succeeding periods. Although the computation of the moving averages is simple and straightforward, this measure is not without theoretical and technical interest. The properties and utility of moving averages can perhaps be brought out more vividly by a hypothetical illustration. Hypothetical data in Table 18–5 are assumed to have a uniform cyclical duration of 5 years and equal amplitude

Table 18–5. Computation of Three-Year and Five-Year Moving Averages for Hypothetical Data of a Five-Year Cycle with Uniform Duration and Amplitude

Year (1)	*Original value* (2)	*Three-year mt* (3)	*Three-year ma* (4)	*Five-year mt* (5)	*Five-year ma* (6)
1	1	—	—	—	—
2	2	6	2.0	—	—
3	3	7	2.3	9	1.8
4	2	6	2.0	10	2.0
5	1	5	1.7	11	2.2
6	2	6	2.0	12	2.4
7	3	9	3.0	13	2.6
8	4	10	3.3	14	2.8
9	3	9	3.0	15	3.0
10	2	8	2.7	16	3.2
11	3	9	3.0	17	3.4
12	4	12	4.0	18	3.6
13	5	13	4.3	19	3.8
14	4	12	4.0	20	4.0
15	3	11	3.7	21	4.2
16	4	12	4.0	22	4.4
17	5	15	5.0	23	4.6
18	6	16	5.3	24	4.8
19	5	15	5.0	—	—
20	4	—	—	—	—

of 2 units. Three-year and five-year moving averages are fitted to the data. The procedure for calculating three-year moving averages is as follows:

(1) Compute the three-year moving totals. This is done by adding up the values of the first three years and centering at the second year. This is the first three-year moving total. Then the value of the first year is dropped and the value of the fourth year is added to form the second three-year moving total, which is centered at the third year. And so the computation moves through the end of the series. The three-year moving totals are entered in column 3 of Table 18–5.

(2) The three-year moving averages are obtained simply by dividing each of the three-year moving totals by 3. These values for our example are entered in column 4 of Table 18–5.

A similar procedure is used to compute the five-year moving averages. In a five-year moving average, the value of each year is replaced by the mean of the values of the five successive years of which two precede and two succeed the given year. Both five-year moving totals and moving averages are, of course, centered in the middle of the respective five-year period, with the first five-year moving total and the moving average entered in the third year. Columns 5 and 6 in Table 18–5 illustrate the computation of five-year moving averages. It may also be pointed out in passing that in computing moving averages for an even number of periods, the procedure becomes more complicated. Such a procedure will be discussed in the next chapter.

The results of our calculations are plotted with the original data in Figure 18–7. Both sets of moving averages may be considered as the statistical expression of secular movement of our hypothetical series.

By studying Table 18–5 and Figure 18–7, we can reach several conclusions about the characteristics of moving averages:

(1) If the data show a periodical fluctuation, a moving average of length equal to the period will completely eliminate the periodic variations.

(2) If the series changes on the average by a constant amount per time unit and its fluctuations are periodic, a moving average of length equal to the period will be linear.

(3) Even if the data show a periodic fluctuation, a moving average of some length other than the length of the period, no matter how small the difference is between the duration of periodicity of the original series and the duration of the moving average, the moving average cannot obliterate completely the periodic variations in the original series. The averaging process then only tends to smooth out somewhat the short-run highs and lows.

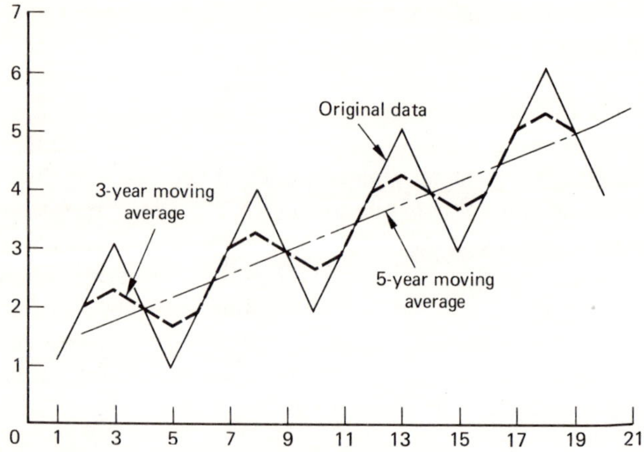

Figure 18–7 Three-Year and Five-Year Moving Average Trends Fitted to Hypothetical Data of a Five-Year Cycle with Uniform Duration and Amplitude

(4) From the preceding paragraph it may be deduced that if the data fluctuate without periodicity, a moving average will have only the effect of eliminating the lesser variations to some extent, thereby smoothing the general appearance of the time series.

With these properties in mind, we can see that the moving average may constitute a satisfactory trend for a series that is basically linear and that is regular in duration and amplitude. But these conditions are seldom met by any economic time series. The use of moving average as a trend measure entails two disadvantages. First, in computing moving averages we lose some years at each end of the series. For instance, we lose two years at each end with the five-year moving average. Had we computed a seven-year moving average, we would have lost three years at each end.

A still more serious drawback is that the moving average is not represented by some mathematical function and, therefore, is not capable of objective projection into the future. Since one of the main objectives of trend analysis is that of forecasting, the moving average, lacking this desirable property, is no longer in wide use as a trend measure.

Nevertheless, the method of moving averages is a very useful technique in analyzing time series. First, in problems in which the trend of the time series is clearly not linear and in which we are concerned only with the general movements of the time series, be it trend, or a cycle, or possibly both, it is customary to study the smoothing behavior of the series by applying a moving average. Second, as will be shown in the next chapter, the characteristic of a moving average is the basis of seasonal analysis. In that context, a 12-month moving average is calculated to eliminate seasonal variations that have a periodicity of one year.

18–7 SHIFTING THE TREND ORIGIN

In computing trends, the middle of the time series is always used as the origin in order to short-cut the computation. But very often we need to change the origin of the trend equation to some other point in the series. This is either to facilitate comparisons of trend values among neighboring years or to convert a trend equation from an annual to a monthly basis. Shifting the origin is a very simple matter. Consider the trend equation for data in Table 18–1.

$$y_t = 1534.07 - 148.04x.$$

(Origin: 1961)

The origin of 1961 is taken to mean that the precise origin is the center of that year: July 1. Now, if we wish to shift the trend equation to 1954, we note that this year precedes the stated origin of 1961 by 7 time units. Consequently, we must deduct 7 times the annual increment—that is, $b(-7)$—from the trend value of 1961:

$$y_t = 1543.07 - 148.04(-7) = 2579.35.$$

The value of 2579.35 becomes the trend value at the new origin, 1954, and the trend equation may now be written as

$$y_t = 2579.35 - 148.04x.$$

(Origin: 1954)

Again, as another example, consider the trend equation for data in Table 18–2:

$$y_t = 118.144 + 7.528x.$$

(Origin: 1960–1961)

The origin of 1960–1961 means that the precise origin is the center of this period: January 1, 1961. Suppose we wish to express the trend equation with an origin at 1961—that is, at July 1, 1961. Since X units is a whole year, the new origin desired is 0.5 units succeeding the old origin. To achieve this we must then add half of the annual increase, or $b(\frac{1}{2})$, to the trend value at the old origin. Then we have

$$y_t = 118.144 + 7.528(\tfrac{1}{2}) = 121.878.$$

The trend equation reads:

$$y_t = 121.878 + 7.528x.$$

(Origin: 1961)

The shifting of the trend origin amounts to finding the new Y intercept—that is, the new value of a. The value of b remains unchanged, since the slope of the trend line is the same, irrespective of the origin.

From the preceding discussion, the procedure of shifting the origin may be generalized by the expression

$$y_t = a + b(x + k), \tag{18–14}$$

where k is the number of time units shifted. If the origin is shifted forward in time, k is positive; if the converse, k is negative. Thus, to change the origin of the trend equation for the radiators and convectors data (Table 18–1), $y_t = 1543 - 148x$, from 1961 to 1954, we have

$$\begin{aligned} y_t &= 1543.07 - 148.04(x - 7) \\ &= 1543.07 - 148.04x + 1036.28 \\ &= 2579.35 - 148.04x. \end{aligned}$$

(Origin: 1954)

Formula (18–14) can also be expanded to cover parabolic trend equations. As an example, we may consider the following second-degree trend equation for data in Table 18–3:

$$y_t = 22.14 + 1.385x + 0.1526x^2.$$

(Origin: 1962–1963)

If we wish to shift the origin for this equation to 1963, we may follow a procedure suggested by formula (18–14):

$$\begin{aligned} y_t &= 22.14 + 1.385(x + 0.5) + 0.1526(x + 0.5)^2 \\ &= 22.14 + 1.385x + 0.6925 + 0.1526(x^2 + 1x + 0.25) \\ &= 22.99 + 1.5376x + 0.1526x^2. \end{aligned}$$

(Origin: 1963)

18–8 REDUCING ANNUAL TREND VALUES TO MONTHLY VALUES

For simplicity and ease of computation, trends are usually fitted to annual rather than quarterly or monthly data. To eliminate trend from a time series so that we may study seasonal or cyclical patterns, we need monthly trend values. Fortunately, monthly trend equations can be derived from annual trend equations without any loss of accuracy. The conversion of annual trend equations to monthly values depends upon the units in which the original data are expressed. These are two possible situations. One is that Y units are annual totals, such as the GNP, annual total production of steel, or total population. The other is that Y units are monthly averages, such as the monthly average synthetic rubber production presented in Table 18–2. The monthly averages of synthetic rubber production, like other monthly average data, are annual production for each year divided by 12. We shall discuss these two situations in turn.

The case where X equals one year and the annual value is a total of monthly values is illustrated by our previous example of United States Merchandise Exports. Since in this situation the a value is expressed as the sum of each 12 monthly values of the year, we must divide it by 12 in order to reduce the a value to terms of monthly Y values. The a value so computed is clearly the arithmetic mean of the monthly values. The origin of the monthly series is one-twelfth of the a value computed from annual totals.

The value of b in an annual trend equation for annual totals measures the annual total change in Y. When b is divided by 12, it has been reduced to annual change in monthly magnitudes. When it is divided by 12 again, it is further reduced to monthly change in monthly magnitudes. The necessity of dividing b twice by 12—that is, by 144 altogether—must be clearly understood. The division of annual change by 12 gives us only the change from some month in a given year to the corresponding month in the following year, or the annual change in monthly magnitudes. However, we are seeking an expression of the change in each and every month—that is, monthly change in monthly magnitudes. Thus, we must divide b by 12 again.

In conclusion, to reduce an annual trend equation to a monthly trend equation when the original data are given as annual totals, a is divided by 12 and b is divided by 144. The annual trend equation for U.S. GNP data may be cited as an illustration. Recall that it is

$$\log y_t = 2.71870 + 0.01666x.$$

(Origin: 1961)

This can be converted to a monthly basis as

$$\log y_t = \frac{2.71870}{12} + \frac{0.01666}{144}x$$
$$= 0.226558 + 0.0001157x.$$

(Origin: July 1, 1961; $X = 1$ month)

Note that the origin for this monthly trend equation is the beginning of

July. In trend analysis we usually have the origin at the center of a period. Thus, we may rewrite the equation in the following form:

$$\log y_t = 0.2265558 + 0.0001157(x + \tfrac{1}{2})$$
$$= 0.226616 + 0.001157x.$$

(Origin: July 1961)

Where data are given as monthly averages per year, the *a* value in the annual trend equation is already the arithmetic mean of the 12-month total. In other words, it is already at the monthly level. The *b* value now represents the annual change in monthly magnitudes. As a result, to convert an annual trend equation when annual data are expressed as monthly averages, *a* remains unchanged and *b* is divided by 12 only once. Consider the annual trend equation for data in Table 18–1:

$$y_t = 1543.07 - 148.04x.$$

(Origin: 1961; Y unit = annual monthly averages)

To convert this equation into monthly terms, we have

$$y_t = 1543.07 - \frac{148.04}{12}(x + 0.5)$$

$$= 1536.9017 - 12.3367x.$$

(Origin: July 1961; X = 1 month)

We may, on occasion, need to convert an annual trend equation into quarterly values. The procedure involved is similar to that which has just been discussed. This is left as an exercise for the student.

18–9 CHOICE OF PERIOD AND MATHEMATICAL EXPRESSIONS FOR TREND

In measuring trend, we must take note of two further analytical considerations. We must choose carefully the period to be studied, and we must choose a mathematical expression to describe the secular movement.

Generally speaking, the longer the period covered the more significant the trend. Secular movements cannot be expected to reveal themselves clearly within a short span of time. When the period is too short, the general drift of the series may be unduly influenced by the cyclical fluctuations. This would make it difficult to separate the various sources of variation in time series. As a minimum safeguard, it may be said that to compute trend the period must cover at least two or three complete cycles.

Furthermore, in order to produce accurate values of the Y intercept and the slope of the trend equation, the period must be so selected that the number of prosperous years is about the same as the number of depressed years, and the first and last years of the series must be on the opposite sides of the cycle; that is, if the series starts with an upswing it should end with a downswing. If the number of prosperous years is greater than the number of depressed years, the Y intercept and, therefore, the level of the trend will be too high. If the situation is reversed, then the level of the trend will be too low. If the series starts and ends with a cyclical upswing, the value of the slope will be too large. If both the first and last years of the series coincide with a cyclical downswing, then the value of the slope will be too small. Correct value of the slope, therefore, can be derived only for a series that starts with an upswing and ends with a downswing, or vice versa. Usually bias in the slope is more damaging than bias in the level, since a trend becomes progressively worse as X increases if b is inaccurate.

The selection of the mathematical expression for a trend is always a difficult matter, because there are really no positive or objective tests that can be applied. In any case, in selecting the mathematical expression to be employed in trend fitting for a given series, the very first thing to do is to establish a hypothesis or theory about the long-run average behavior of the series. Next, we plot the data and subject them to economic analysis in accordance with the hypothesis we have formulated. If observation of the graph seems to support our hypothesis about the data, then a mathematical expression and the method of fitting it may be finally determined in conjunction with the following criteria:

(1) The mathematical expression, subject to modification in view of the other criteria, must provide the best fit.
(2) The mathematical expression must be able to provide reasonable extrapolated values.
(3) The mathematical expression must be able to provide a trend that is free from cyclical and irregular influences.

The selection of a mathematical expression, therefore, cannot be defended by mathematical tests of goodness-of-fit alone. For this reason the more complex curves are avoided in applied work. Such curves may on occasion provide excellent fits, but they are neither free from cyclical influences nor capable of yielding reasonable extrapolated values.

The truth is that trend values are affected by a myriad of economic and social forces, the legal framework, and natural phenomena, and the attempt to express their aggregate impact by a simple formula is often fruitless and frustrating.

EXERCISES

18–1 Variable Y in Table 18–P1 stands for the percentage ratio of agricultural employment to total employment in the United States, 1950–1968. Plot the data on arithmetic paper and explain whether an arithmetic straight-line trend is appropriate.

Table 18–P1

Year	Y	Year	Y
1950	12.55	1960	8.60
1951	11.60	1961	8.20
1952	11.30	1962	7.65
1953	11.60	1963	6.90
1954	10.65	1964	6.55
1955	10.65	1965	6.15
1956	10.15	1966	5.45
1957	9.55	1967	5.15
1958	9.15	1968	5.00
1959	8.90		

SOURCE: Adapted from *Statistical Abstract of the United States*, 1968.

18–2 Variable Y in Table 18–P2 stands for production of pneumatic casings in the United States, 1955–1968 (annual total production in millions). Plot the data on both arithmetic and semi-log papers. From your graphs, what type of general drift does this series seem to possess?

Table 18–P2

Year	Y	Year	Y
1955	112.2	1963	139.1
1956	100.4	1964	158.1
1957	106.9	1965	167.9
1958	96.6	1966	177.2
1959	118.0	1967	163.2
1960	119.8	1968	203.1
1961	116.8		
1962	113.9		

SOURCE: Adapted from *Survey of Current Business*, Department of Commerce.

18–3 Follow the instructions in Exercise 18–2 for the series of new plant and equipment expenditures in the United States, 1955–1968 (annual totals in billions of dollars) in Table 18–P3.

Table 18–P3

Year	Y	Year	Y
1957	37.0	1963	39.2
1958	30.5	1964	44.9
1959	32.5	1965	52.0
1960	35.7	1966	60.6
1961	34.4	1967	61.7
1962	37.3	1968	65.2

SOURCE: *Statistical Abstract of the United States*, 1968.

18–4 Follow the instructions in Exercise 18–2 for the series of crude petroleum production in the United States, 1952–1968 (annual monthly averages in millions of barrels) in Table 20–P4.

Table 18–P4

Year	Y	Year	Y
1952	190.8	1961	218.3
1953	196.4	1962	222.9
1954	192.9	1963	229.0
1955	207.0	1964	232.1
1956	218.1	1965	237.3
1957	218.1	1966	253.0
1958	204.1	1967	268.0
1959	214.1	1968	277.4
1960	214.1		

SOURCE: Adapted from *Survey of Current Business*, Department of Commerce.

18–5 (a) Show that the arithmetic straight-line trend by the method of least squares for data in Table 18–P1 is

$$y_t = 8.72 - 0.42x.$$

(Origin: 1959; $X = 1$ year)

(b) Compute the trend values.

(c) What is the projected trend value of 1970 for this series?

18–6 (a) Show that the least squares second-degree polynomial trend for data in Table 18–P2 is

$$y_t = 127.9 + 7.2x + 0.539x^2.$$

(Origin: 1961–1962; X unit = 1 year; Y unit = annual total production in millions)

(b) Compute the trend values.
(c) Project the trend value for 1970.

18–7 (a) Fit a second-degree polynomial trend to data in Table 18–P3 by the method of least squares.
(b) Compute the trend values.
(c) Project the trend value for 1971.

18–8 (a) Show that a geometric straight-line trend fitted to the data in Table 18–P4 is

$$\log y_t = 2.34622 + 0.00859x.$$

(Origin: 1960; X unit = 1 year; Y unit = annual monthly averages in millions of barrels)

(b) Convert the trend above into the exponential form.
(c) What is the constant rate of increase per year for this series?
(d) What is the projected trend value for 1970?

18–9 (a) Fit an exponential trend to the data in Table 18–P5.
(b) What is the average rate of growth per annum for this series?

Table 18–P5. GNP of the Republic of Utopia

(Average annual total in billions of dollars)

Decade	Y	Decade	Y
1910	2	1950	40
1920	5	1960	90
1930	10	1970	170
1940	18		

SOURCE: Hypothetical.

18–10 (a) Change the origins of all the trend equations obtained in Problems 18–5 to 18–8 inclusive to July 1968.
(b) Reduce the first and fourth resulting trend equations in (a) into monthly magnitudes and center the origin to the middle of July 1968.

19

MEASURES OF SHORT-RUN FLUCTUATIONS

19–1 INTRODUCTION

The measurement of secular trend is but one of the problems connected with time-series analysis. Such series are also subject to short-run fluctuations, seasonal and cyclical in nature, that are of great interest and importance to economic research and formulation of business policy. In this chapter we shall deal with the measurement of both types of variations.

Seasonal variations in time series are all-pervasive, and an understanding of their patterns aids managerial decision making in a variety of fields. Accurate measurements of seasonal changes also allow us to adjust statistically for them, thereby leaving only trends, cycles, and irregular movements in the series. Deseasonalized data are often easier to interpret, since, for example, one is less likely to confuse a seasonal downswing with deterioration of general business activity.

The measurement of cyclical fluctuations is regarded by many as the central task in the analysis of time series. Interest in cycles may involve either a given series—such as cyclical movements of steel production, of retail prices, of volumes of sales—or the complex patterns of cycles in general business activity. Note that cyclical changes in individual series are closely related to business cycles. Various studies have shown that the majority of economic time series that reflect such movements do conform, with varying degrees of lead and lag, to cycles in the economy at large.

Seasonal variations are computed from quarterly or monthly data. These variations, which recur every year, do not show up if data are lumped together by years or by longer time periods. Cyclical fluctuations, however, can be measured from annual, quarterly, or monthly data.

19–2 TYPES OF SEASONALS

Seasonal variations in time series are of two types: those whose pattern remains stable over a number of years, and those whose pattern gradually changes. The measurement of constant seasonals is a simple matter, but the analysis of changing seasonals involves rather complicated theoretical as well as practical considerations.

A *seasonal index* consists of twelve numbers, one for each month of a year, or for each of the months of a number of years, showing the relative amount of activity for a year, or a number of years, that has typically taken place in each month. Thus, a seasonal index may be specific or typical. A *specific seasonal index* refers to the seasonal changes during a particular year. A *typical seasonal index* is obtained by averaging a number of specific seasonals. It is thus a generalized expression of seasonal variations for a series.

By observing specific seasonals one can determine whether the seasonal pattern of a series is stable or is changing gradually or abruptly. The representativeness of a typical index demands that the specific seasonals must be stable. With changing specific seasonals, the construction of a typical seasonal index is highly questionable and of little practical importance.

19–3 CALCULATION OF THE TYPICAL SEASONAL INDEX BY THE METHOD OF RATIO TO MOVING AVERAGE

There are available several methods of isolating seasonal variations from a series, each having merits and defects. Among these, the *method of ratio to moving average* is far superior from a theoretical point of view as well as for practical reasons. We shall consider this method in detail before turning to alternative techniques.

The method of ratio to moving average starts with the multiplicative assumption of time series. That is, $Y = TSCI$. It first attempts to estimate TC by employing a twelve-month moving average. After this, TC is to be eliminated in order to estimate SI. This is done by dividing TC into the original data. That is,

$$SI = \frac{TSCI}{TC}.$$

Finally, irregular movements are eliminated from SI by an averaging process.

To illustrate the application of this method we shall use the series of frozen vegetables (stocks, cold storage, end of month), 1952–1958; these data are entered in column 1 of Table 19–1.

Table 19-1. Frozen Vegetables, Stocks, Cold Storage, End of the Month, 1952–1958

(Millions of pounds)

Computation of ratios to centered 12-month moving averages

Year and month	*Stock* (1)	*12-month moving total* (2)	*12-month moving average* (3)	*Centered 12-month moving average* (4)	*Ratio to moving average (percent)* (5)
1952					
Jan.	444			—	—
Feb.	399	—	—	—	—
Mar.	348	—	—	—	—
Apr.	314	—	—	—	—
May	302	—	—	—	—
June	337	—	—	—	—
		5204	434		
July	385			436.0	88.3
		5255	438		
Aug.	463			440.0	105.2
		5306	442		
Sept.	530			445.0	119.1
		5378	448		
Oct.	577			451.0	127.9
		5448	454		
Nov.	570			456.5	124.9
		5507	459		
Dec.	535			461.0	116.1
		5554	463		
1953					
Jan.	495			466.5	106.1
		5637	470		
Feb.	450			474.5	94.8
		5748	479		
Mar.	420			485.5	86.5
		5906	492		
Apr.	384			499.0	77.0
		6066	506		
May	361			512.0	70.5
		6218	518		
June	384			525.0	73.1
		6388	532		
July	468			538.0	87.0
		6524	544		
Aug.	574			548.5	104.6
		6638	553		
Sept.	688			557.0	123.5
		6731	561		
Oct.	737			564.5	130.6
		6817	568		
Nov.	722			571.5	126.3
		6902	575		
Dec.	705			577.5	122.1
		6962	580		
1954					
Jan.	631			581.0	108.6
		6987	582		
Feb.	564			583.5	96.7
		7015	585		
Mar.	513			585.0	87.7
		7025	585		
Apr.	470			584.0	80.5
		6998	583		
May	446			581.5	76.7
		6965	580		
June	444			578.0	76.8
		6909	576		
July	493			573.5	86.0
		6855	571		

Table 19–1. Frozen Vegetables, Stocks, Cold Storage, End of the Month, 1952–1958 (*Continued*)

(Millions of pounds)

Computation of ratios to centered 12-month moving averages

Year and month	*Stock* (1)	*12-month moving total* (2)	*12-month moving average* (3)	*Centered 12-month moving average* (4)	*Ratio to moving average (percent)* (5)
Aug.	602			568.5	105.9
		6796	566		
Sept.	698			564.0	123.8
		6740	562		
Oct.	710			560.0	126.8
		6697	558		
Nov.	689			556.0	123.9
		6647	554		
Dec.	649			553.0	117.4
		6622	552		
1955					
Jan.	577			551.5	104.6
		6612	551		
Feb.	505			551.0	91.7
		6615	551		
Mar.	457			550.0	83.1
		6590	549		
Apr.	427			548.5	77.8
		6573	548		
May	396			547.0	72.4
		6547	546		
June	419			545.0	76.9
		6522	544		
July	483			543.0	89.0
		6503	542		
Aug.	605			541.5	111.7
		6494	541		
Sept.	673			541.0	124.4
		6487	541		
Oct.	693			540.5	128.2
		6484	540		
Nov.	663			541.0	122.6
		6503	542		
Dec.	624			544.5	114.6
		6565	547		
1956					
Jan.	558			551.5	101.2
		6670	556		
Feb.	496			562.0	88.3
		6816	568		
Mar.	450			576.0	78.1
		7012	584		
Apr.	424			597.0	71.0
		7234	610		
May	415			616.5	67.3
		7476	623		
June	481			633.0	76.0
		7710	643		
July	588			652.5	90.1
		7939	662		
Aug.	751			671.0	111.9
		8165	680		
Sept.	869			689.0	126.1
		8380	698		
Oct.	915			708.0	129.2
		8612	718		
Nov.	905			726.5	124.6
		8822	735		
Dec.	858			742.5	115.6
		8998	750		
1957					
Jan.	787			755.5	104.2
		9137	761		
Feb.	722			766.5	94.2
		9260	772		

Table 19-1. Frozen Vegetables, Stocks, Cold Storage, End of the Month, 1952–1958 (*Concluded*)

(Millions of pounds)

Computation of ratios to centered 12-month moving averages

Year and month	*Stock* (1)	*12-month moving total* (2)	*12-month moving average* (3)	*Centered 12-month moving average* (4)	*Ratio to moving average (percent)* (5)
1957					
Mar.	665			776.5	85.6
		9376	781		
Apr.	656			784.5	83.6
		9454	788		
May	625			790.0	79.1
		9506	792		
June	657			793.0	82.8
		9530	794		
July	727			794.0	91.6
		9524	794		
Aug.	874			793.0	110.2
		9500	792		
Sept.	985			790.0	124.7
		9457	788		
Oct.	993			784.5	126.6
		9378	781		
Nov.	957			777.5	123.1
		9289	774		
Dec.	882			769.5	114.6
		9182	765		
1958					
Jan.	781			762.0	102.5
		9106	759		
Feb.	698			755.5	92.4
		9025	752		
Mar.	622			747.0	83.3
		8901	742		
Apr.	577			738.0	78.2
		8812	734		
May	536			732.0	73.2
		8754	730		
June	550			728.5	75.5
		8719	727		
July	651			—	—
		—	—		
Aug.	793			—	—
		—	—		
Sept.	861			—	—
		—	—		
Oct.	904			—	—
		—	—		
Nov.	899			—	—
		—	—		
Dec.	847			—	—
		—	—		

SOURCE: Various issues of *Business Statistics*.

Estimate of Trend and Cycles

The first step of the ratio to moving average method is to estimate TC. To do this we may note again that seasonal variations have a rigid periodicity of twelve months. Also, random and sporadic factors usually produce changes in data that last less than a year. Furthermore, a moving average will smooth out fluctuations that are uniform in duration and amplitude if the term of the moving average is equal to, or is an integral multiple of, the period of the fluctuation. There is then reason to believe that a 12-month moving average will eliminate nearly all the seasonal and most of the irregular influences in the series. It cannot, however, remove all the seasonal and irregular forces. This is so because seasonal changes are periodicities, but their amplitudes may vary somewhat from year to year. For instance, during a warm winter in Miami Beach the hotel business there should be better than during a rather cold winter. Furthermore, the periodicity of irregular movements is always unknown but it is very unlikely that it would be an integral multiple of 12 months. Remembering this limitation, we may say that a 12-month moving average, which attempts to eliminate seasonal and irregular variations, represents the remaining components—trend and cycle—in the series. Thus the 12-month moving average may be called the approximate *TC* curve.

Despite the limitation mentioned above, it is interesting to note that the 12-month moving average for the frozen-vegetable data seems to have ironed out almost completely the seasonal and irregular movements in the series. It can be seen in Figure 19–1 that not only does the 12-month moving average form a smooth curve; it also seems no longer to follow the seasonal peaks and troughs in the original data. Indeed, this smooth curve seems to reflect the trends and cycles of the series quite well.

The calculation of a 12-month moving average starts, as does the calculation of any moving average, with adding up the first 12 monthly values in the series to form a 12-month moving total. The second moving total is obtained by dropping the value of the first month from, and adding

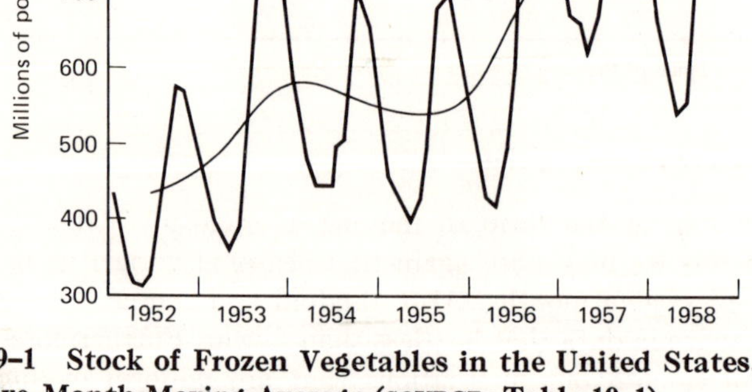

Figure 19–1 Stock of Frozen Vegetables in the United States, 1952–1958, and Twelve-Month Moving Average (SOURCE: Table 19–1)

the value of the thirteenth month to, the first moving total, and so on until all the moving totals have been obtained from the series. Then each moving total is divided by 12 to get the 12-month moving averages. The moving totals and averages so obtained fall in between two months. In our example, the first moving total and average fall between June and July of 1952, the second pair between July and August of the same year, and so on. However, data that are typical of a month should be centered at the middle of the month. Thus, to compute moving averages when an even number of time units is used, an additional step is required to center the average at the middle of each time unit. For a 12-month period, centered averages are obtained by adding the averages two at a time and dividing each sum by 2. The first centered 12-month moving average would fall on the seventh month of the series, the second on the eighth month, and so on. The procedure of computing a 12-month moving average is illustrated in columns 2 through 4 in Table 19–1.

Estimate of Seasonal and Irregular Movements

With the estimate of TC, the next step involves their elimination from original data so that SI can be isolated. The estimate of SI, as already mentioned, is made by dividing the original data by the centered 12-month moving averages; that is, $SI = TSCI/TC$. The resulting ratios are usually expressed in percentage form, as in column 5 of Table 19–1. From this step the name "method of ratio to moving average" is derived.

In computing the centered 12-month moving average, 6 months are lost at each end of the series. For each remaining month there is a ratio of original data to moving average. The 12 monthly ratios of each year are the specific seasonal indices for that year, which are theoretically free from trend and cyclical effects. They are, therefore, estimates of SI. These estimates, however, are slightly biased, mainly because the moving averages may have smoothed out some of the cyclical changes at peaks and troughs. Also, as mentioned previously, the movement of the moving averages may still contain some irregular influences. Consequently, just as we consider the centered 12-month moving average an approximate estimate of TC, so we consider the ratios to moving averages an approximation of SI.

Before we attempt to isolate S from SI, or to construct a typical seasonal index, we must examine the specific seasonals to see whether such an attempt is justified, which depends upon their stability. If the seasonals change through time, it would be of little practical value to have a generalized expression for them.

To examine the stability of specific seasonals, we first put them in tabular form and then observe them together with their graphs. The specific seasonals for our illustrative problem are recorded in Table 19–2.

Table 19–2. Specific Seasonals for End-of-Month Stock of Frozen Vegetables, 1952–1958

	Year						
Month	1952	1953	1954	1955	1956	1957	1958
Jan.		106.1	108.6	104.6	101.2	104.2	102.5
Feb.		94.8	96.7	91.7	88.3	94.2	92.4
Mar.		86.5	87.7	83.1	78.1	85.6	83.3
Apr.		77.0	80.5	77.8	71.0	83.6	78.2
May		70.5	76.7	72.4	67.3	79.1	73.2
June		73.1	76.8	76.9	76.0	82.8	75.5
July	88.3	87.0	86.0	89.0	90.1	91.6	
Aug.	105.2	104.6	105.9	111.7	111.9	110.2	
Sept.	119.1	123.5	123.8	124.4	126.1	124.7	
Oct.	127.9	130.6	126.8	128.2	129.2	126.6	
Nov.	124.9	126.3	123.9	122.6	124.6	123.1	
Dec.	116.1	122.1	117.4	114.6	115.6	114.6	

SOURCE: Table 19–1.

As graphic aid to the study of the data in Table 19–2, the ratios may be either drawn as a continuous series or represented by a *tier graph* as in Figure 19–2. In a tier graph, the specific seasonal indices are piled up one after the other. The horizontal scale is identical for all years, but the vertical scale is a movable one. That is, while the vertical scales for all the years are of the same size, the scales for adjacent years are separated by a constant vertical distance. The actual vertical scales need not be shown, since we are interested in studying only the general movements and timing of the specific seasonal patterns.

From Figure 19–2 we can see that all the specific seasonal indices have identical turning points: the month of May is their seasonal low and October is their seasonal high. Also, although their amplitudes seem to differ slightly, they do give the general impression of a high degree of uniformity within months. These mild differences in amplitude should actually be expected, since the specific seasonals contain irregular movements as well. Finally, if we read Table 19–2 horizontally month by month, we find that even though values of the same month for various years vary from each other, these variations seem to follow a random pattern without any noticeable trend. That is, all the January values, all the February values, and so on, do not seem to increase or decrease on the average through time. From these observations we may conclude that the seasonal pattern of frozen vegetable stock may be considered a stable one.

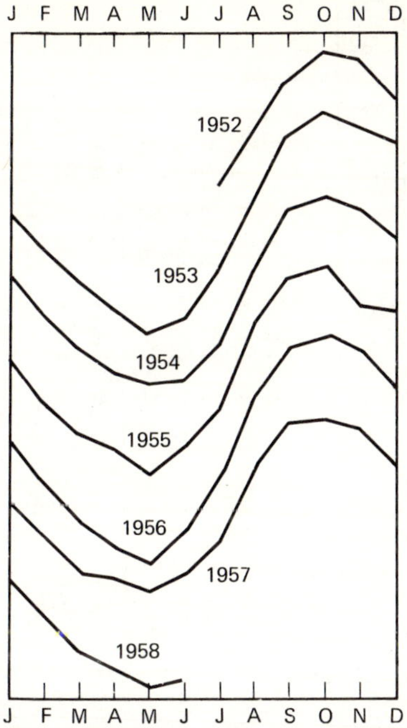

Figure 19-2 Tier Graph Showing Specific Seasonal Patterns for End-of-Month Stock of Frozen Vegetables (SOURCE: Table 19-2)

As a result we decide that the construction of a typical seasonal index for the data is justified.

Eliminating the Irregular Movements

Under the assumption that the irregular component of *SI* is a normally distributed random variable with mean zero, then the irregular movements will be averaged out by monthly means of the *SI* values, leaving only the seasonal component. This assumption would be true if the irregular variations were caused by random forces alone, but we know otherwise. Irregular fluctuations may also be influenced by sporadic forces that often produce extreme values. The conventional arithmetic mean, being affected by the extreme observations all the time, cannot then be expected to eliminate from *SI* all the influences unrelated to seasonal fluctuation. One way out of this difficulty is to use the median of the monthly values as estimates of *S*. A still better alternative is to use a modified arithmetic mean that combines the advantages of the mean and the median.

To compute the modified monthly means we first array the monthly *SI* values as shown in Table 19-3. Next we disregard the largest and the smallest values of each array and include only the central values in the

Table 19–3. Computation of Seasonal Index for Stock of Frozen Vegetables, 1952–1958

Rank	Jan.	Feb.	Mar.	Apr.	May	June	July	Aug.	Sept.	Oct.	Nov.	Dec.
1	101.2	88.3	78.1	71.0	67.3	73.1	86.0	104.6	111.9	126.6	112.6	114.6
2	102.5	91.7	83.1	77.0	70.5	75.5	87.0	105.2	123.5	126.8	123.1	114.6
3	104.2	92.4	83.3	77.8	72.4	76.0	88.3	105.9	123.8	127.9	123.9	115.6
4	104.6	94.2	85.6	78.2	73.2	76.8	89.0	110.2	124.4	128.2	124.6	116.1
5	106.1	94.8	86.5	80.5	76.7	76.9	90.1	111.7	124.7	129.2	124.9	117.4
6	108.6	96.7	87.7	83.6	79.1	82.8	91.6	111.9	126.1	130.6	126.3	122.1
Total of four central items	417.4	373.1	338.5	313.5	292.8	305.2	354.4	433.0	496.4	512.1	496.5	463.7
Modified means	104.4	93.3	84.6	78.4	73.2	76.3	88.6	108.2	124.1	128.0	124.1	115.9
Adjusted seasonal index	104.5	93.4	84.7	78.4	73.2	76.3	88.7	108.3	124.2	128.1	124.2	116.0

SOURCE: Table 19–1.

calculation of the monthly means. The number of values to be excluded from both ends depends upon the data on hand. In our illustration we note that for some months the SI values are quite close to each other; other months have one extreme high and/or one low entry. Therefore, it is sufficient to discard only one value from each end of the arrays. In other words, the modified means for our present illustration may be computed from the central four percentages for each month. These averages are the remaining effects in SI and thus they are estimates of S—the typical seasonal index, or simply the seasonal index. The sum of the twelve values in a seasonal index should add up to 1200, or an average 100.0. This total in our illustration is 1199.1, which is close enough for most practical and theoretical analyses of seasonal variations. Nevertheless, many people are trained to believe that everything should attain the highest accuracy. To accommodate this, we adjust the 12 modified means by multiplying each by a ratio of 1200 to the sum of the modified means. This adjusting factor for our illustration is

$$\frac{1200}{1199.1} = 1.00075.$$

The adjusted seasonal index for our example appears in the last row of Table 19–3, which, after rounding, adds up to 1200.

A graph of the seasonal index appears in Figure 19–3, which shows that the seasonal pattern of our illustrative series is quite pronounced, with a seasonal low associated with May that is slightly less than 27 percent below, and a seasonal high in October that is more than 28 percent above, an average of 100. A moment's reflection will show that the seasonal pattern for frozen vegetable stock is closely associated with the rate of

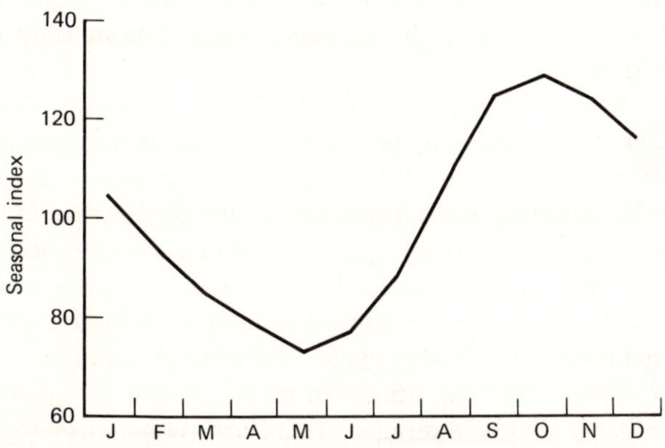

Figure 19–3 Seasonal Index of Frozen Vegetable Stock in the United States, 1952–1958

fresh vegetable production and the rate of vegetable consumption. The production of fresh vegetables normally reaches its seasonal peak in early summer and its seasonal trough in late fall. On the other hand, the rate of consumption of vegetables remains relatively constant throughout the year. Consequently, during the months of winter and spring, when harvests of fresh vegetables continue to decline but the level of vegetable consumption remains stable, frozen vegetable stocks are bound to shrink progressively. Throughout the summer and fall, when the production of vegetables is increasing without a corresponding rise in vegetable consumption, the stock of frozen vegetables should increase.

19–4 ALTERNATIVE METHODS OF MEASURING THE SEASONAL INDEX

Two other methods of computing the seasonal index merit some mention, especially because one of them brings out indirectly the advantages of the ratio to moving average method.

The simplest possible method of computing a seasonal index is to calculate the average value for each month (for January, for February, and so on) and express averages as percentages so that the percentages of the average for all 12 months can add up to 1200, or average 100.

This method is indeed simple but very crude. It assumes there is no trend component in the series; that is, $Y = CSI$. This, however, is an unjustified assumption. Most economic series have trends, and, therefore, the seasonal index computed by this method is actually an index of trends and seasonals. Furthermore, the effects of cycles on the original values may or may not be eliminated by the averaging process. This depends on the duration of the cycle and the term of the average—that is, on the number of months included in the average. Thus, this method is of little value and is seldom used.

The second alternative is the method of *ratio to trend.* This method assumes that seasonal variation for a given month is a constant fraction of trend. Starting with the model $Y = TSCI$, some argue that trend can be eliminated by dividing each observation by its corresponding trend value. The ratios resulting from this operation compose SCI. Each of these ratios to trend is a one-based relative—that is, a pure number with a unity base. Next, an average is computed for the ratios for each month. This averaging process eliminates cycle and random influences from the ratios to trend. Thus, the 12 averages of ratios to trend contain only the seasonal component. These averages, therefore, constitute the seasonal index. As with the method of ratio to moving average, some slight corrections are necessary in order to adjust these ratios to average unity.

The ratio to trend method is certainly a more logical procedure than the crude measurement of seasonals mentioned above. It has an advantage over the moving average procedure too, for it has a ratio to trend value for each month for which data are available. Thus, unlike the case of moving averages, no loss of data occurs. This is a distinct advantage, especially when the period covered by the time series is very short.

The main defect of the ratio to trend method is that if there are pronounced cyclical swings in the series, the trend—whether a straight line or a curve—can never follow the actual data as closely as a 12-month moving average does. As a result, the ratio to trend will tend to be more variable than the ratios to a 12-month moving average. Under these circumstances, it is likely that the averaging process will have a greater ability to remove irregular movements from the ratios to moving averages than it has in removing irregular and cyclical movements from the ratios to trend. This is another way of saying that a seasonal index computed by the ratio to moving average method may be less biased than one calculated by the ratio to trend method.

In conclusion, because of its theoretical and practical advantages, the ratio to moving average method is probably the one most widely used. Its application is not confined to periodic fluctuations of 12-month duration. It can be used equally well for any periodic fluctuation, such as hourly variations of temperature within the day or daily variations in sales within the week.

19–5 CHANGING SEASONAL

As mentioned before, the use of the conventional methods to measure typical variations is appropriate and significant only if the variations are stable from year to year. Unfortunately, seasonal patterns tend to change through time. Changes in seasonals may show up in the amplitude or—and this is more serious—in timing. These changes are sometimes abrupt and sometimes gradual. We are concerned here with the appropriate ways to measure such changing seasonals.

In certain situations the seasonal variations of a series may remain stable for a number of years and then change suddenly into patterns quite different from the old, which in turn may remain stable for years before they change again. Such changes are caused either by powerful sporadic factors or by deliberate managerial policies. During World War II, which necessitated an all-out effort in many lines of production, seasonals in these lines disappeared but reestablished themselves afterward. Changes in the seasonal variations of automobile sales were observed when the industry altered the design of cars to make them less dependent on weather and when it changed from early spring to fall the introduction of new models.

A simple but satisfactory method of dealing with this kind of seasonal variation is to analyze the variations independently for the two periods separated by the abrupt change. The seasonal index for each separated period may be computed either by the ratio to moving average method or by the ratio to trend device. When we use such indices in analysis, each index should be used for the period for which it is constructed. In synthesis, we would ignore the index computed before the abrupt change and employ only the one established after the change.

Gradual changes in seasonals are fairly common. They are brought about by modifications of custom, technology, business practices, and other factors that change slowly but steadily. New designs of automobiles have greatly reduced the amplitude of gasoline consumption. The emergence of air conditioning has increased the summer use of electricity. Changes in consumers' tastes, encouraged by intensive advertising, have gradually eliminated the pronounced seasonal fluctuations in ice-cream production. An elastic credit supply and the central banking reserves of the federal reserve banks have progressively reduced seasonal swings in short-term interest rates almost to the vanishing point.

Gradual changes in seasonal patterns are usually secular in nature. In such cases, no average can be a typical representation of the seasonal pattern, since errors that run in one direction cannot be removed by averaging. When the change in a seasonal is gradual, instead of deriving a typical seasonal index illogically, we study the steady and progressive change in the specific seasonal indices month by month and obtain a measure of change in the index for each month. When the specific seasonal indexes—that is, ratios to moving averages—have been obtained, the ratios of each month are plotted as an independent time-series graph. Twelve such graphs, with the X axis scaled in years, characterize the changes in seasonal pattern and therefore enable us to measure them. We reason that if there is a changing seasonal pattern, then progressive change will be reflected in the movements of the ratios for each month in the form of a discernible general drift from year to year. That is, the movement in the values of each month for the year can be appropriately defined by a trend line. The trend line here serves, as usual, the purpose of an average. The trend value for each month in each year is the first approximation of the seasonal index for that month in that year. We therefore end our analysis of gradually changing seasonal patterns with as many specific indexes as there are years in the period under examination. These specific index values are trend values. The preliminary specific seasonal indices for each year, as before, must be adjusted to average 100. To adjust data for seasonal variations, each monthly original observation is divided by the adjusted seasonal index for that month in the same year. For forecasting, we use the projected trend values by extending the trend line for the desired month.

Whenever the values of the same month for the years tend to change gradually by a constant absolute amount, straight-line trends are appropriate. In many cases, however, values for each given month may not behave linearly. One frequently used device is to draw a freehand trend line through plotted points for a single month; the trend values are then read from the chart. This method is actually used by the Federal Reserve System in dealing with changing seasonals for the construction of its industrial production index. Another widely used method is to compute a moving average of five or seven terms. This procedure is used by the National Bureau of Economic Research. Here, as elsewhere in statistical analysis, the trend line to be determined depends upon the nature of the data and the judgment and experience of the investigator.

19–6 MEASURING CYCLES FROM ANNUAL DATA

Cyclical fluctuations do not repeat themselves periodically as do seasonal variations. Neither do they behave fortuitously as do the irregular movements. The cycles of a specific series usually embrace a certain broad pattern that shows some similarity but always contains some differences in duration and intensity. Because of this lack of uniformity in cycles, it is possible to isolate them but impossible to project them into the future. This, as will be seen later, constitutes the most difficult part of forecasting with time series.

Isolating cycles from annual data is a relatively simple matter. In annual data there are only two components: the trend and the cycle. Seasonal variations, being monthly changes that complete their pattern within a year, do not show up in annual data. Irregular movements, which are also short-run influences compared to a year, produce negligible influences upon annual data, and their positive and negative effects tend to offset each other during the course of a year. With annual data, the trend value is considered as the *statistical normal*—the value that represents the normal annual growth if there had been no cyclical influences. Conversely, then, we may consider cyclical fluctuations as the deviations from the normal values—the trend.

Thus the problem of isolating cycles from annual data is the same as the elimination of trend. Trend can be eliminated by subtracting the trend value from the original data if we assume an additive model, $Y = T + I$. The results are *cyclical residuals*. Or, if we assume a multiplicative model, $Y = TC$, cycles are isolated by dividing original data by the trend values, obtaining *cyclical relatives.* In order to facilitate comparison of cyclical fluctuations among different time series, both cyclical residuals and relatives are expressed in terms of percentage deviations from the trend. The percentage deviations can be obtained by either of the following two formulas:

$$\text{percentage deviations of cyclical residuals} = \left(\frac{y - y_t}{y_t}\right)(100) \text{ percent,} \tag{19–1}$$

$$\text{percentage deviations of cyclical relatives} = \left(\frac{y}{y_t} - 1\right)(100) \text{ percent.} \tag{19–2}$$

It may be observed that values of percentage deviations derived from both formulas are identical.

To illustrate the measuring of cyclical fluctuations of annual data under the additive as well as multiplicative models, we shall use the synthetic rubber production series in Table 18–2. Computations of cyclical residuals, cyclical relatives, and percentage deviations for this series are given in Table 19–4. These results are portrayed in Figure 19–4.

Table 19–4. Computation of Cyclical Fluctuations of Synthetic Rubber Production in the United States, 1953–1968

Year	*Production* y	*Trend* y_t*	*Cyclical residuals* $y - y_t$	*Cyclical relatives* $(y/y_t)100$	*Percentage deviations*
1953	70.7	61.69	+9.01	114.6	+14.6
1954	51.9	69.22	−17.32	75.0	−25.0
1955	80.9	76.75	+4.15	105.4	+5.4
1956	90.0	84.27	+5.73	106.8	+6.8
1957	93.2	91.80	+1.40	101.5	+1.5
1958	87.9	99.33	−11.43	88.5	−11.5
1959	114.9	106.85	+8.05	107.5	+7.5
1960	119.5	114.38	+5.12	104.5	+4.5
1961	117.0	121.91	−4.91	96.0	−4.0
1962	131.1	129.43	+1.67	101.3	+1.3
1963	134.1	136.96	−2.68	97.9	−2.1
1964	147.0	144.49	+2.51	101.7	+1.7
1965	151.1	152.02	−0.92	99.4	−0.6
1966	164.1	159.54	+4.56	102.9	+2.9
1967	159.3	167.07	−7.79	95.3	−4.7
1968	177.6	174.60	+3.00	101.7	+1.7

* $y_t = 118.144 + 7.527x$ ($X = 0$: 1960–1961).

Table 19-5. Cyclical Irregular Movements for United States Factory Production of Cheese, 1947–1955

Year and month	*Production* y *(millions of pounds)* (1)	*Trend** y_t (T) (2)	*Seasonal*† *index* S *(as ratios)* (3)	*Statistical normal* TS (4)	*Cyclical-irregular movements* y/TS *(percent)* (5)
1947					
Jan.	74.3	93.3	0.780	72.774	102.1
Feb.	78.3	93.5	0.770	71.995	108.8
Mar.	100.2	93.6	0.968	90.605	110.6
Apr.	114.1	93.8	1.099	103.086	110.7
May	140.6	94.0	1.410	132.540	106.1
June	148.1	94.1	1.416	133.246	111.1
July	133.3	94.3	1.233	116.272	114.6
Aug.	104.0	94.4	1.096	103.462	100.5
Sept.	87.7	94.6	0.933	88.262	99.4
Oct.	80.9	94.8	0.839	79.537	101.7
Nov.	60.9	94.9	0.712	67.569	90.1
Dec.	60.4	95.1	0.744	70.754	85.4
1948					
Jan.	63.7	95.3	0.780	74.334	85.7
Feb.	65.2	95.4	0.770	73.458	88.8
Mar.	80.3	95.6	0.968	92.541	86.8
Apr.	94.0	95.8	1.099	105.284	89.3
May	129.8	95.9	1.410	135.219	96.0
June	131.9	96.1	1.416	136.078	96.9
July	114.7	96.3	1.233	118.738	96.6
Aug.	108.5	96.4	1.096	105.654	102.7
Sept.	89.4	96.6	0.933	90.128	99.2
Oct.	81.5	96.8	0.839	81.215	100.4
Nov.	67.3	96.9	0.712	68.993	97.5
Dec.	72.1	97.1	0.744	72.242	99.8
. . .					

SOURCE: *Business Statistics.*

* $y_t = 93.29 + 0.165x$ (origin, January 1947).

† Modified mean of four central ratios of actual values to centered 12-month moving averages for 1947–1955.

2 contains the trend values for the series, which are derived by an arithmetic straight-line trend equation, $y_t = 93.29 + 0.165x$, with origin at January 1947. The seasonal pattern of this series is stable, and a typical seasonal index is computed by the ratio to moving average method, the

monthly index values being modified means of the four central items. These index values are recorded in column 3 in ratio form. Column 4 presents the values of statistical normal values for the series. These values are obtained by multiplying trend values by seasonal index. When we divide the *TS* values into original observations in column 1 and multiply the results by 100, we obtain the estimates of cyclical-irregular fluctuations in percentage form, as shown in column 5. The cyclical-irregular percentages are plotted in the lower chart in Figure 19–5. These percentages

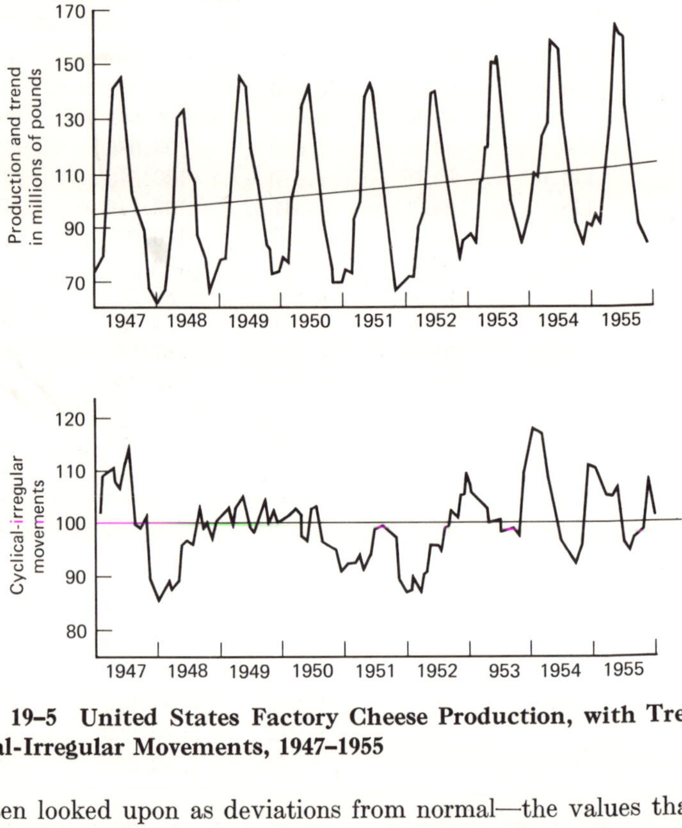

Figure 19–5 United States Factory Cheese Production, with Trend and Cyclical-Irregular Movements, 1947–1955

are often looked upon as deviations from normal—the values that could have been had there been no cyclical and irregular fluctuations.

Sometimes we wish to eliminate most of the irregular movements from the *CI* curve. This is often done by smoothing the curve with freehand drawing or by the calculation of a moving average. To estimate *C* by taking a moving average of *CI*, the terms used are often 3 months or 5 months. Odd terms make computation easier, and short durations avoid the tendency to smooth out too much of the cycles.

The isolation of cyclical fluctuations by a moving average is actually not a satisfactory method. A 3-month moving average may smooth out only a negligible part of the cycle, and the resulting curve may not be very smooth. A 5-month moving average may produce a smoother estimate of cycles, but the portion of cycles smoothed out may be considerable.

The fact that there is no satisfactory method of removing irregular movements, coupled with some theoretical considerations—on which we shall here comment briefly—has led to the practice of analyzing cyclical and irregular fluctuations as a single entity. Irregular movements caused by random forces are relatively unimportant. Furthermore, random forces occur quite frequently, usually from unknown sources. Thus it is neither possible nor necessary to identify them: it is best to treat them as part of the cyclical swing that we actually observe. It seems reasonable to consider the slight sawtooth effect of cyclical movement as a part of the cycles, but can we also ignore the short-term sporadic changes that rise and fall with considerable magnitude just as the cycles do?

Part of the answer lies in the basic difference between these two types of variation. It is the main property of cycles that they are cumulative and self-generating. Thus, if the fluctuations in the short run are such that they rise abruptly and then immediately fall back toward the previous level, then they are not cumulative and should be considered as irregular variations. Sometimes sporadic forces are quite capable of generating fluctuations that are cumulative for a number of months. Prolonged strikes, speculative booms, and war scares are examples of such irregular forces. Under these circumstances we may insist that, to be qualified as cycles, the cumulative expansions and contractions must last a minimum number of months, say longer than 12 as used by the National Bureau of Economic Research. Here we come to the realization that cyclical causes are diversified, complicated, and dynamic. They can never be expected to produce smooth cycles as portrayed by the sine and cosine functions of the time scale. Irregular movements are part of the aggregate of factors that create the ragged cyclical course we observe in reality. Our attention to irregular movements should warn us that not every rise and fall comprises a cycle. When we mark off the turning points of cycles, we must be careful that they reveal faithfully the underlying cyclical forces at work.

In examining the cyclical and irregular movements of our illustrative series as plotted in Figure 19–5, it is interesting to observe first the upper chart, which shows the original data with trend. The seasonal variations of cheese production are so pronounced that the original data reveal hardly any cyclical movements. But the lower chart, for *CI*, tells quite a different story. Here we see the necessity of eliminating seasonals from the original data before we can profitably study cycles.

19–8 COMMENTS ON THE ESTIMATE OF *CI*

The irregular nature of cyclical movements defies any attempt to use an "average" cycle to represent their effects on a time series. The best—or the least unsatisfactory—approach is to measure these fluctuations indirectly by removing effects of trend and seasonal forces. The remaining variations are considered cyclical variations (though we must remember that they also reflect irregular influences). This method is admittedly imperfect, not only because a cycle estimated by it cannot be projected into the future but also because the estimate is to some degree biased.

To estimate CI, we must first estimate T and S. The actual behavior of the cyclical and irregular movements depends on the trend line as well as on the seasonal index. When we determine the trend line, we usually have two purposes in mind: (1) extrapolation, and (2) adjustment of the original data by subtraction or division. But these two functions of the trend are often contradictory. We may find that a given trend equation isolates cycles satisfactorily but is inadequate for forecasting, or vice versa. Furthermore, the same mathematical trend function usually yields different trend values when the initial and terminal years selected differ and when the period covered varies. Consequently, the cyclical relatives computed from deviations from trend are affected by the choice of trend function and the choice of the period to which it is fitted.

The adjustment for seasonals is also rather arbitrary. The seasonal patterns of many series are subject to change without notice, and even if they remain relatively constant, the values of typical seasonal indices always differ with different methods of computation. This means that the appearance of deseasonalized data is affected by whether we use changing or typical seasonal indices and on the method of computation.

The resulting CI is influenced, finally, by the assumption we make for decomposition of the series. Have we used the multiplicative or the additive model? In choosing one, we may be unable to determine whether the other is closer to the relationship of the components of the series.

From these considerations, we may say that the estimate of CI includes not only the cyclical and irregular forces but also errors in defining the trend, in determining the seasonal pattern, and in decomposing the original series. Since these errors can never be precisely determined, there must always be some uncertainty and speculation in the interpretation and use of the statistical results that we call "cycles."

Secular trend, seasonal variations, and cyclical fluctuations are all real and significant in many economic and social time series, even though our measures for them are only approximations. These measures, approximations though they may be, can be of great use in both economic research and business management. They can enable us to forecast the future behavior of time series.

19–9 TIME SERIES AND FORECASTING

A Note on Forecasting Techniques

One of the basic aims of time-series analysis is forecasting. Forecasting, we should now observe, is much more than projecting a series mechanically into the future. When a firm, an industry, or a government formulates economic plans, it makes assumptions about the future course of business activity on the basis of observations of the past. Here lies the basic difficulty in attempting to forecast future business conditions. The future may be an extension of the past, but it can hardly be expected to be an exact replica. Forces responsible for economic change are numerous and complex. They are often difficult to discover and to measure. They may appear in all kinds of combinations. Furthermore, they may be constantly changing. Even if the response mechanism of an industry or of the economy remains essentially the same through time, it is not always easy to predict how business conditions will respond to new stimuli whose effects we have failed to observe before.

The fact that the past can never be a perfect guide to the future warns us that forecasting cannot be a routine application of techniques or theoretical ideas to a list of unchanging variables. Successful forecasting requires expert blending of economic theory and thorough familiarity with the relevant statistical data. It should utilize both quantitative and qualitative information. The forecaster must be able to distinguish between new facts that are important and those that are not. He must be competent to decide when past relationships can be relied upon and when they cannot. He must be able to appreciate the effects of nonmeasurable socioeconomic and political forces upon business activities. In other words, forecasting is, and probably will remain, more an art than a science.

Numerous forecasting techniques with varying degrees of complexity have been devised during the past several decades. Most fall into one of three broad categories: the naive method, the barometric method, and the analytical method.

A forecasting method is said to be *naive* if it lacks rigid theoretical basis. As a method of forecasting, it ranges from simple coin tossing to decide on an upward or a downward movement to mechanical projection of a time series into the future. A frequently used naive device simply assumes that things will not change. Plans are made on the assumption that, insofar as the events in question are concerned, the future will resemble the present. It is predicted, for example, that sales for the next quarter will be the same as those for this quarter, or that sales for the next quarter will increase by the same amount as they did in the last quarter. As a matter of fact, for most short-term decisions this is exactly what is done. Of course, this method becomes more questionable as the period of forecast becomes longer. The naive method in general, and time-series analysis in particular, assume that the future is an extension of the past.

The *barometric method* implies that historical patterns tend to repeat themselves and that the future can be predicted from certain happenings of the present. Thus, past statistical behavior that seems to be associated regularly with fluctuations in particular series or general business conditions is discovered and used as the basis for forecasting. Foreshadowing, or leading, series are examined to provide an advance reading of what will follow in the series to be forecasted. The search for foreshadowing series is not based only on the leads actually revealed by historical data. Theoretical considerations as to the leads and lags in various series are also employed. Cross-checking between empiricism and theory in the search for leading series is the best hope for establishing the thesis that the future does not represent a break from the past but that changes are largely determined by present conditions. The leading series selected serve as barometers of future changes in specific series or general business conditions. The barometric method, therefore, involves the use of a statistical indicator or indicators that suggest the direction in which a specific time series or the aggregate economy is heading.

The *analytical method* of forecasting entails detailed analyses of causative forces operating currently on the variable to be predicted. Relationships of cause and effect disclosed by analyzing current data are used to judge the future course of the causative forces and their effects on the variable's future behavior. Analytical techniques may be nonmathematical, such as the rather naive factor-listing method or the more sophisticated method of opinion polling; or mathematical, such as econometric models. In any case, this method calls for careful analyses of forces at work in order to establish useful relationships and their quantification. The employment of past statistical relationships is, however, by no means excluded. As a matter of fact, historical patterns are often used to estimate the constants contained in the mathematical models. Although the analytical method can be used to forecast the activity of a firm or an industry, its greatest advance in recent years has come from its use in forecasting the framework of the national income accounts.

We shall now comment on the forecasting with time series.

Forecasting a Series by Itself

The method of forecasting a time series by itself is sometimes called the *economic rhythm method*. Often used by an individual firm or industry to forecast sales, production, plant facilities, or the like, it involves simply the projection into the future of its measurable components: the trend, the seasonal, and the cycle. In projecting a series we may merely wish to project each of the measurable components independently, or we may wish to use these projections to arrive at a synthesis of the components as the forecast of the actual value. Forecasting by synthesis of trend, seasonal, and cycle is just the reverse of decomposing the time series. That is, we first forecast values of the three components for a given time period and then combine these projected values into a single forecast by a process of multiplication.

A firm or industry experiencing steady and substantial growth must forecast future trends so that it can plan the construction of new facilities. Such forecasts will enable business managers to avoid the unprofitable practice of basing expansion decisions on the current demand for their product. Forecasts will help them determine whether an increase in demand is permanent (secular) or temporary (cyclical). When a cyclical expansion is mistaken for secular growth, unneeded capacity may be built. When secular growth is incorrectly interpreted as a temporary expansion, a share of the market may be lost forever to competitors. A forecast of secular trend aids not only in the expansion of capacity in accordance with the growth of demand but also in the purchase or construction of equipment and plant at the most advantageous time, which is usually when the demand for the firm's output is below its peak or when the producers of fixed assets are temporarily in a depressed stage of their activity.

In the forecast of secular trend, the first step is the measurement of the trend in the immediate past. The forecast would be made by extrapolation—that is, projection of the fitted trend into the future. Such projection assumes that the forces behind the trend will continue to have the same effects in the future as in the past; consequently, the further a fitted trend is projected into the future, the more questionable the forecast becomes. However, since secular trend is a type of change that is quite persistent, it is fairly safe to project a series several years into the future.

Forecasts in seasonals are necessary for such activities as budgeting and production scheduling, which are planned on a monthly basis.

Where it is appropriate to construct a typical seasonal index, it will serve as an adequate forecast of the seasonal variations for the next year. If the seasonal is undergoing a gradual change, the specific seasonal index of one year will often suffice as the forecast for the seasonal of the next year. Or, if the seasonal changes assume a clearly defined trend, the projected monthly trend values may serve as the forecasts. With an abruptly changing seasonal, the seasonal index computed after the change should be used as the forecast. Forecasts based on seasonal indexes are made by taking an annual forecast based on trend or some other device and combining it with seasonal indexes, producing monthly forecasts that take account of seasonal fluctuations. For instance, if it is forecast that sales of a firm in the next year will be, say, 1,200,000 units and if there were no seasonal variations in the firm's sales, the monthly sales would be 100,000 units for each month. But if January is expected to be only 80 percent of the monthly average sales, then the forecast for January should be only 80,000 units. In other words, the forecast of total annual sales can be transformed into monthly forecasts by multiplying the average monthly forecasted sales by the appropriate monthly seasonal index values.

Sometimes a useful approximate forecast of actual values can be made by synthesis of trend and seasonal variation. When the forecast is made on an annual basis, it is done by simply projecting the trend line. If the forecast is needed on a monthly basis, the projected monthly trend values are multiplied by the appropriate seasonal indices. A forecast of actual values merely on the extension of trend line and the seasonal indexes may be adequate for many types of decisions. Nevertheless, we must recognize certain limitations of such a forecast:

(1) The type of trend equation selected may not be appropriate.
(2) The trend constants and, therefore, the trend values are subject to sampling errors.
(3) The forces influencing the trend may change after the trend equation has been selected.
(4) The seasonal may not remain the same as in the past, or it may have been undergoing a change that has not been adequately described.
(5) The forecast is made under the assumption that it is not influenced by cyclical and irregular effects.

The most important of these limitations is the last. Nevertheless, it may be discovered that many products, such as motor fuel, shoes, or paperboard for food packaging (Figure 19–6), have been experiencing an ever-expanding market with only mild response to cyclical forces. Except for irregular disturbances, as by wars and strikes, these products have grown fairly steadily through time, despite shifts in their growth rates. For such products, projections of the statistical normal can be considered as useful forecasts for both short- and long-run planning.

For series, such as consumer durable goods and investment goods, that are greatly affected by the cyclical factor, projection of cyclical movement becomes the most important part of forecasting a series by itself. Nearly all types of short-run activities of such firms—ordering raw materials,

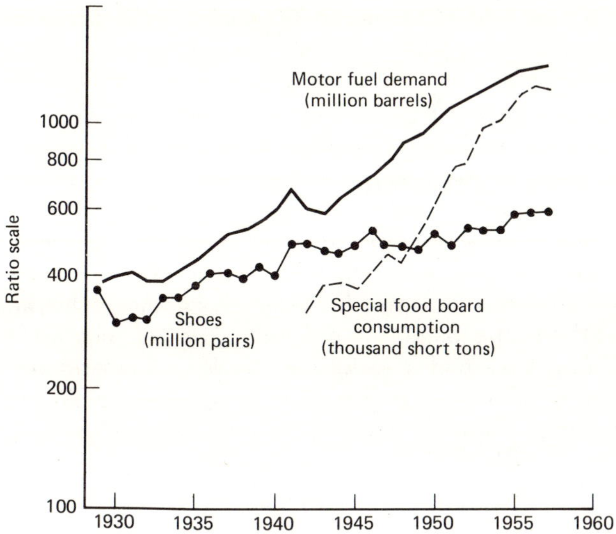

Figure 19–6 Demands for Motor Fuel, Shoes, and Special Food Paper-Board, 1929–1957 (SOURCE: *Survey of Current Business,* November 1957)

scheduling production or advertising, hiring workers, arranging for a line of credit—are influenced by cyclical fluctuations. Unfortunately, accurate forecasting of both cyclical turning points and amplitude is extremely difficult if not impossible, because there is simply no objective way of extending cyclical relatives or residuals into the future. Nearly all of the other methods of forecasting mentioned earlier are concerned with cycles, but they are primarily for predicting general business conditions rather than specific cycles.

Sometimes, in the attempt to project cyclical movement mathematically, a cycle is fitted to data by some function of the sines or cosines of the time scale. Such procedures assume periodicities in both duration and amplitude of the cyclical pattern or that the pattern is changing in accordance with some known law—assumptions that almost no economic time series can meet.

A device often used in forecasting cyclical movement is the subjective extension of a cycle curve. The estimate of CI in the recent past is first smoothed by a moving average or by freehand, and both the CI values and the resulting C values are plotted in the same graph. After careful study of the relationship between the CI and C curves, and the behavior of the C curve in the recent past, the C curve is extended a few months to a year into the future by freehand. Projections of the C values are then read from the chart. Such projection of the cycle is not a mechanical procedure as in the case of trend and seasonal projections; it involves a great deal of subjective judgment. This method, therefore, can be used only by an investigator who is thoroughly familiar with the business under consideration and who proceeds carefully and conscientiously.

Forecasting a Series by Other Series

Although there is no satisfactory method of forecasting the cycle of a series by the series itself, a promising method does exist for predicting the cyclical fluctuations of one series by those of another. This is possible because cyclical turning points of different series do not occur at the same time. Thus, if we find a series that leads (anticipates the turning points in) another series rather persistently, we then employ regression technique to predict the turning points, or even the cyclical magnitudes, of the lagging series, Y, by the leading series, X.

A simple model that uses a related series X to forecast the series of Y is

$$\bar{y}_t = a + bx_t.$$

Also, the independent variable may be lagged:

$$\bar{y}_t = a + bx_{t-1},$$

where the lead-and-lag relationship may be determined on theoretical grounds, empirical observations, or both. Often one can easily determine the optimal lag by studying the simple correlation coefficients for each lag.

Several points must be stressed in the application of regression analysis to time-series data. First, we should use CI values for both the dependent and independent variables. The reason is that our main purpose is to forecast cyclical fluctuations, and two series may appear to be highly associated when data with the trend components are used but may have a low degree of association with detrended data.

Second, more accurate results in forecasting can be obtained with monthly data than with quarterly or annual data. It is conceivable too that very reliable leads and lags can be established as data are made available on weekly or more frequent bases. The scarcity of time series on a basis shorter than a month limits sharply the utility of this method.

Finally, the forecaster (the independent variable) selected must be one that precedes with some degree of regularity the one to be forecasted. We should not select as the forecaster, call it A, a variable that may lead B during the upswing of cycle but lag it during the downswing. Thus, for example, we should not select business activity as the forecaster for the rate of interest, since the expansion of business activity may lead to a rise of interest, but rise in interest may bring about a contraction in business activity.

EXERCISES

Problems 19–1 through 19–8 are based on the following data:

Volume of Exports from the Republic of Independence

Year	*Quarter*	*Y value (millions of dollars)*	*Year*	*Quarter*	*Y value (millions of dollars)*
1960	1	100	1964	1	140
	2	110		2	145
	3	105		3	142
	4	120		4	150
1961	1	116	1965	1	148
	2	125		2	160
	3	128		3	158
	4	138		4	170
1962	1	122	1966	1	168
	2	126		2	175
	3	120		3	170
	4	134		4	180
1963	1	131			
	2	153			
	3	150			
	4	160			

SOURCE: Hypothetical.

19–1 Compute an arithmetic linear trend for data, shift the origin to the first quarter of 1960, and compute all the trend values.

19–2 Compute centered four-quarter moving averages and plot the result against the original data.

19–3 Determine the ratios to moving averages.

19–4 Tabulate the specific seasonals and present them graphically by a tier chart. Comment on the behaviors of these specific seasonals.

19–5 Compute the typical seasonal index for data as the modified means of the two central values. If necessary, compute the adjusted seasonals.

19–6 Give a graphic representation of the typical seasonal index.

19–7 Use the results obtained previously to determine the statistical normal for the series: *TS*.

19–8 Determine the cyclical irregular movements of the series. Graph the results.

19–9 The monthly trend equation for production (in thousands of pounds) of a certain firm is

$$y_t = 100 + 0.2x \qquad (X = 0, \text{Jan. } 1969),$$

and the typical seasonal variations for the same series are estimated as follows:

J	F	M	A	M	J	J	A	S	O	N	D
80	85	110	115	130	110	100	90	90	115	95	80

(a) Forecast the monthly trend values of the series for 1970.
(b) Forecast the statistical normal of this series for 1970.

19–10 Department store sales (Y) in billions of dollars in the United States are related to the previous year's disposable income (X), also in billions, by the following regression equation:

$$\bar{y}_t = 10 + 0.05x_{t-1}.$$

If it has been estimated that in 1975 the disposable income will be 560 billions, what is the forecast of department store sales for 1976?

20

FURTHER TOPICS AND REFERENCES

20–1 THE END OF A BEGINNING

We now have come to the end of a beginning. This text has presented many standard statistical methods, widely and successfully used for many years. The obvious limitations of an introductory course, however, have prevented our covering many other topics, elementary as well as advanced, that are equally important. It seems appropriate, therefore, at the end of our beginning, that we briefly introduce the reader to some of these topics and suggest some references for his further study, in the hope that we have not closed but have rather opened up for him the subject of statistical reasoning.

As the reader explores statistical analysis further, he may encounter technical difficulties from time to time if he lacks a basic knowledge in calculus and matrix algebra. There are many good texts covering these topics, such as the programmed basic text by E. W. Martin, Jr., entitled *Mathematics for Decision Making* (Irwin, 1969). Volume I of this text covers linear mathematics and Volume II covers calculus. The reader may also benefit by consulting Serge Lang's *A Complete Course in Calculus*, 2d ed., (Addison-Wesley, 1968). A text by Clopper Almon, Jr., presents a uniform, connected exposition of linear algebra: *Matrix Methods in Economics* (Addison-Wesley, 1967).

20–2 BAYESIAN VERSUS CLASSICAL STATISTICS

As by now the reader should be fully aware, statistics is a method of decision making under uncertainty, embracing two major alternative approaches—the Bayesian and the classical. Though Bayesian and classical statistics are agreed to be quite different, it is almost impossible to draw a definite boundary between them. The main reason is that both schools of thought are based on probability theory, although the Bayesians use subjective probability while the classicists accept only the mathematical and relative-frequency definitions. The two main differences are: (1) Classical decision procedures are based on sample information alone, while Bayesian decision procedures can be applied on subjective prior information alone or on both prior judgment and additional information. (2) The Bayesian procedures take explicit account of specific loss functions for wrong decisions, whereas the classical methods are sometimes quite vague on this point. Thus, as a decision theory, the Bayesian approach is more flexible and complete than the classical approach.

An attempt has been made in this text to give a reasonably balanced presentation of Bayesian and classical statistics, reflecting the author's feeling that they are complementary to, rather than substitutes for, each other. Basic texts with a similar orientation include Peters and Summers, *Statistical Analysis for Business Decisions* (Prentice-Hall, 1968) and Spurr and Bonini, *Statistical Analysis for Business Decisions* (Irwin, 1967). A more advanced text that treats both the classical and Bayesian methods quite extensively is *Statistics: Probability, Inference and Decision* by Hays and Winkler (Holt, Rinehart and Winston). This book was published in two volumes in 1970 and then combined into a single text in its 1971 printing.

For broader coverage of classical analysis at the elementary level than the present volume offers, the reader is referred to this author's *Statistical Analysis* (Holt, Rinehart and Winston, 1969); E. C. Bryant's *Statistical Analysis*, 2d ed. (McGraw-Hill, 1966); S. B. Richmond's *Statistical Analysis*, 2d ed. (Ronald, 1969); and Taro Yamane's *Statistics*, 2d ed. (Harper & Row, 1967).

Texts with Bayesian orientation at the elementary level include Howard Raiffa, *Decision Analysis* (Addison-Wesley, 1968); R. Schlaifer, *Probability and Statistics for Business Decisions* (McGraw-Hill, 1959); K. Sasaki, *Statistics for Modern Decision Making* (Wadsworth, 1968); Dyckman, Smidt, and McAdams, *Management Decision Making under Uncertainty* (Macmillan, 1969); and R. Schlaifer, *Analysis of Decisions under Uncertainty* (McGraw-Hill, 1965). A more advanced text is Pratt, Raiffa, and Schlaifer, *Introduction to Statistical Decision Theory* (McGraw-Hill, 1965). An advanced and standard text on Bayesian statistics is *Applied Statistical Decision Theory* under the joint authorship of H. Raiffa and R. Schlaifer (Harvard University Press, 1961).

20-3 MATHEMATICAL STATISTICS

Mathematical statistics is concerned with the derivation of probability models and the underlying mathematical formulations for various methods of statistical inference and decision making. Knowledge in this area will also enhance the reader's understanding of the merits and limitations of statistical methods as applied in actual situations.

Books on mathematical statistics and probability theory readily appreciable by the reader with a background in elementary calculus include: E. Borel, *Elements of the Theory of Probability* (Prentice-Hall, 1965); J. E. Freund, *Mathematical Statistics* (Prentice-Hall, 1962); P. Hoel, *Introduction to Mathematical Statistics*, 3d ed. (Wiley, 1963); W. B. Lindgren, *Statistical Theory*, 2d ed. (Macmillan, 1968); and J. Neyman, *First Course in Probability and Statistics* (Holt, Rinehart and Winston, 1950). Some more advanced, well-known texts in this field are W. Feller, *An Introduction to Probability and Its Applications*, Vols. 1 and 2 (Wiley, 1957, 1966); and E. Parzen, *Modern Probability Theory and Its Applications* (Wiley, 1960).

20–4 METHODS OF SAMPLING

Throughout our discussion of statistical inference and decision making we have assumed that we are dealing with simple random samples. There are, however, many procedures of sampling, each producing a different kind of sample. Designs other than simple random sampling have been introduced for use in circumstances in which random sampling may be impossible or another method may be more desirable.

Broadly speaking, sampling designs can be grouped as nonrandom and random. *Nonrandom sampling* is a process of sample selection on considerations other than those of probability, such as expert judgment or convenience. *Random sampling* is also referred to as *probability sampling*, since if the sampling process is random, the laws of probability can be applied. We note that randomness can enter a process of sampling in a number of ways, and therefore, random samples may be of many kinds. In addition to simple, or unrestricted, random sampling procedures, with which we are now quite familiar, the more frequently employed random sampling procedures are the following:

(1) *Stratified random sampling*. This process requires that the population be divided into homogeneous groups, called *strata*, from each of which a subsample is drawn by the simple random procedure. A stratified sample results when these subsamples are combined. A stratified sample is usually proportional; that is, the number of items drawn from each group is proportional to the size of that group relative to the whole population.

(2) *Cluster sampling*. Under this plan, the population is divided into a number of groups, called *clusters*, and then a sample of groups is

selected at random to represent the population. These chosen clusters may be considered as the final sample, or a subsample may be drawn from each cluster to form the final sample. When the clusters consist of geographical areas, this procedure is often called *area sampling*.

(3) *Systematic sampling*. This is a procedure of selecting the kth item from a listing of the population, with a random start. Here, the number k is merely the sampling ratio, n/N. For instance, if 100 items are to be included in the sample for a population of 1000 items, then $k = 10$. Next, we select a number from 1 to 10 at random as the random start. If this number turns out to be, say, 3, then we begin with 3 and select every tenth item thereafter—the 13th, the 23rd, and so on—until all 100 items have been obtained.

(4) *Sequential sampling*. This concept has already been discussed in connection with acceptance sampling.

For detailed presentation of alternative sampling plans and their appropriate uses, the reader may consult the following standard texts: W. G. Cochran, *Sampling Techniques*, 2d ed. (Wiley, 1963); W. E. Deming, *Sample Design for Business Research* (Wiley, 1960); Hanson, Hurwitz, and Madow, *Sample Survey Methods and Theory* (Wiley, 1953); and A. Stuart, *Basic Ideas of Scientific Sampling* (Hafner, 1962).

20-5 NONPARAMETRIC METHODS

In introducing inferential procedures in this text thus far, we have usually assumed that samples have originated from populations having known distributions. Each theoretical distribution function depends upon one or more parameters. These procedures can be appropriately called the *parametric* case.

In many situations, however, it is impossible to specify the form of population distribution; we can merely assume that it is continuous. The process of drawing conclusions directly from sample observations without formal assumptions about the mathematical form of the population distribution is called the *nonparametric*, or *distribution-free*, theory. Actually, the reader is not completely ignorant of this branch of statistics. The chi-square tests on frequencies, goodness-of-fit, and so on introduced in Chapter 14, and the rank-correlation analysis presented in Chapter 17, are all nonparametric techniques. Again, the Chebyshev inequality, although not strictly nonparametric since it involves μ and σ, is distribution-free since it holds for any probability distribution.

In this section we shall introduce two other frequently employed nonparametric tests and then comment on the importance and limitations of nonparametric statistics in general.

The Runs Above and Below the Median Test for Randomness

The parametric theory of statistical inference rests on randomness. For parametric procedures to apply, the sample must be a random sample: a sample observation must be identically distributed with the population. This assumption may easily be violated, either because the population changes with respect to one or more characteristics while the sample is being drawn, or because dependence has somehow crept into the sample observations. In other words, the population may not be in statistical control or the sample may not be random. When this happens, the classical theory fails. Clearly, methods that can evaluate the randomness of a sample are of supreme importance in justifying the use of parametric inference procedures.

A simple nonparametric method for randomness based on the concept of "runs" is called the *runs above and below the median* test. Suppose we have a sample of, say, 26 observations gathered in the order listed below:

97, 89, 25, 81, 11, 83, 16, 96, 44, 32, 98, 19, 68,
33, 25, 54, 74, 82, 17, 49, 33, 22, 62, 20, 92, 80.

A moment's study of this set of data reveals that the value of the median is $(54 + 49)/2 = 51.5$. We proceed to observe now whether each observation is above or below the median. If above, we mark it as an a, and if below, as a b. For the sample above, we have the following sequence of a's and b's:

aa-b-a-b-a-b-a-bb-a-b-a-bb-aaa-bbbb-a-b-aa.

Each sequence of a's or b's, uninterrupted by the other letter, is a run; we have, therefore, 17 runs for our data.

Now, there is reason to believe that if the sample is random, then the number of runs above and below the median should also be a random variable. Furthermore, if the population is continuous, n is even, and $n \geq 25$, the sampling distribution of this number of runs, R, is asymptotically normal with

$$E(R) = \frac{n + 2}{2} \tag{20–1}$$

and

$$V(R) = \frac{n(n - 2)}{4(n - 1)}. \tag{20–2}$$

The null hypothesis to be tested here is usually stated as: "The population is in statistical control," or "The sample is a random sample." The alternative is often of the left-tail variety. This is especially appropriate if there are too few runs, since such a situation usually indicates a shifting population mean. We may occasionally wish to conduct a right-tail test when there are too many runs, revealing a systematic swing above and below the median from one observation to another, thus making us suspect that the data might have been manipulated to yield too perfect a result. When we have no specific ideas, except that H_1 should be the opposite of H_0, we may then proceed to make a two-sided test. In any event, when the normal theory can be applied to the distribution of R, the testing statistic is

$$Z = \frac{R - E(R)}{\sqrt{V(R)}} . \tag{20–3}$$

Suppose we wish to have a two-sided test for our illustrative data at $\alpha = 0.01$; we would have

$$E(R) = \frac{26 + 2}{2} = 14,$$

$$V(R) = \frac{26(26 - 2)}{4(26 - 1)} = 6.24,$$

and

$$Z = \frac{17 - 14}{\sqrt{6.24}} = 1.0.$$

Since the observed value of Z falls between -2.58 and $+2.58$, H_0 cannot be rejected at the 1 percent level of significance. We conclude therefore that the sample is a random sample.

In closing our discussion of the test of randomness with runs above and below the median, let us note three things. First, when the sample is small, exact probabilities can be determined with the binomial theorem with $p = \frac{1}{2}$, since R is a binominal variable. Second, the condition that n be even can be violated to yield satisfactory results when n is large enough for normal approximation. Third, when there are sample values that are tied with (equal to) the median value, either they may be disregarded so that number of runs and the sample size are reduced, or they may be assigned a's or b's by a random process, such as the toss of a coin.

The Wilcoxon Test for Two Matched Samples

In Chapter 15 we noted that the significance of the difference between two sample means derived from two dependent (matched) samples can be evaluated by a t test. We shall now see that a corresponding nonparametric method is called the Wilcoxon, or *signed-rank*, test. This is perhaps the most powerful design among all nonparametric tests for the identity between two population distributions with *matched pairs*. Its mechanics are quite simple. When observations are given in n pairs, we first observe their differences, d_i, and then rank these differences according to their absolute magnitudes. Next, we attach the signs of these ranks for the corresponding differences. Finally, we find the testing statistic, T, defined as the sum of the ranks with the less frequent sign.

When the null hypothesis, $f(x) = g(y)$, is true, then each of the 2^n possible sets of signed ranks (obtained by arbitrarily assigning + and − signs 1 through n) is equally likely to occur. On this basis, the exact distribution of T can be found.

To test $f(x) = g(y)$ against $f(x) \neq g(y)$, we may obtain the approximate significance levels for the smaller rank total by the following expressions.

Denoting the 5 percent level of the lower rank total as $T_{0.05}$, we have

$$T_{0.05} = \frac{n^2 - 7n + 10}{5}. \tag{20–4}$$

We reject H_0 at $\alpha = 0.05$ if $T \leq T_{0.05}$.

Denoting the 1 percent level for the lower rank total as $T_{0.01}$, we have

$$T_{0.01} = \frac{11n^2}{60} + 5 - 2n. \tag{20–5}$$

We reject H_0 at $\alpha = 0.01$ if $T \leq T_{0.01}$.

Now, an example is in order. Suppose that two auditing procedures are to be compared with ten pairs of auditors, matched by sex, and the number of accounts audited per hour are as given in columns two and three of Table 20–1; is there a significant difference in the two methods of auditing at $\alpha = 0.05$? at $\alpha = 0.01$?

The less frequent sign in this case is minus, and the sum of these minus

Table 20–1

Pair	*Procedure* 1	*Procedure* 2	d_i	*Rank of* d_i
1	109	61	48	10
2	100	78	22	9
3	77	57	20	8
4	69	64	5	5
5	59	60	−1	−1
6	69	75	−6	−6
7	87	85	2	2
8	88	84	4	4
9	88	85	3	3
10	91	83	8	7

ranks is

$$T = 1 + 6 = 7,$$

$$T_{0.05} = \frac{(10)^2 - 7(10) + 10}{5} = 8,$$

$$T_{0.01} = \frac{11(10)^2}{60} + 5 - 2(10) = 3.33.$$

Thus, H_0 is rejected at $\alpha = 0.05$, but it cannot be rejected at $\alpha = 0.01$.

When $n \geq 10$, the sampling distribution of T is approximately normal with

$$E(T) = \frac{n(n+1)}{4} \tag{20–6}$$

and

$$V(T) = \frac{n(n+1)(2n+1)}{24}. \tag{20–7}$$

Hence, for large samples we use

$$Z = \frac{T - E(T)}{\sigma_T} \tag{20–8}$$

as the testing statistic in the signed-rank test. For example, with data in Table 20–1, we have

$$E(T) = \frac{(10)(10+1)}{4} = 27.5,$$

$$V(T) = \frac{10(10+1)(2)(10)+1)}{24} = 96,$$

$$Z = \frac{7 - 27.5}{\sqrt{96}} = -2.09.$$

Thus, $P(T \leq 7) = N(-2.09) = 0.0183$. This result indicates again that H_0 can be rejected at $\alpha = 0.05$ but not at $\alpha = 0.01$. It is interesting to note that, according to Wilcoxon's table, the exact probability, $P(T \leq 7)$, for $n = 10$ is 0.0186. The normal value is an excellent approximation for this exact probability.

When ties occur in this test, we assign to each the mean of the ranks they would have had if all were different. Note also that equations for T_α are used for two-sided tests only, but (20–8) can be employed for both one-sided and two-sided tests.

Further Remarks

The nonparametric theory arose as early as the middle of the nineteenth century, and the well-known nonparametric method of testing independence and goodness-of-fit by the chi-square test was formulated by Karl Pearson in 1900. The true beginning of this theory, however, was Spearman's rank-correlation coefficient test for covariability, popularized by Harold Hotelling and Margaret R. Pabst in 1936. Even then there was little interest in nonparametric statistics until 1945, when Wilcoxon proposed a test for the two-sample case that was distinguished by both its simplicity and its excellent results, even in cases where the population distribution is known to be normal.

There are many reasons for the wide adoption and the continuous growth of the nonparametric theory since 1945. Theoretically, it is always desirable to discover procedures of statistical inference that have the minimum number of restrictions imposed by the underlying assumptions. Practically, it is important for scientific investigators to have at their disposal simple and yet widely applicable techniques.

Nonparametric statistics, being primarily concerned with order, or categorical, relations, provides us two important motives for its study and development. First, data may occur naturally in the form of ranks, or just categories. Such being the case, arithmetic differences between quantities provide no meaningful significance, and so direct ordinal analysis would suffice for reaching a judgment about the independence of underlying variables. Moreover, when the order relations in data are considered alone, the reasoning behind each nonparametric method requires relatively simple probability theory, so that the discrete sampling distributions can be easily found.

The foregoing remarks should not suggest that nonparametric methods are somehow superior to parametric decision procedures. In most situations where both the parametric and nonparametric techniques apply, the former is in fact distinctly more desirable than the latter on at least two counts.

First, in general, the nonparametric tests have less than 100 percent power efficiency as compared to the parametric tests. The concept "power efficiency" may be explained by a simple example. Suppose that we have two alternative testing procedures for a given H_0, A and B. If A requires $n = 25$ but B requires $n = 50$ in order to reach a power of, say, 0.95 for a given true alternative at $\alpha = 0.05$, then we say that the power efficiency of B is only 50 percent of that of A. It can be shown, for example, for normal populations, that the signed-rank test requires about 5 percent more scores than the t test in order to be equally powerful. Of course, when the assumption of normality is not met, the reverse may even be true.

Second, the null hypothesis tested by a nonparametric method is seldom equivalent to that tested by a parametric procedure. In general, with nonparametric tests, hypotheses are less precise and yield less information in conclusions. For example, when the usual assumptions for the linear model in the analysis of variance are met, we test the hypothesis that the treatment means are equal by the F ratio. In accepting this H_0, we have not only accepted the statement of equal means, but we have also implied that the treatment populations are identically distributed. When this H_0 is rejected, we know precisely that the populations are not all identical with respect to their central tendencies. The order method of comparing several samples, on the other hand, is concerned with the null hypothesis that the population distributions are the same. When this H_0 is rejected, we have a rather vague conclusion that the population distributions are not all identical. But are they different in terms of central tendency or in terms of variability? This question is left unanswered by the nonparametric tests. Thus, the departure from identity makes the nonparametric procedures highly insensitive. To be sure, this insensitivity can be reduced or even eliminated if we are willing to make some mild assumptions about population distributions. Such assumptions, however, may be just as offensive as those that the order statistics have been designed to avoid.

Some recent texts on nonparametric statistics include Kraft and Eeden, *A Nonparametric Introduction to Statistics* (Macmillan, 1968); G. E. Noether, *Elements of Nonparametric Statistics* (Wiley, 1967); and S. Siegel, *Nonparametric Statistics for the Behavioral Sciences* (McGraw-Hill, 1956).

20–6 ANALYSIS OF VARIANCE AND EXPERIMENTAL DESIGN

This text has only briefly introduced the analysis of variance—a technique that is amazingly versatile. It can be employed to analyze experimental designs with three or more factors. Furthermore, these designs may take the form of *fixed-effects*, *random-effects*, or *mixed-effects* models. The central focus of the analysis of variance is that it enables us to evaluate the merits of, say, three or more makes of machines, by designing the experiments in such a way that the possible effects of different types of raw materials and/or different machine operators on the performance of the machines can also be revealed.

Many references are available for this branch of statistics. The interested reader may benefit by starting with C. C. Li's simple and quite readable *Introduction to Experimental Statistics* (McGraw-Hill, 1964). Charles R. Hicks offers simple and realistic numerical examples in *Fundamental Concepts in the Design of Experiments* (Holt, Rinehart and Winston, 1964).

A systematic account of the methods from the viewpoint of the user and the consultant is given by O. Kempthorne's *The Design and Analysis of Experiments* (Wiley, 1952). A comprehensive treatment of the underlying principles of experimental design for behavioral scientists can be found in B. J. Winer's *Statistical Principles in Experimental Design* (McGraw-Hill, 1962).

20–7 MULTIPLE LINEAR REGRESSION ANALYSIS

The simple linear regression analysis introduced in Chapter 17 dealt with two variables. The main objectives were to determine the type and degree of association between the two variables and to arrive at an equation that might be used to predict the value of the dependent variable, Y, based on the value of the independent variable, X.

The simple linear regression model is very useful, but it ignores the possible effects of any other independent variable that might be significantly associated with the dependent variable. For example, if a simple linear regression analysis yields a coefficient of determination of, say, $r^2 = 0.65$, this means that only 65 percent of the variation in the values of the dependent variable is associated with the variation in the single independent variable. Thirty-five percent of the total variability in Y remains unexplained. Clearly, the total variability in Y can be explained more fully if appropriate additional independent variables are introduced.

When two or more independent variables are introduced in the regression analysis and if the dependent variable can be assumed to be linearly related to them, then we have what is called a multiple linear regression model.

If a dependent variable, Y, is associated with two independent variables, X_2 and X_3, the sample multiple regression equation would be given as

$$\bar{y}_c = a + bx_2 + cx_3, \tag{20–9}$$

which represents, geometrically, a fitted regression plane to the scatter in a three-dimensional space. In this expression, a is the Y intercept or the elevation of the fitted regression plane, and b and c are called *net*, or *partial, regression coefficients*. Precisely, b measures the average influence of X_2 on Y when X_3 is held constant, and c measures the average effect of X_3 on Y when X_2 is held constant.

To compute the constants in (20–9) by the method of least squares, we must solve the following set of normal equations simultaneously:

$$\text{(I)} \qquad na + b\sum x_2 + c\sum x_3 = \sum y,$$

$$\text{(II)} \qquad a\sum x_2 + b\sum x_2^2 + c\sum x_2x_3 = \sum yx_2,$$

$$\text{(III)} \qquad a\sum x_3 + b\sum x_2x_3 + c\sum x_3^2 = \sum yx_3.$$

Now, if we shift the origin from (0, 0, 0) to $(\bar{y}, \bar{x}_2, \bar{x}_3)$ by defining $d_1 = y - \bar{y}$, $d_2 = x_2 - \bar{x}_2$, and $d_3 = x_3 - \bar{x}_3$, the preceding set of equations will be reduced to

$$\text{(II)} \qquad b \sum d_2^2 + c \sum d_2 d_3 = \sum d_1 d_2,$$

$$\text{(III)} \qquad b \sum d_2 d_3 + c \sum d_3^2 = \sum d_1 d_3.$$

From these, we have

$$b = \frac{(\sum d_1 d_2)(\sum d_3^2) - (\sum d_1 d_3)(\sum d_2 d_3)}{(\sum d_2^2)(\sum d_3^2) - (\sum d_2 d_3)^2}, \qquad (20\text{–}10)$$

$$c = \frac{(\sum d_1 d_3)(\sum d_2^2) - (\sum d_1 d_2)(\sum d_2 d_3)}{(\sum d_2^2)(\sum d_3^2) - (\sum d_2 d_3)^2}. \qquad (20\text{–}11)$$

With the partial regression coefficients determined, the Y intercept of the fitted regression plane can be obtained by

$$a = \bar{y} - b\bar{x}_2 - c\bar{x}_2. \qquad (20\text{–}12)$$

As an example, suppose the sales manager of a certain firm believes that sales ability, among other factors, might be associated with salesmen's verbal reasoning ability and vocational interest. To verify this, ten salesmen are selected at random from his staff and given two sets of tests, one for verbal reasoning ability and the other for vocational interest. The

Table 20–2

Y	X_2	X_3	Y^2	X_2^2	X_3^2	YX_2	YX_3	X_2X_3
1	1	2	1	1	4	1	2	2
1	1	1	1	1	1	1	1	1
1	2	1	1	4	1	2	1	2
2	2	3	4	4	9	4	6	6
2	3	2	4	9	4	6	4	6
4	3	4	16	9	16	12	16	12
3	4	3	9	16	9	12	9	12
5	4	5	25	16	25	20	25	20
6	5	4	36	25	16	30	24	20
6	5	6	36	25	36	30	36	30
31	30	31	133	110	121	118	124	111

$\bar{y} = 3.1, \quad \bar{x}_2 = 3.0, \quad \bar{x}_3 = 3.1$

results are given in the first three columns of Table 20–2, where

Y = monthly average gross sales of salesman in thousands of dollars,
X_2 = score on verbal reasoning ability test, ranging from 1, very poor, to 5, excellent, and
X_3 = score on vocational interest test, ranging from 1, nil, to 6, exceedingly strong.

The sums of squares and of cross products are given in the rest of the table.

To estimate the population regression equation of Y on X_2 and X_3 for data on hand, we first compute the sums of squares and of cross products in terms of deviations from their respective mean values:

$$\begin{aligned}
\sum d_1^2 &= \sum (y - \bar{y})^2 = \sum y^2 - \bar{y} \sum y \\
&= 133 - 3.1(31) = 36.9;
\end{aligned}$$

$$\begin{aligned}
\sum d_2^2 &= \sum x_2^2 - \bar{x}_2 \sum x_2 \\
&= 110 - 3.0(30) = 20.0;
\end{aligned}$$

$$\begin{aligned}
\sum d_3^2 &= \sum x_3^2 - \bar{x}_3 \sum x_3 \\
&= 121 - 3.1(31) = 24.9;
\end{aligned}$$

$$\begin{aligned}
\sum d_1 d_2 &= \sum (y - \bar{y})(x_2 - \bar{x}_2) \\
&= \sum xy_2 - \bar{y} \sum x_2 \\
&= 118 - 3.1(30) = 25.0;
\end{aligned}$$

$$\begin{aligned}
\sum d_1 d_3 &= \sum yx_3 - \bar{y} \sum x_3 \\
&= 124 - 3.1(31) \\
&= 27.9;
\end{aligned}$$

$$\begin{aligned}
\sum d_2 d_3 &= \sum x_2 x_3 - \bar{x}_2 \sum x_3 \\
&= 111 - 3.0(31) = 18.0.
\end{aligned}$$

Next, by (20–10) and (20–11), we have

$$b = \frac{(25.0)(24.9) - (27.9)(18.0)}{(20.0)(24.9) - (18.0)^2}$$

= 0.6914 increase in sales in thousands of dollars per point increase in verbal reasoning ability test score when vocational interest score is held constant;

$$c = \frac{(27.9)(20.0) - (25.0)(18.0)}{(20.0)(24.9) - (18.0)^2}$$

= 0.6207 increase in sales in thousands of dollars per point increase in vocational interest test score when verbal reasoning ability score is held constant.

Finally,

$$a = 3.1 - 0.6914(3.0) - 0.6207(3.1)$$
$$= -0.8984 \text{ thousands of dollars.}$$

The estimated regression equation of sales ability on verbal reasoning ability and vocational interest can now be written as

$$\bar{y}_c = -0.8984 + 0.6914x_2 + 0.6207x_3.$$

This equation can be used by the sales manager to predict a new applicant's sales ability in terms of his test scores in verbal reasoning ability and vocational interest. For example, if an applicant's test scores are $X_2 = 3$ and $X_3 = 4$, his predicted sales ability would be

$$\bar{y}_c = -0.8984 + 0.6914(3) + 0.6207(4)$$
$$= 3.6586.$$

This means that given $X_2 = 3$ and $X_3 = 4$, a monthly average gross sales of \$3658.60 can be predicted for the salesman in question.

Let us mention also that the percentage of total variation in Y that is explained by variations in X_2 and X_3 is measured by the *coefficient of multiple determination*, denoted by R^2 and defined as

$$R^2 = \frac{\text{explained variation}}{\text{total variation}}$$

$$= \frac{\sum (y - \bar{y}_c)^2}{\sum (y - \bar{y})^2}$$

$$= \frac{b \sum d_1 d_2 + c \sum d_1 d_3}{\sum d_1^2}. \quad (20\text{–}13)$$

For our example, we have

$$R^2 = \frac{0.6914(25.0) + 0.6207(27.9)}{36.9}$$

$$= 0.9377.$$

Thus, more than 93 percent of the variation in sales ability is explained by verbal reasoning ability and vocational interest of salesmen. Compared to $r_{12}^2 = 0.8469$ and $r_{13}^2 = 0.8472$ for our illustrative data, multiple regression analysis is clearly an improvement over the simple linear regression of Y on X_2 or of Y on X_3 alone.

References for linear regression analysis are given together with those for nonlinear models at the end of the next section.

20-8 NONLINEAR REGRESSION ANALYSIS

While linear regression models may characterize a large class of associations among variables, sometimes the relationship between two or more variables can be adequately described only by some curvilinear function. From theory, we may know in advance that the relationship between two variables can be adequately represented only by some curvilinear mathematical function. For example, the liquidity preference theory suggests that the demand for money to hold as cash balances is a decreasing and nonlinear function of the interest rate. Again, the law of diminishing returns dictates that total production is positively and nonlinearly related to the variable input or inputs. Occasions may also arise in which theoretical considerations enable us to judge only whether the association between two variables should be positive or negative but not whether it is linear or otherwise. The laws of supply and demand illustrate this possibility. Here the usual procedure is to inspect the scatter diagram constructed with sample data; the general pattern of the scatter may suggest an appropriate form of regression, whether a linear or some type of nonlinear function.

As in linear regression analysis, nonlinear relationships may be simple as well as multiple. Among the simple curvilinear regression models are the second-degree polynomial function, for which the sample regression equation is given as

$$\bar{y}_c = a + bx + cx^2, \tag{20-14}$$

and the exponential function, which is of the form

$$\bar{y}_c = ab^x. \tag{20-15}$$

Both these functions were discussed in detail in Chapter 18. Two other simple curvilinear regression models, the reciprocal and the power functions, are also of great interest, and we shall now comment on them briefly.

Scatter diagrams for, say, sample data with reference to the demand for and supply of a commodity, Y, may indicate that the appropriate regression equations should be, respectively,

$$\bar{y}_c = a + \frac{b}{x} \tag{20-16}$$

and

$$\bar{y}_c = a - \frac{b}{x}, \tag{20-17}$$

where X may stand for the price of Y. We note that in the case of (20-16) the slope of the function is everywhere negative and decreases in absolute value as X increases. As X approaches zero, Y approaches ∞; as X approaches ∞, Y approaches a. For case (20–17), the function is everywhere positive, and Y approaches a as X approaches ∞.

For the functions above, a *reciprocal transformation* would make the determination of a and b quite simple. If we define a new variable $Z = 1/X$, then (20–16) and (20–17) become

$$\bar{y}_c = a + bZ \tag{20–18}$$

and

$$\bar{y}_c = a - bZ, \tag{20–19}$$

respectively. These are now simple linear regressions of Y on Z. Procedures developed in Chapter 17 can be applied to obtain a and b. After a and b are found, the regression of Y on Z can easily be converted to the original forms by noting that $Z = 1/X$.

Sometimes a power function such as

$$\bar{y}_c = ax^b \tag{20–20}$$

may represent the relationship between Y and X in the sample. Here, we would like to find the unknown power b. For this function, if $b > 1$, Y increases at an increasing rate when X increases. If $0 < b < 1$, the rate of increase of this function decreases continuously. For either value of b, Y approaches ∞ as X approaches ∞. When $b = -1$, this function is represented by a rectangular hyperbola—a curve for which the locus of a point is such that the product XY is a constant.

To facilitate the fit of a power regression curve, we take logarithms of both sides of (20–20), obtaining

$$\log \bar{y}_c = \log a + b(\log x). \tag{20–21}$$

This is the result of a double-log transformation, since both X and Y are transformed to logarithms. Now, if we let

$$\log \bar{y}_c = \bar{z}_c,$$

$$\log a = c,$$

and

$$\log x = w,$$

we can rewrite the preceding expression as

$$\bar{z}_c = c + bw, \tag{20–22}$$

which is a simple linear regression equation of Z on W. Hence, for actual computations, we take logarithms of both X and Y values and proceed to determine the regression coefficients c and b in the usual manner.

A multiple sample regression may take the form

$$\bar{y}_c = ax_2^b x_3^c. \tag{20–23}$$

With this regression equation, we may compute the sample partial regression coefficients again by applying the principle of least squares to the logarithmic values of the variables. Namely, taking logarithms of both sides of (20–23), we may transform the original multiplicative expression into an additive model:

$$\log \bar{y}_c = \log a + b(\log x_2) + c(\log x_3). \tag{20–24}$$

If we let

$$\log \bar{y}_c = \bar{z}_c, \quad \log x_2 = z_2, \quad \log x_3 = z_3, \quad \text{and} \quad \log a = d,$$

then (20–24) can be written as

$$\bar{z}_c = d + bz_2 + cz_3, \tag{20–25}$$

which is linear in all variables. As a consequence, we can apply the linear methods developed in the last section to the logarithms of Y, X_2, and X_3 to find the values of the constants in (20–25). Once d, b, and c are calculated, the sample regression equation in the form of (20–25) can be converted back to the original form of (20–23) by noting the definitions of Z's and that $d = \log a$.

Double-log transformations are frequently used by economists because they correspond to the assumption of constant elasticity between the dependent and independent variables, and the simple application of linear methods to logarithms of the variables gives a direct estimate or estimates of elasticities. For example, the production function of the United States as a whole, estimated by Paul H. Douglas and C. W. Cobb, is

$$P = 1.10L^{0.75}C^{0.25},$$

where P is the production index of total output per year, L is the index of labor inputs, and C is the index of capital inputs. Sample indexes for this estimate ran from 1900 to 1922, with 1899 = 100. We note here that the estimated exponents are the elasticities of production with respect to labor and capital. The first tells us that if there is a 1 percent increase in labor inputs with capital held constant, there will be 0.75 percent increase in production. Similarly, the second indicates that there will be 0.25 percent increase in production if there is a 1 percent increase in capital with labor held constant.

We see that in treating curvilinear relationships we usually search for an initial transformation of data, by way of either reciprocals or logarithms, in such a manner that the relationship between the transformed variables appears to be approximately linear. Such transformations are made mainly for two reasons. First, as the reader may have already guessed, this procedure is to make calculations easier, since computational problems in nonlinear systems may be quite formidable. Second, transformations are desirable because statistical theory is developed primarily under the assumption of linearity.

A lucid discussion on the important topics of linear as well as nonlinear association analysis can be found in J. Johnston's *Econometric Methods* (McGraw-Hill, 1963), Chapters 1–4. For more comprehensive coverage the reader may consult Draper and Smith, *Applied Regression Analysis* (Wiley, 1966), and Ezekiel and Fox, *Methods of Correlation and Regression Analysis*, 3d ed. (Wiley, 1959).

20–9 INDEX NUMBERS

Conspicuously absent from our presentation thus far are index numbers, which are popular and useful in measuring the quantity of some heterogeneous aggregate. The aggregate may be one of physical quantity, such as stocks in the balance sheet or flows in the income statement. It may also be a list of prices, such as prices paid for the purchase of various types of inputs by a firm or prices received by a department store from its sales. In each case, the problem is to derive a single figure that is descriptive of the volume of, or change in, a given aggregate through time or from one place to another. *Index* numbers, in effect, relate a variable or variables in a given period to the same variable or variables in another period, called the *base period.* An index, the simplified name for index numbers, computed from a single variable, is called a *simple index,* whereas an index constructed from a group of variables is a *composite index.*

A simple index is easy to construct and its meaning is readily apparent. Consider now the hypothetical data in Table 20–3. Index A is computed

Table 20–3. Illustration of Simple Production Index

Year	*Production (thousands of pounds)*	*Index A* (1966 = 100)	*Index B* (1960–1961 = 100)
1960	8.5	63	81
1961	12.5	93	119
1962	9.4	70	90
1963	10.7	79	102
1964	13.6	101	130
1965	15.3	113	146
1966	13.5	100	129
1967	12.8	95	122
1968	14.7	109	140
1969	16.7	124	159
1970	18.0	133	171

by selecting 1966 as the base period. The index numbers are then found by dividing each year's production figure by that for 1966 and multiplying by 100. Thus, the index numbers are presented in the form of percentages. They can, if we wish, be given also as proportions. Notice that the index number of each year is interpreted in terms of the base period. For example, the 1970 index means that production in that year is 133 percent of 1966 production. Index B in the same table is computed by using the average production of 1960 and 1961 (that is, 10.5 thousand pounds) as the base. This base is divided into each year's production, and the results are multiplied by 100.

Serious problems arise when we attempt to combine many variables in a meaningful way to arrive at a composite index. In general, there are two alternative methods for the construction of composite indices: aggregate and relative. An *aggregate index* is one constructed by pricing out quantities of goods and services. Namely, it is the result of dividing an aggregate (in dollars) at one date by an aggregate at another date. For an index constructed with an *average of relatives* a number of separate index numbers are computed and then combined by weighted averages.

For the construction of composite indices two basic formulas are available: Laspeyres and Paasche. Most index numbers currently in use have been constructed with one of these methods, with or without some modifications. For example, an aggregate Laspeyres price index is defined as

$$L = \frac{\sum p_n q_0}{\sum p_0 q_0}, \tag{20–26}$$

and the corresponding Paasche index is

$$P = \frac{\sum p_n q_n}{\sum p_0 q_n}, \tag{20–27}$$

where p and q stand for the prices and quantities of the variables under consideration and the subscripts 0 and n refer to the base-period and given-period quantities, respectively.

Notice that L and P differ, owing to different weighting systems. The Laspeyres index uses base-period quantities as weights, while the weights for the Paasche index are given-period quantities. P and L not only produce different results but also provide different types of information. Specifically, the Laspeyres index seeks to answer the question: What is the change in the aggregate value of the base period's list of goods and services when valued at given-period prices? With reference to the consumer price index, for example, L measures the change in the cost of the family's base-period way of life. In contrast, the Paasche index attempts to answer the question: What would be the value of the given period's list of goods and services when valued at base-period prices? Again, referring to a consumer price index, P measures the base-period cost of the family's given-period way of life. Thus, L prices the base-period budget; P prices the given-period budget. Also, as can be seen from the computations given below, $L > P$. In general, it may be said that L is a maximum estimate of the change in the price level from the base period to the given period; P is a minimum estimate. This difference, however, is not fortuitous. It makes no sense to ask which formula is more accurate or better. Each is meaningful in the sense that it has a simple and precise physical interpretation. Indeed, we can actually consider the difference as a meaningful range of values delimiting the true change in the level of prices.

Now, suppose we wish to construct a price index for data given in Table 20–4 with 1965 = 100. We have

$$L_{1970} = \frac{15(100) + 4(250) + 2(50)}{10(100) + 5(250) + 1(50)}$$

$$= 1.13 \text{ or } 113 \text{ percent}$$

and

$$P_{1970} = \frac{15(90) + 4(300) + 2(45)}{10(90) + 5(300) + 1(45)}$$

$$= 1.08 \text{ or } 108 \text{ percent.}$$

Table 20–4. Hypothetical Data for Constructing Price Indices

	Price		*Quantity*	
Item	1965	1970	1965	1970
A	$10.00	$15.00	100	90
B	5.00	4.00	250	300
C	1.00	2.00	50	45

Though somewhat abused, index numbers, especially price indices, are high on the list of statistical descriptive measures in everyday use. One important employment of price indices is for the deflation of value series. For instance, when GNP is divided by an appropriate price index, we have what is called GNP in constant dollars; when money income is divided by an appropriate price index, we have real income; and so on. In each instance a deflated value series represents the real change in that series from the base period, since changes in the series due to changes in prices are eliminated.

A person who must construct index numbers needs a good reference on the subject. A short but highly instructive book on the methods of constructing index numbers in W. M. Persons' *The Construction of Index*

Numbers (Houghton-Mifflin, 1928). This volume gives special attention to the problems of bias resulting from the type of average used and from correlation between quantities and prices. The definitive work on index numbers in Irving Fisher's *The Making of Index Numbers* (Houghton-Mifflin, 1927), which discusses tests of reliability for many different formulas for computing index numbers. A more recent text is B. D. Mudgett's *Index Numbers* (Wiley, 1951). A good discussion on the difficulty and limitations of constructing price indices can be found in Joint Economic Committee, 87th Congress, 1st Session: *Government Price Statistics: Hearings before the Subcommittee on Economic Statistics* (Government Printing Office, 1961). If one wants merely an introductory knowledge of the significance of index numbers and some of their limitations, either Chapter 11 in Yamane's *Statistics* or Chapter 15 in this author's *Statistical Analysis* (both cited earlier) should suffice. Also, business and economics students should become acquainted with the common index numbers currently in use by consulting *Survey of Current Business*, a monthly publication of the U. S. Department of Commerce.

20–10 ADDITIONAL REFERENCE WORKS

To complete our list of references, we might cite some sources for operations research, game theory, and time-series analysis and forecasting.

Operations research is concerned with the application of mathematical and statistical methods to the study of the operations of large, complex organizations or activities. Its objective is to provide top-level administrators with a quantitative basis for decisions that will increase the effectiveness of such organizations in carrying out their basic objectives. A book recommended for its rigor and comprehensive coverage is H. M. Wagner, *Principles of Operations Research* (Prentice-Hall, 1970). This volume requires a considerable knowledge of mathematics. Wagner also lists further references in this field.

Statistical decision theory is concerned with problems where the consequences of various acts taken by the decision maker are beyond his control; these consequences are treated as the states of nature. Methods for dealing with decision situations involving the competitive actions of rational opponents are part of game theory. Both kinds of problems can be viewed as different aspects of a large set of situations to which the overall title "decision theory" is often applied. There are several good texts on game theory. One that surveys the central ideas and results of game theory and related decision-making models without using excessive mathematical detail is *Games and Decisions*, co-authored by R. D. Luce and H. Raiffa (Wiley, 1957).

A comprehensive, clear presentation of the classical approach to time-series analysis can be found in Croxton, Cowden, and Klein, *Applied General Statistics*, 3d ed. (Prentice-Hall, 1967). Chapter 22 in this volume also discusses correlation in time series. An excellent but rather advanced treatment of some topics in time-series analysis is given in Part 3 of Gerhard Tintner's *Econometrics* (Wiley, 1952). On forecasting, the reader may consult R. D. Brown, *Smoothing, Forecasting, and Prediction* (Prentice-Hall, 1963), and Spencer, Clark, and Hoguet, *Business and Economic Forecasting* (Irwin, 1961). Additional references can be found in these two volumes.

To keep up with current developments, the reader may wish to study regularly some journals in statistics, such as the *Journal of the American Statistical Association*, *Econometrica*, and *Review of Economics and Statistics*.

The reader may also find the following set of general reference works and tables a valuable addition to his personal library: Buckland and Fox, *Bibliography of Basic Texts on Statistical Methods* (Hafner, 1963); Hauser and Leonard, *Government Statistics for Business Use*, 2d ed. (Wiley, 1956); E. B. Cox, *Basic Tables in Business and Economics* (McGraw-Hill, 1966); Freund and Williams, *Dictionary/Outline of Basic Statistics* (McGraw-Hill, 1966); and Rand Corporation, *A Million Random Digits with 100,000 Normal Deviates* (Macmillan, 1955).

APPENDIX A

SETS, PERMUTATIONS, COMBINATIONS, AND SUMMATIONS

A–1 THE NOTATION OF A SET AND WAYS OF SPECIFYING A SET

A "set" is such a fundamental concept that the whole field of contemporary mathematics can be derived from it and from certain rules of logic. A *set* is a well-defined collection of things, objects, people, or symbols. "Well-defined" means that there must be no doubt whatsoever whether or not a given item belongs to the set under consideration.

The objects that belong to a set are called the *members*, or *elements*, or *elementary events*, of that set. A set is said to be *finite* if the number of its members is a positive integer or zero; otherwise the set is *infinite*.

We will use the convention of denoting different sets by capital letters, such as A, B, C. Furthermore, if x is some object that is a member of the set, say, S, we then write

$$x \in S,$$

which is read "x is an element of the set S" or "x belongs to the set S." The notation "$\in$" is referred to as an inclusive symbol. To denote that some object x is not a member of the set, say, A, we write

$$x \notin A.$$

There are, in practice, two ways to specify a set: the "roster" and the "rule" methods.

The *roster method* specifies a set by listing the names of all members of the set. For example, by this rule the set C—the outcomes of tossing a coin—may be presented as

$$C = \{H, T\}.$$

Again, the set D, consisting of the number of dots on the face of a die, is

$$D = \{1, 2, 3, 4, 5, 6\}.$$

This method is clear-cut and convenient when the number of elements in a set is small. It becomes cumbersome, however, when the number of elements in the set is at all large.

The *rule method* specifies a set by stating the requirements that any item must meet in order to be a member of the set. A rule of membership may be given in words. For example,

"D is the set of all elements x such that x is an integer from one to six."

In a more precise and compact form,

$$D = \{x\colon x \text{ is an integer and } 1 \leq x \leq 6\},$$

where the ":" is read "such that"; on its left side is a symbol for an arbitrary member of the set, and on the right the rule that defines membership in the set.

A coin and a die are tossed. Let x stand for the face of the coin and y for that of the die; then S, the set of all possible pairs (x, y) that the coin and die may show, can be specified either as

$$S = \{(H, 1), (H, 2), \ldots, (H, 6), (T, 1), \ldots, (T, 6)\},$$

by the roster method, or as

$$S = \{(x, y) : x \text{ is } H \text{ or } T \text{ and } y \text{ is an integer from 1 to 6}\},$$

by the rule method.

A-2 UNIVERSAL SET AND SUBSETS

The *universal set* in any discussion is the totality of all members under consideration. From the universal set we may select special subsets of interest. The set A is said to be a subset of the set B if every element of A is an element of B. Consider, for example,

$$U = \{x\colon x \text{ is a letter in the English alphabet}\};$$

$$A = \{e\colon e \text{ is a vowel in the English alphabet}\};$$

$$B = \{x, y, z\}.$$

Sets A and B are subsets of the universal set U, and we write

$$A \subset U \quad \text{and} \quad B \subset U.$$

The notation "$\subset$" is read "is a subset of" and is an inclusion symbol. Be sure to distinguish between the two statements

$$e \in A \quad \text{and} \quad A \subset B.$$

The former reveals that e is an element contained in A; the latter indicates that A is a set with all its elements contained in B; that is, $A = \{e\} \subset B$.

It is often instructive to give a pictorial representation of a set and the relationship between the universal set and its subsets. Usually an arbitrary plane figure represents a set. All elements of the set are thus considered to be points contained within the boundary of the figure. See Figure A–1(a). A subset can be represented by a closed region in the plane figure that represents the universe. See Figure A–1(b). Such pictures are called *Venn diagrams* after the English logician John Venn (1834–1883).

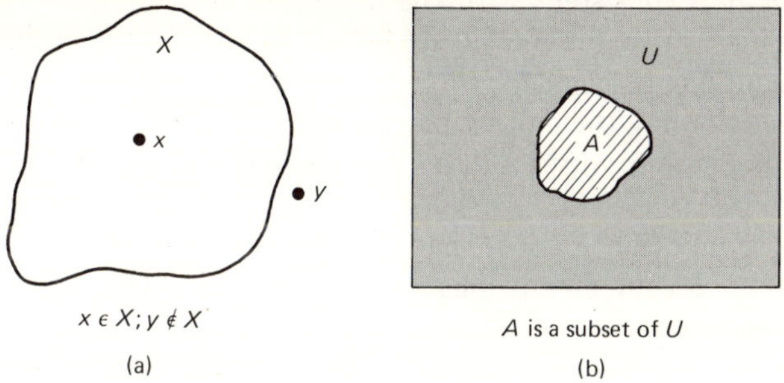

Figure A–1 Venn Diagrams for Membership and Subset

A number of implications and relationships between sets are of interest in statistical analysis. These are listed below.

(1) The meaning of a universal set is consistent with our definition of a statistical population in Chapter 1, or with the "sample space" introduced in Chapter 2.

(2) A subset can be defined not only on a universal set but also on another subset. For example, consider

U: the set of American citizens,
A: the set of all naturalized American citizens, and
B: the set of all Oriental-born American citizens.

Clearly, A is a subset of U, and B is a subset of A. The Venn diagram for these sets is given in Figure A–2.

(3) By definition, every set may be considered as a subset of itself. In particular, a universe is a subset of the universe.

(4) The *null set*, defined as the set that contains no element and denoted as $\varnothing$, is a subset of any set. There are two reasons for this convention.

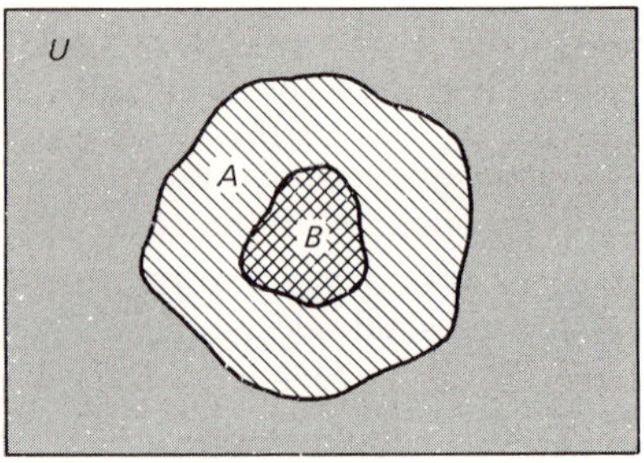

Figure A–2 Venn Diagrams Showing that *B* is a Subset of *A* and that *A* is a Subset of *U*

First, it is consistent with the definition of a subset, which requires that every element of a subset belong to another set S. Since the null set contains no element, there is no element in $\varnothing$ to violate this condition. Second, the null set plays a role analogous to that of zero in the number system: if the null set is combined with any set S, the result is S.

It is important to note that $\varnothing$ is not the same as $\{0\}$. The former is empty—devoid of any elements, while the latter does contain an element—zero.

(5) Two sets are said to be *equal* if they contain identical elements. For example, the sets

$$A = \{1, 2, 3, a, b, c\}$$

and

$$B = \{1, a, 3, b, 2, c\}$$

are equal; $A = B$. Note that the order of appearance of the elements in a set is immaterial. Note also that two sets of the same size are not necessarily equal, but two equal sets must be of the same size.

(6) Two sets A and B are said to be *disjoint*, or *mutually exclusive*, if they have no elements in common. For example, the sets consisting of elements in both

$$A = \{a, e, i, o, u\}$$

and

$$B = \{x, y, z\}$$

is $\varnothing$.

(7) Sets, in statistical analysis, may also be called events. In particular, a universal set is a *certain event*, or *certainty*, and the null set is an *impossible event*.

(8) It can be shown that if U is a finite set with N elements, then there are 2^N different subsets of U, including U and $\varnothing$.

(9) All sets, including U and $\varnothing$, are called events in probability theory.

A–3 OPERATIONS ON SETS

Given a universal set U and its subsets, $A, B, C, \ldots$, we can perform on these subsets certain operations that produce other, or perhaps the same, subsets. Among all the operations on sets, we are interested only in "complementation," "union," and "intersection."

Complement: The *complement* of any set A with respect to any stated or understood universal set is the set of elements of the universal set that do not belong to A. Denoting the complement of A by $\bar{A}$, we have

$$\bar{A} = \{x : x \in U \text{ but } x \notin A\}.$$

The complement of A is shown by the shaded area in Figure A–3.

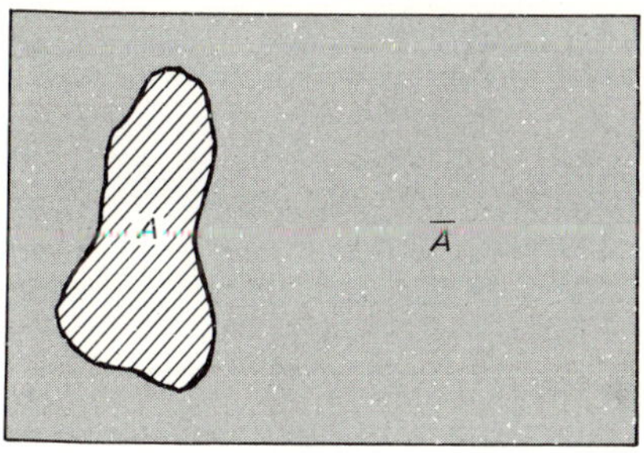

Figure A–3 Venn Diagram Showing that $\bar{A}$ is the Complement of A

Let us return to the example in which a coin and a die are tossed. Here, the universe is

$$U = \{(x, y) : x \text{ is } H \text{ or } T \text{ and } y \text{ is an integer from 1 to 6}\}.$$

Now, if we define

$$A = \{(x, y) : x \text{ is } H \text{ and } y \text{ is an integer from 1 to 6}\},$$

then $$\bar{A} = \{(x, y) : x \text{ is } T \text{ and } y \text{ is an integer from 1 to 6}\}.$$

Again, let U consist of the numbers 1, 2, 3, . . . , 9 and the 26 letters of the alphabet, $a, b, \ldots, z$. If

$$B = \{1, 2, 3, 4, 5, a, b, c, d, e, f\},$$

then $$\bar{B} = \{6, 7, 8, 9, g, h, i, \ldots, z\}.$$

In the case of the complement, it is particularly crucial to make the universal set explicitly known.

Union: The *union* of two sets A and B is the set of elements that belong to at least one of the sets A and B—that is, to A or B. We symbolize this

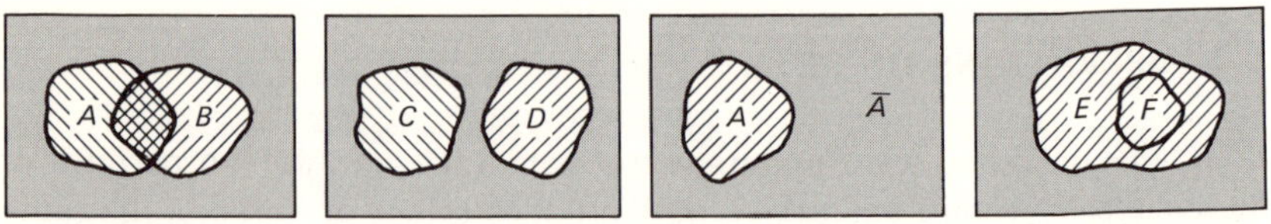

$A \cup B$ $C \cup D$ $A \cup \bar{A}$ $E \cup F$

Figure A–4 Venn Diagrams for Unions

operation by $A \cup B$ (read "A union B" or "A cup B"), and we write

$$A \cup B = \{x: x \in A \text{ or } x \in B \text{ or } x \text{ belongs to both } A \text{ and } B\}.$$

The union of A and B is indicated in the shaded area in Figure A–4. Consider, for example,

$$A = \{a, b, c, d, e, f\}$$

and $$B = \{a, e, i, o, u\},$$

both of which are subsets of the set of all letters in the English alphabet. The set whose elements are either in A or in B or in both is the set

$$C = A \cup B = \{a, b, c, d, e, f, i, o, u\}.$$

Again if,

$$A = \{x: x \text{ is a positive odd integer}\}$$

and $$B = \{x: x \text{ is a positive odd integer less than } 100\},$$

then $$A \cup B = \{x: x \text{ is a positive odd integer}\}.$$

Intersection: The intersection of two sets A and B is the set of elements that belong to both A and B. We symbolize this operation as $A \cap B$ (read "A intersection B" or "A cap B"), and we write

$$A \cap B = \{x: x \in A \text{ and } x \in B\}.$$

The intersection of A and B is indicated by the shaded area in Figure A–5.

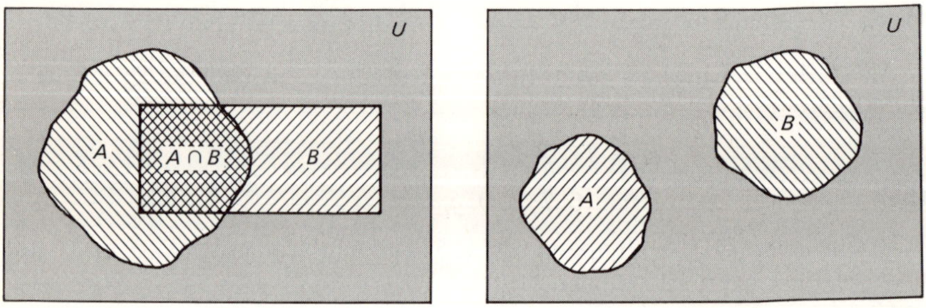

$A \cap B = \emptyset$

(a) (b)

Figure A–5 The Intersection of *A* and *B*

Associated with the operation of union is the notion of partitioning, which is extremely useful in probability theory. The sets $S_1, S_2, \ldots, S_n$ are said to *partition* (or form a *partition* of) the set A if

(a) $S_i \subset U; i = 1, 2, \ldots, n;$
(b) $S_i \cap S_j = \varnothing; i, j = 1, 2, \ldots, n$ and $i \neq j$; and
(c) $S_1 \cup S_2 \cup S_3 \cup \cdots \cup S_n = A.$

In other words, a group of disjoint (mutually exclusive) subsets partition the set that is their union, which may be the universal set. See Figure A–6.

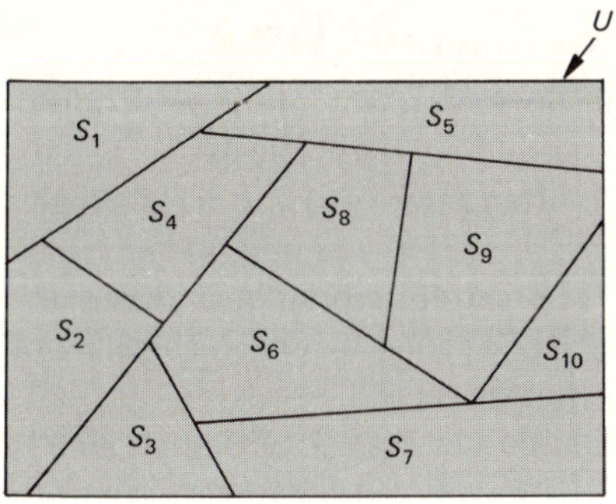

Figure A–6 Partition of a Universal Set

A–4 ALGEBRA OF SETS

Having defined the operations of complement, union, and intersection for sets, we may formulate an "algebra of sets." Many of the basic laws of set theory are analogous with the rules of algebra of real numbers. Some of the important ones are given here without proof, but most of these are intuitively plausible, and the reader, as an exercise, can verify their validity by applying them to sets of their own choices.

Commutative laws: Unions and interestion of sets are commutative; that is, for any sets A and B

$$A \cup B = B \cup A \tag{A–1}$$

and

$$A \cap B = B \cap A. \tag{A–2}$$

Associative laws: Unions and intersection of sets are *associative*; that is, for any sets A, B, and C we have

$$(A \cup B) \cup C = A \cup (B \cup C) \tag{A–3}$$

and
$$(A \cap B) \cap C = A \cap (B \cap C). \tag{A–4}$$

Distributive laws: In the algebra of sets, intersection can be distributed over union and union can be distributed over intersection; that is,

$$A \cap (B \cup C) = (A \cap B) \cup (A \cap C) \tag{A–5}$$

and
$$A \cup (B \cap C) = (A \cup B) \cap (A \cup C). \tag{A–6}$$

Identity laws: The union of any set S and the null set $\varnothing$ is the same set S; that is,

$$S \cup \varnothing = S. \tag{A–7}$$

The intersection of any set S and the universal set U is the same set S; that is,

$$S \cap U = S. \tag{A–8}$$

We note the similarities of addition and multiplication in the algebra of real numbers to union and intersection in the algebra of sets. Because of these similarities, we often refer to $A \cup B$ as the logical sum, and to $A \cap B$ as the logical product.

Idempotent laws: The careful reader may have already noted from the distributive laws that the analogy between the algebra of real numbers and the algebra of sets is not perfect. This distinction can be further reinforced by considering the idempotent laws. If A is any real number, then $A + A = 2A$ and $A(A) = A^2$. However, if A is any set, then we have two idempotent laws:

$$A \cup A = A \tag{A–9}$$

and
$$A \cap A = A. \tag{A–10}$$

Complement laws: For every subset S of a universal set U there is one and only one complement of S, namely, $\bar{S}$. The union and intersection of S and $\bar{S}$ have the following three properties:

$$S \cup \bar{S} = U, \tag{A–11}$$

$$S \cap \bar{S} = \varnothing, \tag{A–12}$$

and
$$\bar{\bar{S}} = S, \tag{A–13}$$

where $\bar{\bar{S}}$ stands for the complement of the complement of S.

De Morgan's laws: These laws state that

$$\overline{(A \cup B)} = \bar{A} \cap \bar{B} \tag{A–14}$$

and
$$\overline{(A \cap B)} = \bar{A} \cup \bar{B}. \tag{A–15}$$

Verbally, (A–14) says that the complement of $A \cup B$ is equal to the intersection of the complement of A with the complement of B, and (A–15) says that the complement of $A \cap B$ is equal to the union of the complement of A and the complement of B.

A-5 THE MULTIPLICATIVE PRINCIPLE

In discussing the theory of sets, we noted that the number of subsets that can be formed from a universe of N elements is 2^N. But we did not explain how to reach this conclusion. Also, in discussing probability theory we need an efficient way of counting the number of ways of arranging a set of objects, especially when the objects are numerous. For these problems we must consider what are called permutations and combinations, both of which are based upon a simple concept—the principle of multiplication.

The principle of multiplication states: If an operation can be performed in n_1 ways, and then, after it is performed in any one of these ways, a second operation can be performed in n_2 ways, and, after it is performed in any one of these ways, a third operation can be performed in n_3 ways, and so on for k operations, then the k operations can be performed in

$$(n_1)(n_2)(n_3)\cdots(n_k) \text{ ways.} \qquad \text{(A–16)}$$

For example, in how many ways can three letters A, B, and C be arranged in order?

One way to solve this problem is to list the possible arrangements and count them. A tree diagram, Figure A–7, provides an organized way of listing the arrangements so that none is missed. Here the initial point, or origin, is denoted as O. If we follow all branches starting with O from left to right of the tree, we set the six possible arrangements listed in the column on the extreme right. Or, by the principle of multiplication, we have

$$(n_1)(n_2)(n_3) = (3)(2)(1) = 6 \text{ ways.}$$

A coin and die are tossed; in how many ways can they fall?

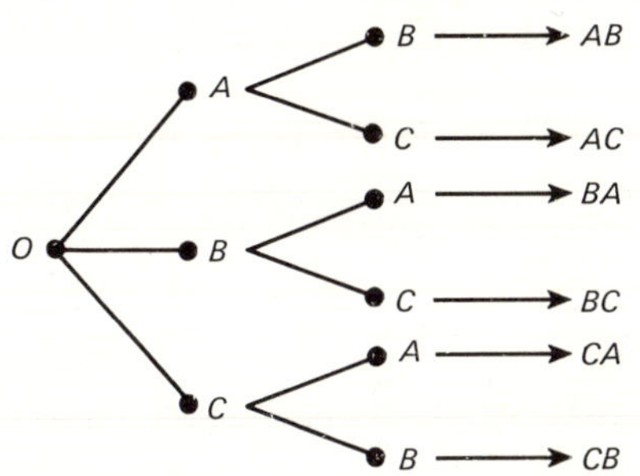

Figure A–7 Tree Diagram for Number of Ways of Arranging Three Letters, *A*, *B*, and *C*

There are two ways for a coin to fall, T or H, and for each outcome of the coin, the die can fall six different ways. Hence,

$$(n_1)(n_2) = (2)(6) = 12 \text{ ways.}$$

If a department store has five entrances and six exits, in how many ways can one enter and leave the store?

$$(n_1)(n_2) = (5)(6) = 30 \text{ ways.}$$

Now, given N elements in the universal set, for each there are two ways to be selected—to take it or to leave it. Hence, by the principle of multiplication, we have

$$\underbrace{(2)(2)\cdots(2)}_{N \text{ terms}} = 2^N.$$

A–6 PERMUTATIONS

A permutation of a number of objects is an arrangement of these objects in a definite order. The number of permutations of a set of N things, taken all together, is $N!$. Denoting this number by ${}_NP_N$, we have

$$ {}_NP_N = N!, \tag{A–17}$$

where $N!$ is read "N factorial" and is the product of all whole integers from 1 to N; that is,

$$N! = N(N-1)(N-2)\cdots(3)(2)(1). \tag{A–18}$$

In particular, we see that

$$1! = 1,$$

$$2! = (2)(1) = 2,$$

$$3! = (3)(2)(1) = 6,$$

$$4! = (4)(3)(2)(1) = 24.$$

We also note that

$$10! = 10(9!),$$

$$100! = 100(99!),$$

$$(N+1)! = (N+1)N!,$$

or, in general, if $N \geq r$, then

$$N! = r!(r+1)(r+2)\cdots(N-1)(N). \tag{A–19}$$

For example,

$$\begin{aligned} 100! &= 99!(100) \\ &= 98!(99)(100) \\ &= 90!(91)(92)\cdots(99)(100) \\ &= 11!(12)(13)\cdots(99)(100). \end{aligned}$$

Finally, we also define $0! = 1$.

Equation (A–17) is the consequence of the direct application of the multiplicative principle. We have, in permutation, N spaces to fill. The first space can be filled with any one of the N objects, and so in N ways. After this has been done (in any of the N ways), the second space can be filled with any of the remaining $N - 1$ objects. Likewise, the third place can be filled in $N - 2$ ways, the fourth space in $N - 3$ ways, and so forth. Therefore, by the multiplicative principle and the definition of the factorial sign, the number of ways of filling N spaces is $N!$.

In how many different ways can the ten digits from 0 to 9 be arranged by taking all of them at a time? Answer:

$${}_{10}P_{10} = (10)(9)(8)\cdots(2)(1) = 3{,}628{,}800.$$

The total number of arrangements of objects taken n at a time from N objects, with $n \leq N$, denoted as ${}_NP_n$, is

$${}_NP_n = \frac{N!}{(N-n)!}. \qquad \text{(A–20)}$$

Again, by the multiplicative principle, we see that n spaces can be filled in

$${}_NP_n = N(N-1)(N-2)\cdots(N-n+1) \text{ ways.}$$

The righthand side of this expression consists of n factors; if this expression is multiplied by $(N-n)!/(N-n)!$, we obtain the convenient expression (A–20) for ${}_NP_n$.

Note that if $N = n$, we have

$${}_NP_n = \frac{N!}{0!} = N!.$$

Four signal flags are to be run up, one above the other, on a flagpole. How many different signals can be run up using six different flags four at a time?

$$\begin{aligned} {}_6P_4 &= \frac{6!}{(6-4)!} = \frac{2!(6)(5)(4)(3)}{2!} \\ &= 360. \end{aligned}$$

A-7 COMBINATIONS

A *combination* is a selection of objects considered without regard to their order. The *total number of combinations* of a set of N things taken n at a time, $N \geq n$, usually denoted by ${}_NC_n$ or by $\binom{N}{n}$, is

$$ {}_NC_n = \binom{N}{n} = \frac{N!}{n!(N-n)!}. \tag{A-21} $$

To evaluate (A-21), we may consider the list of possible selections of two letters at a time from three letters, A, B, and C. By rearranging, we get two permutations from each of the three selections

$$ AB, \quad AC, \quad \text{and} \quad BC, $$

since each 2-subset can be arranged $2!$ ways. This operation yields a total of $3(2!) = 6$ permutations, as listed below.

Combinations ${}_3C_2$	*Permutations* ${}_3P_2$
AB	AB, BA
AC	AC, CA
BC	BC, CB

Clearly, all 6 permutations of 3 letters, taken 2 at a time, are obtained by arranging the 3 combinations. That is,

$$ (\text{number of combinations})\,(2!) = (\text{number of permutations}), $$

or, in symbols,

$$ {}_3C_2(2!) = {}_3P_2. $$

Hence,

$$ {}_3C_2 = \frac{{}_3P_2}{2!} = \frac{3!/(3-2)!}{2!} $$

$$ = \frac{3!}{2!(3-2)!} = 3. $$

It is important to repeat that in a permutation, order counts; in a combination, order does not count. The equivalent general expression is

$$ {}_NP_n = {}_NC_n({}_nP_n). \tag{A-22} $$

A traveling basketball squad has 10 players. The coach must select a starting team for the forthcoming game. How many different teams of 5 players can be designed for this purpose? Here we are not interested in the positions each of the five players in each team is to take. It is therefore a problem of combination, and

$$
\begin{aligned}
{}_{10}C_5 &= \frac{10!}{5!(10-5)!} \\
&= \frac{5!(6)(7)(8)(9)(10)}{5!(5)(4)(3)(2)(1)} \\
&= 252 \text{ combinations.}
\end{aligned}
$$

If, in selecting the team, the coach also has to designate positions, then order counts and the problem is one of permutation. Hence,

$$
\begin{aligned}
{}_{10}P_5 &= \frac{10!}{(10-5)!} \\
&= \frac{5!}{5!}(6)(7)(8)(9)(10) \\
&= 30{,}240 \text{ permutations.}
\end{aligned}
$$

Note that, when a team has been selected, the number of ways the 5 players can be assigned is

$${}_5P_5 = (5)(4)(3)(2)(1) = 120 \text{ ways.}$$

Thus,

$$30{,}240 \text{ assignments} = (252 \text{ teams})(120 \text{ assignments per team}).$$

In closing our discussion on combinations, we may observe the following relationships:

(a) $$\binom{N}{1} = \frac{N!}{1!(N-1)!} = N;$$

(b) $$\binom{N}{N} = \frac{N!}{N!(N-N)!} = \frac{N!}{N!0!} = 1;$$

(c) $$\binom{N}{N-n} = \frac{N!}{n!(N-n)!} = \binom{N}{n}.$$

The last relationship states that the number of combinations of objects taken $(N - n)$ at a time is the same as the number of combinations of N objects taken n at a time. This is not surprising, since whenever we select n objects from the N, we leave $(N - n)$ behind. Thus, if a sample of 20 is taken at random without replacement from a population of 100 items, how many possible samples of 20 are there? Answer:

$$\binom{100}{20} = \binom{100}{80}.$$

A–8 THE $\sum$ NOTATION

The Greek capital letter sigma, $\sum$, is a shorthand notation for designating a sum. For example, the sum

$$x_1 + x_2 + \cdots + x_n$$

may be conveniently written with the single summation sign as

$$\sum_{i=1}^{n} x_i, \tag{A–23}$$

which is read as "the summation of x_i from $i = 1$ to $i = n$." Here, i is called the *index of summation* and is a variable ranging over the integers $1, 2, \ldots, n$. The expression $i = 1$ below the $\sum$ sign indicates that 1 is the initial value taken on by i, and the n above the $\sum$ sign indicates the terminal value of i. The symbol x_i is called the *summand*; it is a function of i that takes on values of $x_1, x_2, \ldots, x_n$, respectively, as i takes on values $1, 2, \ldots, n$. The $\sum$ sign says that the values taken by the summand are to be added.

The symbol used for the index of summation is sometimes called a *dummy index*, because it is completely arbitrary. For example,

$$x_1 + x_2 + \cdots + x_n = \sum_{i=1}^{n} x_i = \sum_{j=1}^{n} x_j = \sum_{k=1}^{n} x_k = \cdots.$$

As illustrations, we see that

$$\sum_{i=1}^{3} i = 1 + 2 + 3 = 6,$$

$$\sum_{i=1}^{3} i^2 = 1^2 + 2^2 + 3^2 = 14,$$

$$\sum_{i=0}^{r} \frac{1}{2^i} = \frac{1}{2^0} + \frac{1}{2^1} + \frac{1}{2^2} + \cdots + \frac{1}{2^r},$$

$$\sum_{i=1}^{5} x_i = x_1 + x_2 + \cdots + x_5,$$

$$\sum_{i=6}^{9} x_i = x_6 + x_7 + x_8 + x_9.$$

From the last two examples we can readily appreciate that

$$x_1 + x_2 + \cdots + x_9 = \sum_{i=1}^{5} x_i + \sum_{i=6}^{9} x_i = \sum_{i=1}^{9} x_i.$$

As we saw in Chapter 10, a familiar function of n values $x_1, x_2, \ldots, x_n$ is their arithmetic mean (their sum divided by n), denoted by $\bar{x}$. Using the $\sum$ notation, we can write

$$\bar{x} = \frac{x_1 + x_2 + \cdots + x_n}{n}$$

$$= \frac{1}{n} \sum_{i=1}^{n} x_i.$$

A-9 FINITE DOUBLE SUMS

The values of a variable X, the summand, are sometimes arranged as a two-way table as follows:

$$\begin{array}{cccc} x_{11} & x_{12} & \cdots & x_{1c} \\ x_{21} & x_{22} & \cdots & x_{2c} \\ \vdots & & & \\ x_{r1} & x_{r2} & \cdots & x_{rc} \end{array}$$

where the individual values may be denoted as x_{ij}, $i = 1, 2, \ldots, r$, $j = 1, 2, \ldots, c$. The sum of the x_{ij} can be obtained by first adding the various rows and then adding the row totals to get the desired result. This is simply the application of the single summation twice to get a double summation sign as derived below:

$$\sum_{j=1}^{c} x_{1j} + \sum_{j=1}^{c} x_{2j} + \cdots + \sum_{j=1}^{c} x_{rj} = \sum_{i=1}^{r} \sum_{j=1}^{c} x_{ij}.$$

The same result can be obtained if we add the columns first and then add the column sums to get the grand sum. Thus,

$$\sum_{i=1}^{r} \sum_{j=1}^{c} x_{ij} = \sum_{j=1}^{c} \sum_{i=1}^{r} x_{ij}. \qquad \text{(A-24)}$$

If the range of summation is completely clear from the context, the index of summation can be omitted. Thus, we can use

$$\sum x \quad \text{to replace} \quad \sum_{i=1}^{n} x_i;$$

$$\frac{1}{n}\sum x \quad \text{to replace} \quad \frac{1}{n}\sum_{i=1}^{n} x_i;$$

$$\sum\sum x_{ij} \quad \text{to replace} \quad \sum_{j=1}^{c}\sum_{i=1}^{r} x_{ij};$$

$$\sum x^2 - (\sum x)^2 \quad \text{to replace} \quad \sum_{i=1}^{n} x_i^2 - (\sum_{i=1}^{n} x_i)^2;$$

and so on.

A–10 THEOREMS OF SUMMATION

Theorem I

If a is some constant value over the n diffrrent observations i, then

$$\sum_{i=1}^{n} a = na. \tag{A–25}$$

The reason is that each of the x's is equal to a constant quantity a, and thus

$$\sum_{i=1}^{n} a = a + a + \cdots + a = na.$$

Extending this theorem to double sums, we have

$$\sum_{i=1}^{r}\sum_{j=1}^{c} a = \sum_{i=1}^{r} ca. \tag{A–26}$$

Theorem II

Given the value a, which is a constant over all individual values entering into the summation, then

$$\sum_{i=1}^{n} ax_i = a\sum_{i=1}^{n} x_i, \tag{A–27}$$

since

$$\begin{aligned}\sum_{i=1}^{n} ax_i &= ax_1 + ax_2 + \cdots + ax_n \\ &= a(x_1 + x_2 + \cdots + x_n) \\ &= a\sum_{i=1}^{n} x_i.\end{aligned}$$

This rule can also be applied to double sums. That is,

$$\sum_i \sum_j a x_{ij} = a \sum_i \sum_j x_{ij}. \qquad \text{(A–28)}$$

Theorem III

If the only operation to be carried out before a sum is taken is itself a sum (or a difference), then the summation may be distributed; namely,

$$\sum_i (x_i + y_i - z_i) = \sum_i x_i + \sum_i y_i - \sum_i z_i. \qquad \text{(A–29)}$$

This is true because

$$\begin{aligned}\sum_i (x_i + y_i - z_i) &= (x_1 + y_1 - z_1) + (x_2 + y_2 - z_2) \\ &\qquad + \cdots + (x_n + y_n - z_n) \\ &= (x_1 + \cdots + x_n) + (y_1 + \cdots + y_n) \\ &\qquad - (z_1 + \cdots + z_n) \\ &= \sum_i x_i + \sum_i y_i - \sum_i z_i.\end{aligned}$$

Theorem III is a very handy tool for algebraic manipulations on summation signs. For example,

$$\begin{aligned}\sum_i^r \sum_j^c (x_{ij} - a)^2 &= \sum_i \sum_j (x_{ij}^2 - 2a x_{ij} + a^2) \\ &= \sum_i \sum_j x_{ij}^2 - \sum_i \sum_j 2a x_{ij} + \sum_i \sum_j a^2 \\ &= \sum_i \sum_j x_{ij}^2 - 2a \sum_i \sum_j x_{ij} + \sum_i^r c a^2 \\ &= \sum_i \sum_j x_{ij}^2 - 2a \sum_i \sum_j x_{ij} + r c a^2.\end{aligned}$$

Theorem IV

If some operation is to be carried out on the individual values of X before the summation, this is indicated by mathematical punctuations, or symbols having the force of punctuations. Unless the summation notation is included within this punctuation, the summation is to be made after the other operation. For example,

$$\sum x^2 = \sum (x_1^2 + x_2^2 + \cdots + x_n^2),$$

whereas

$$(\sum x)^2 = (x_1 + x_2 + \cdots + x_n)^2.$$

Also,

$$\sum \sqrt{x_i} = \sqrt{x_1} + \sqrt{x_2} + \cdots + \sqrt{x_n},$$

whereas

$$\sqrt{\sum x} = \sqrt{x_1 + x_2 + \cdots + x_n}.$$

This rule is especially important for evaluating double sums. For instance,

$$\sum_i \left(\sum_j x_{ij}\right)^2 = \sum_i (x_{i1} + x_{i2} + x_{i3} + \cdots + x_{ic})^2.$$

APPENDIX B

STATISTICAL TABLES

B–1 Squares and Square Roots 588
B–2 Common Logarithms 597
B–3 The Cumulative Binomial Distribution 599
B–4 The Hypergeometric Distribution 603
B–5 The Cumulative Poisson Distribution 605
B–6 Exponential Functions 607
B–7 Values of the Standard Normal Distribution Function 608
B–8 Unit Normal Loss Function 610
B–9 The Chi-Square Distribution 611
B–10 Student's t Distribution 612
B–11 F Distribution 613
B–12 Values of $z = \left(\frac{1}{2}\right) \ln \left(\frac{1+r}{1-r}\right)$ 616
B–13 Equidistributed Random Numbers 618

Table B–1. Squares and Square Roots

n	n^2	$\sqrt{n}$	$\sqrt{10n}$
1.00	1.0000	1.00000	3.16228
1.01	1.0201	1.00499	3.17805
1.02	1.0404	1.00995	3.19374
1.03	1.0609	1.01489	3.20936
1.04	1.0816	1.01980	3.22490
1.05	1.1025	1.02470	3.24037
1.06	1.1236	1.02956	3.25576
1.07	1.1449	1.03441	3.27109
1.08	1.1664	1.03923	3.28634
1.09	1.1881	1.04403	3.30151
1.10	1.2100	1.04881	3.31662
1.11	1.2321	1.05357	3.33167
1.12	1.2544	1.05830	3.34664
1.13	1.2769	1.06301	3.36155
1.14	1.2996	1.06771	3.37639
1.15	1.3225	1.07238	3.39116
1.16	1.3456	1.07703	3.40588
1.17	1.3689	1.08167	3.42053
1.18	1.3924	1.08628	3.43511
1.19	1.4161	1.09087	3.44964
1.20	1.4400	1.09545	3.46410
1.21	1.4641	1.10000	3.47851
1.22	1.4884	1.10454	3.49285
1.23	1.5129	1.10905	3.50714
1.24	1.5376	1.11355	3.52136
1.25	1.5625	1.11803	3.53553
1.26	1.5876	1.12250	3.54965
1.27	1.6129	1.12694	3.56371
1.28	1.6384	1.13137	3.57771
1.29	1.6641	1.13578	3.59166
1.30	1.6900	1.14018	3.60555
1.31	1.7161	1.14455	3.61939
1.32	1.7424	1.14891	3.63318
1.33	1.7689	1.15326	3.64692
1.34	1.7956	1.15758	3.66060
1.35	1.8225	1.16190	3.67423
1.36	1.8496	1.16619	3.68782
1.37	1.8769	1.17047	3.70135
1.38	1.9044	1.17473	3.71484
1.39	1.9321	1.17898	3.72827
1.40	1.9600	1.18322	3.74166
1.41	1.9881	1.18743	3.75500
1.42	2.0164	1.19164	3.76829
1.43	2.0449	1.19583	3.78153
1.44	2.0736	1.20000	3.79473
1.45	2.1025	1.20416	3.80789
1.46	2.1316	1.20830	3.82099
1.47	2.1609	1.21244	3.83406
1.48	2.1904	1.21655	3.84708
1.49	2.2201	1.22066	3.86005

n	n^2	$\sqrt{n}$	$\sqrt{10n}$
1.50	2.2500	1.22474	3.87298
1.51	2.2801	1.22882	3.88587
1.52	2.3104	1.23288	3.89872
1.53	2.3409	1.23693	3.91152
1.54	2.3716	1.24097	3.92428
1.55	2.4025	1.24499	3.93700
1.56	2.4336	1.24900	3.94968
1.57	2.4649	1.25300	3.96232
1.58	2.4964	1.25698	3.97492
1.59	2.5281	1.26095	3.98748
1.60	2.5600	1.26491	4.00000
1.61	2.5921	1.26886	4.01248
1.62	2.6244	1.27279	4.02492
1.63	2.6569	1.27671	4.03733
1.64	2.6896	1.28062	4.04969
1.65	2.7225	1.28452	4.06202
1.66	2.7556	1.28841	4.07431
1.67	2.7889	1.29228	4.08656
1.68	2.8224	1.29615	4.09878
1.69	2.8561	1.30000	4.11096
1.70	2.8900	1.30384	4.12311
1.71	2.9241	1.30767	4.13521
1.72	2.9584	1.31149	4.14729
1.73	2.9929	1.31529	4.15933
1.74	3.0276	1.31909	4.17133
1.75	3.0625	1.32288	4.18330
1.76	3.0976	1.32665	4.19524
1.77	3.1329	1.33041	4.20714
1.78	3.1684	1.33417	4.21900
1.79	3.2041	1.33791	4.23084
1.80	3.2400	1.34164	4.24264
1.81	3.2761	1.34536	4.25441
1.82	3.3124	1.34907	4.26615
1.83	3.3489	1.35277	4.27785
1.84	3.3856	1.35647	4.28952
1.85	3.4225	1.36015	4.30116
1.86	3.4596	1.36382	4.31277
1.87	3.4969	1.36748	4.32435
1.88	3.5344	1.37113	4.33590
1.89	3.5721	1.37477	4.34741
1.90	3.6100	1.37840	4.35890
1.91	3.6481	1.38203	4.37035
1.92	3.6864	1.38564	4.38178
1.93	3.7249	1.38924	4.39318
1.94	3.7636	1.39284	4.40454
1.95	3.8025	1.39642	4.41588
1.96	3.8416	1.40000	4.42719
1.97	3.8809	1.40357	4.43847
1.98	3.9204	1.40712	4.44972
1.99	3.9601	1.41067	4.46094

Table B-1. Squares and Square Roots (*Continued*)

n	n^2	$\sqrt{n}$	$\sqrt{10n}$
2.00	4.0000	1.41421	4.47214
2.01	4.0401	1.41774	4.48330
2.02	4.0804	1.42127	4.49444
2.03	4.1209	1.42478	4.50555
2.04	4.1616	1.42829	4.51664
2.05	4.2025	1.43178	4.52769
2.06	4.2436	1.43527	4.53872
2.07	4.2849	1.43875	4.54973
2.08	4.3264	1.44222	4.56070
2.09	4.3681	1.44568	4.57165
2.10	4.4100	1.44914	4.58258
2.11	4.4521	1.45258	4.59347
2.12	4.4944	1.45602	4.60435
2.13	4.5369	1.45945	4.61519
2.14	4.5796	1.46287	4.62601
2.15	4.6225	1.46629	4.63681
2.16	4.6656	1.46969	4.64758
2.17	4.7089	1.47309	4.65833
2.18	4.7524	1.47648	4.66905
2.19	4.7961	1.47986	4.67974
2.20	4.8400	1.48324	4.69042
2.21	4.8841	1.48661	4.70106
2.22	4.9284	1.48997	4.71169
2.23	4.9729	1.49332	4.72229
2.24	5.0176	1.49666	4.73286
2.25	5.0625	1.50000	4.74342
2.26	5.1076	1.50333	4.75395
2.27	5.1529	1.50665	4.76445
2.28	5.1984	1.50997	4.77493
2.29	5.2441	1.51327	4.78539
2.30	5.2900	1.51658	4.79583
2.31	5.3361	1.51987	4.80625
2.32	5.3824	1.52315	4.81664
2.33	5.4289	1.52643	4.82701
2.34	5.4756	1.52971	4.83735
2.35	5.5225	1.53297	4.84768
2.36	5.5696	1.53623	4.85798
2.37	5.6169	1.53948	4.86826
2.38	5.6644	1.54272	4.87852
2.39	5.7121	1.54596	4.88876
2.40	5.7600	1.54919	4.89898
2.41	5.8081	1.55242	4.90918
2.42	5.8564	1.55563	4.91935
2.43	5.9049	1.55885	4.92950
2.44	5.9536	1.56205	4.93964
2.45	6.0025	1.56525	4.94975
2.46	6.0516	1.56844	4.95984
2.47	6.1009	1.57162	4.96991
2.48	6.1504	1.57480	4.97996
2.49	6.2001	1.57797	4.98999

n	n^2	$\sqrt{n}$	$\sqrt{10n}$
2.50	6.2500	1.58114	5.00000
2.51	6.3001	1.58430	5.00999
2.52	6.3504	1.58745	5.01996
2.53	6.4009	1.59060	5.02991
2.54	6.4516	1.59374	5.03984
2.55	6.5025	1.59687	5.04975
2.56	6.5536	1.60000	5.05964
2.57	6.6049	1.60312	5.06952
2.58	6.6564	1.60624	5.07937
2.59	6.7081	1.60935	5.08920
2.60	6.7600	1.61245	5.09902
2.61	6.8121	1.61555	5.10882
2.62	6.8644	1.61864	5.11859
2.63	6.9169	1.62173	5.12835
2.64	6.9696	1.62481	5.13809
2.65	7.0225	1.62788	5.14782
2.66	7.0756	1.63095	5.15752
2.67	7.1289	1.63401	5.16720
2.68	7.1824	1.63707	5.17687
2.69	7.2361	1.64012	5.18652
2.70	7.2900	1.64317	5.19615
2.71	7.3441	1.64621	5.20577
2.72	7.3984	1.64924	5.21536
2.73	7.4529	1.65227	5.22494
2.74	7.5076	1.65529	5.23450
2.75	7.5625	1.65831	5.24404
2.76	7.6176	1.66132	5.25357
2.77	7.6729	1.66433	5.26308
2.78	7.7284	1.66733	5.27257
2.79	7.7841	1.67033	5.28205
2.80	7.8400	1.67332	5.29150
2.81	7.8961	1.67631	5.30094
2.82	7.9524	1.67929	5.31037
2.83	8.0089	1.68226	5.31977
2.84	8.0656	1.68523	5.32917
2.85	8.1225	1.68819	5.33854
2.86	8.1796	1.69115	5.34790
2.87	8.2369	1.69411	5.35724
2.88	8.2944	1.69706	5.36656
2.89	8.3521	1.70000	5.37587
2.90	8.4100	1.70294	5.38516
2.91	8.4681	1.70587	5.39444
2.92	8.5264	1.70880	5.40370
2.93	8.5849	1.71172	5.41295
2.94	8.6436	1.71464	5.42218
2.95	8.7025	1.71756	5.43139
2.96	8.7616	1.72047	5.44059
2.97	8.8209	1.72337	5.44977
2.98	8.8804	1.72627	5.45894
2.99	8.9401	1.72916	5.46809

Table B–1. Squares and Square Roots (*Continued*)

n	n^2	$\sqrt{n}$	$\sqrt{10n}$
3.00	9.0000	1.73205	5.47723
3.01	9.0601	1.73494	5.48635
3.02	9.1204	1.73781	5.49545
3.03	9.1809	1.74069	5.50454
3.04	9.2416	1.74356	5.51362
3.05	9.3025	1.74642	5.52268
3.06	9.3636	1.74929	5.53173
3.07	9.4249	1.75214	5.54076
3.08	9.4864	1.75499	5.54977
3.09	9.5481	1.75784	5.55878
3.10	9.6100	1.76068	5.56776
3.11	9.6721	1.76352	5.57674
3.12	9.7344	1.76635	5.58570
3.13	9.7969	1.76918	5.59464
3.14	9.8596	1.77200	5.60357
3.15	9.9225	1.77482	5.61249
3.16	9.9856	1.77764	5.62139
3.17	10.0489	1.78045	5.63028
3.18	10.1124	1.78326	5.63915
3.19	10.1761	1.78606	5.64801
3.20	10.2400	1.78885	5.65685
3.21	10.3041	1.79165	5.66569
3.22	10.3684	1.79444	5.67450
3.23	10.4329	1.79722	5.68331
3.24	10.4976	1.80000	5.69210
3.25	10.5625	1.80278	5.70088
3.26	10.6276	1.80555	5.70964
3.27	10.6929	1.80831	5.71839
3.28	10.7584	1.81108	5.72713
3.29	10.8241	1.81384	5.73585
3.30	10.8900	1.81659	5.74456
3.31	10.9561	1.81934	5.75326
3.32	11.0224	1.82209	5.76194
3.33	11.0889	1.82483	5.77062
3.34	11.1556	1.82757	5.77927
3.35	11.2225	1.83030	5.78792
3.36	11.2896	1.83303	5.79655
3.37	11.3569	1.83576	5.80517
3.38	11.4244	1.83848	5.81378
3.39	11.4921	1.84120	5.82237
3.40	11.5600	1.84391	5.83095
3.41	11.6281	1.84662	5.83952
3.42	11.6964	1.84932	5.84808
3.43	11.7649	1.85203	5.85662
3.44	11.8336	1.85472	5.86515
3.45	11.9025	1.85742	5.87367
3.46	11.9716	1.86011	5.88218
3.47	12.0409	1.86279	5.89067
3.48	12.1104	1.86548	5.89915
3.49	12.1801	1.86815	5.90762

n	n^2	$\sqrt{n}$	$\sqrt{10n}$
3.50	12.2500	1.87083	5.91608
3.51	12.3201	1.87350	5.92453
3.52	12.3904	1.87617	5.93296
3.53	12.4609	1.87883	5.94138
3.54	12.5316	1.88149	5.94979
3.55	12.6025	1.88414	5.95819
3.56	12.6736	1.88680	5.96657
3.57	12.7449	1.88944	5.97495
3.58	12.8164	1.89209	5.98331
3.59	12.8881	1.89473	5.99166
3.60	12.9600	1.89737	6.00000
3.61	13.0321	1.90000	6.00833
3.62	13.1044	1.90263	6.01664
3.63	13.1769	1.90526	6.02495
3.64	13.2496	1.90788	6.03324
3.65	13.3225	1.91050	6.04152
3.66	13.3956	1.91311	6.04979
3.67	13.4689	1.91572	6.05805
3.68	13.5424	1.91833	6.06630
3.69	13.6161	1.92094	6.07454
3.70	13.6900	1.92354	6.08276
3.71	13.7641	1.92614	6.09098
3.72	13.8384	1.92873	6.09918
3.73	13.9129	1.93132	6.10737
3.74	13.9876	1.93391	6.11555
3.75	14.0625	1.93649	6.12372
3.76	14.1376	1.93907	6.13188
3.77	14.2129	1.94165	6.14003
3.78	14.2884	1.94422	6.14817
3.79	14.3641	1.94679	6.15630
3.80	14.4400	1.94936	6.16441
3.81	14.5161	1.95192	6.17252
3.82	14.5924	1.95448	6.18061
3.83	14.6689	1.95704	6.18870
3.84	14.7456	1.95959	6.19677
3.85	14.8225	1.96214	6.20484
3.86	14.8996	1.96469	6.21289
3.87	14.9769	1.96723	6.22093
3.88	15.0544	1.96977	6.22896
3.89	15.1321	1.97231	6.23699
3.90	15.2100	1.97484	6.24500
3.91	15.2881	1.97737	6.25300
3.92	15.3664	1.97990	6.26099
3.93	15.4449	1.98242	6.26897
3.94	15.5236	1.98494	6.27694
3.95	15.6025	1.98746	6.28490
3.96	15.6816	1.98997	6.29285
3.97	15.7609	1.99249	6.30079
3.98	15.8404	1.99499	6.30872
3.99	15.9201	1.99750	6.31664

Table B-1. Squares and Square Roots (*Continued*)

n	n^2	$\sqrt{n}$	$\sqrt{10n}$
4.00	16.0000	2.00000	6.32456
4.01	16.0801	2.00250	6.33246
4.02	16.1604	2.00499	6.34035
4.03	16.2409	2.00749	6.34823
4.04	16.3216	2.00998	6.35610
4.05	16.4025	2.01246	6.36396
4.06	16.4836	2.01494	6.37181
4.07	16.5649	2.01742	6.37966
4.08	16.6464	2.01990	6.38749
4.09	16.7281	2.02237	6.39531
4.10	16.8100	2.02485	6.40312
4.11	16.8921	2.02731	6.41093
4.12	16.9744	2.02978	6.41872
4.13	17.0569	2.03224	6.42651
4.14	17.1396	2.03470	6.43428
4.15	17.2225	2.03715	6.44205
4.16	17.3056	2.03961	6.44981
4.17	17.3889	2.04206	6.45755
4.18	17.4724	2.04450	6.46529
4.19	17.5561	2.04695	6.47302
4.20	17.6400	2.04939	6.48074
4.21	17.7241	2.05183	6.48845
4.22	17.8084	2.05426	6.49615
4.23	17.8929	2.05670	6.50384
4.24	17.9776	2.05913	6.51153
4.25	18.0625	2.06155	6.51920
4.26	18.1476	2.06398	6.52687
4.27	18.2329	2.06640	6.53452
4.28	18.3184	2.06882	6.54217
4.29	18.4041	2.07123	6.54981
4.30	18.4900	2.07364	6.55744
4.31	18.5761	2.07605	6.56506
4.32	18.6624	2.07846	6.57267
4.33	18.7489	2.08087	6.58027
4.34	18.8356	2.08327	6.58787
4.35	18.9225	2.08567	6.59545
4.36	19.0096	2.08806	6.60303
4.37	19.0969	2.09045	6.61060
4.38	19.1844	2.09284	6.61816
4.39	19.2721	2.09523	6.62571
4.40	19.3600	2.09762	6.63325
4.41	19.4481	2.10000	6.64078
4.42	19.5364	2.10238	6.64831
4.43	19.6249	2.10476	6.65582
4.44	19.7136	2.10713	6.66333
4.45	19.8025	2.10950	6.67083
4.46	19.8916	2.11187	6.67832
4.47	19.9809	2.11424	6.68581
4.48	20.0704	2.11660	6.69328
4.49	20.1601	2.11896	6.70075

n	n^2	$\sqrt{n}$	$\sqrt{10n}$
4.50	20.2500	2.12132	6.70820
4.51	20.3401	2.12368	6.71565
4.52	20.4304	2.12603	6.72309
4.53	20.5209	2.12838	6.73053
4.54	20.6116	2.13073	6.73795
4.55	20.7025	2.13307	6.74537
4.56	20.7936	2.13542	6.75278
4.57	20.8849	2.13776	6.76018
4.58	20.9764	2.14009	6.76757
4.59	21.0681	2.14243	6.77495
4.60	21.1600	2.14476	6.78233
4.61	21.2521	2.14709	6.78970
4.62	21.3444	2.14942	6.79706
4.63	21.4369	2.15174	6.80441
4.64	21.5296	2.15407	6.81175
4.65	21.6225	2.15639	6.81909
4.66	21.7156	2.15870	6.82642
4.67	21.8089	2.16102	6.83374
4.68	21.9024	2.16333	6.84105
4.69	21.9961	2.16564	6.84836
4.70	22.0900	2.16795	6.85565
4.71	22.1841	2.17025	6.86294
4.72	22.2784	2.17256	6.87023
4.73	22.3729	2.17486	6.87750
4.74	22.4676	2.17715	6.88477
4.75	22.5625	2.17945	6.89202
4.76	22.6576	2.18174	6.89928
4.77	22.7529	2.18403	6.90652
4.78	22.8484	2.18632	6.91375
4.79	22.9441	2.18861	6.92098
4.80	23.0400	2.19089	6.92820
4.81	23.1361	2.19317	6.93542
4.82	23.2324	2.19545	6.94262
4.83	23.3289	2.19773	6.94982
4.84	23.4256	2.20000	6.95701
4.85	23.5225	2.20227	6.96419
4.86	23.6196	2.20454	6.97137
4.87	23.7169	2.20681	6.97854
4.88	23.8144	2.20907	6.98570
4.89	23.9121	2.21133	6.99285
4.90	24.0100	2.21359	7.00000
4.91	24.1081	2.21585	7.00714
4.92	24.2064	2.21811	7.01427
4.93	24.3049	2.22036	7.02140
4.94	24.4036	2.22261	7.02851
4.95	24.5025	2.22486	7.03562
4.96	24.6016	2.22711	7.04273
4.97	24.7009	2.22935	7.04982
4.98	24.8004	2.23159	7.05691
4.99	24.9001	2.23383	7.06399

Table B–1. Squares and Square Roots (*Continued*)

n	n^2	$\sqrt{n}$	$\sqrt{10n}$
5.00	25.0000	2.23607	7.07107
5.01	25.1001	2.23830	7.07814
5.02	25.2004	2.24054	7.08520
5.03	25.3009	2.24277	7.09225
5.04	25.4016	2.24499	7.09930
5.05	25.5025	2.24722	7.10634
5.06	25.6036	2.24944	7.11337
5.07	25.7049	2.25167	7.12039
5.08	25.8064	2.25389	7.12741
5.09	25.9081	2.25610	7.13442
5.10	26.0100	2.25832	7.14143
5.11	26.1121	2.26053	7.14843
5.12	26.2144	2.26274	7.15542
5.13	26.3169	2.26495	7.16240
5.14	26.4196	2.26716	7.16938
5.15	26.5225	2.26936	7.17635
5.16	26.6256	2.27156	7.18331
5.17	26.7289	2.27376	7.19027
5.18	26.8324	2.27596	7.19722
5.19	26.9361	2.27816	7.20417
5.20	27.0400	2.28035	7.21110
5.21	27.1441	2.28254	7.21803
5.22	27.2484	2.28473	7.22496
5.23	27.3529	2.28692	7.23187
5.24	27.4576	2.28910	7.23878
5.25	27.5625	2.29129	7.24569
5.26	27.6676	2.29347	7.25259
5.27	27.7729	2.29565	7.25948
5.28	27.8784	2.29783	7.26636
5.29	27.9841	2.30000	7.27324
5.30	28.0900	2.30217	7.28011
5.31	28.1961	2.30434	7.28697
5.32	28.3024	2.30651	7.29383
5.33	28.4089	2.30868	7.30068
5.34	28.5156	2.31084	7.30753
5.35	28.6225	2.31301	7.31437
5.36	28.7296	2.31517	7.32120
5.37	28.8369	2.31733	7.32803
5.38	28.9444	2.31948	7.33485
5.39	29.0521	2.32164	7.34166
5.40	29.1600	2.32379	7.34847
5.41	29.2681	2.32594	7.35527
5.42	29.3764	2.32809	7.36206
5.43	29.4849	2.33024	7.36885
5.44	29.5936	2.33238	7.37564
5.45	29.7025	2.33452	7.38241
5.46	29.8116	2.33666	7.38918
5.47	29.9209	2.33880	7.39594
5.48	30.0304	2.34094	7.40270
5.49	30.1401	2.34307	7.40945

n	n^2	$\sqrt{n}$	$\sqrt{10n}$
5.50	30.2500	2.34521	7.41620
5.51	30.3601	2.34734	7.42294
5.52	30.4704	2.34947	7.42967
5.53	30.5809	2.35160	7.43640
5.54	30.6916	2.35372	7.44312
5.55	30.8025	2.35584	7.44983
5.56	30.9136	2.35797	7.45654
5.57	31.0249	2.36008	7.46324
5.58	31.1364	2.36220	7.46994
5.59	31.2481	2.36432	7.47663
5.60	31.3600	2.36643	7.48331
5.61	31.4721	2.36854	7.48999
5.62	31.5844	2.37065	7.49667
5.63	31.6969	2.37276	7.50333
5.64	31.8096	2.37487	7.50999
5.65	31.9225	2.37697	7.51665
5.66	32.0356	2.37908	7.52330
5.67	32.1489	2.38118	7.52994
5.68	32.2624	2.38238	7.53658
5.69	32.3761	2.38537	7.54321
5.70	32.4900	2.38747	7.54983
5.71	32.6041	2.38956	7.55645
5.72	32.7184	2.39165	7.56307
5.73	32.8329	2.39374	7.56968
5.74	32.9476	2.39583	7.57628
5.75	33.0625	2.39792	7.58288
5.76	33.1776	2.40000	7.58947
5.77	33.2929	2.40208	7.59605
5.78	33.4084	2.40416	7.60263
5.79	33.5241	2.40624	7.60920
5.80	33.6400	2.40832	7.61577
5.81	33.7561	2.41039	7.62234
5.82	33.8724	2.41247	7.62889
5.83	33.9889	2.41454	7.63544
5.84	34.1056	2.41661	7.64199
5.85	34.2225	2.41868	7.64853
5.86	34.3396	2.42074	7.65506
5.87	34.4569	2.42281	7.66159
5.88	34.5744	2.42487	7.66812
5.89	34.6921	2.42693	7.67463
5.90	34.8100	2.42899	7.68115
5.91	34.9281	2.43105	7.68765
5.92	35.0464	2.43311	7.69415
5.93	35.1649	2.43516	7.70065
5.94	35.2836	2.43721	7.70714
5.95	35.4025	2.43926	7.71362
5.96	35.5216	2.44131	7.72010
5.97	35.6409	2.44336	7.72658
5.98	35.7604	2.44540	7.73305
5.99	35.8801	2.44745	7.73951

Table B-1. Squares and Square Roots (*Continued*)

n	n^2	$\sqrt{n}$	$\sqrt{10n}$
6.00	36.0000	2.44949	7.74597
6.01	36.1201	2.45153	7.75242
6.02	36.2404	2.45357	7.75887
6.03	36.3609	2.45561	7.76531
6.04	36.4816	2.45764	7.77174
6.05	36.6025	2.45967	7.77817
6.06	36.7236	2.46171	7.78460
6.07	36.8449	2.46374	7.79102
6.08	36.9664	2.46577	7.79744
6.09	37.0881	2.46779	7.80385
6.10	37.2100	2.46982	7.81025
6.11	37.3321	2.47184	7.81665
6.12	37.4544	2.47386	7.82304
6.13	37.5769	2.47588	7.82943
6.14	37.6996	2.47790	7.83582
6.15	37.8225	2.47992	7.84219
6.16	37.9456	2.48193	7.84857
6.17	38.0689	2.48395	7.85493
6.18	38.1924	2.48596	7.86130
6.19	38.3161	2.48797	7.86766
6.20	38.4400	2.48998	7.87401
6.21	38.5641	2.49199	7.88036
6.22	38.6884	2.49399	7.88670
6.23	38.8129	2.49600	7.89303
6.24	38.9376	2.49800	7.89937
6.25	39.0625	2.50000	7.90569
6.26	39.1876	2.50200	7.91202
6.27	39.3129	2.50400	7.91833
6.28	39.4384	2.50599	7.92465
6.29	39.5641	2.50799	7.93095
6.30	39.6900	2.50998	7.93725
6.31	39.8161	2.51197	7.94355
6.32	39.9424	2.51396	7.94984
6.33	40.0689	2.51595	7.95613
6.34	40.1956	2.51794	7.96241
6.35	40.3225	2.51992	7.96869
6.36	40.4496	2.52190	7.97496
6.37	40.5769	2.52389	7.98123
6.38	40.7044	2.52587	7.98749
6.39	40.8321	2.52784	7.99375
6.40	40.9600	2.52982	8.00000
6.41	41.0881	2.53180	8.00625
6.42	41.2164	2.53377	8.01249
6.43	41.3449	2.53574	8.01873
6.44	41.4736	2.53772	8.02496
6.45	41.6025	2.53969	8.03119
6.46	41.7316	2.54165	8.03741
6.47	41.8609	2.54362	8.04363
6.48	41.9904	2.54558	8.04984
6.49	42.1201	2.54755	8.05605

n	n^2	$\sqrt{n}$	$\sqrt{10n}$
6.50	42.2500	2.54951	8.06226
6.51	42.3801	2.55147	8.06846
6.52	42.5104	2.55343	8.07465
6.53	42.6409	2.55539	8.08084
6.54	42.7716	2.55734	8.08703
6.55	42.9025	2.55930	8.09321
6.56	43.0336	2.56125	8.09938
6.57	43.1649	2.56320	8.10555
6.58	43.2964	2.56515	8.11172
6.59	43.4281	2.56710	8.11788
6.60	43.5600	2.56905	8.12404
6.61	43.6921	2.57099	8.13019
6.62	43.8244	2.57294	8.13634
6.63	43.9569	2.57488	8.14248
6.64	44.0896	2.57682	8.14862
6.65	44.2225	2.57876	8.15475
6.66	44.3556	2.58070	8.16088
6.67	44.4889	2.58263	8.16701
6.68	44.6224	2.58457	8.17313
6.69	44.7561	2.58650	8.17924
6.70	44.8900	2.58844	8.18535
6.71	45.0241	2.59037	8.19146
6.72	45.1584	2.59230	8.19756
6.73	45.2929	2.59422	8.20366
6.74	45.4276	2.59615	8.20975
6.75	45.5625	2.59808	8.21584
6.76	45.6976	2.60000	8.22192
6.77	45.8329	2.60192	8.22800
6.78	45.9684	2.60384	8.23408
6.79	46.1041	2.60576	8.24015
6.80	46.2400	2.60768	8.24621
6.81	46.3761	2.60960	8.25227
6.82	46.5124	2.61151	8.25833
6.83	46.6489	2.61343	8.26438
6.84	46.7856	2.61534	8.27043
6.85	46.9225	2.61725	8.27647
6.86	47.0596	2.61916	8.28251
6.87	47.1969	2.62107	8.28855
6.88	47.3344	2.62298	8.29458
6.89	47.4721	2.62488	8.30060
6.90	47.6100	2.62679	8.30662
6.91	47.7481	2.62869	8.31264
6.92	47.8864	2.63059	8.31865
6.93	48.0249	2.63249	8.32466
6.94	48.1636	2.63439	8.33067
6.95	48.3025	2.63629	8.33667
6.96	48.4416	2.63818	8.34266
6.97	48.5809	2.64008	8.34865
6.98	48.7204	2.64197	8.35464
6.99	48.8601	2.64386	8.36062

Table B–1. Squares and Square Roots (*Continued*)

n	n^2	$\sqrt{n}$	$\sqrt{10n}$
7.00	49.0000	2.64575	8.36660
7.01	49.1401	2.64764	8.37257
7.02	49.2804	2.64953	8.37854
7.03	49.4209	2.65141	8.38451
7.04	49.5616	2.65330	8.39047
7.05	49.7025	2.65518	8.39643
7.06	49.8436	2.65707	8.40238
7.07	49.9849	2.65895	8.40833
7.08	50.1264	2.66083	8.41427
7.09	50.2681	2.66271	8.42021
7.10	50.4100	2.66458	8.42615
7.11	50.5521	2.66646	8.43208
7.12	50.6944	2.66833	8.43801
7.13	50.8369	2.67021	8.44393
7.14	50.9796	2.67208	8.44985
7.15	51.1225	2.67395	8.45577
7.16	51.2656	2.67582	8.46168
7.17	51.4089	2.67769	8.46759
7.18	51.5524	2.67955	8.47349
7.19	51.6961	2.68142	8.47939
7.20	51.8400	2.68328	8.48528
7.21	51.9841	2.68514	8.49117
7.22	52.1284	2.68701	8.49706
7.23	52.2729	2.68887	8.50294
7.24	52.4176	2.69072	8.50882
7.25	52.5625	2.69258	8.51469
7.26	52.7076	2.69444	8.52056
7.27	52.8529	2.69629	8.52643
7.28	52.9984	2.69815	8.53229
7.29	53.1441	2.70000	8.53815
7.30	53.2900	2.70185	8.54400
7.31	53.4361	2.70370	8.54985
7.32	53.5824	2.70555	8.55570
7.33	53.7289	2.70740	8.56154
7.34	53.8756	2.70924	8.56738
7.35	54.0225	2.71109	8.57321
7.36	54.1696	2.71293	8.57904
7.37	54.3169	2.71477	8.58487
7.38	54.4644	2.71662	8.59069
7.39	54.6121	2.71846	8.59651
7.40	54.7600	2.72029	8.60233
7.41	54.9081	2.72213	8.60814
7.42	55.0564	2.72397	8.61394
7.43	55.2049	2.72580	8.61974
7.44	55.3536	2.72764	8.62554
7.45	55.5025	2.72947	8.63134
7.46	55.6516	2.73130	8.63713
7.47	55.8009	2.73313	8.64292
7.48	55.9504	2.73496	8.64870
7.49	56.1001	2.73679	8.65448

n	n^2	$\sqrt{n}$	$\sqrt{10n}$
7.50	56.2500	2.73861	8.66025
7.51	56.4001	2.74044	8.66603
7.52	56.5504	2.74226	8.67179
7.53	56.7009	2.74408	8.67756
7.54	56.8516	2.74591	8.68332
7.55	57.0025	2.74773	8.68907
7.56	57.1536	2.74955	8.69483
7.57	57.3049	2.75136	8.70057
7.58	57.4564	2.75318	8.70632
7.59	57.6081	2.75500	8.71206
7.60	57.7600	2.75681	8.71780
7.61	57.9121	2.75862	8.72353
7.62	58.0644	2.76043	8.72926
7.63	58.2169	2.76225	8.73499
7.64	58.3696	2.76405	8.74071
7.65	58.5225	2.76586	8.74643
7.66	58.6756	2.76767	8.75214
7.67	58.8289	2.76948	8,75785
7.68	58.9824	2.77128	8.76356
7.69	59.1361	2.77308	8.76926
7.70	59.2900	2.77489	8.77496
7.71	59.4441	2.77669	8.78066
7.72	59.5984	2.77849	8.78635
7.73	59.7529	2.78029	8.79204
7.74	59.9076	2.78209	8.79773
7.75	60.0625	2.78388	8.80341
7.76	60.2176	2.78568	8.80909
7.77	60.3729	2.78747	8.81476
7.78	60.5284	2.78927	8.82043
7.79	60.6841	2.79106	8.82610
7.80	60.8400	2.79285	8.83176
7.81	60.9961	2.79464	8.83742
7.82	61.1524	2.79643	8.84308
7.83	61.3089	2.79821	8.84873
7.84	61.4656	2.80000	8.85438
7.85	61.6225	2.80179	8.86002
7.86	61.7796	2.80357	8.86566
7.87	61.9369	2.80535	8.87130
7.88	62.0944	2.80713	8.87694
7.89	62.2521	2.80891	8.88257
7.90	62.4100	2.81069	8.88819
7.91	62.5681	2.81247	8.89382
7.92	62.7264	2.81425	8.89944
7.93	62.8849	2.81603	8.90505
7.94	63.0436	2.81780	8.91067
7.95	63.2025	2.81957	8.91628
7.96	63.3616	2.82135	8.92188
7.97	63.5209	2.82312	8.92749
7.98	63.6804	2.82489	8.93308
7.99	63.8401	2.82666	8.93868

Table B–1. Squares and Square Roots (*Continued*)

n	n^2	$\sqrt{n}$	$\sqrt{10n}$
8.00	64.0000	2.82843	8.94427
8.01	64.1601	2.83019	8.94986
8.02	64.3204	2.83196	8.95545
8.03	64.4809	2.83373	8.96103
8.04	64.6416	2.83549	8.96660
8.05	64.8025	2.83725	8.97218
8.06	64.9636	2.83901	8.97775
8.07	65.1249	2.84077	8.98332
8.08	65.2864	2.84253	8.98888
8.09	65.4481	2.84429	8.99444
8.10	65.6100	2.84605	9.00000
8.11	65.7721	2.84781	9.00555
8.12	65.9344	2.84956	9.01110
8.13	66.0969	2.85132	9.01665
8.14	66.2596	2.85307	9.02219
8.15	66.4225	2.85482	9.02774
8.16	66.5856	2.85657	9.03327
8.17	66.7489	2.85832	9.03881
8.18	66.9124	2.86007	9.04434
8.19	67.0761	2.86182	9.04986
8.20	67.2400	2.86356	9.05539
8.21	67.4041	2.86531	9.06091
8.22	67.5684	2.86705	9.06642
8.23	67.7329	2.86880	9.07193
8.24	67.8976	2.87054	9.07744
8.25	68.0625	2.87228	9.08295
8.26	68.2276	2.87402	9.08845
8.27	68.3929	2.87576	9.09395
8.28	68.5584	2.87750	9.09945
8.29	68.7241	2.87924	9.10494
8.30	68.8900	2.88097	9.11043
8.31	69.0561	2.88271	9.11592
8.32	69.2224	2.88444	9.12140
8.33	69.3889	2.88617	9.12688
8.34	69.5556	2.88791	9.13236
8.35	69.7225	2.88964	9.13783
8.36	69.8896	2.89137	9.14330
8.37	70.0569	2.89310	9.14877
8.38	70.2244	2.89482	9.15423
8.39	70.3921	2.89655	9.15969
8.40	70.5600	2.89828	9.16515
8.41	70.7281	2.90000	9.17061
8.42	70.8964	2.90172	9.17606
8.43	71.0649	2.90345	9.18150
8.44	71.2336	2.90517	9.18695
8.45	71.4025	2.90689	9.19239
8.46	71.5716	2.90861	9.19783
8.47	71.7409	2.91033	9.20326
8.48	71.9104	2.91204	9.20869
8.49	72.0801	2.91376	9.21412

n	n^2	$\sqrt{n}$	$\sqrt{10n}$
8.50	72.2500	2.91548	9.21954
8.51	72.4201	2.91719	9.22497
8.52	72.5904	2.91890	9.23038
8.53	72.7609	2.92062	9.23580
8.54	72.9316	2.92233	9.24121
8.55	73.1025	2.92404	9.24662
8.56	73.2736	2.92575	9.25203
8.57	73.4449	2.92746	9.25743
8.58	73.6164	2.92916	9.26283
8.59	73.7881	2.93087	9.26823
8.60	73.9600	2.93258	9.27362
8.61	74.1321	2.93428	9.27901
8.62	74.3044	2.93598	9.28440
8.63	74.4769	2.93769	9.28978
8.64	74.6496	2.93939	9.29516
8.65	74.8225	2.94109	9.30054
8.66	74.9956	2.94279	9.30591
8.67	75.1689	2.94449	9.31128
8.68	75.3424	2.94618	9.31665
8.69	75.5161	2.94788	9.32202
8.70	75.6900	2.94958	9.32738
8.71	75.8641	2.95127	9.33274
8.72	76.0384	2.95296	9.33809
8.73	76.2129	2.95466	9.34345
8.74	76.3876	2.95635	9.34880
8.75	76.5625	2.95804	9.35414
8.76	76.7376	2.95973	9.35949
8.77	76.9129	2.96142	9.36483
8.78	77.0884	2.96311	9.37017
8.79	77.2641	2.96479	9.37550
8.80	77.4400	2.96648	9.38083
8.81	77.6161	2.96816	9.38616
8.82	77.7924	2.96985	9.39149
8.83	77.9689	2.97153	9.39681
8.84	78.1456	2.97321	9.40213
8.85	78.3225	2.97489	9.40744
8.86	78.4996	2.97658	9.41276
8.87	78.6769	2.97825	9.41807
8.88	78.8544	2.97993	9.42338
8.89	79.0321	2.98161	9.42868
8.90	79.2100	2.98329	9.43398
8.91	79.3881	2.98496	9.43928
8.92	79.5664	2.98664	9.44458
8.93	79.7449	2.98831	9.44987
8.94	79.9236	2.98998	9.45516
8.95	80.1025	2.99166	9.46044
8.96	80.2816	2.99333	9.46573
8.97	80.4609	2.99500	9.47101
8.98	80.6404	2.99666	9.47629
8.99	80.8201	2.99833	9.48156

Table B–1. Squares and Square Roots (*Concluded*)

n	n^2	$\sqrt{n}$	$\sqrt{10n}$
9.00	81.0000	3.00000	9.48683
9.01	81.1801	3.00167	9.49210
9.02	81.3604	3.00333	9.49737
9.03	81.5409	3.00500	9.50263
9.04	81.7216	3.00666	9.50789
9.05	81.9025	3.00832	9.51315
9.06	82.0836	3.00998	9.51840
9.07	82.2649	3.01164	9.52365
9.08	82.4464	3.01330	9.52890
9.09	82.6281	3.01496	9.53415
9.10	82.8100	3.01662	9.53939
9.11	82.9921	3.01828	9.54463
9.12	83.1744	3.01993	9.54987
9.13	83.3569	3.02159	9.55510
9.14	83.5396	3.02324	9.56033
9.15	83.7225	3.02490	9.56556
9.16	83.9056	3.02655	9.57079
9.17	84.0889	3.02820	9.57601
9.18	84.2724	3.02985	9.58123
9.19	84.4561	3.03150	9.58645
9.20	84.6400	3.03315	9.59166
9.21	84.8241	3.03480	9.59687
9.22	85.0084	3.03645	9.60208
9.23	85.1929	3.03809	9.60729
9.24	85.3776	3.03974	9.61249
9.25	85.5625	3.04138	9.61769
9.26	85.7476	3.04302	9.62289
9.27	85.9329	3.04467	9.62808
9.28	86.1184	3.04631	9.63328
9.29	86.3041	3.04795	9.63846
9.30	86.4900	3.04959	9.64365
9.31	86.6761	3.05123	9.64883
9.32	86.8624	3.05287	9.65401
9.33	87.0489	3.05450	9.65919
9.34	87.2356	3.05614	9.66437
9.35	87.4225	3.05778	9.66954
9.36	87.6096	3.05941	9.67471
9.37	87.7969	3.06105	9.67988
9.38	87.9844	3.06268	9.68504
9.39	88.1721	3.06431	9.69020
9.40	88.3600	3.06594	9.69536
9.41	88.5481	3.06757	9.70052
9.42	88.7364	3.06920	9.70567
9.43	88.9249	3.07083	9.71082
9.44	89.1136	3.07246	9.71597
9.45	89.3025	3.07409	9.72111
9.46	89.4916	3.07571	9.72625
9.47	89.6809	3.07734	9.73139
9.48	89.8704	3.07896	9.73653
9.49	90.0601	3.08058	9.74166

n	n^2	$\sqrt{n}$	$\sqrt{10n}$
9.50	90.2500	3.08221	9.74679
9.51	90.4401	3.08383	9.75192
9.52	90.6304	3.08545	9.75705
9.53	90.8209	3.08707	9.76217
9.54	91.0116	3.08869	9.76729
9.55	91.2025	3.09031	9.77241
9.56	91.3936	3.09192	9.77753
9.57	91.5849	3.09354	9.78264
9.58	91.7764	3.09516	9.78775
9.59	91.9681	3.09677	9.79285
9.60	92.1600	3.09839	9.79796
9.61	92.3521	3.10000	9.80306
9.62	92.5444	3.10161	9.80816
9.63	92.7369	3.10322	9.81326
9.64	92.9296	3.10483	9.81835
9.65	93.1225	3.10644	9.82344
9.66	93.3156	3.10805	9.82853
9.67	93.5089	3.10966	9.83362
9.68	93.7024	3.11127	9.83870
9.69	93.8961	3.11288	9.84378
9.70	94.0900	3.11448	9.84886
9.71	94.2841	3.11609	9.85393
9.72	94.4784	3.11769	9.85901
9.73	94.6729	3.11929	9.86408
9.74	94.8676	3.12090	9.86914
9.75	95.0625	3.12250	9.87421
9.76	95.2576	3.12410	9.87927
9.77	95.4529	3.12570	9.88433
9.78	95.6484	3.12730	9.88939
9.79	95.8441	3.12890	9.89444
9.80	96.0400	3.13050	9.89949
9.81	96.2361	3.13209	9.90454
9.82	96.4324	3.13369	9.90959
9.83	96.6289	3.13528	9.91464
9.84	96.8256	3.13688	9.91968
9.85	97.0225	3.13847	9.92472
9.86	97.2196	3.14006	9.92975
9.87	97.4169	3.14166	9.93479
9.88	97.6144	3.14325	9.93982
9.89	97.8121	3.14484	9.94485
9.90	98.0100	3.14643	9.94987
9.91	98.2081	3.14802	9.95490
9.92	98.4064	3.14960	9.95992
9.93	98.6049	3.15119	9.96494
9.94	98.8036	3.15278	9.96995
9.95	99.0025	3.15436	9.97497
9.96	99.2016	3.15595	9.97998
9.97	99.4009	3.15753	9.98499
9.98	99.6004	3.15911	9.98999
9.99	99.8001	3.16070	9.99500

Table B-2. Common Logarithms

N	0	1	2	3	4	5	6	7	8	9
10	0000	0043	0086	0128	0170	0212	0253	0294	0334	0374
11	0414	0453	0492	0531	0569	0607	0645	0682	0719	0755
12	0792	0828	0864	0899	0934	0969	1004	1038	1072	1106
13	1139	1173	1206	1239	1271	1303	1335	1367	1399	1430
14	1461	1492	1523	1553	1584	1614	1644	1673	1703	1732
15	1761	1790	1818	1847	1875	1903	1931	1959	1987	2014
16	2041	2068	2095	2122	2148	2175	2201	2227	2253	2279
17	2304	2330	2355	2380	2405	2430	2455	2480	2504	2529
18	2553	2577	2601	2625	2648	2672	2695	2718	2742	2765
19	2788	2810	2833	2856	2878	2900	2923	2945	2967	2989
20	3010	3032	3054	3075	3096	3118	3139	3160	3181	3201
21	3222	3243	3263	3284	3304	3324	3345	3365	3385	3404
22	3424	3444	3464	3483	3502	3522	3541	3560	3579	3598
23	3617	3636	3655	3674	3692	3711	3729	3747	3766	3784
24	3802	3820	3838	3856	3874	3892	3909	3927	3945	3962
25	3979	3997	4014	4031	4048	4065	4082	4099	4116	4133
26	4150	4166	4183	4200	4216	4232	4249	4265	4281	4298
27	4314	4330	4346	4362	4378	4393	4409	4425	4440	4456
28	4472	4487	4502	4518	4533	4548	4564	4579	4594	4609
29	4624	4639	4654	4669	4683	4698	4713	4728	4742	4757
30	4771	4786	4800	4814	4829	4843	4857	4871	4886	4900
31	4914	4928	4942	4955	4969	4983	4997	5011	5024	5038
32	5051	5065	5079	5092	5105	5119	5132	5145	5159	5172
33	5185	5198	5211	5224	5237	5250	5263	5276	5289	5302
34	5315	5328	5340	5353	5366	5378	5391	5403	5416	5428
35	5441	5453	5465	5478	5490	5502	5514	5527	5539	5551
36	5563	5575	5587	5599	5611	5623	5635	5647	5658	5670
37	5682	5694	5705	5717	5729	5740	5752	5763	5775	5786
38	5798	5809	5821	5832	5843	5855	5866	5877	5888	5899
39	5911	5922	5933	5944	5955	5966	5977	5988	5999	6010
40	6021	6031	6042	6053	6064	6075	6085	6096	6107	6117
41	6128	6138	6149	6160	6170	6180	6191	6201	6212	6222
42	6232	6243	6253	6263	6274	6284	6294	6304	6314	6325
43	6335	6345	6355	6365	6375	6385	6395	6405	6415	6425
44	6435	6444	6454	6464	6474	6484	6493	6503	6513	6522
45	6532	6542	6551	6561	6571	6580	6590	6599	6609	6618
46	6628	6637	6646	6656	6665	6675	6684	6693	6702	6712
47	6721	6730	6739	6749	6758	6767	6776	6785	6794	6803
48	6812	6821	6830	6839	6848	6857	6866	6875	6884	6893
49	6902	6911	6920	6928	6937	6946	6955	6964	6972	6981
50	6990	6998	7007	7016	7024	7033	7042	7050	7059	7067
51	7076	7084	7093	7101	7110	7118	7126	7135	7143	7152
52	7160	7168	7177	7185	7193	7202	7210	7218	7226	7235
53	7243	7251	7259	7267	7275	7284	7292	7300	7308	7316
54	7324	7332	7340	7348	7356	7364	7372	7380	7388	7396
N	0	1	2	3	4	5	6	7	8	9

Table B-2. Common Logarithms (*Concluded*)

N	0	1	2	3	4	5	6	7	8	9
55	7404	7412	7419	7427	7435	7443	7451	7459	7466	7474
56	7482	7490	7497	7505	7513	7520	7528	7536	7543	7551
57	7559	7566	7574	7582	7589	7597	7604	7612	7619	7627
58	7634	7642	7649	7657	7664	7672	7679	7686	7694	7701
59	7709	7716	7723	7731	7738	7745	7752	7760	7767	7774
60	7782	7789	7796	7803	7810	7818	7825	7832	7839	7846
61	7853	7860	7868	7875	7882	7889	7896	7903	7910	7917
62	7924	7931	7938	7945	7952	7959	7966	7973	7980	7987
63	7993	8000	8007	8014	8021	8028	8035	8041	8048	8055
64	8062	8069	8075	8082	8089	8096	8102	8109	8116	8122
65	8129	8136	8142	8149	8156	8162	8169	8176	8182	8189
66	8195	8202	8209	8215	8222	8228	8235	8241	8248	8254
67	8261	8267	8274	8280	8287	8293	8299	8306	8312	8319
68	8325	8331	8338	8344	8351	8357	8363	8370	8376	8382
69	8388	8395	8401	8407	8414	8420	8426	8432	8439	8445
70	8451	8457	8463	8470	8476	8482	8488	8494	8500	8506
71	8513	8519	8525	8531	8537	8543	8549	8555	8561	8567
72	8573	8579	8585	8591	8597	8603	8609	8615	8621	8627
73	8633	8639	8645	8651	8657	8663	8669	8675	8681	8686
74	8692	8698	8704	8710	8716	8722	8727	8733	8739	8745
75	8751	8756	8762	8768	8774	8779	8785	8791	8797	8802
76	8808	8814	8820	8825	8831	8837	8842	8848	8854	8859
77	8865	8871	8876	8882	8887	8893	8899	8904	8910	8915
78	8921	8927	8932	8938	8943	8949	8954	8960	8965	8971
79	8976	8982	8987	8993	8998	9004	9009	9015	9020	9025
80	9031	9036	9042	9047	9053	9058	9063	9069	9074	9079
81	9085	9090	9096	9101	9106	9112	9117	9122	9128	9133
82	9138	9143	9149	9154	9159	9165	9170	9175	9180	9186
83	9191	9196	9201	9206	9212	9217	9222	9227	9232	9238
84	9243	9248	9253	9258	9263	9269	9274	9279	9284	9289
85	9294	9299	9304	9309	9315	9320	9325	9330	9335	9340
86	9345	9350	9355	9360	9365	9370	9375	9380	9385	9390
87	9395	9400	9405	9410	9415	9420	9425	9430	9435	9440
88	9445	9450	9455	9460	9465	9469	9474	9479	9484	9489
89	9494	9499	9504	9509	9513	9518	9523	9528	9533	9538
90	9542	9547	9552	9557	9562	9566	9571	9576	9581	9586
91	9590	9595	9600	9605	9609	9614	9619	9624	9628	9633
92	9638	9643	9647	9652	9657	9661	9666	9671	9675	9680
93	9685	9689	9694	9699	9703	9708	9713	9717	9722	9727
94	9731	9736	9741	9745	9750	9754	9759	9763	9768	9773
95	9777	9782	9786	9791	9795	9800	9805	9809	9814	9818
96	9823	9827	9832	9836	9841	9845	9850	9854	9859	9863
97	9868	9872	9877	9881	9886	9890	9894	9899	9903	9908
98	9912	9917	9921	9926	9930	9934	9939	9943	9948	9952
99	9956	9961	9965	9969	9974	9978	9983	9987	9991	9996
N	0	1	2	3	4	5	6	7	8	9

Table B-3. The Cumulative Binomial Distribution

n	*r*	*p* = .10	*p* = .20	*p* = .25	*p* = .30	*p* = .40	*p* = .50
5	0	.59049	.32768	.23730	.16807	.07776	.03125
	1	.91854	.73728	.63281	.52822	.33696	.18750
	2	.99144	.94208	.89648	.83692	.68256	.50000
	3	.99954	.99328	.98437	.96922	.91296	.81250
	4	.99999	.99968	.99902	.99757	.98976	.96875
	5	1.00000	1.00000	1.00000	1.00000	1.00000	1.00000
10	0	.34868	.10737	.05631	.02825	.00605	.00098
	1	.73610	.37581	.24403	.14931	.04636	.01074
	2	.92981	.67780	.52559	.38278	.16729	.05469
	3	.98720	.87913	.77588	.64961	.38228	.17187
	4	.99837	.96721	.92187	.84973	.63310	.37695
	5	.99985	.99363	.98027	.95265	.83376	.62305
	6	.99999	.99914	.99649	.98941	.94524	.82812
	7	1.00000	.99992	.99958	.99841	.98771	.94531
	8		1.00000	.99997	.99986	.99832	.98926
	9			1.00000	.99999	.99990	.99902
	10				1.00000	1.00000	1.00000
15	0	.20589	.03518	.01336	.00475	.00047	.00003
	1	.54904	.16713	.08018	.03527	.00517	.00049
	2	.81594	.39802	.23609	.12683	.02711	.00369
	3	.94444	.64816	.46129	.29687	.09050	.01758
	4	.98728	.83577	.68649	.51549	.21728	.05923
	5	.99775	.93895	.85163	.72162	.40322	.15088
	6	.99969	.98194	.94338	.86886	.60981	.30362
	7	.99997	.99576	.98270	.94999	.78690	.50000
	8	1.00000	.99921	.99581	.98476	.90495	.69638
	9		.99989	.99921	.99635	.96617	.84912
	10		.99999	.99988	.99933	.99065	.94077
	11		1.00000	.99999	.99991	.99807	.98242
	12			1.00000	.99999	.99972	.99631
	13				1.00000	.99997	.99951
	14					1.00000	.99997
	15						1.00000
20	0	.12158	.01153	.00317	.00080	.00004	.00000
	1	.39175	.06918	.02431	.00764	.00052	.00002
	2	.67693	.20608	.09126	.03548	.00361	.00020
	3	.86705	.41145	.22516	.10709	.01596	.00129
	4	.95683	.62965	.41484	.23751	.05095	.00591
	5	.98875	.80421	.61717	.41637	.12560	.02069
	6	.99761	.91331	.78578	.60801	.25001	.05766
	7	.99958	.96786	.89819	.77227	.41589	.13159
	8	.99994	.99002	.95907	.88667	.59560	.25172
	9	.99999	.99741	.98614	.95204	.75534	.41190
	10	1.00000	.99944	.99606	.98286	.87248	.58810
	11		.99990	.99906	.99486	.94347	.74828
	12		.99998	.99982	.99872	.97897	.86841
	13		1.00000	.99997	.99974	.99353	.94234

Table B–3. The Cumulative Binomial Distribution (*Continued*)

n	r	$p = .10$	$p = .20$	$p = .25$	$p = .30$	$p = .40$	$p = .50$
20	14			1.00000	.99996	.99839	.97931
	15				.99999	.99968	.99409
	16				1.00000	.99995	.99871
	17					.99999	.99980
	18					1.00000	.99998
	19						1.00000
25	0	.07179	.00378	.00075	.00013	.00000	.00000
	1	.27121	.02739	.00702	.00157	.00005	.00000
	2	.53709	.09823	.03211	.00896	.00043	.00001
	3	.76359	.23399	.09621	.03324	.00237	.00008
	4	.90201	.42067	.21374	.09047	.00947	.00046
	5	.96660	.61669	.37828	.19349	.02936	.00204
	6	.99052	.78004	.56110	.34065	.07357	.00732
	7	.99774	.89088	.72651	.51185	.15355	.02164
	8	.99954	.95323	.85056	.67693	.27353	.05388
	9	.99992	.98267	.92867	.81056	.42462	.11476
	10	.99999	.99445	.97033	.90220	.58577	.21218
	11	1.00000	.99846	.98027	.95575	.73228	.34502
	12		.99963	.99663	.98253	.84623	.50000
	13		.99992	.99908	.99401	.92220	.65498
	14		.99999	.99979	.99822	.96561	.78782
	15		1.00000	.99996	.99955	.98683	.88524
	16			.99999	.99990	.99567	.94612
	17			1.00000	.99998	.99879	.97836
	18				1.00000	.99972	.99268
	19					.99995	.99796
	20					.99999	.99954
	21					1.00000	.99992
	22						.99999
	23						1.00000
50	0	.00515	.00001	.00000	.00000		
	1	.03379	.00019	.00001	.00000		
	2	.11173	.00129	.00009	.00000		
	3	.25029	.00566	.00050	.00003		
	4	.43120	.01850	.00211	.00017		
	5	.61612	.04803	.00705	.00072	.00000	
	6	.77023	.10340	.01939	.00249	.00001	
	7	.87785	.19041	.04526	.00726	.00006	
	8	.94213	.30733	.09160	.01825	.00023	
	9	.97546	.44374	.16368	.04023	.00076	.00000
	10	.99065	.58356	.26220	.07885	.00220	.00001
	11	.99678	.71067	.38162	.13904	.00569	.00005
	12	.99900	.81394	.51099	.22287	.01325	.00015
	13	.99971	.88941	.63704	.32788	.02799	.00047
	14	.99993	.93928	.74808	.44683	.05396	.00130
	15	.99998	.96920	.83692	.56918	.09550	.00330
	16	1.00000	.98556	.90169	.68388	.15609	.00767
	17		.99374	.94488	.78219	.23688	.01642

Table B–3. The Cumulative Binomial Distribution (*Continued*)

n	r	p = .10	p = .20	p = .25	p = .30	p = .40	p = .50
50	18		.99749	.97127	.85944	.33561	.03245
	19		.99907	.98608	.91520	.44648	.05946
	20		.99968	.99374	.95224	.56103	.10132
	21		.99990	.99738	.97491	.67014	.16112
	22		.99997	.99898	.98772	.76602	.23994
	23		.99999	.99963	.99441	.84383	.33591
	24		1.00000	.99988	.99763	.90219	.44386
	25			.99996	.99907	.94266	.55614
	26			.99999	.99966	.96859	.66409
	27			1.00000	.99988	.98397	.76006
	28				.99996	.99238	.83888
	29				.99999	.99664	.89868
	30				1.00000	.99863	.94054
	31					.99948	.96755
	32					.99982	.98358
	33					.99994	.99233
	34					.99998	.99670
	35					1.00000	.99870
	36						.99953
	37						.99985
	38						.99995
	39						.99999
	40						1.00000
100	0	.00003					
	1	.00032					
	2	.00194					
	3	.00784					
	4	.02371	.00000				
	5	.05758	.00002				
	6	.11716	.00008				
	7	.20605	.00028	.00000			
	8	.32087	.00086	.00001			
	9	.45129	.00233	.00004			
	10	.58316	.00570	.00014	.00000		
	11	.70303	.01257	.00039	.00001		
	12	.80182	.02533	.00103	.00002		
	13	.87612	.04691	.00246	.00006		
	14	.92743	.08044	.00542	.00016		
	15	.96011	.12851	.01108	.00040		
	16	.97940	.19234	.02111	.00097		
	17	.98999	.27119	.03763	.00216		
	18	.99542	.36209	.06301	.00452	.00000	
	19	.99802	.46016	.09953	.00889	.00001	
	20	.99919	.55946	.14883	.01646	.00002	
	21	.99969	.65403	.21144	.02883	.00004	
	22	.99989	.73893	.28637	.04787	.00011	
	23	.99996	.81091	.37018	.07553	.00025	
	24	.99999	.86865	.46167	.11357	.00056	
	25	1.00000	.91252	.55347	.16313	.00119	

Table B-3. The Cumulative Binomial Distribution (*Concluded*)

n	*r*	*p* = .10	*p* = .20	*p* = .25	*p* = .30	*p* = .40	*p* = .50
100	26		.94417	.64174	.22440	.00240	
	27		.96585	.72238	.29637	.00460	.00000
	28		.97998	.79246	.37678	.00843	.00001
	29		.98875	.85046	.46234	.01478	.00002
	30		.99394	.89621	.54912	.02478	.00004
	31		.99687	.93065	.63311	.03985	.00009
	32		.99845	.95540	.71072	.06150	.00020
	33		.99926	.97241	.77926	.09125	.00044
	34		.99966	.98357	.83714	.13034	.00089
	35		.99985	.99059	.88392	.17947	.00176
	36		.99994	.99482	.92012	.23861	.00332
	37		.99998	.99725	.94695	.30681	.00602
	38		.99999	.99860	.96602	.38219	.01049
	39		1.00000	.99931	.97901	.46208	.01760
	40			.99968	.98750	.54329	.02844
	41			.99985	.99283	.62253	.04431
	42			.99994	.99603	.69674	.06661
	43			.99997	.99789	.76347	.09667
	44			.99999	.99891	.82110	.13563
	45			1.00000	.99946	.86891	.18410
	46				.99974	.90702	.24206
	47				.99988	.93621	.30865
	48				.99995	.95770	.38218
	49				.99998	.97290	.46021
	50				.99999	.98324	.53979
	51				1.00000	.98999	.61782
	52					.99424	.69135
	53					.99680	.79794
	54					.99829	.81590
	55					.99912	.86437
	56					.99956	.90333
	57					.99979	.93339
	58					.99990	.95569
	59					.99996	.97156
	60					.99998	.98240
	61					.99999	.98951
	62					1.00000	.99398
	63						.99668
	64						.99824
	65						.99911
	66						.99956
	67						.99980
	68						.99991
	69						.99996
	70						.99998
	71						.99999
	72						1.00000

Table B–4. The Hypergeometric Distribution

N	n	k	r or x	$P(r)$	$p(x)$	N	n	k	r or x	$P(r)$	$p(x)$
10	1	1	0	0.900000	0.900000	10	5	3	0	0.083333	0.083333
10	1	1	1	1.000000	0.100000	10	5	3	1	0.500000	0.416667
10	2	1	0	0.800000	0.800000	10	5	3	2	0.916667	0.416667
10	2	1	1	1.000000	0.200000	10	5	3	3	1.000000	0.083333
10	2	2	0	0.622222	0.622222	10	5	4	0	0.023810	0.023810
10	2	2	1	0.977778	0.355556	10	5	4	1	0.261905	0.238095
10	2	2	2	1.000000	0.022222	10	5	4	2	0.738095	0.476190
10	3	1	0	0.700000	0.700000	10	5	4	3	0.976190	0.238095
10	3	2	1	1.000000	0.300000	10	5	4	4	1.000000	0.023810
10	3	2	0	0.466667	0.466667	10	5	5	0	0.003968	0.003968
10	3	2	1	0.933333	0.466667	10	5	5	1	0.103175	0.099206
10	3	2	2	1.000000	0.066667	10	5	5	2	0.500000	0.396825
10	3	3	0	0.291667	0.291667	10	5	5	3	0.896825	0.396825
10	3	3	1	0.816667	0.525000	10	5	5	4	0.996032	0.099206
10	3	3	2	0.991667	0.175000	10	5	5	5	1.000000	0.003968
10	3	3	3	1.000000	0.008333	10	6	1	0	0.400000	0.400000
10	4	1	0	0.600000	0.600000	10	6	1	1	1.000000	0.600000
10	4	1	1	1.000000	0.400000	10	6	2	0	0.133333	0.133333
10	4	2	0	0.333333	0.333333	10	6	2	1	0.666667	0.533333
10	4	2	1	0.866667	0.533333	10	6	2	2	1.000000	0.333333
10	4	2	2	1.000000	0.133333	10	6	3	0	0.033333	0.033333
10	4	3	0	0.166667	0.166667	10	6	3	1	0.333333	0.300000
10	4	3	1	0.666667	0.500000	10	6	3	2	0.833333	0.500000
10	4	3	2	0.966667	0.300000	10	6	3	3	1.000000	0.166667
10	4	3	3	1.000000	0.033333	10	6	4	0	0.004762	0.004762
10	4	4	0	0.071429	0.071429	10	6	4	1	0.119048	0.114286
10	4	4	1	0.452381	0.380952	10	6	4	2	0.547619	0.428571
10	4	4	2	0.880952	0.428571	10	6	4	3	0.928571	0.380952
10	4	4	3	0.995238	0.114286	10	6	4	4	1.000000	0.071429
10	4	4	4	1.000000	0.004762	10	6	5	1	0.023810	0.023810
10	5	1	0	0.500000	0.500000	10	6	5	2	0.261905	0.238095
10	5	1	1	1.000000	0.500000	10	6	5	3	0.738095	0.476190
10	5	2	0	0.222222	0.222222	10	6	5	4	0.976190	0.238095
10	5	2	1	0.777778	0.555556	10	6	5	5	1.000000	0.023810
10	5	2	2	1.000000	0.222222	10	6	6	2	0.071429	0.071429

Extracted with permission from Gerald J. Lieberman and Donald B. Owen, *Tables of the Hypergeometric Probability Distribution*, Stanford University Press, Stanford, Calif., 1961.

Table B–4. The Hypergeometric Distribution (*Concluded*)

N	n	k	r or x	$P(r)$	$p(x)$	N	n	k	r or x	$P(r)$	$p(x)$
10	6	6	3	0.452381	0.380952	10	8	3	2	0.533333	0.466667
10	6	6	4	0.880952	0.428571	10	8	3	3	1.000000	0.466667
10	6	6	5	0.995238	0.114286	10	8	4	2	0.133333	0.133333
10	6	6	6	1.000000	0.004762	10	8	4	3	0.666667	0.533333
10	7	1	0	0.300000	0.300000	10	8	4	4	1.000000	0.333333
10	7	1	1	1.000000	0.700000	10	8	5	3	0.222222	0.222222
10	7	2	0	0.066667	0.066667	10	8	5	4	0.777778	0.555556
10	7	2	1	0.533333	0.466667	10	8	5	5	1.000000	0.222222
10	7	2	2	1.000000	0.466667	10	8	6	4	0.333333	0.333333
10	7	3	0	0.008333	0.008333	10	8	6	5	0.866667	0.533333
10	7	3	1	0.183333	0.175000	10	8	6	6	1.000000	0.133333
10	7	3	2	0.708333	0.525000	10	8	7	5	0.466667	0.466667
10	7	3	3	1.000000	0.291667	10	8	7	6	0.933333	0.466667
10	7	4	1	0.033333	0.033333	10	8	7	7	1.000000	0.066667
10	7	4	2	0.333333	0.300000	10	8	8	6	0.622222	0.622222
10	7	4	3	0.833333	0.500000	10	8	8	7	0.977778	0.355556
10	7	4	4	1.000000	0.166667	10	8	8	8	1.000000	0.022222
10	7	5	2	0.083333	0.083333	10	9	1	0	0.100000	0.100000
10	7	5	3	0.500000	0.416667	10	9	1	1	1.000000	0.900000
10	7	5	4	0.916667	0.416667	10	9	2	1	0.200000	0.200000
10	7	5	5	1.000000	0.083333	10	9	2	2	1.000000	0.800000
10	7	6	3	0.166667	0.166667	10	9	3	2	0.300000	0.300000
10	7	6	4	0.666667	0.500000	10	9	3	3	1.000000	0.700000
10	7	6	5	0.966667	0.300000	10	9	4	3	0.400000	0.400000
10	7	6	6	1.000000	0.033333	10	9	4	4	1.000000	0.600000
10	7	7	4	0.291667	0.291667	10	9	5	4	0.500000	0.500000
10	7	7	5	0.816667	0.525000	10	9	5	5	1.000000	0.500000
10	7	7	6	0.991667	0.175000	10	9	6	5	0.600000	0.600000
10	7	7	7	1.000000	0.008333	10	9	6	6	1.000000	0.400000
10	8	1	0	0.200000	0.200000	10	9	7	6	0.700000	0.700000
10	8	1	1	1.000000	0.800000	10	9	7	7	1.000000	0.300000
10	8	2	0	0.022222	0.022222	10	9	8	7	0.800000	0.800000
10	8	2	1	0.377778	0.355556	10	9	8	8	1.000000	0.200000
10	8	2	2	1.000000	0.622222	10	9	9	8	0.900000	0.900000
10	8	3	1	0.066667	0.066667	10	9	9	9	1.000000	0.100000

Table B–5. The Cumulative Poisson Distribution

r	μ = .1	μ = .2	μ = .3	μ = .4	μ = .5
0	.90484	.81873	.74082	.67302	.60653
1	.99532	.98248	.96306	.93845	.90980
2	.99985	.99885	.99640	.99207	.98561
3	1.00000	.99994	.99973	.99922	.99825
4		1.00000	.99998	.99994	.99983
5			1.00000	1.00000	.99999
6					1.00000

r	μ = .6	μ = .7	μ = .8	μ = .9	μ = 1.0
0	.54881	.49658	.44933	.40657	.36788
1	.87810	.84419	.80879	.77248	.73576
2	.97688	.96586	.95258	.93714	.91970
3	.99664	.99425	.99092	.98654	.98101
4	.99961	.99921	.99859	.99766	.99634
5	.99996	.99991	.99982	.99966	.99941
6	1.00000	.99999	.99998	.99996	.99992
7		1.00000	1.00000	1.00000	.99999
8					1.00000

r	μ = 2	μ = 3	μ = 4	μ = 5	μ = 6
0	.13534	.04979	.01832	.00674	.00248
1	.40601	.19915	.09158	.04043	.01735
2	.67668	.42319	.23810	.12465	.06197
3	.85712	.64723	.43347	.26503	.15120
4	.94735	.81526	.62884	.44049	.28506
5	.98344	.91608	.78513	.61596	.44568
6	.99547	.96649	.88933	.76218	.60630
7	.99890	.98810	.94887	.86663	.74398
8	.99976	.99620	.97864	.93191	.84724
9	.99995	.99890	.99187	.96817	.91608
10	.99999	.99971	.99716	.98630	.95738
11	1.00000	.99993	.99908	.99455	.97991
12		.99998	.99973	.99798	.99117
13		1.00000	.99992	.99930	.99637
14			.99998	.99977	.99860
15			1.00000	.99993	.99949
16				.99998	.99982
17				1.00000	.99994
18					.99998
19					1.00000

Table B–5. The Cumulative Poisson Distribution (*Concluded*)

r	$\mu = 7$	$\mu = 8$	$\mu = 9$	$\mu = 10$
0	.00091	.00033	.00012	.00004
1	.00730	.00302	.00123	.00050
2	.02964	.01375	.00623	.00277
3	.08176	.04238	.02123	.01034
4	.17299	.09963	.05496	.02925
5	.30071	.19124	.11569	.06709
6	.44971	.31337	.20678	.13014
7	.59871	.45296	.32390	.22022
8	.72909	.59255	.45565	.33282
9	.83050	.71662	.58741	.45793
10	.90148	.81589	.70599	.58304
11	.94665	.88808	.80301	.69678
12	.97300	.93620	.87577	.79156
13	.98719	.96582	.92615	.86446
14	.99428	.98274	.95853	.91654
15	.99759	.99177	.97796	.95126
16	.99904	.99628	.98889	.97296
17	.99964	.99841	.99468	.98572
18	.99987	.99935	.99757	.99281
19	.99996	.99975	.99894	.99655
20	.99999	.99991	.99956	.99841
21	1.00000	.99997	.99982	.99930
22		.99999	.99993	.99970
23		1.00000	.99998	.99988
24			.99999	.99995
25			1.00000	.99998
26				.99999
27				1.00000

Table B–6. Exponential Functions

x	e^{-x}	e^{x}
.01	.9900	1.0101
.02	.9802	1.0202
.03	.9704	1.0305
.04	.9608	1.0408
.05	.9512	1.0513
.06	.9418	1.0618
.07	.9324	1.0725
.08	.9231	1.0833
.09	.9139	1.0942
.10	.9048	1.1052
.20	.8187	1.2214
.30	.7408	1.3499
.40	.6703	1.4918
.50	.6065	1.6487
.60	.5488	1.8221
.70	.4966	2.0138
.80	.4493	2.2255
.90	.4066	2.4596
1.00	.3679	2.7183
2.00	.1353	7.3891
3.00	.04979	20.0886
4.00	.01832	54.598
5.00	.00674	148.41
6.00	.00248	403.43
7.00	.000912	1096.6
8.00	.000335	2981.0
9.00	.000123	8103.1
10.00	.000045	22026.0

Table B–7. Values of the Standard Normal Distribution Function

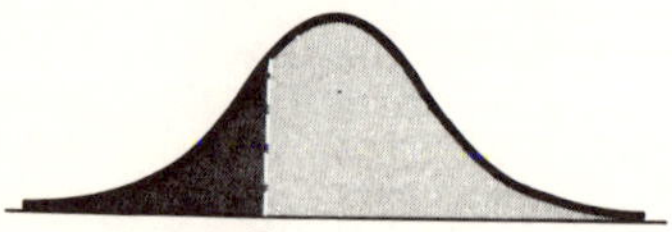

z

z	0	1	2	3	4	5	6	7	8	9
−3.	.0013	.0010	.0007	.0005	.0003	.0002	.0002	.0001	.0001	.0000
−2.9	.0019	.0018	.0017	.0017	.0016	.0016	.0015	.0015	.0014	.0014
−2.8	.0026	.0025	.0024	.0023	.0023	.0022	.0021	.0021	.0020	.0019
−2.7	.0035	.0034	.0033	.0032	.0031	.0030	.0029	.0028	.0027	.0026
−2.6	.0047	.0045	.0044	.0043	.0041	.0040	.0039	.0038	.0037	.0036
−2.5	.0062	.0060	.0059	.0057	.0055	.0054	.0052	.0051	.0049	.0048
−2.4	.0082	.0080	.0078	.0075	.0073	.0071	.0069	.0068	.0066	.0064
−2.3	.0107	.0104	.0102	.0099	.0096	.0094	.0091	.0089	.0087	.0084
−2.2	.0139	.0136	.0132	.0129	.0126	.0122	.0119	.0116	.0113	.0110
−2.1	.0179	.0174	.0170	.0166	.0162	.0158	.0154	.0150	.0146	.0143
−2.0	.0228	.0222	.0217	.0212	.0207	.0202	.0197	.0192	.0188	.0183
−1.9	.0287	.0281	.0274	.0268	.0262	.0256	.0250	.0244	.0238	.0233
−1.8	.0359	.0352	.0344	.0336	.0329	.0322	.0314	.0307	.0300	.0294
−1.7	.0446	.0436	.0427	.0418	.0409	.0401	.0392	.0384	.0375	.0367
−1.6	.0548	.0537	.0526	.0516	.0505	.0495	.0485	.0475	.0465	.0455
−1.5	.0668	.0655	.0643	.0630	.0618	.0606	.0594	.0582	.0570	.0559
−1.4	.0808	.0793	.0778	.0764	.0749	.0735	.0722	.0708	.0694	.0681
−1.3	.0968	.0951	.0934	.0918	.0901	.0885	.0869	.0853	.0838	.0823
−1.2	.1151	.1131	.1112	.1093	.1075	.1056	.1038	.1020	.1003	.0985
−1.1	.1357	.1335	.1314	.1292	.1271	.1251	.1230	.1210	.1190	.1170
−1.0	.1587	.1562	.1539	.1515	.1492	.1469	.1446	.1423	.1401	.1379
− .9	.1841	.1814	.1788	.1762	.1736	.1711	.1685	.1660	.1635	.1611
− .8	.2119	.2090	.2061	.2033	.2005	.1977	.1949	.1922	.1894	.1867
− .7	.2420	.2389	.2358	.2327	.2297	.2266	.2236	.2206	.2177	.2148
− .6	.2743	.2709	.2676	.2643	.2611	.2578	.2546	.2514	.2483	.2451
− .5	.3085	.3050	.3015	.2981	.2946	.2912	.2877	.2843	.2810	.2776
− .4	.3446	.3409	.3372	.3336	.3300	.3264	.3228	.3192	.3156	.3121
− .3	.3821	.3783	.3745	.3707	.3669	.3632	.3594	.3557	.3520	.3483
− .2	.4207	.4168	.4129	.4090	.4052	.4013	.3974	.3936	.3897	.3859
− .1	.4602	.4562	.4522	.4483	.4443	.4404	.4364	.4325	.4286	.4247
− .0	.5000	.4960	.4920	.4880	.4840	.4801	.4761	.4721	.4681	.4641

Table B–7. Values of the Standard Normal Distribution Function (*Concluded*)

z	0	1	2	3	4	5	6	7	8	9
.0	.5000	.5040	.5080	.5120	.5160	.5199	.5239	.5279	.5319	.5359
.1	.5398	.5438	.5478	.5517	.5557	.5596	.5636	.5675	.5714	.5753
.2	.5793	.5832	.5871	.5910	.5948	.5987	.6026	.6064	.6103	.6141
.3	.6179	.6217	.6255	.6293	.6331	.6368	.6406	.6443	.6480	.6517
.4	.6554	.6591	.6628	.6664	.6700	.6736	.6772	.6808	.6844	.6879
.5	.6915	.6950	.6985	.7019	.7054	.7088	.7123	.7157	.7190	.7224
.6	.7257	.7291	.7324	.7357	.7389	.7422	.7454	.7486	.7517	.7549
.7	.7580	.7611	.7642	.7673	.7703	.7734	.7764	.7794	.7823	.7852
.8	.7881	.7910	.7939	.7967	.7995	.8023	.8051	.8078	.8106	.8133
.9	.8159	.8186	.8212	.8238	.8264	.8289	.8315	.8340	.8365	.8389
1.0	.8413	.8438	.8461	.8485	.8508	.8531	.8554	.8577	.8599	.8621
1.1	.8643	.8665	.8686	.8708	.8729	.8749	.8770	.8790	.8810	.8830
1.2	.8849	.8869	.8888	.8907	.8925	.8944	.8962	.8980	.8997	.9015
1.3	.9032	.9049	.9066	.9082	.9099	.9115	.9131	.9147	.9162	.9177
1.4	.9192	.9207	.9222	.9236	.9251	.9265	.9278	.9292	.9306	.9319
1.5	.9332	.9345	.9357	.9370	.9382	.9394	.9406	.9418	.9430	.9441
1.6	.9452	.9463	.9474	.9484	.9495	.9505	.9515	.9525	.9535	.9545
1.7	.9554	.9564	.9573	.9582	.9591	.9599	.9608	.9616	.9625	.9633
1.8	.9641	.9648	.9656	.9664	.9671	.9678	.9686	.9693	.9700	.9706
1.9	.9713	.9719	.9726	.9732	.9738	.9744	.9750	.9756	.9762	.9767
2.0	.9772	.9778	.9783	.9788	.9793	.9798	.9803	.9808	.9812	.9817
2.1	.9821	.9826	.9830	.9834	.9838	.9842	.9846	.9850	.9854	.9857
2.2	.9861	.9864	.9868	.9871	.9874	.9878	.9881	.9884	.9887	.9890
2.3	.9893	.9896	.9898	.9901	.9904	.9906	.9909	.9911	.9913	.9916
2.4	.9918	.9920	.9922	.9925	.9927	.9929	.9931	.9932	.9934	.9936
2.5	.9938	.9940	.9941	.9943	.9945	.9946	.9948	.9949	.9951	.9952
2.6	.9953	.9955	.9956	.9957	.9959	.9960	.9961	.9962	.9963	.9964
2.7	.9965	.9966	.9967	.9968	.9969	.9970	.9971	.9972	.9973	.9974
2.8	.9974	.9975	.9976	.9977	.9977	.9978	.9979	.9979	.9980	.9981
2.9	.9981	.9982	.9982	.9983	.9984	.9984	.9985	.9985	.9986	.9986
3.	.9987	.9990	.9993	.9995	.9997	.9998	.9998	.9999	.9999	1.0000

***Note 1:** If a random variable X is not "standard," its values must be "standardized": $Z = (X - \mu)/\sigma$. That is,*

$$P(X \leq x) = N\left(\frac{x-\mu}{\sigma}\right).$$

***Note 2:** For $z \geq 4$, $N(z) = 1$ to four decimal places; for $z \leq -4$, $N(z) = 0$ to four decimal places.*

Table B–8. Unit Normal Loss Function

The value $L_N(D)$ is the expected opportunity loss (or EVPI) for a linear loss function with slope one and a unit normal distribution. The value D represents the relative position of the break-even point.

When using $L_N(D)$ for a general normal distribution, the value D represents the deviation of the break-even point K from the mean μ, expressed in standard deviation, σ, units.

D	.00	.01	.02	.03	.04	.05	.06	07	.08	.09
.0	.3989	.3940	.3890	.3841	.3793	.3744	.3697	.3649	.3602	.3556
.1	.3509	.3464	.3418	.3373	.3328	.3284	.3240	.3197	.3154	.3111
.2	.3069	.3027	.2986	.2944	.2904	.2863	.2824	.2784	.2745	.2706
.3	.2668	.2630	.2592	.2555	.2518	.2481	.2445	.2409	.2374	.2339
.4	.2304	.2270	.2236	.2203	.2169	.2137	.2104	.2072	.2040	.2009
.5	.1978	.1947	.1917	.1887	.1857	.1828	.1799	.1771	.1742	.1714
.6	.1687	.1659	.1633	.1606	.1580	.1554	.1528	.1503	.1478	.1453
.7	.1429	.1405	.1381	.1358	.1334	.1312	.1289	.1267	.1245	.1223
.8	.1202	.1181	.1160	.1140	.1120	.1100	.1080	.1061	.1042	.1023
.9	.1004	.09860	.09680	.09503	.09328	.09156	.08986	.08819	.08654	.08491
1.0	.08332	.08174	.08019	.07866	.07716	.07568	.07422	.07279	.07138	.06999
1.1	.06862	.06727	.06595	.06465	.06336	.06210	.06086	.05964	.05844	.05726
1.2	.05610	.05496	.05384	.05274	.05165	.05059	.04954	.04851	.04750	.04650
1.3	.04553	.04457	.04363	.04270	.04179	.04090	.04002	.03916	.03831	.03748
1.4	.03667	.03587	.03508	.03431	.03356	.03281	.03208	.03137	.03067	.02998
1.5	.02931	.02865	.02800	.02736	.02674	.02612	.02552	.02494	.02436	.02380
1.6	.02324	.02270	.02217	.02165	.02114	.02064	.02015	.01967	.01920	.01874
1.7	.01829	.01785	.01742	.01699	.01658	.01617	.01578	.01539	.01501	.01464
1.8	.01428	.01392	.01357	.01323	.01290	.01257	.01226	.01195	.01164	.01134
1.9	.01105	.01077	.01049	.01022	$.0^{2}9957$	$.0^{2}9698$	$.0^{2}9445$	$.0^{2}9198$	$.0^{2}8957$	$.0^{2}8721$
2.0	$.0^{2}8491$	$.0^{2}8266$	$.0^{2}8046$	$.0^{2}7832$	$.0^{2}7623$	$.0^{2}7418$	$.0^{2}7219$	$.0^{2}7024$	$.0^{2}6835$	$.0^{2}6649$
2.1	$.0^{2}6468$	$.0^{2}6292$	$.0^{2}6120$	$.0^{2}5952$	$.0^{2}5788$	$.0^{2}5628$	$.0^{2}5472$	$.0^{2}5320$	$.0^{2}5172$	$.0^{2}5028$
2.2	$.0^{2}4887$	$.0^{2}4750$	$.0^{2}4616$	$.0^{2}4486$	$.0^{2}4358$	$.0^{2}4235$	$.0^{2}4114$	$.0^{2}3996$	$.0^{2}3882$	$.0^{2}3770$
2.3	$.0^{2}3662$	$.0^{2}3556$	$.0^{2}3453$	$.0^{2}3352$	$.0^{2}3255$	$.0^{2}3159$	$.0^{2}3067$	$.0^{2}2977$	$.0^{2}2889$	$.0^{2}2804$
2.4	$.0^{2}2720$	$.0^{2}2640$	$.0^{2}2561$	$.0^{2}2484$	$.0^{2}2410$	$.0^{2}2337$	$.0^{2}2267$	$.0^{2}2199$	$.0^{2}2132$	$.0^{2}2067$
2.5	$.0^{2}2005$	$.0^{2}1943$	$.0^{2}1883$	$.0^{2}1826$	$.0^{2}1769$	$.0^{2}1715$	$.0^{2}1662$	$.0^{2}1610$	$.0^{2}1560$	$.0^{2}1511$
3.0	$.0^{3}3822$	$.0^{3}3689$	$.0^{3}3560$	$.0^{3}3436$	$.0^{3}3316$	$.0^{3}3199$	$.0^{3}3087$	$.0^{3}2978$	$.0^{3}2873$	$.0^{3}2771$
3.5	$.0^{4}5848$	$.0^{4}5620$	$.0^{4}5400$	$.0^{4}5188$	$.0^{4}4984$	$.0^{4}4788$	$.0^{4}4599$	$.0^{4}4417$	$.0^{4}4242$	$.0^{4}4073$
4.0	$.0^{5}7145$	$.0^{5}6835$	$.0^{5}6538$	$.0^{5}6253$	$.0^{5}5980$	$.0^{5}5718$	$.0^{5}5468$	$.0^{5}5227$	$.0^{5}4997$	$.0^{5}4777$

Table B–9. The Chi-Square Distribution

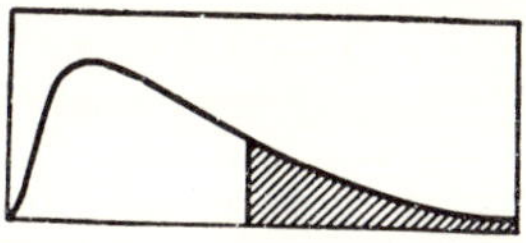

df	Probability that chi-square value will be exceeded							
	.995	.990	.975	.950	.050	.025	.010	.005
1	---	---	---	.004	3.84	5.02	6.63	7.88
2	.01	.02	.05	.10	5.99	7.38	9.21	10.60
3	.07	.11	.22	.35	7.81	9.35	11.34	12.84
4	.21	.30	.48	.71	9.49	11.14	13.28	14.86
5	.41	.55	.83	1.15	11.07	12.83	15.09	16.75
6	.68	.87	1.24	1.64	12.59	14.45	16.81	18.55
7	.99	1.24	1.69	2.17	14.07	16.01	18.48	20.28
8	1.34	1.65	2.18	2.73	15.51	17.53	20.09	21.96
9	1.73	2.09	2.70	3.33	16.92	19.02	21.67	23.59
10	2.16	2.56	3.25	3.94	18.31	20.48	23.21	25.19
11	2.60	3.05	3.82	4.57	19.68	21.92	24.72	26.76
12	3.07	3.57	4.40	5.23	21.03	23.34	26.22	28.30
13	3.57	4.11	5.01	5.89	22.36	24.74	27.69	29.82
14	4.07	4.66	5.63	6.57	23.68	26.12	29.14	31.32
15	4.60	5.23	6.26	7.26	25.00	27.49	30.58	32.80
16	5.14	5.81	6.91	7.96	26.30	28.85	32.00	34.27
17	5.70	6.41	7.56	8.67	27.59	30.19	33.41	35.72
18	6.26	7.01	8.23	9.39	28.87	31.53	34.81	37.16
19	6.84	7.63	8.91	10.12	30.14	32.85	36.19	38.58
20	7.43	8.26	9.59	10.85	31.41	34.17	37.57	40.00
21	8.03	8.90	10.28	11.59	32.67	35.48	38.93	41.40
22	8.64	9.54	10.98	12.34	33.92	36.78	40.29	42.80
23	9.26	10.20	11.69	13.09	35.17	38.08	41.64	44.18
24	9.89	10.86	12.40	13.85	36.42	39.36	42.98	45.56
25	10.52	11.52	13.12	14.61	37.65	40.65	44.31	46.93
26	11.16	12.20	13.84	15.38	38.89	41.92	45.64	48.29
27	11.81	12.88	14.57	16.15	40.11	43.19	46.96	49.64
28	12.46	13.56	15.31	16.93	41.34	44.46	48.28	50.99
29	13.12	14.26	16.05	17.71	42.56	45.72	49.59	52.34
30	13.79	14.95	16.79	18.49	43.77	46.98	50.89	53.67
40	20.71	22.16	24.43	26.51	55.76	59.34	63.69	66.77
50	27.99	29.71	32.36	34.76	67.50	71.42	76.15	79.49
60	35.53	37.48	40.48	43.19	79.08	83.30	88.38	91.95
70	43.28	45.44	48.76	51.74	90.53	95.02	100.43	104.22
80	51.17	53.54	57.15	60.39	101.88	106.63	112.33	116.32
90	59.20	61.75	65.65	69.13	113.14	118.14	124.12	128.30
100	67.33	70.06	74.22	77.93	124.34	129.56	135.81	140.17

Table B–10. Student's *t* Distribution

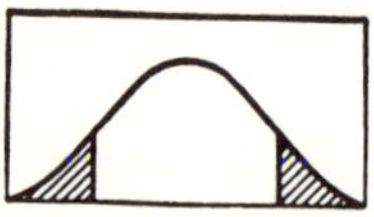

Degrees of freedom	Probability												
	0.9	0.8	0.7	0.6	0.5	0.4	0.3	0.2	0.1	0.05	0.02	0.01	0.001
1	0.158	0.325	0.510	0.727	1.000	1.376	1.963	3.078	6.314	12.706	31.821	63.657	636.619
2	0.142	0.289	0.445	0.617	0.816	1.061	1.386	1.886	2.920	4.303	6.965	9.925	31.598
3	0.137	0.277	0.424	0.584	0.765	0.978	1.250	1.638	2.353	3.182	4.541	5.841	12.924
4	0.134	0.271	0.414	0.569	0.741	0.941	1.190	1.533	2.132	2.776	3.747	4.604	8.610
5	0.132	0.267	0.408	0.559	0.727	0.920	1.156	1.476	2.015	2.571	3.365	4.032	6.869
6	0.131	0.265	0.404	0.553	0.718	0.906	1.134	1.440	1.943	2.447	3.143	3.707	5.959
7	0.130	0.263	0.402	0.549	0.711	0.896	1.119	1.415	1.895	2.365	2.998	3.499	5.408
8	0.130	0.262	0.399	0.546	0.706	0.889	1.108	1.397	1.860	2.306	2.896	3.355	5.041
9	0.129	0.261	0.398	0.543	0.703	0.883	1.100	1.383	1.833	2.262	2.821	3.250	4.781
10	0.129	0.260	0.397	0.542	0.700	0.879	1.093	1.372	1.812	2.228	2.764	3.169	4.587
11	0.129	0.260	0.396	0.540	0.697	0.876	1.088	1.363	1.796	2.201	2.718	3.106	4.437
12	0.128	0.259	0.395	0.539	0.695	0.873	1.083	1.356	1.782	2.179	2.681	3.055	4.318
13	0.128	0.259	0.394	0.538	0.694	0.870	1.079	1.350	1.771	2.160	2.650	3.012	4.221
14	0.128	0.258	0.393	0.537	0.692	0.868	1.076	1.345	1.761	2.145	2.624	2.977	4.140
15	0.128	0.258	0.393	0.536	0.691	0.866	1.074	1.341	1.753	2.131	2.602	2.947	4.073
16	0.128	0.258	0.392	0.535	0.690	0.865	1.071	1.337	1.746	2.120	2.583	2.921	4.015
17	0.128	0.257	0.392	0.534	0.689	0.863	1.069	1.333	1.740	2.110	2.567	2.898	3.965
18	0.127	0.257	0.392	0.534	0.688	0.862	1.067	1.330	1.734	2.101	2.552	2.878	3.922
19	0.127	0.257	0.391	0.533	0.688	0.861	1.066	1.328	1.729	2.093	2.539	2.861	3.883
20	0.127	0.257	0.391	0.533	0.687	0.860	1.064	1.325	1.725	2.086	2.528	2.845	3.850
21	0.127	0.257	0.391	0.532	0.686	0.859	1.063	1.323	1.721	2.080	2.518	2.831	3.819
22	0.127	0.256	0.390	0.532	0.686	0.858	1.061	1.321	1.717	2.074	2.508	2.819	3.792
23	0.127	0.256	0.390	0.532	0.685	0.858	1.060	1.319	1.714	2.069	2.500	2.807	3.767
24	0.127	0.256	0.390	0.531	0.685	0.857	1.059	1.318	1.711	2.064	2.492	2.797	3.745
25	0.127	0.256	0.390	0.531	0.684	0.856	1.058	1.316	1.708	2.060	2.485	2.787	3.725
26	0.127	0.256	0.390	0.531	0.684	0.856	1.058	1.315	1.706	2.056	2.479	2.779	3.707
27	0.127	0.256	0.389	0.531	0.684	0.855	1.057	1.314	1.703	2.052	2.473	2.771	3.690
28	0.127	0.256	0.389	0.530	0.683	0.855	1.056	1.313	1.701	2.048	2.467	2.763	3.674
29	0.127	0.256	0.389	0.530	0.683	0.854	1.055	1.311	1.699	2.045	2.462	2.756	3.659
30	0.127	0.256	0.389	0.530	0.683	0.854	1.055	1.310	1.697	2.042	2.457	2.750	3.646
40	0.126	0.255	0.388	0.529	0.681	0.851	1.050	1.303	1.684	2.021	2.423	2.704	3.551
60	0.126	0.254	0.387	0.527	0.679	0.848	1.046	1.296	1.671	2.000	2.390	2.660	3.460
120	0.126	0.254	0.386	0.526	0.677	0.845	1.041	1.289	1.658	1.980	2.358	2.617	3.373
∞	0.126	0.253	0.385	0.524	0.674	0.842	1.036	1.282	1.645	1.960	2.326	2.576	3.291

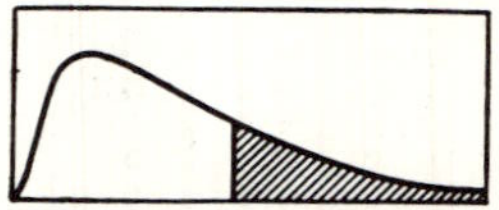

Table B–11. *F* Distribution

Upper 10% points

δ_2 \ δ_1	1	2	3	4	5	6	7	8	9	10	12	15	20	24	30	40	60	120	∞
1	39·86	49·50	53·59	55·83	57·24	58·20	58·91	59·44	59·86	60·19	60·71	61·22	61·74	62·00	62·26	62·53	62·79	63·06	63·33
2	8·53	9·00	9·16	9·24	9·29	9·33	9·35	9·37	9·38	9·39	9·41	9·42	9·44	9·45	9·46	9·47	9·47	9·48	9·49
3	5·54	5·46	5·39	5·34	5·31	5·28	5·27	5·25	5·24	5·23	5·22	5·20	5·18	5·18	5·17	5·16	5·15	5·14	5·13
4	4·54	4·32	4·19	4·11	4·05	4·01	3·98	3·95	3·94	3·92	3·90	3·87	3·84	3·83	3·82	3·80	3·79	3·78	3·76
5	4·06	3·78	3·62	3·52	3·45	3·40	3·37	3·34	3·32	3·30	3·27	3·24	3·21	3·19	3·17	3·16	3·14	3·12	3·10
6	3·78	3·46	3·29	3·18	3·11	3·05	3·01	2·98	2·96	2·94	2·90	2·87	2·84	2·82	2·80	2·78	2·76	2·74	2·72
7	3·59	3·26	3·07	2·96	2·88	2·83	2·78	2·75	2·72	2·70	2·67	2·63	2·59	2·58	2·56	2·54	2·51	2·49	2·47
8	3·46	3·11	2·92	2·81	2·73	2·67	2·62	2·59	2·56	2·54	2·50	2·46	2·42	2·40	2·38	2·36	2·34	2·32	2·29
9	3·36	3·01	2·81	2·69	2·61	2·55	2·51	2·47	2·44	2·42	2·38	2·34	2·30	2·28	2·25	2·23	2·21	2·18	2·16
10	3·29	2·92	2·73	2·61	2·52	2·46	2·41	2·38	2·35	2·32	2·28	2·24	2·20	2·18	2·16	2·13	2·11	2·08	2·06
11	3·23	2·86	2·66	2·54	2·45	2·39	2·34	2·30	2·27	2·25	2·21	2·17	2·12	2·10	2·08	2·05	2·03	2·00	1·97
12	3·18	2·81	2·61	2·48	2·39	2·33	2·28	2·24	2·21	2·19	2·15	2·10	2·06	2·04	2·01	1·99	1·96	1·93	1·90
13	3·14	2·76	2·56	2·43	2·35	2·28	2·23	2·20	2·16	2·14	2·10	2·05	2·01	1·98	1·96	1·93	1·90	1·88	1·85
14	3·10	2·73	2·52	2·39	2·31	2·24	2·19	2·15	2·12	2·10	2·05	2·01	1·96	1·94	1·91	1·89	1·86	1·83	1·80
15	3·07	2·70	2·49	2·36	2·27	2·21	2·16	2·12	2·09	2·06	2·02	1·97	1·92	1·90	1·87	1·85	1·82	1·79	1·76
16	3·05	2·67	2·46	2·33	2·24	2·18	2·13	2·09	2·06	2·03	1·99	1·94	1·89	1·87	1·84	1·81	1·78	1·75	1·72
17	3·03	2·64	2·44	2·31	2·22	2·15	2·10	2·06	2·03	2·00	1·96	1·91	1·86	1·84	1·81	1·78	1·75	1·72	1·69
18	3·01	2·62	2·42	2·29	2·20	2·13	2·08	2·04	2·00	1·98	1·93	1·89	1·84	1·81	1·78	1·75	1·72	1·69	1·66
19	2·99	2·61	2·40	2·27	2·18	2·11	2·06	2·02	1·98	1·96	1·91	1·86	1·81	1·79	1·76	1·73	1·70	1·67	1·63
20	2·97	2·59	2·38	2·25	2·16	2·09	2·04	2·00	1·96	1·94	1·89	1·84	1·79	1·77	1·74	1·71	1·68	1·64	1·61
21	2·96	2·57	2·36	2·23	2·14	2·08	2·02	1·98	1·95	1·92	1·87	1·83	1·78	1·75	1·72	1·69	1·66	1·62	1·59
22	2·95	2·56	2·35	2·22	2·13	2·06	2·01	1·97	1·93	1·90	1·86	1·81	1·76	1·73	1·70	1·67	1·64	1·60	1·57
23	2·94	2·55	2·34	2·21	2·11	2·05	1·99	1·95	1·92	1·89	1·84	1·80	1·74	1·72	1·69	1·66	1·62	1·59	1·55
24	2·93	2·54	2·33	2·19	2·10	2·04	1·98	1·94	1·91	1·88	1·83	1·78	1·73	1·70	1·67	1·64	1·61	1·57	1·53
25	2·92	2·53	2·32	2·18	2·09	2·02	1·97	1·93	1·89	1·87	1·82	1·77	1·72	1·69	1·66	1·63	1·59	1·56	1·52
26	2·91	2·52	2·31	2·17	2·08	2·01	1·96	1·92	1·88	1·86	1·81	1·76	1·71	1·68	1·65	1·61	1·58	1·54	1·50
27	2·90	2·51	2·30	2·17	2·07	2·00	1·95	1·91	1·87	1·85	1·80	1·75	1·70	1·67	1·64	1·60	1·57	1·53	1·49
28	2·89	2·50	2·29	2·16	2·06	2·00	1·94	1·90	1·87	1·84	1·79	1·74	1·69	1·66	1·63	1·59	1·56	1·52	1·48
29	2·89	2·50	2·28	2·15	2·06	1·99	1·93	1·89	1·86	1·83	1·78	1·73	1·68	1·65	1·62	1·58	1·55	1·51	1·47
30	2·88	2·49	2·28	2·14	2·05	1·98	1·93	1·88	1·85	1·82	1·77	1·72	1·67	1·64	1·61	1·57	1·54	1·50	1·46
40	2·84	2·44	2·23	2·09	2·00	1·93	1·87	1·83	1·79	1·76	1·71	1·66	1·61	1·57	1·54	1·51	1·47	1·42	1·38
60	2·79	2·39	2·18	2·04	1·95	1·87	1·82	1·77	1·74	1·71	1·66	1·60	1·54	1·51	1·48	1·44	1·40	1·35	1·29
120	2·75	2·35	2·13	1·99	1·90	1·82	1·77	1·72	1·68	1·65	1·60	1·55	1·48	1·45	1·41	1·37	1·32	1·26	1·19
∞	2·71	2·30	2·08	1·94	1·85	1·77	1·72	1·67	1·63	1·60	1·55	1·49	1·42	1·38	1·34	1·30	1·24	1·17	1·00

Table 18 from *Biometrika Tables for Statisticians*. Volume 1, edited by E. S. Pearson and H. O. Hartley.
Used with the kind permission of E. S. Pearson and the trustees of *Biometrika*.

Table B–11. *F* Distribution (*Continued*)

Upper 5% points

δ_2 \ δ_1	1	2	3	4	5	6	7	8	9	10	12	15	20	24	30	40	60	120	∞
1	161·4	199·5	215·7	224·6	230·2	234·0	236·8	238·9	240·5	241·9	243·9	245·9	248·0	249·1	250·1	251·1	252·2	253·3	254·3
2	18·51	19·00	19·16	19·25	19·30	19·33	19·35	19·37	19·38	19·40	19·41	19·43	19·45	19·45	19·46	19·47	19·48	19·49	19·50
3	10·13	9·55	9·28	9·12	9·01	8·94	8·89	8·85	8·81	8·79	8·74	8·70	8·66	8·64	8·62	8·59	8·57	8·55	8·53
4	7·71	6·94	6·59	6·39	6·26	6·16	6·09	6·04	6·00	5·96	5·91	5·86	5·80	5·77	5·75	5·72	5·69	5·66	5·63
5	6·61	5·79	5·41	5·19	5·05	4·95	4·88	4·82	4·77	4·74	4·68	4·62	4·56	4·53	4·50	4·46	4·43	4·40	4·36
6	5·99	5·14	4·76	4·53	4·39	4·28	4·21	4·15	4·10	4·06	4·00	3·94	3·87	3·84	3·81	3·77	3·74	3·70	3·67
7	5·59	4·74	4·35	4·12	3·97	3·87	3·79	3·73	3·68	3·64	3·57	3·51	3·44	3·41	3·38	3·34	3·30	3·27	3·23
8	5·32	4·46	4·07	3·84	3·69	3·58	3·50	3·44	3·39	3·35	3·28	3·22	3·15	3·12	3·08	3·04	3·01	2·97	2·93
9	5·12	4·26	3·86	3·63	3·48	3·37	3·29	3·23	3·18	3·14	3·07	3·01	2·94	2·90	2·86	2·83	2·79	2·75	2·71
10	4·96	4·10	3·71	3·48	3·33	3·22	3·14	3·07	3·02	2·98	2·91	2·85	2·77	2·74	2·70	2·66	2·62	2·58	2·54
11	4·84	3·98	3·59	3·36	3·20	3·09	3·01	2·95	2·90	2·85	2·79	2·72	2·65	2·61	2·57	2·53	2·49	2·45	2·40
12	4·75	3·89	3·49	3·26	3·11	3·00	2·91	2·85	2·80	2·75	2·69	2·62	2·54	2·51	2·47	2·43	2·38	2·34	2·30
13	4·67	3·81	3·41	3·18	3·03	2·92	2·83	2·77	2·71	2·67	2·60	2·53	2·46	2·42	2·38	2·34	2·30	2·25	2·21
14	4·60	3·74	3·34	3·11	2·96	2·85	2·76	2·70	2·65	2·60	2·53	2·46	2·39	2·35	2·31	2·27	2·22	2·18	2·13
15	4·54	3·68	3·29	3·06	2·90	2·79	2·71	2·64	2·59	2·54	2·48	2·40	2·33	2·29	2·25	2·20	2·16	2·11	2·07
16	4·49	3·63	3·24	3·01	2·85	2·74	2·66	2·59	2·54	2·49	2·42	2·35	2·28	2·24	2·19	2·15	2·11	2·06	2·01
17	4·45	3·59	3·20	2·96	2·81	2·70	2·61	2·55	2·49	2·45	2·38	2·31	2·23	2·19	2·15	2·10	2·06	2·01	1·96
18	4·41	3·55	3·16	2·93	2·77	2·66	2·58	2·51	2·46	2·41	2·34	2·27	2·19	2·15	2·11	2·06	2·02	1·97	1·92
19	4·38	3·52	3·13	2·90	2·74	2·63	2·54	2·48	2·42	2·38	2·31	2·23	2·16	2·11	2·07	2·03	1·98	1·93	1·88
20	4·35	3·49	3·10	2·87	2·71	2·60	2·51	2·45	2·39	2·35	2·28	2·20	2·12	2·08	2·04	1·99	1·95	1·90	1·84
21	4·32	3·47	3·07	2·84	2·68	2·57	2·49	2·42	2·37	2·32	2·25	2·18	2·10	2·05	2·01	1·96	1·92	1·87	1·81
22	4·30	3·44	3·05	2·82	2·66	2·55	2·46	2·40	2·34	2·30	2·23	2·15	2·07	2·03	1·98	1·94	1·89	1·84	1·78
23	4·28	3·42	3·03	2·80	2·64	2·53	2·44	2·37	2·32	2·27	2·20	2·13	2·05	2·01	1·96	1·91	1·86	1·81	1·76
24	4·26	3·40	3·01	2·78	2·62	2·51	2·42	2·36	2·30	2·25	2·18	2·11	2·03	1·98	1·94	1·89	1·84	1·79	1·73
25	4·24	3·39	2·99	2·76	2·60	2·49	2·40	2·34	2·28	2·24	2·16	2·09	2·01	1·96	1·92	1·87	1·82	1·77	1·71
26	4·23	3·37	2·98	2·74	2·59	2·47	2·39	2·32	2·27	2·22	2·15	2·07	1·99	1·95	1·90	1·85	1·80	1·75	1·69
27	4·21	3·35	2·96	2·73	2·57	2·46	2·37	2·31	2·25	2·20	2·13	2·06	1·97	1·93	1·88	1·84	1·79	1·73	1·67
28	4·20	3·34	2·95	2·71	2·56	2·45	2·36	2·29	2·24	2·19	2·12	2·04	1·96	1·91	1·87	1·82	1·77	1·71	1·65
29	4·18	3·33	2·93	2·70	2·55	2·43	2·35	2·28	2·22	2·18	2·10	2·03	1·94	1·90	1·85	1·81	1·75	1·70	1·64
30	4·17	3·32	2·92	2·69	2·53	2·42	2·33	2·27	2·21	2·16	2·09	2·01	1·93	1·89	1·84	1·79	1·74	1·68	1·62
40	4·08	3·23	2·84	2·61	2·45	2·34	2·25	2·18	2·12	2·08	2·00	1·92	1·84	1·79	1·74	1·69	1·64	1·58	1·51
60	4·00	3·15	2·76	2·53	2·37	2·25	2·17	2·10	2·04	1·99	1·92	1·84	1·75	1·70	1·65	1·59	1·53	1·47	1·39
120	3·92	3·07	2·68	2·45	2·29	2·17	2·09	2·02	1·96	1·91	1·83	1·75	1·66	1·61	1·55	1·50	1·43	1·35	1·25
∞	3·84	3·00	2·60	2·37	2·21	2·10	2·01	1·94	1·88	1·83	1·75	1·67	1·57	1·52	1·46	1·39	1·32	1·22	1·00

Table B–11. *F* Distribution (*Concluded*)

Upper 1% points

δ_2 \ δ_1	1	2	3	4	5	6	7	8	9	10	12	15	20	24	30	40	60	120	∞
1	4052	4999·5	5403	5625	5764	5859	5928	5982	6022	6056	6106	6157	6209	6235	6261	6287	6313	6339	6366
2	98·50	99·00	99·17	99·25	99·30	99·33	99·36	99·37	99·39	99·40	99·42	99·43	99·45	99·46	99·47	99·47	99·48	99·49	99·50
3	34·12	30·82	29·46	28·71	28·24	27·91	27·67	27·49	27·35	27·23	27·05	26·37	26·69	26·60	26·50	26·41	26·32	26·22	26·13
4	21·20	18·00	16·69	15·98	15·52	15·21	14·98	14·80	14·66	14·55	14·37	14·20	14·02	13·93	13·84	13·75	13·65	13·56	13·46
5	16·26	13·27	12·06	11·39	10·97	10·67	10·46	10·29	10·16	10·05	9·89	9·72	9·55	9·47	9·38	9·29	9·20	9·11	9·02
6	13·75	10·92	9·78	9·15	8·75	8·47	8·26	8·10	7·98	7·87	7·72	7·56	7·40	7·31	7·23	7·14	7·06	6·97	6·88
7	12·25	9·55	8·45	7·85	7·46	7·19	6·99	6·84	6·72	6·62	6·47	6·31	6·16	6·07	5·99	5·91	5·82	5·74	5·65
8	11·26	8·65	7·59	7·01	6·63	6·37	6·18	6·03	5·91	5·81	5·67	5·52	5·36	5·28	5·20	5·12	5·03	4·95	4·86
9	10·56	8·02	6·99	6·42	6·06	5·80	5·61	5·47	5·35	5·26	5·11	4·96	4·81	4·73	4·65	4·57	4·48	4·40	4·31
10	10·04	7·56	6·55	5·99	5·64	5·39	5·20	5·06	4·94	4·85	4·71	4·56	4·41	4·33	4·25	4·17	4·08	4·00	3·91
11	9·65	7·21	6·22	5·67	5·32	5·07	4·89	4·74	4·63	4·54	4·40	4·25	4·10	4·02	3·94	3·86	3·78	3·69	3·60
12	9·33	6·93	5·95	5·41	5·06	4·82	4·64	4·50	4·39	4·30	4·16	4·01	3·86	3·78	3·70	3·62	3·54	3·45	3·36
13	9·07	6·70	5·74	5·21	4·86	4·62	4·44	4·30	4·19	4·10	3·96	3·82	3·66	3·59	3·51	3·43	3·34	3·25	3·17
14	8·86	6·51	5·56	5·04	4·69	4·46	4·28	4·14	4·03	3·94	3·80	3·66	3·51	3·43	3·35	3·27	3·18	3·09	3·00
15	8·68	6·36	5·42	4·89	4·56	4·32	4·14	4·00	3·89	3·80	3·67	3·52	3·37	3·29	3·21	3·13	3·05	2·96	2·87
16	8·53	6·23	5·29	4·77	4·44	4·20	4·03	3·89	3·78	3·69	3·55	3·41	3·26	3·18	3·10	3·02	2·93	2·84	2·75
17	8·40	6·11	5·18	4·67	4·34	4·10	3·93	3·79	3·68	3·59	3·46	3·31	3·16	3·08	3·00	2·92	2·83	2·75	2·65
18	8·29	6·01	5·09	4·58	4·25	4·01	3·84	3·71	3·60	3·51	3·37	3·23	3·08	3·00	2·92	2·84	2·75	2·66	2·57
19	8·18	5·93	5·01	4·50	4·17	3·94	3·77	3·63	3·52	3·43	3·30	3·15	3·00	2·92	2·84	2·76	2·67	2·58	2·49
20	8·10	5·85	4·94	4·43	4·10	3·87	3·70	3·56	3·46	3·37	3·23	3·09	2·94	2·86	2·78	2·69	2·61	2·52	2·42
21	8·02	5·78	4·87	4·37	4·04	3·81	3·64	3·51	3·40	3·31	3·17	3·03	2·88	2·80	2·72	2·64	2·55	2·46	2·36
22	7·95	5·72	4·82	4·31	3·99	3·76	3·59	3·45	3·35	3·26	3·12	2·98	2·83	2·75	2·67	2·58	2·50	2·40	2·31
23	7·88	5·66	4·76	4·26	3·94	3·71	3·54	3·41	3·30	3·21	3·07	2·93	2·78	2·70	2·62	2·54	2·45	2·35	2·26
24	7·82	5·61	4·72	4·22	3·90	3·67	3·50	3·36	3·26	3·17	3·03	2·89	2·74	2·66	2·58	2·49	2·40	2·31	2·21
25	7·77	5·57	4·68	4·18	3·85	3·63	3·46	3·32	3·22	3·13	2·99	2·85	2·70	2·62	2·54	2·45	2·36	2·27	2·17
26	7·72	5·53	4·64	4·14	3·82	3·59	3·42	3·29	3·18	3·09	2·96	2·81	2·66	2·58	2·50	2·42	2·33	2·23	2·13
27	7·68	5·49	4·60	4·11	3·78	3·56	3·39	3·26	3·15	3·06	2·93	2·78	2·63	2·55	2·47	2·38	2·29	2·20	2·10
28	7·64	5·45	4·57	4·07	3·75	3·53	3·36	3·23	3·12	3·03	2·90	2·75	2·60	2·52	2·44	2·35	2·26	2·17	2·06
29	7·60	5·42	4·54	4·04	3·73	3·50	3·33	3·20	3·09	3·00	2·87	2·73	2·57	2·49	2·41	2·33	2·23	2·14	2·03
30	7·56	5·39	4·51	4·02	3·70	3·47	3·30	3·17	3·07	2·98	2·84	2·70	2·55	2·47	2·39	2·30	2·21	2·11	2·01
40	7·31	5·18	4·31	3·83	3·51	3·29	3·12	2·99	2·89	2·80	2·66	2·52	2·37	2·29	2·20	2·11	2·02	1·92	1·80
60	7·08	4·98	4·13	3·65	3·34	3·12	2·95	2·82	2·72	2·63	2·50	2·35	2·20	2·12	2·03	1·94	1·84	1·73	1·60
120	6·85	4·79	3·95	3·48	3·17	2·96	2·79	2·66	2·56	2·47	2·34	2·19	2·03	1·95	1·86	1·76	1·66	1·53	1·38
∞	6·63	4·61	3·78	3·32	3·02	2·80	2·64	2·51	2·41	2·32	2·18	2·04	1·88	1·79	1·70	1·59	1·47	1·32	1·00

Table B–12. Values of $z = \left(\frac{1}{2}\right) \ln \left(\frac{1 + r}{1 - r}\right)$

r	*r* (3rd decimal)					*r*	*r* (3rd decimal)				
	.000	**.002**	**.004**	**.006**	**.008**		**.000**	**.002**	**.004**	**.006**	**.008**
.00	.0000	.0020	.0040	.0060	.0080	**.35**	.3654	.3677	.3700	.3723	.3746
1	.0100	.0120	.0140	.0160	.0180	**6**	.3769	.3792	.3815	.3838	.3861
2	.0200	.0220	.0240	.0260	.0280	**7**	.3884	.3907	.3931	.3954	.3977
3	.0300	.0320	.0340	.0360	.0380	**8**	.4001	.4024	.4047	.4071	.4094
4	.0400	.0420	.0440	.0460	.0480	**9**	.4118	.4142	.4165	.4189	.4213
.05	.0500	.0520	.0541	.0561	.0581	**.40**	.4236	.4260	.4284	.4308	.4332
6	.0601	.0621	.0641	.0661	.0681	**1**	.4356	.4380	.4404	.4428	.4453
7	.0701	.0721	.0741	.0761	.0782	**2**	.4477	.4501	.4526	.4550	.4574
8	.0802	.0822	.0842	.0862	.0882	**3**	.4599	.4624	.4648	.4673	.4698
9	.0902	.0923	.0943	.0963	.0983	**4**	.4722	.4747	.4772	.4797	.4822
.10	.1003	.1024	.1044	.1064	.1084	**.45**	.4847	.4872	.4897	4922	.4948
1	.1104	.1125	.1145	.1165	.1186	**6**	.4973	.4999	.5024	.5049	.4075
2	.1206	.1226	.1246	.1267	.1287	**7**	.5101	.5126	.5152	.5178	.5204
3	.1307	.1328	.1348	.1368	.1389	**8**	.5230	.5256	.5282	.5308	.5334
4	.1409	.1430	.1450	.1471	.1491	**9**	.5361	.5387	.5413	.5440	.5466
.15	.1511	.1532	.1552	.1573	.1593	**.50**	.5493	.5520	.5547	.5573	.5600
6	.1614	.1634	.1655	.1676	.1696	**1**	.5627	.5654	.5682	.5709	.5736
7	.1717	.1737	.1758	.1779	.1799	**2**	.5763	.5791	.5818	.5846	.5874
8	.1820	.1841	.1861	.1882	.1903	**3**	.5901	.5929	.5957	.5985	.6013
9	.1923	.1944	.1965	.1986	.2007	**4**	.6042	.6070	.6098	.6127	.6155
.20	.2027	.2048	.2069	.2090	.2111	**.55**	.6194	.6213	.6241	.6270	.6299
1	.2132	.2153	.2174	.2195	.2216	**6**	.6328	.6358	.6387	.6416	.6446
2	.2237	.2258	.2279	.2300	.2321	**7**	.6475	.6505	.6535	.6565	.6595
3	.2342	.2363	.2384	.2405	.2427	**8**	.6625	.6655	.6685	.6716	.6746
4	.2448	.2469	.2490	.2512	.2533	**9**	.6777	.6807	.6838	.6869	.6900
.25	.2554	.2575	.2597	.2618	.2640	**.60**	.6931	.6963	.6994	.7026	.7057
6	.2661	.2683	.2704	.2726	.2747	**1**	.7089	.7121	.7153	.7185	.7218
7	.2769	.2790	.2812	.2833	.2855	**2**	.7250	.7283	.7315	.7348	.7381
8	.2877	.2899	.2920	.2942	.2964	**3**	.7414	.7447	.7481	.7514	.7548
9	.2986	.3008	.3029	.3051	.3073	**4**	.7582	.7616	.7650	.7684	.7718
.30	.3095	.3117	.3139	.3161	.3183	**.65**	.7753	.7788	.7823	.7858	.7893
1	.3205	.3228	.3250	.3272	.3294	**6**	.7928	.7964	.7999	.8035	.8071
2	.3316	.3339	.3361	.3383	.3406	**7**	.8107	.8144	.8180	.8217	.8254
3	.3428	.3451	.3473	.3496	.3518	**8**	.8291	.8328	.8366	.8404	.8441
4	.3541	.3564	.3586	.3609	.3632	**9**	.8480	.8518	.8556	.8595	.8634

Table B–12. Values of $z = \left(\frac{1}{2}\right) \ln \left(\frac{1+r}{1-r}\right)$ *(Continued)*

r	*r* (3rd decimal)					*r*	*r* (3rd decimal)				
	.000	**.002**	**.004**	**.006**	**.008**		**.000**	**.002**	**.004**	**.006**	**.008**
.70	.8673	.8712	.8752	.8792	.8832	**.85**	1.256	1.263	1.271	1.278	1.286
1	.8872	.8912	.8953	.8994	.9035	**6**	1.293	1.301	1.309	1.317	1.325
2	.9076	.9118	.9160	.9202	.9245	**7**	1.333	1.341	1.350	1.358	1.367
3	.9287	.9330	.9373	.9417	.9461	**8**	1.376	1.385	1.394	1.403	1.412
4	.9505	.9549	.9549	.9639	.9684	**9**	1.422	1.432	1.442	1.452	1.462
.75	0.973	0.978	0.982	0.987	0.991	**.90**	1.472	1.483	1.494	1.505	1.516
6	0.996	1.001	1.006	1.011	1.015	**1**	1.528	1.539	1.551	1.564	1.576
7	1.020	1.025	1.030	1.035	1.040	**2**	1.589	1.602	1.616	1.630	1.644
8	1.045	1.050	1.056	1.061	1.066	**3**	1.658	1.673	1.689	1.705	1.721
9	1.071	1.077	1.082	1.088	1.093	**4**	1.738	1.756	1.774	1.792	1.812
.80	1.099	1.104	1.110	1.116	1.121	**.95**	1.832	1.853	1.874	1.897	1.921
1	1.127	1.133	1.139	1.145	1.151	**6**	1.946	1.972	2.000	2.029	2.060
2	1.157	1.163	1.169	1.175	1.182	**7**	2.092	2.127	2.165	2.205	2.249
3	1.188	1.195	1.201	1.208	1.214	**8**	2.298	2.351	2.410	2.477	2.555
4	1.221	1.228	1.235	1.242	1.249	**9**	2.647	2.759	2.903	3.106	3.453

This table is abridged from Table 14 of the *Biometrika Tables for Statisticians*, Vol. 1 (1st ed.), edited by E. S. Pearson and H. O. Hartley. Used with the kind permission of E. S. Pearson and the trustees of *Biometrika*.

Table B–13. Equidistributed Random Numbers

03 47 43 73 86	36 96 47 36 61	46 98 63 71 62	33 26 16 80 45	60 11 14 10 95
97 74 24 67 62	42 81 14 57 20	42 53 32 37 32	27 07 36 07 51	24 51 79 89 73
16 76 62 27 66	56 50 26 71 07	32 90 79 78 53	13 55 38 58 59	88 97 54 14 10
12 56 85 99 26	96 96 68 27 31	05 03 72 93 15	57 12 10 14 21	88 26 49 81 76
55 59 56 35 64	38 54 82 46 22	31 62 43 09 90	06 18 44 32 53	23 83 01 30 30
16 22 77 94 39	49 54 43 54 82	17 37 93 23 78	87 35 20 96 43	84 26 34 91 64
84 42 17 53 31	57 24 55 06 88	77 04 74 47 67	21 76 33 50 25	83 92 12 96 76
63 01 63 78 59	16 95 55 67 19	98 10 50 71 75	12 86 73 58 07	44 39 52 38 79
33 21 12 34 29	78 64 56 07 82	52 42 07 44 38	15 51 00 13 42	99 66 02 79 54
57 60 86 32 44	09 47 27 96 54	49 17 46 09 62	90 52 84 77 27	08 02 73 43 28
18 18 07 92 46	44 17 16 58 09	79 83 86 19 62	06 76 50 03 10	55 23 64 05 05
25 62 38 97 75	84 16 07 44 99	83 11 46 32 24	20 14 85 88 45	10 93 72 88 71
23 42 40 64 74	82 97 77 77 81	07 45 32 14 08	32 98 94 07 72	93 85 79 10 75
52 36 28 19 95	50 92 26 11 97	00 56 76 31 38	80 22 02 53 53	86 60 42 04 53
37 85 94 35 12	83 39 50 08 30	42 34 07 96 88	54 42 06 87 98	35 85 29 48 39
70 29 17 12 13	40 33 20 38 26	13 89 51 03 74	17 76 37 13 04	07 74 21 19 30
56 62 18 37 35	96 83 50 87 75	97 12 25 93 47	70 33 24 03 54	97 77 46 44 80
99 49 55 22 77	88 42 95 45 72	16 64 36 16 00	04 43 18 66 79	94 77 24 21 90
16 08 15 04 72	33 27 14 34 09	45 59 34 68 49	12 72 07 34 45	99 27 72 95 14
31 16 93 32 43	50 27 89 87 19	20 15 37 00 49	52 85 66 60 44	38 68 88 11 80
68 34 30 13 70	55 74 30 77 40	44 22 78 84 26	04 33 46 09 52	68 07 97 06 57
74 57 25 65 76	59 29 97 68 60	71 91 38 67 54	13 58 18 24 76	15 54 55 95 52
27 42 37 86 53	48 55 90 65 72	96 57 69 36 10	96 46 92 42 45	97 60 49 04 91
00 39 68 29 61	66 37 32 20 30	77 84 57 03 29	10 45 65 04 26	11 04 96 67 24
29 94 98 94 24	68 49 69 10 82	53 75 91 93 30	34 25 20 57 27	40 48 73 51 92
16 90 82 66 59	83 62 64 11 12	67 19 00 71 74	60 47 21 29 68	02 02 37 03 31
11 27 94 75 06	06 09 19 74 66	02 94 37 34 02	76 70 90 30 86	38 45 94 30 38
35 24 10 16 20	33 32 51 26 38	79 78 45 04 91	16 92 53 56 16	02 75 50 95 98
38 23 16 86 38	42 38 97 01 50	87 75 66 81 41	40 01 74 91 62	48 51 84 08 32
31 96 25 91 47	96 44 33 49 13	34 86 82 53 91	00 52 43 48 85	27 55 26 89 62
66 67 40 67 14	64 05 71 95 86	11 05 65 09 68	76 83 20 37 90	57 16 00 11 66
14 90 84 45 11	75 73 88 05 90	52 27 41 14 86	22 98 12 22 08	07 52 74 95 80
68 05 51 18 00	33 96 02 75 19	07 60 62 93 55	59 33 82 43 90	49 37 38 44 59
20 46 78 73 90	97 51 40 14 02	04 02 33 31 08	39 54 16 49 36	47 95 93 13 30
64 19 58 97 79	15 06 15 93 20	01 90 10 75 06	40 78 78 89 62	02 67 74 17 33
05 26 93 70 60	22 35 85 15 13	92 03 51 59 77	59 56 78 06 83	52 91 05 70 74
07 97 10 88 23	09 98 42 99 64	61 71 62 99 15	06 51 29 16 93	58 05 77 09 51
68 71 86 85 85	54 87 66 47 54	73 32 08 11 12	44 95 92 63 16	29 56 24 29 48
26 99 61 65 53	58 37 78 80 70	42 10 50 67 42	32 17 55 85 74	94 44 67 16 94
14 65 52 68 75	87 59 36 22 41	26 78 63 06 55	13 08 27 01 50	15 29 39 39 43

Index

Acceptance region, 324
Acceptance sampling, 159–163
Act, action, 123
Aggregative price index, 656
All-or-nothing loss function, 138
Alpha (α) error, 321
Alternative hypothesis, 319
Analysis of variance, 401, 556
Arithmetic mean, 53, 242, 256
Attribute, 9
Average deviation, 249
Averages, *see* Central tendency

Bayes' Theorem, 194–201
Bayesian inference: estimation, 412–430
  testing, 430–434
Bayesian strategy, 127
  special forms of: for equivalent-state problems, 134
    for nonequivalent-state problems, 165
Bernoulli probability model, 76
Beta (β) error, 321, 339
Binomial probability model: assumptions of, 76–77
  normal approximations of, 273
  properties of, 80
Bivariate data, 216, 447
Bivariate normal distribution, 468

Central limit theorem, 268
Central tendency, measures of: arithmetic mean, 242, 256
  geometric mean, 243, 258
  harmonic mean, 245, 259
  median, 239, 253
  mode, 240, 255
Changing seasonals, 529
Chebyshev inequality, 60
Chi-square, efficient formulas for, 375–376
Chi-square distributions: properties of, 357
  table, 611
Chi-square tests on: contingency tables (independence), 370
  frequencies, 362
  goodness-of-fit, 365
  homogeneity, 376
  variance, 358
Class interval, 225
Class limits (boundaries), 222
Cluster sampling, 549
Coefficient of: confidence, 304
  multiple regression, 557
  simple correlation, 467, 470
  simple determination, 475, 477
  simple regression, 452
Combination, 580
Compound (joint) events, 23
Conditional probability, 25
Confidence coefficient, 203
Confidence limits for: correlation coefficient, 472
  difference between two means, 311
  difference between two proportions, 312
  mean, 305
  proportion, 308
  regression coefficient B, 463
  regression variable μ_{XY}, 463
  regression variable Y, 464
  standard deviation, 361
  variance, 360

Contingency tables, 370
Continuity correction factor, 362
Continuous probability distributions, 62
Continuous variables, 9, 61
Correlation coefficient of: population, 468
  sample, 470
Covariance, 468
Cramer-Rao inequality, 298
Critical region, 324
Critical value, 324
Curvilinear regression: exponential, 561
  multiple, 563
  power, 562
  reciprocal, 561
  second-degree polynomial, 561
Cyclical fluctuations, 491
  measures of: from annual data, 531
    from monthly data, 533

Data, bivariate, 216, 447
  cross-section, 215
  multivariate, 216
  qualitative, 216
  quantitative, 216
  time series, 216
  univariate, 216
Deciles, 289
Decision criteria: coefficient of optimism, 125
  expected value, 127
  maximax, 125
  maximin, 125
  minimax regret, 126
Decision making: with Bayesian posterior analysis, 201
  with Bayesian preposterior analysis, 204
  with Bayesian prior analysis, 129, 198
  with classical procedures, *see* Estimation; Testing hypotheses
Decision tree, 143
Degrees of freedom, 355
Density function, 62
Dependent events, 29
Descriptive statistics, 3
Discrete probability models, 76–92
Dispersion (variation), measures of: average deviation, 249
  range, 248
  relative, 253
  standard deviation, 250
  variance, 250
Distribution function, 51

Elementary events, 14
Elementary units, 5
Elimination theorem of probability, 35
Empirical probability, 18
Error, sampling, 327
  Type I, 321
  Type II, 321
Estimation of: arithmetic mean, 305
  difference between two means, 311
  difference between two proportions, 312
  parameter, 303
  proportion, 308
  standard deviation, 361
  variance, 360
Estimators, properties of: consistency, 290
  efficiency, 298
  sufficiency, 299
  unbiasedness, 296
Events, complementary, 25
  compound (joint), 23
  dependent, 29
  disjoint (mutually exclusive), 22
  elementary, 14
  independent, 31
Expected value (expectation) of: perfect information, 132
  random variables, 55
  sample information, 437
Experiment, random, 13
Explained error, 475
Explained sum of squares, 475
Exponential functions, table, 607
Exponential probability model, 101
Exponential regression, 561
Exponential trend, 502

F distributions, definition, 390
 properties of, 391
 table, 613
F tests of equality: among several means, 401
 between two variances, 399
Finite population correction factor, 282
Forecasting with times series, 539
Frequency density, 225
Frequency distributions, categorical, 217
 class boundaries, 222
 class frequency, 219
 class interval, 225
 classes, number of, 219
 cumulative, 219
 graphical presentation of, 226
 leptokurtic, 260
 numerical, 218, 220
 open classes, 225
 peakedness, 261
 platykurtic, 260
 relative, 217
 skewness, 71, 260
 tabular presentation of, 217
 unequal classes, 225
 uniform classes, 225

Gauss-Markov theorem, 452
Geometric mean, 243
Goodness-of-fit tests, 365

Harmonic mean, 245
Histogram, 228
Homogeneity, tests of, 376
Hypergeometric probability model, 81
 table, 603
Hypotheses, formulation of, 319

Index numbers, 564
Inductive statistics, 4
Infinite population, 7
Interval estimate, *see* Confidence limits
Irregular component, 491

K, as critical values, 325

Kurtosis, 260

Large numbers, law of, 267
Least significant difference, 411
Least squares, method of, 453
Leptokurtic, 260
Level of significance, 321
Likelihood function, 300
Linear correlation model, 467
Linear regression model: multiple, 557
 simple, 452
Linear trend, 494
Logarithms, table, 597

Mathematical expectation, 53
Maximin strategy, 125
Maximum likelihood, method of, 300
Mean squares, 406
Median, 239, 253
Mid-point, 222
Minimax strategy, 126
Minimum variance estimator, 298
Mode, 240, 255
Moments, 259
Monte Carlo simulation, 179
Moving-average trend, 506
Multiple linear regression, 557
Multiple nonlinear regression, 563
Multiplicative law of probability for: dependent events, 29
 independent events, 31
Mutually exclusive events, 22

Nonparametric tests, 550
 goodness-of-fit test, 365
 homogeneity test, 367
 independence test, 370
 versus parametric statistics, 555
 randomness test, 551
 Wilcoxon test, 552
Normal loss integral, 170
 table, 610
Normal probability models, general, 105
 standard, 109
 table, 608
Null hypothesis, 319

OC curve, 161, 339–343
OC function, 160, 339–343
Odds, 20
Ogive, 233
Open-end classes, 225

Parameter, 11
Partial regression coefficient, 557
Payoff, 121
Payoff table, 124
Peakedness, 261
Pearsonian coefficient of skewness, 260
Permutations, 578
Piecewise loss function, 136
Piecewise payoff function, 135
Poisson probability model, 87
as limit to the binomial, 93
relation to exponential model, 101
table, 605
Polygon, frequency, 231
Pooled variance, 356
Population, definition, 5
finite, 7
infinite, 7
Power curve, 339–347
Power function, 339–347
Prior probability, 194
Probability, axioms of, 15
Bayes' theorem, 195
conditional, 25
definition of: classical, 16
objective, 18
subjective, 19
elimination theorem, 35
law of addition: for mutually exclusive events, 22
for overlapping events, 23
law of complementation, 25
law of multiplication: for dependent events, 29
for independent events, 31
likelihood, 195
posterior, 195
prior, 194
Probability distributions, 48
Probability sampling, 549

Quadratic loss function, 138
Qualitative data, 216
Quantitative data, 216
Quartiles, 289
Queuing theory: analytical approach, 173
Monte Carlo approach, 179

Random experiments, 13
Random numbers, 21
table, 618
Random samples, simple, 20
Random variables, 47
Range, 248
Rank correlation, 479
Ratio-to-moving-average method, 518
Reference contract, 147
Region of acceptance, 324
Region of rejection, 324
Regression: population standard deviation of, 451
population variance of, 451
Regression analysis: curvilinear, 561
multiple linear, 557
simple linear, 449
Regrets (opportunity losses), 126
Relative frequency, 217
Residual variance, 451

Sample, 3
Sample points, 14
Sample size, determination of, 306, 310, 439
Sample space, 14
Sampling distribution of: arithmetic mean, 471
difference between two means, 286
difference between two proportions, 285
median and related measures, 288
number of successes, 273
proportion, 275
Scatter diagram, 454
Seasonal index: alternative methods of, 528
construction of, 518

Sequential decision, 140
Sets, theory of, 569
Shifting the origin, 509
Significance, level of, 321
Simple random sampling, 20
Simulation, 179
Single-channel queuing problems, 173
Single-tail tests, 320
Skewness, 260
Small samples, 354
Squares and square roots, table of, 588
Standard error, 273
Standard normal probability model, 109
Standardized random variables, 59
State of nature, 121
Statistic, 3
Statistical hypotheses, 319
Statistical investigation, stages of, 2
Statistical tables, 588–618
Statistics, 1
  descriptive, 3
  inductive, 4
Stochastic process, 13
Student's distributions, *see* t distributions
Subjective probability, 19
Subpopulations, 450
Sufficiency, 299
Sum of squares, 405
Summation sign, $\sum$, 582
Systematic sampling, 550

t distributions, 388
  properties of, 389
  table, 612
t tests for: difference between two means, 397
  means, 392
  rank correlation, 479
  simple correlation, 471
  simple regression coefficient, 463
Testing hypotheses about: arithmetic mean, 382
  correlation coefficient, 471
  difference between means, 335, 397
  difference between proportions, 337
  equality between variances, 319
Theoretical (expected) frequencies, 362
Time series, models of, 492
Total sum of squares, 405, 474
Trends, exponential, 502
  geometric straight line, 502
  linear, 494
  moving averages, 506
  second-degree parabola, 499
12-month moving averages, 571
Type I error, 321
Type II error, 321, 339

Unbiasedness, 296
Unexplained error, 474
Unexplained sum of squares, 475
Uniform probability models: continuous, 100
  discrete, 84
Uniformly most powerful test, 343
Unions, axiom of, 15
Universe, *see* Population
Utility, assumptions for, 145
  definition of, 144
  expected, 145
  functions of money, 149
  index, 144

Variables, random, 47
Variance, 56
Variance, analysis of, 401, 556
Variation, *see* Dispersion
Venn diagram, 571

Waiting-line theory, *see* Queuing theory

Z, critical ratios, 325
Z-transformation, 59